Optimal State Estimation for Process Monitoring, Fault Diagnosis and Control

Optimal State Estimation for Process Monitoring, Fault Diagnosis and Control

Ch. Venkateswarlu
Chief Scientist (Retd.), Indian Institute of Chemical Technology (CSIR-IICT), Hyderabad, India

Rama Rao Karri
Petroleum and Chemical Engineering, Faculty of Engineering, Universiti Teknologi Brunei, Gadong, Brunei Darussalam

ELSEVIER

Elsevier
Radarweg 29, PO Box 211, 1000 AE Amsterdam, Netherlands
The Boulevard, Langford Lane, Kidlington, Oxford OX5 1GB, United Kingdom
50 Hampshire Street, 5th Floor, Cambridge, MA 02139, United States

Notices
Knowledge and best practice in this field are constantly changing. As new research and experience broaden our understanding, changes in research methods, professional practices, or medical treatment may become necessary.

Practitioners and researchers must always rely on their own experience and knowledge in evaluating and using any information, methods, compounds, or experiments described herein. In using such information or methods they should be mindful of their own safety and the safety of others, including parties for whom they have a professional responsibility.

To the fullest extent of the law, neither the Publisher nor the authors, contributors, or editors, assume any liability for any injury and/or damage to persons or property as a matter of products liability, negligence or otherwise, or from any use or operation of any methods, products, instructions, or ideas contained in the material herein.

British Library Cataloguing-in-Publication Data
A catalogue record for this book is available from the British Library

Library of Congress Cataloging-in-Publication Data
A catalog record for this book is available from the Library of Congress

ISBN: 978-0-323-85878-6

For Information on all Elsevier publications
visit our website at https://www.elsevier.com/books-and-journals

Publisher: Susan Dennis
Editorial Project Manager: Sam Young
Production Project Manager: Sruthi Satheesh
Cover Designer: Greg Harris

Typeset by MPS Limited, Chennai, India

Working together
to grow libraries in
developing countries

www.elsevier.com • www.bookaid.org

Contents

About the authors

Dr. Ch. Venkateswarlu has formerly worked as a scientist, senior principal scientist, and chief scientist at Indian Institute of Chemical Technology (IICT), Hyderabad, a premier research and development (R&D) institute of Council of Scientific and Industrial Research (CSIR), India. Recently, he retired as the Director R&D at BV Raju Institute of Technology (BVRIT), Narsapur, Greater Hyderabad. Prior to this, he has also worked as a professor, principal, and head of Chemical Engineering Department of the same institute. He completed his graduation from Andhra University, India, and from Indian Institute of Chemical Engineers, and postgraduation and PhD in Chemical Engineering from Osmania University, Hyderabad, India. He holds 35 years of R&D and industry experience along with 20 years of teaching experience. His research interests lie in the areas of dynamic process modeling and simulation; process identification and dynamic optimization; process monitoring and fault diagnosis; state estimation and soft sensing; statistical process control and advanced process control; applied engineering mathematics and evolutionary computing; artificial intelligence and expert systems; and bioprocess engineering and bioinformatics. He has published more than 120 research papers in peer journals of repute along with some more international and national proceeding publications. He is also credited with 150 technical paper presentations and invited lectures. He has authored a book published by Elsevier along with a few book chapters. He is also in editorial boards of a few international journals. He has executed several R&D projects sponsored by DST and Industry. He is a reviewer of several international research journals and many national and international research project proposals. He has guided several postgraduate and PhD students. He has served as a long-term guest faculty for premier institutes such as Bhaba Atomic Research Centre Scientific Officers Training, BITS Pilani MS (off-campus), and IICT-CDAC Bioinformatics Programs. He is a fellow of Andhra Pradesh Akademi of Sciences and Telangana State Academy of Sciences. He received various awards in recognition to his R&D and academic contributions.

Dr. Rama Rao Karri working as a professor (Sr. Asst.) in Petroleum and Chemical Engineering, Faculty of Engineering, Universiti Teknologi Brunei (UTB), Brunei Darussalam. He has obtained his PhD in chemical engineering from Indian Institute of Technology (IIT) Delhi, Masters from IIT Kanpur, and Bachelors from Andhra University College of Engineering, Visakhapatnam, India. He has worked as a postdoctoral research fellow at NUS, Singapore, for about 6 years and has over 18 years of working experience in academics, industry, and research. He has led the water resources cluster team at UTB (2015−17), which has focused on water−energy nexus, smart water, and environmental management. His research activities are mainly focused in the development of modeling methods for systems with inherent uncertainties in multidisciplinary fields. He has experience of working in multidisciplinary fields and has expertise in various evolutionary optimization techniques and process modeling. He has published around 100+ research articles in reputed journals, book chapters, and conference proceedings with a combined impact factor of 312.2 and has an h-index of 23 (Scopus and Google Scholar). He is an editorial board member in 6 renowned journals and a peer-review member for more than 81 reputed journals from Elsevier (22), Springer (18), Taylor & Francis (9), Scientific Reports, and other reputed journals (32), and has peer reviewed more than 300 manuscripts. He is a recipient of Publons Peer Reviewer Award as top 1% of global peer reviewers for Environment & Ecology and Crossfield categories for the year 2019. He is also delegating as an advisory board member for many international conferences. He held a position as Editor-in-Chief in *International Journal of Chemoinformatics and Chemical Engineering*, IGI Global, USA (April 2019−August 2020). He is also a Managing Guest Editor for special issues (1) "Magnetic nano composites and emerging applications," in *Journal of Environmental Chemical Engineering*

(IF: 5.909), and (2) "Novel CoronaVirus (COVID-19) in Environmental Engineering Perspective," in *Journal of Environmental Science and Pollution Research* (IF: 4.223), Springer. He has also been a guest editor for the special issue "Nanocomposites for the Sustainable Environment" in *Applied Sciences Journal* (IF: 3.056), MDPI. He is the associate editor in *Scientific Reports* (IF: 4.379), Springer Nature, and *International Journal of Energy and Water Resources (IJEWR)*, Springer Inc. He is also the coeditor and managing editor for seven Elsevier, one Springer, and one CRC edited books. Recently he has been listed in the 2021 world 2% scientists report compiled by Stanford University Researchers.

Preface

This book is addressed to students, researchers, and industry professionals in multiple domains of science, engineering, and technology. This book covers the fundamentals and advanced topics of state estimation with a number of solved examples and case studies, which are beneficial to the personnel of different disciplines to gain knowledge and apply to the problems encountered in their respective domains. General readers of initial chapters that cover fundamentals, algorithms, basic problems, and solved examples on state estimation are expected to have familiarity with the basics of mathematics. Readers of lateral chapters that deal with the design and real applications of state estimators are expected to have familiarity with the fundamentals of mathematics, mathematical modeling, and control concepts along with their basic domain knowledge in engineering and science.

State estimation or soft sensing is an approach that deals with the methods that are used to estimate the unmeasured process variables through the use of known measurements and process knowledge. The enhanced degree of process automation and the growing demand for higher performance in industrial systems make state estimation an integral part of various operational strategies associated with the process systems engineering field. It occupies a greater role in the design, modeling, monitoring, fault diagnosis, optimization, control and operation of all kinds of physical, chemical, biological, and other engineering and technological processes through the use of systematic computer-aided approaches. The task of state estimation has become more challenging in the face of growing demand to meet the requirements of process safety, environmental regulations, energy efficiency, better product quality, and optimum utilization of available resources.

The contents of this book are organized such that the readers in a step-by-step approach can learn the basics of state estimation, acquaint themselves with the relevant algorithms, and understand the problem-solving approaches. As they go further, they will be getting familiar with the design concepts and real applications of state estimation towards process monitoring, fault diagnosis, online optimization, and control. Based on the knowledge gained from the real applications, the audience can create or apply the concepts and strategies of state estimation to solve the problems in their respective domains of science and engineering. Therefore this book covers the pedagogical aspects such as knowledge, understanding, application, analysis, evaluation, and creation.

This book is presented in five parts, wherein the details regarding the description, algorithms, examples, procedures, real system applications, and case studies concerning different approaches and methods of state estimation are elaborated in detail. The first part of this book consists of six chapters covering the basic details, a variety of state estimation algorithms in different approaches, and their implementation procedures, along with solved examples. Moreover, various methods of sensor configuration for state estimation and their implementation procedures are covered in this part. The second part comprises seven chapters covering mechanistic model-based filtering and observation techniques as well as data-driven model-based state estimation methods. The design and implementation of various state estimation strategies for monitoring applications of several real processes are presented and discussed thoroughly in this part. The third part of the book contains three chapters emphasizing the design and application of various quantitative model-based state estimation methods for process fault detection and diagnosis in various nonlinear processes. The fourth part consists of three chapters dealing with the methods and applications of state estimation toward process control. The algorithmic implementation, design, and evaluation of different state estimation methods with several real applications are presented in detail in this part. The fifth part consists of a chapter explaining the importance of state estimation in online optimization. Different state estimation methods with design and implementation procedures for online optimization of real processes are presented in this part. The last chapter at the end of this book briefs an overview, opportunities, challenges, and future directions in state estimation.

Formulation, design, and implementation of various state estimation strategies to solve a wide variety of base cases as well as real engineering problems make this book more beneficial to researchers working in multiple domains. Many references with a variety of state estimation and related theme titles are included at the end of each chapter of this book. These references will be immensely useful to the readers to advance their general knowledge and domain knowledge in the field of state estimation.

Finally, the first author takes this opportunity to thank his colleague Dr. K. Yamuna Rani, beloved PhD students Dr. Rama Rao Karri, Dr. C. Sumana, Dr. N. Anil, Dr. P. Anand, Dr. J.S. Eswari, and several graduate and postgraduate students at the Indian Institute of Chemical Technology (CSIR-IICT) for their contributory research support to this book. Also, thanks to CSIR-IICT management for providing this opportunity to pursue the research and publish its major outcomes.

Ch. Venkateswarlu and Rama Rao Karri

Part I

Basic details and state estimation algorithms

Chapter 1

Optimal state estimation and its importance in process systems engineering

1.1 Introduction

Monitoring, diagnosis, and control of process plants heavily rely on accurate and quickly accessible real-time information of the states that describe the operating behavior of the process. The variables that provide a complete representation of the internal status of the system at a given instant of time are referred to the states of the system. In several processes, the full state of the plant is not directly measurable, and the process measurements are usually corrupted with noise. Therefore, reducing the noise in the measurements and reconstructing the state of the plant is a crucial issue from the process monitoring, fault diagnosis and control point of view. Filtering and prediction are elementary concepts to state estimation. The aim of the filtering and estimation problem is to establish the best estimate for the true variable of a system from an incomplete and noisy set of observations of the system. Optimal estimation of unmeasured process states is attractive, when conventional hardware sensors are not available, or the cost and technical limitations prohibit their usage. The methods that are employed to estimate the unmeasured process variables are called state estimators/soft sensors. These soft sensors use known measurements and process knowledge to provide the estimates of the unmeasured process variables, which are incorporated in process operation schemes to achieve economic optimum. The advancement in computer technology and cheap computing power leads to new developments in soft sensor technology.

1.2 Significance of state estimation

State estimators or soft sensors can be built by using currently available online, offline, and historical information of the process. The behavior of any process is indicated by the states of the output variables, which are dependent on the inputs to the process. A subset of these process outputs represents the primary variables that quantify the productivity in terms of the purity, quality, or properties of the product. These so-called primary variables are often the ones that are difficult to measure online. The other outputs (e.g., temperature, flow, and pressure) are called secondary variables, which can be easily measured online. Because of the intrinsic feature of the process, the states of the secondary variables have a certain dependency on the states of the primary variables. For example, pressures and temperatures can define the liquid composition in a system. It is not easy to access the measurements in time for states like concentrations and the parameters like reaction rates and heat transfer coefficients in chemical and biochemical processes. If such states and parameters are estimated using the known measurements of temperatures and pressures, these estimated variables can be used in process operation schemes to improve plant performance. For instance, separation systems like multicomponent distillation columns are to be operated as precisely as possible to meet the desired product purity specifications. If the current compositions of the columns are available in time, they can form a base for improving the process performance through operator decision-making or in developing an automatic closed-loop control scheme. An automatic sensor system for composition measurement is not economical since component analyzers like gas chromatograph involve large measurement delays and high investment and maintenance costs. As an alternative, component concentrations can be estimated from the temperature measurement using the state estimators. Thus, it should be possible to use the readily available secondary variables to infer the primary variable through the use of a soft sensor. State estimation has become an integral part of various schemes such a process design, modeling, monitoring, fault diagnosis, optimization, control, and operation of all kinds of processes.

Optimal State Estimation for Process Monitoring, Fault Diagnosis and Control. DOI: https://doi.org/10.1016/B978-0-323-85878-6.00013-0

1.3 Role of state estimation in process systems engineering

The technological advancements have made the process industry undergo significant changes emphasizing optimally running the plants with due consideration to safety, monitoring, and control with minimal manual intervention. The increased degree of automation and the growing demand for higher performance, efficiency, and reliability in industrial systems makes the task of state estimation an integral part of various operational strategies such as monitoring, diagnosis, and control concerning the field of process systems engineering. State estimation is a subject of immense interest with applications in conventional and advanced process control, supervisory control, process monitoring, online optimization, product quality monitoring, data reconciliation, fault detection, and diagnosis, etc. The state estimation-based applications have become more crucial in the face of growing demand to meet process safety requirements, environmental regulations, energy efficiency, better product quality, and optimum utilization of available resources. To meet various goals concerning plant operations, detailed information about the number of process variables and parameters is required. Installation of additional instrumentation and sensors to monitor the plant's operation can be expensive both in terms of the capital cost as well as maintenance cost. With the use of state estimator/soft sensor, it is possible to estimate many other variables and process parameters by strategically measuring few key variables of the process. However, to gain the maximum benefit from the state estimation techniques, the sensors that provide vital information about the measured variables are to be placed at optimal locations in the process plant. Thus, optimal selection of measurements has become an important prerequisite for the successful estimation of process states. Various sensor configuration methods have been introduced for the purpose of process state estimation. Thus, state estimation with optimal sensors selection has attained greater significance in monitoring, diagnosis, and control of high dimensional systems.

1.4 Outline of this book

Approaches for state estimation can be broadly classified into mechanistic/first principle model-based approach and data-driven model-based approach. The mechanistic model-based approach includes various stochastic model-based filtering and observation techniques such as Kalman filter, extended Kalman filter, unscented Kalman filter, square root unscented Kalman filter, particle filter, and ensemble Kalman filter. The data-driven model-based approach includes the state estimators derived based on principal component analysis, partial least squares, artificial neural networks, and radial basis function networks. Further various linear and nonlinear observers are also applied for optimal state estimation. The details regarding the description, algorithms, examples, applications, and case studies concerning different approaches and methods of state estimation are elaborated in this book to benefit a wider audience of different disciplines.

The book is presented in five sections highlighting several applications of state estimation in process systems engineering field. The first section covers the fundamental details about state estimation, where in a wide variety of state estimation algorithms in different approaches and their implementation procedures, along with solved examples are presented in-detail. Also, various methods of sensor configuration for state estimation and their implementation procedures are covered. The second section covers process monitoring applications of state estimation, where in design and implementation of various state estimation strategies for monitoring applications of several real processes are discussed. The third section of the book, emphasizes the methods and approaches of state estimation towards process fault detection and diagnosis. The significance of state estimation towards this direction is dealt with different real applications of fault detection and diagnosis in design, execution and performance evaluation. The fourth section of this book presents the methods and approaches of state estimation towards process control with algorithmic implementation, design, and evaluation procedures. The importance of state estimation towards process control is described with several real applications. The fifth section explains the importance of state estimation in online optimization of engineering systems. In this section, different state estimation methods are presented with design and implementation procedures for online optimization of real processes.

1.5 Summary

A state estimator or soft sensor is a mathematical technique used to infer important process variables that are not physically measured. Optimal estimation of the states of a plant from its model in conjunction with its measured outputs constitutes a dominant role in the process systems engineering domain. State estimation is highly useful and supportive to various process operation schemes such as advanced process control, fault detection, process monitoring, and product quality control that aim to achieve higher process performance. This introductory chapter explains the significance and importance of state estimation and briefs about the theme of this book that deals with the details of fundamentals, algorithms, examples, applications, and case studies concerning different approaches and several methods of state estimation.

Chapter 2

Introduction to stochastic processes and state estimation filtering

2.1 Introduction

In many autonomous systems, the knowledge of the system state is essentially required to monitor the status of the process. In a realistic situation, the state of a process is often not directly obtainable but it is usually inferred or estimated based on the system outputs measured by instruments (sensors) along with the support of a dynamic model representing the system. In most cases, building a perfect model to capture all the dynamic phenomenon is not possible. To compensate for the unmodelled dynamics, process noise is often added to the dynamic model. Moreover, to account the measurement errors in realistic situation, proper measurement noise is added to the measurement model. Processes that are associated with such a random noise phenomenon are defined as stochastic or random processes. For most engineering applications, the process noise and measurement noise are assumed to follow zero-mean Gaussian or normal distribution. The general idea of filtering and estimation problem is to establish the best estimate for the true variable of a system from an incomplete and noisy set of observations of the system.

This chapter describes and elaborates the basic concepts relating to stochastic variables, noise, probability, probability distributions, random processes, stochastic representation of general dynamic models, filtering, prediction, and estimation. The concepts and definitions presented in this chapter provide a basic framework for other chapters of this book.

2.2 Probability and stochastic variables

A probability variable is a random variable, whereas a stochastic variable is a chance variable. The probability variable is usually denoted by a capital letter X or Y. The name suggests that the variable has something to do with the concept of probabilities. Suppose, a die is rolled and X is the outcome, then the outcome varies for every turn. Thus X represents a random variable. The possible values of X are 1, 2, 3, 4, 5, and 6. Occurrence of each of these values has a probability of 1/6. In a statistical sense, probability represents the relative frequency of an event, when it is observed for more number of times. Assume that a discrete random variable X takes on the values $x_1, x_2, \ldots, x_k$ as a result of an experiment. If X takes the value x_i for m observations and the total number of trials is n, then the ratio m/n is called the relative frequency of the event $X = x_i$. The relative frequency m/n itself is a random number and changes according to the number of trials performed. As an example of relative frequency, consider the tossing of a coin. Here the events are heads and tails. Considering both the events occur equally likely, that implies the coin is unbiased and it is tossed a large number of times such that the event heads appear exactly half the number of tails. The relative frequency of occurrence of heads is thus 0.5. Similarly, the probability of tails is 0.5.

In general, the probability is a nonnegative number and its values lie between 0 and 1. A value of "zero" indicates that the event will not occur and "one" signifies that the event certainly occurs. Suppose we have a case of n possible events, $x_1, x_2, \ldots, x_n$, which are mutually exclusive in the sense that the occurrence of one event excludes the other. Consider the probabilities of occurrence of these events as $p_1, p_2, \ldots, p_n$, respectively, such that the combined probability of all the events is unity, that is, $p_1 + p_2 + \cdots + p_n = 1$.

Probability measure: The probability of an event is measured by assigning a numerical value for each event. Consider a sample space S with $n(S)$ outcomes in which the event A occurs $n(A)$ times. The probability measure is expressed as

$$\text{Probability of an event} = \frac{\text{Number of favorable outcomes}}{\text{Number of possible outcomes}}$$

$$p(A) = \frac{n(A)}{n(S)} \tag{2.1}$$

Optimal State Estimation for Process Monitoring, Fault Diagnosis and Control. DOI: https://doi.org/10.1016/B978-0-323-85878-6.00020-8

If the sample space S has n equally liked outcomes, then

$$p_1 = p_2 = \ldots = p_n = 1/n \tag{2.2}$$

This is called a uniform distribution on S.

Example 1: Consider rolling a fair die 3 times, find the probability when all the numbers are the same in each trail.

Solution

When the die is rolled three times, the sample space has 216 ordered triplets as (i, j, k). Since the die is fair, the event of interest is the equal probability of the same numbers with six outcomes:

$$A = \{(1, 1, 1), (2, 2, 2), \ldots, (6, 6, 6)\}$$

Thus we have a uniform probability distribution. Here $n(A) = 6$ and $n(S) = 216$.
Thus, $P(A) = 6/216 = 1/36$.

2.2.1 Probability theorems

The addition and multiplication theorems of probability are briefed as follows:

Addition theorem: Let A and B are two independent events in the sample space of an experiment.

$$P(A \cup B) = P(A) + P(B) - P(A \cap B) \tag{2.3}$$

If A and B are mutually exclusive events, $A \cap B = $ null set. Thus

$$P(A \cup B) = P(A) + P(B) \tag{2.4}$$

If the sample space consists of n events such that

$$S = A_1 \cup A_2 \cup \ldots \cup A_n \tag{2.5}$$

Thus, according to the addition theorem,

$$P(S) = P(A_1) + P(A_2) + \cdots + P(A_n) = \sum_{i=1}^{n} P(A_i) \tag{2.6}$$

Multiplication theorem: Let A and B are two independent events in the sample space of an experiment. According to the multiplication theorem,

$$P(A \cap B) = P(A).P(B) \tag{2.7}$$

If $A_1, A_2, \ldots, A_n$ are independent events in the sample space, then

$$P(A_1 \cap A_2 \cap \cdots \cap A_n) = P(A_1) \times P(A_2) \cdots \times P(A_n) \tag{2.8}$$

Example 2: Consider throwing of a single die with sample space $S = \{1, 2, 3, 4, 5, 6\}$. What is the probability of getting an odd number <3 and even number >3?

Solution
The sample space S is represented by

$$S = \{1,\ 2,\ 3,\ 4,\ 5,\ 6\}$$

Let A and B are mutually exclusive events in the sample space.

$$A = \{\text{odd number} < 3\} = \{1\}$$
$$B = \{\text{even number} > 3\} = \{4, 6\}$$

$$P(A \cup B) = P(A) + P(B)$$
$$= \frac{1}{6} + \frac{2}{6} = \frac{1}{2}$$

2.2.2 Conditional probability

It is a probability measure that deals with dependent events. Suppose A and B are two dependent events; the $P(A/B)$ defines the probability of event A after the occurrence of B. Then $P(A/B)$ can be interpreted as a probability of A given B. Similarly, $P(B/A)$ defines the vice versa. This concept is known as conditional probability since a condition on the occurrence of B or A is specified.

The measure for conditional probability is given as follows:

$$P(A/B) = \frac{P(A \cap B)}{P(B)}, \quad \text{provided } P(B) \neq 0 \tag{2.9}$$

or

$$P(A/B) = \frac{P(B \cap A)}{P(A)}, \quad \text{provided } P(A) \neq 0 \tag{2.10}$$

If A and B are independent events, then

$$P(A/B) = \frac{P(A) \times P(B)}{P(B)} = P(A) \tag{2.11}$$

It is often required to find the probability of event B after the occurrence of event A. This probability is called the conditional probability of B given A and denoted as $P(B/A)$. The probability measure for this case is given by

$$P(B/A) = \frac{P(A \cap B)}{P(A)} \tag{2.12}$$

In case if the events A and B occur in a sample space S, and $P(A) \neq 0$, $P(B) \neq 0$, then

$$P(A \cap B) = P(A) \times P(B/A) = P(B) \times P(A/B) \tag{2.13}$$

If A and B are independent events, then

$$P(A \cap B = P(A) \times P(B) \tag{2.14}$$

If $P(A) \neq 0$, $P(B) \neq 0$, then

$$P(A/B) = P(A); \quad P(B/A) = P(B) \tag{2.15}$$

This means that the probability of A does not depend on the occurrence or nonoccurrence of the probability of B, and vice versa.

Example 3: A problem in an engineering subject is given to three students A, B, and C. The chances of solving the problem by the students are 1/3, 1/4, and 1/5, respectively. What is the probability that the problem is solved?

Solution

The probability that A can solve the problem, $P(A) = 1/3$.
The probability that A cannot solve the problem $= 1 - 1/3 = 2/3$
The probability that B can solve the problem, $P(B) = 1/4$.
The probability that B cannot solve the problem $= 1 - 1/4 = 3/4$
The probability that C can solve the problem, $P(C) = 1/5$.
The probability that C cannot solve the problem $= 1 - 1/5 = 4/5$
The probability that A, B, and C cannot solve the problem $= 2/3 \times 3/4 \times 4/5 = 2/5$
The probability that the problem is solved by at least one student $= 1 - 2/5 = 3/5$

2.3 Probability distributions and distribution functions

The probability functions and the probability distributions are described as follows:

2.3.1 Discrete random variables and discrete probability distributions

Random variables can be discrete or continuous. If a random variable takes a finite set of countable values, it is called a discrete random variable. Let X be a discrete random variable. Our interest is to compute the probabilities of the form $P(X = x_k)$ for various values of x_k in the range of X. As x_k varies as $x_1, x_2, \ldots$, etc., in the range of X, the probability $P(X = x_k)$ also varies. Thus $P(X = x_k)$ is a function of x_k. The probability function is represented by

$$P(X = x_k) = f(x_k), \quad k = 1, 2, \ldots \tag{2.16}$$

The probability function is also called a *probability distribution* which is given by

$$P(X = x) = f(x) \tag{2.17}$$

For $x = x_k$, this function follows Eq. (2.16); while for other values of x, $f(x) = 0$.
In general, $f(x)$ is a probability function if

(i) $f(x) \geq 0$
(ii) $\sum f(x) = 1$, where the sum is taken as the overall possible values of x.

Example 4: Suppose that a coin of the head (H) and tail (T) is tossed twice, the sample space becomes $S = \{HH, HT, TH, TT\}$. Let X be the random variable representing the number of heads that can come up in the sample space. Find the probability function corresponding to the random variable X.

Solution
We have X is a random variable representing each of the sample point HH, HT, TH, and TT in the sample space. Thus, we have

$$P(HH) = 1/4, \quad P(HT) = 1/4, \quad P(TH) = 1/4, \quad \text{and } P(TT) = 1/4.$$

Then

$$P(X = 0) = P(TT) = 1/4$$
$$P(X = 1) = P(HT \cup TH) = P(HT) + P(TH) = 1/4 + 1/4 = 1/2$$
$$P(X = 2) = P(HH) = 1/4$$

The probability function is given in Table 2.1.

Example 5: Find whether the following functions $f_1(x)$ and $f_2(x)$ represent the probability distribution functions for discrete random variables.

$$f_1(x) = \begin{cases} 3/4; & x = -2 \\ 1/4; & x = 3 \\ 0; & \text{elsewhere} \end{cases}, \quad f_2(x) = \begin{cases} 2/3; & x = -1 \\ -1/3; & x = 4 \\ 0; & \text{elsewhere} \end{cases}$$

Solution
The function $f_1(x)$ takes nonzero values 3/4 and 1/4 at the points $x = -2$ and 3, respectively, and x takes all other values with zero probability, that is, $f_1(-2) = 3/4$, $f_1(3) = 1/4$, and $f_1(x) = 0$, elsewhere. Condition (i) is satisfied since $f_1(x) \geq 0$ for all values of x. Condition (ii) is also satisfied because $3/4 + 1/4 + 0 = 1$. Hence $f_1(x)$ represents a probability function for a discrete random variable x.

TABLE 2.1 Probability function for example 4.

x	0	1	2
f(x)	1/4	½	1/4

Similarly, in the case of $f_2(x)$, $\sum_x f_2(x) = 1$, thus condition (ii) is satisfied. However, $f_2(x)$ at $x = 4$, that is, $f_2(4) = -1/3$, which is negative. Hence condition (i) is violated. Therefore $f_2(x)$ cannot be the probability function of any random variable.

Cumulative distribution function: The distribution function or cumulative distribution function for a discrete random variable X can be obtained from its probability function by noting that, for all x in $(-\infty, \infty)$,

$$F(x) = P(X \leq x) = \sum_{u \leq x} f(u) \tag{2.18}$$

where the sum is the overall values of u taken on by X for which $u \leq x$.

If X takes on only a finite number of values $x_1, x_2, \ldots, x_n$, then the distribution function is given by

$$F(x) = \begin{cases} 0 & -\infty < x < x_1 \\ f(x_1) & x_1 \leq x < x_2 \\ f(x_1) + f(x_2) & x_2 \leq x < x_3 \\ \vdots & \vdots \\ f(x_1) + f(x_2) + \cdots + f(x_n) & x_n \leq x < \infty \end{cases} \tag{2.19}$$

Example 6: Find the distribution function for the random variable x in Example 4.

Solution
The distribution function is

$$F(x) = \begin{cases} 0 & ; & -\infty \leq x \leq 0 \\ 1/4 & ; & 0 \leq x < 1 \\ 3/4 & ; & 1 \leq x < 2 \\ 1 & ; & 2 \leq x < \infty \end{cases}$$

The distribution function [1] is sketched, as shown in Fig. 2.1.

From the distribution function (Fig. 2.1), it can be observed that the magnitude of jumps at 0, 1, and 2 are the probabilities given in Table 2.1. This shows that the probability function can be obtained from the distribution function. Accordingly, the probability function of a discrete random variable obtained from the distribution function is given by:

$$f(x) = F(x) - \lim_{u \to x^-} F(u) \tag{2.20}$$

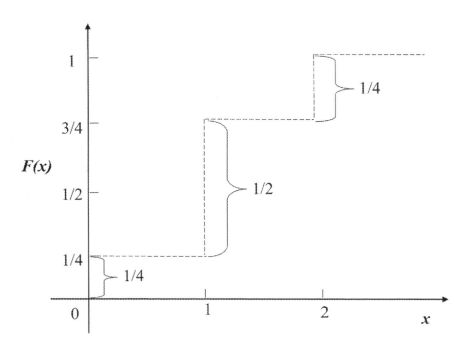

FIGURE 2.1 Plot of the distribution function.

2.3.2 Continuous random variables and continuous probability distributions

A random variable that assumes an infinite number of values is called a continuous random variable. The distribution function of a continuous random variable X is defined as $F(x) = P(X \leq x)$.

$$F(x) = P(X \leq x) \tag{2.21}$$

The $F(x)$ is called the Probability Distribution Function of the random variable X. The following are the properties of the probability distribution function:

$$
\begin{aligned}
&F(x) \in [0, 1]\\
&F(-\infty) = 0\\
&F(\infty) = 1\\
&F(a) \leq F(b) \text{ if } a \leq b\\
&P(a < X \leq b) = F(b) - F(a)
\end{aligned}
\tag{2.22}
$$

The conditional probability distribution function of the random variable X for event A is defined as

$$
\begin{aligned}
F(x/A) &= P(X \leq x/A)\\
&= \frac{P(X \leq x/A)}{P(A)}
\end{aligned}
\tag{2.23}
$$

The distribution function $F(x) = P(X \leq x)$ can be considered as a monotonically increasing function which increases from 0 to 1 and is represented as shown in Fig. 2.2.

On the basis of integral calculus, the probability distribution function can be defined as the derivative of $F(x)$ as

$$\frac{dF(x)}{dx} = f(x) \tag{2.24}$$

The function $f(x)$ has the properties

$$
\begin{aligned}
&f(x) \geq 0\\
&\int_{-\infty}^{\infty} f(x)dx = 1\\
&P(a < X \leq b) = \int_{a}^{b} f(x)dx
\end{aligned}
\tag{2.25}
$$

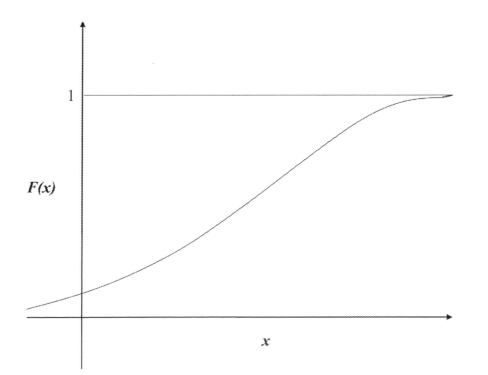

FIGURE 2.2 Monotonically increasing function from 0 to 1.

The function $f(x)$ is also called a *probability density function(PDF)* or simply *density function* of a continuous random variable. If $f(x)$ is the density function for a random variable X, then we can represent $y = f(x)$ graphically by a curve as in Fig. 2.3. Since $f(x) \geq 0$, the curve cannot fall below the x-axis. The entire area bounded by the curve and the x-axis must be 1. Geometrically the probability that X is between a and b, that is, $P(a < X < b)$, is then represented by the area shown shaded in Fig. 2.3.

The conditional PDF of the random variable X for event A is defined as

$$f(x/A) = \frac{dF(x/A)}{dx} \tag{2.26}$$

Suppose we have two random variables X_1 and X_2, the conditional probability density of the first given that the second takes on the variable x_2 as

$$f(x_1, x_2) = \frac{f(x_1, x_2)}{f(x_2)} \tag{2.27}$$

where $f(x_1, x_2)$ and $f(x_2)$ are the joint probability density of the two random variables and the marginal density of the second variable.

Example 7: The probability density is represented by an exponential decay function of the form

$$f(t) = \begin{cases} \dfrac{1}{3}e^{-t/3}; & 0 \leq t < \alpha \\ 0; & \text{elsewhere} \end{cases}$$

Compute the probabilities (i) $P(0 \leq t \leq 2)$, and (ii) $P(3 \leq t \leq 8)$.
Solution
Probabilities are calculated as the areas under the density curve between the corresponding ordinates or the integral of the density over the given interval.

(i) The probability $P(0 \leq t \leq 2)$ is given by

$$\int_0^2 \frac{1}{3}e^{-t/3}dt = -e^{-t/3}\Big|_0^2 = 1 - e^{-2/3}$$

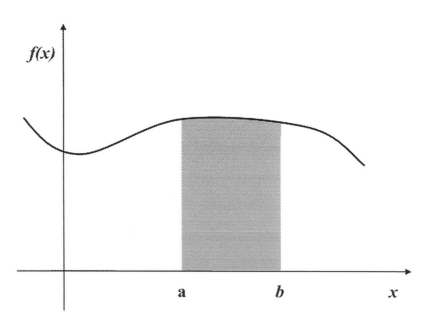

FIGURE 2.3 Probability that X is between a and b.

(ii) The probability $P(3 \leq t \leq 8)$ is given by

$$\int_3^8 \frac{1}{3} e^{-t/3} dt = -e^{-t/3}\Big|_3^8 = e^{-1} - e^{-8/3}$$

The probabilities are shown in the shaded areas covered by dotted lines as shown in Fig. 2.4.

Example 8: Consider the following density function

$$f(x) = \begin{cases} kx^3; & 0 < x < 4 \\ 0; & \text{otherwise} \end{cases}$$

(a) Find the constant k

(b) Compute $P(2 < X < 3)$

Solution

The function $f(x)$ has the properties such as $f(x) \geq 0$, and $\int_{-\infty}^{\infty} f(x)dx = 1$.

(a) $f(x) \geq 0$ if $k > 0$

$$\int_{-\infty}^{\infty} f(x)dx = \int_0^4 kx^3 dx$$
$$= \frac{kx^4}{4}\Big|_0^4 = 64k$$

Since this must be equal to 1, which means $64k = 1 \Rightarrow k = 1/64$.

(b) $P(2 < X < 3) = \int_2^3 \frac{1}{64} x^3 dx = \frac{x^4}{256}\Big|_2^3 = \frac{81}{256} - \frac{16}{256} = \frac{65}{256}$

If $f(x)$ is continuous, we know the probability of X is equal to any particular value is zero. In such a case, we can replace either or both of the signs $<$ in Eq. (2.25) by $\leq$. Thus, for this example,

$$P(2 \leq X \leq 3) = \frac{65}{256}$$

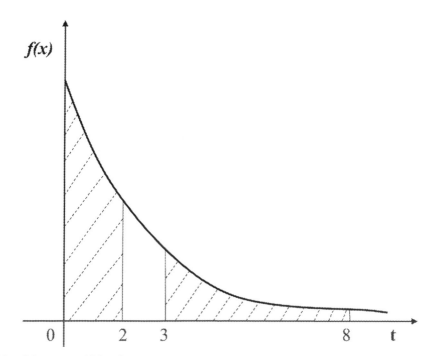

FIGURE 2.4 Probabilities of the exponential function.

Example 9: If two random variables have the joint probability density

$$f(x_1, x_2) = \begin{cases} \dfrac{3}{2}(x_1 + 3x_2); & \text{for } 0 < x_1 < 1 \ \& \ 0 < x_2 < 1 \\ 0; & \text{otherwise} \end{cases}$$

Find the conditional probability density of the first variable, given that the second variable takes on the value x_2.
Solution
First, we find the marginal density of the second random variable by integrating the function with respect to x_1.

$$f(x_2) = \int_0^1 \frac{3}{2}(x_1 + 3x_2)dx_1$$

$$= \frac{3}{4}(1 + 6x_2); \quad \text{for } (0 < x_2 < 1)$$

and $f(x_2) = 0$, otherwise.

Hence, by definition, the conditional probability of the first random variable given that the second takes on the value x_2 is

$$f(x_1, x_2) = \frac{\frac{3}{2}(x_1 + 3x_2)}{\frac{3}{4}(1 + 6x_2)} = \frac{2x_1 + 6x_2}{1 + 6x_2} \quad \text{for } 0 < x_1 < 1, 0 < x_2 < 1)$$

and $f(x_1, x_2) = 0$, otherwise.

2.4   White Gaussian noise and colored noise

In filtering and estimation problems, most of the physical systems under consideration are noisy. The noise may arise in a number of ways. For example, the inputs to the system may associate with noise which is unknown/unpredictable. The outputs from the system may be noisy due to sensor signal measurement inaccuracies. White noise and colored noise are important signals in stochastic systems.

White noise: In signal processing, white noise is a random signal having equal intensity at different frequencies, giving it a constant power spectral density. White noise draws its name from white light, although light that appears white, generally it does not have a flat power spectral density over the visible band. In discrete-time, white noise is a discrete signal whose samples are regarded as a sequence of serially uncorrelated random variables with zero mean and finite variance. If each sample has a normal distribution with zero mean, the signal is said to be additive white Gaussian noise.

Consider the discrete model of a typical system is represented by the following equations:

$$\begin{aligned} x_{k+1} &= f(x_k, \theta) + Gw \\ y_k &= Hx_k + v \end{aligned} \tag{2.28}$$

where $x \in R^n$ denote the state variables, θ denote system parameters, and $y \in R^m$ denotes the outputs. H is $m \times n$ measurement matrix, G is state noise coefficient matrix, and the symbols w and v refer to Gaussian white noises. White noises are some uncertainties that are independent time series and distributed identically, which means no autocorrelation between them. In a particular case, the "Gaussian white noise" has a normal distribution with zero mean and standard variation σ.

For the system in Eq. (2.28), the Gaussian white noise $\{w\}$ on states and Gaussian white noise $\{v\}$ on output has the following statistical expectations with respect to their mean and autocorrelations. The mean vector and autocorrelation matrix of w and v are:

$$\begin{aligned} \mu_w &= E\{w\} = 0, R_{ww} = E\{ww^T\} = \sigma^2 I, \\ \mu_v &= E\{v\} = 0, R_{vv} = E\{vv^T\} = \sigma^2 I \end{aligned} \tag{2.29}$$

where E is the expected value operator and I is an identity matrix.

Colored noise: Colored noises are some uncertainties that are dependent on their past states and have an autocorrelation. In particular, passing a "Gaussian white noise" from first-order filter results colored noise.

The model of a first-order filter, which results in colored noise $\{w\}$ on states is

$$w_{k+1} = cw_k + e_1 \tag{2.30}$$

where $\{e_1\}$ is Gaussian white noise and c is a real constant.

Similarly, the realization $\{v\}$ for colored noise on measured output is defined as

$$v_{k+1} = dv_k + e_2 \tag{2.31}$$

where $\{e_2\}$ is Gaussian white noise and d is a real constant.

2.5 Stochastic/random processes

A stochastic process is any process describing the evolution in time of a random phenomenon. A *stochastic process* represents a system for which there are observations at certain times, and the observed value at each time is a random variable. This means that, at a specific time for each observation, there is a certain probability to get a specific outcome. In general, probability depends on what has been obtained in the previous observations. In a probabilistic sense, a stochastic process $X(t)$ consists of an experiment with a probability measure P defined on a sample space S and a function that assigns a time function $x(t, s)$ associated with outcomes of an experiment. $X(t)$ is a discrete value process if the set of all possible values of $X(t)$ at all times t is a countable set S_X; otherwise, $X(t)$ is a continuous value process. The notation $X(t)$ refers to either the process or the variable that exhibits randomness at time t.

How the stochastic representation of a process/variable/signal is different from its deterministic representation is illustrated as follows. Certain variables representing physical quantities such as voltage, current, distance, temperature, etc. are called signals. They are real variables. Time is generally an independent variable and a variable/signal is said to be deterministic, if it is exactly predictable for the time span of interest. Examples include

$$x(t) = 10 \sin 2\pi f \quad \text{(sine wave)} \tag{2.32}$$

$$x(t) = \begin{cases} 1, & t \geq 0 \\ 0, & t \leq 0 \end{cases} \quad \text{(unit step)} \tag{2.33}$$

$$x(t) = \begin{cases} 1 - e^{-t}, & t \geq 0 \\ 0, & t < 0 \end{cases} \quad \text{(exponential response)} \tag{2.34}$$

These signals are described by functions in the usual mathematical sense. By specifying a numerical value of t, the corresponding value of x is determined from these functions. Thus, there exists a functional relationship between the variables, and there is nothing chancy about any of these variables.

In contrast to a deterministic signal, a random signal has always some element of chance associated with it. Thus, it is not predictable in a deterministic sense. Example of random signals are:

$$x(t) = 10 \sin(2\pi t + \theta) \tag{2.35}$$

where θ is a random variable uniformly distributed between 0 and 2π.

$$x(t) = A \sin(2\pi t + \theta) \tag{2.36}$$

where θ and A are independent random variables with known dimensions. As these signals have some element of chance associated with them, they are called random signals. These relations are known as random/stochastic representations.

Let us consider the description of a random signal in more detail. It might be the common audible radio noise. If we note from an oscillographic recording of the radio speaker current, it might appear as shown in Fig. 2.5

Such a signal might have some kind of spectral description. Yet, the precise mathematical description of such a signal is remarkably elusive. Imagine sampling the noise shown in the above figure at a particular point in time, say t_1. The numerical value obtained would be governed largely by chance, which suggests it might be considered to be a random variable. Each of these noise-like signals, so generated is called a sample realization of the process. Samples of individual signals at time t_1 would then be sample realizations of the random variable $x(t_1)$. Four of these are shown in Fig. 2.6 as $x_A(t_1)$, $x_B(t_1)$, $x_C(t_1)$, and $x_D(t_1)$. If the signal is sampled at a different time, say t_2, a different random variable $x(t_2)$ is obtained. Thus, in this example, an infinite set of random variables is generated by the random process.

The radio experiment just described is an example of a continuous random process in which time evolves in a continuous manner. The probability density function describing the amplitude of radio signal can also be continuous. However, random processes can also be discrete.

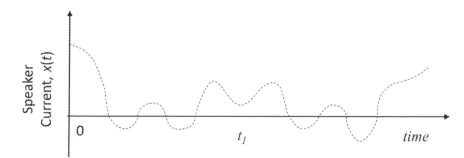

FIGURE 2.5   Typical radio noise signal.

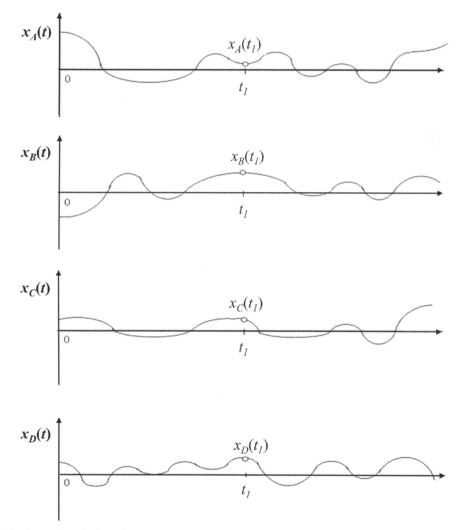

FIGURE 2.6   Ensemble of sample realizationsof a random process.

2.5.1   Stochastic model scheme

A system is deterministic if its inputs together with certain initial conditions and time uniquely specify the output. Otherwise, it is stochastic. We can explain the nonuniqueness response to input signals by supposing that the system has a "random noise" input in addition to the possible control inputs. Thus, denoting the input, output, and noise by u, y, and w respectively, we can represent the system as shown in Fig. 2.7.

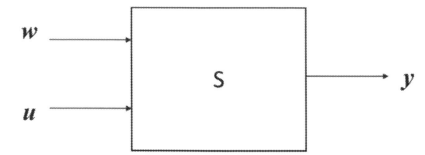

FIGURE 2.7 Stochastic representation of a system with inputs and outputs.

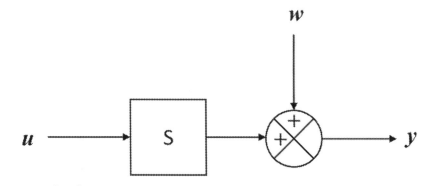
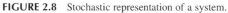

FIGURE 2.8 Stochastic representation of a system.

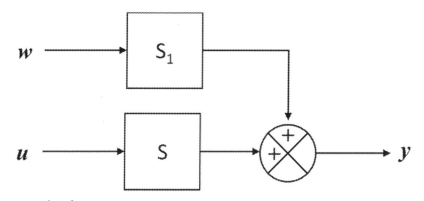
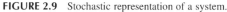

FIGURE 2.9 Stochastic representation of a system.

The actual noise in a system may be well generated internally, say by thermal noise in electronic components. If the fundamental input/output relationship is linear and the noise is additive, then the noise can be regarded as being added at the output, giving us somewhat a simplified model structure as depicted in Fig. 2.8. A full description of the system model with noise has to specify the input/output behavior of the system as well as the statistical characteristics of the noise w. One can specify that w is zero-mean and can be modeled as a part of the system S.

In terms of parameterization, the whole model is reduced to two linear systems, S and S_1; this configuration is shown in Fig. 2.9.

2.5.2 Stochastic representation of real processes

Consider a continuous stirred tank reactor (CSTR) with jacketed vessel (shown in Fig. 2.10) as an example of a stochastic representation of a real process.

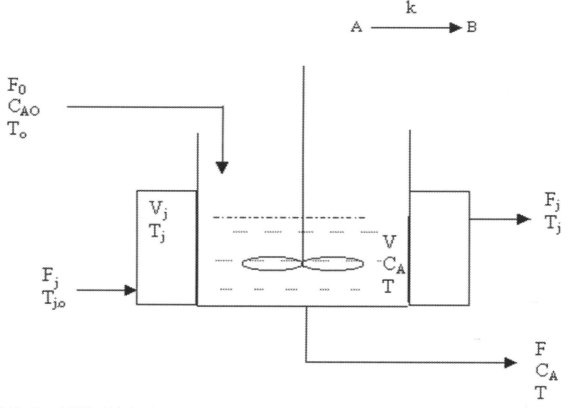

FIGURE 2.10   General CSTR with jacketed system.

Assuming a first-order exothermic reaction in the reactor, the dynamic model of the system is expressed by the following mass and energy balance relations:

$$V\frac{dC_A}{dt} = F(C_{A0} - C_A) - VkC_A$$

$$V\rho C_p\frac{dT}{dt} = F\rho C_p(T_0 - T) - \lambda VkC_A - UA(T - T_j) \tag{2.37}$$

$$k = k_0 e^{-E/RT}$$

Inputs: C_{A0}, T_0, T_j, F_j
Outputs: C_A, T

Here C_A is concentration of reactant A at any time t, C_{A0} is initial reactant concentration, k is reaction rate constant, V is volume of reactor, F is reactant flow rate, T is reaction temperature, T_0 is initial reactant temperature, ρ is density, C_p is specific heat, λ is heat of reaction, T_j is jacket fluid temperature, F_j is jacket fluid flow rate, Vj is jacket volume, U is overall heat transfer coefficient, and A is cross-sectional area of the reactor.

The above model becomes a stochastic representation with the addition of process noise w_1 for the mass balance equation, and w_2 for the energy balance equation:

$$V\frac{dC_A}{dt} = F(C_{A0} - C_A) - VkC_A + w_1$$

$$V\rho C_p\frac{dT}{dt} = F\rho C_p(T_0 - T) - \lambda VkC_A - UA(T - T_j) + w_2 \tag{2.38}$$

If C_A and T are measurable quantities, the random measurements are represented by
$C_A + v_1$ and $T + v_2$, where v_1 and v_2 are measurement noises. These noises can be additive random Gaussian noises.

2.5.3 Stochastic representation of general dynamic models

General dynamic stochastic and deterministic models are given as follows.

Continuous-time linear model:

Deterministic:
process $\quad \dot{x}(t) = Ax(t) + Bu(t)$
measurement $\quad y = Hx(t)$
Noise:
process $\quad \dot{x}(t) = Ax(t) + Bu(t) + w(t)$
measurement $\quad y = Hx(t) + \nu(t)$

Discrete-time linear model

$$
\begin{aligned}
&\text{Deterministic:}\\
&\text{process} \quad x_{k+1} = Ax_k + Bu_k\\
&\text{measurement} \quad y_k = Hx_k\\
&\text{Noise:}\\
&\text{process} \quad x_{k+1} = Ax_k + Bu_k + w_k\\
&\text{measurement} \quad y_k = Hx_k + \nu_k
\end{aligned}
\tag{2.39}
$$

Continuous-time nonlinear model

Deterministic :
process $\quad \dot{x}(t) = f(x,\, u,\, \theta,\, t)$
measurement $\quad y = Hx(t)$
Noise :
process $\quad \dot{x}(t) = f(x,\, u,\, \theta,\, t) + w(t)$
measurement $\quad y = Hx(t) + \nu(t)$

In the above equations, the notation is in standard terminology where x is the state, u is the input, y is the output, θ is a parameter, f is a nonlinear function, w is process noise, v is measurement noise, H is measurement matrix, A and B are coefficient matrices, t is continuous-time, and k is discrete-time.

2.6 Filtering, estimation, and prediction problem

Fundamentally, state estimation is the problem of determining the values of the state variables from the knowledge of the process outputs and inputs. Filtering and prediction are elementary concepts to state estimation.

2.6.1 Filtering, prediction, and smoothing

The basic definitions are given as follows.

Filtering is an operation that involves the extraction of information about a quantity of interest at time, t by using data measured up to and including t. In discussing the filtering problem, it is implied that the system under consideration is noisy. The noise may arise in a number of ways. For example, the inputs to the system may be unknown or unpredictable and may associate with noise, or the outputs of the system derived from the sensors may be noisy, causing some inaccuracy to the measurements. Filtering implies the recovery of the signal at time t of some system information about $s(t)$ using measurements up till time t. A simple example of filtering application in everyday life is in radio reception. Here the signal of interest is the voice signal. This signal is used to modulate a high-frequency carrier that is transmitted to a radio receiver. The received signal is inevitably corrupted by noise, and so, when demodulated, it is filtered to recover the original signal to the possible extent.

Prediction is the forecasting side of information processing. Its aim is to derive information about what the quantity of interest will be like at some time $t + \tau$ in the future ($\tau > 0$) by using data measured up to and including time t. Prediction is mostly referred to as one-step-ahead prediction in estimation problems.

Smoothing is a posteriori form of estimation in which data measured after the time of interest are used for the estimation. Specifically, the smoothed estimate at time t' is obtained by using data measured over the interval $[0, t]$, where $t' < t$.

The Bayes filtering approach and the stochastic filtering approach are the main approaches for filtering and estimation.

2.6.2   Bayes filtering approach for state estimation

In the Bayes approach, probability theory-based recursive Bayesian estimation known as Bayes filter is used for state estimation [2]. Bayes filter is a general probabilistic approach for estimating an unknown probability density function (PDF) recursively over time using incoming measurements and a mathematical model depicting randomness in its description. In continuous-discrete Bayesian filtering, the state is assumed to evolve according to a continuous-time stochastic process and the measurements are considered to be the samples of a discrete-time stochastic process. In discrete Bayes filtering, a discrete-time stochastic process and discrete-time measurements are considered.

Consider a state space model represented by the following state and measurement equations:

$$\begin{aligned} x_k &= f_k(x_{k-1}, w_{k-1}) \\ y_k &= h_k(x_k, v_k) \end{aligned} \tag{2.40}$$

Here k is the discrete-time index, and f_k is a nonlinear function describing the evolution of the state with identically distributed process noise, w_{k-1}. The h_k is a nonlinear function mapping the state space to the measurements with independent and identically distributed noise v_k. The elementary Bayesian strategy for estimating a random variable is briefed as follows. Let $y_{1:k} = (y_i, i = 1, \ldots, k)$ be the measurement sequence. On assuming the known prior distribution $p(x_0)$, the posterior probability can be obtained sequentially by prediction as follows:

$$p(x_k|y_{1:k-1}) = \int p(x_k|x_{k-1})p(x_{k-1}|y_{1:k-1})dx_{k-1} \tag{2.41}$$

This updating step amounts to multiplying the prior distribution by some likelihood function $p(y|x)$ that measures the adequacy of the prior initial guesses with respect to the observations according to the relation $p(x|y) \propto p(y|x)p(x)$:

$$p(x_k|y_{1:k}) = \frac{p(y_{1:k}|x_k)p(x_k|x_{k-1})}{p(y_k|y_{1:k-1})} \tag{2.42}$$

The above equations are the optimal solution from a Bayesian perspective to the nonlinear state estimation problem. The main challenge in Bayesian state estimation for nonlinear systems is that these basic equations cannot be readily evaluated to obtain $p(x_k/y_{1:k})$ because they involve high dimensional integrations.

2.6.3   Stochastic filtering approach for state estimation

Stochastic filtering theory was first established in and around 1940s [3,4], which led to the introduction of the classic Kalman filter (KF) in 1960 [5] and the subsequent Kalman–Bucy filter in 1961 [6]. Stochastic filtering theory-based KF and its variants have dominated with numerous applications in various engineering and scientific areas. KF is a stochastic method used for state estimation in linear systems. The further extended version of the KF, namely, the extended Kalman filter (EKF), has been developed for state estimation in nonlinear systems [2,7,8]. Subsequently, other variations of the KF, such as particle filters and the unscented KF, are found to exhibit great promise for nonlinear state estimation [9–12]. Also, various elegant and computationally efficient observers based on nonlinear systems theory were introduced for state estimation in certain classes of nonlinear systems [13–19].

The mathematical formulation of the stochastic filtering problem requires a dynamic system with measurements. The system is like a physical object, and its behavior can generally be described by equations. The system operates in real-time so that the independent variable in the equations is time. The system may operate in discrete or continuous time, with the underlying equations either difference equations or differential equations. The output of the system may change at discrete instants of time or on a continuous-time basis. Let us consider the general stochastic problem in a dynamic state-space form as given by

$$\dot{x}(t) = f(x(t), u(t), w(t), t) \tag{2.43}$$

$$y(t) = h(x(t), u(t), v(t), t) \tag{2.44}$$

Eq. (2.43) is called the state equation and Eq. (2.44) is called the measurement equation. In the above equations, $x(t)$ represents the state vector, $y(t)$ denotes the measurement vector, $u(t)$ represents the input vector, and $w(t)$ and $v(t)$ refer to the process and measurement noises, respectively. In a simple sense, the stochastic filtering problem is to begin the estimation with initial state $x(0)$ at time t_0 and estimate the optimal current state $x(t)$ at time t given the observations $y(t)$ up to time t. The estimation algorithms such as KF, EKF, and its variants can be used in conjunction with the dynamic model of the system and its noise corrupted measurements for optimal state estimation. These techniques, along with the algorithms, are elaborated in the forthcoming Chapters 3, 4 and 5.

2.7 Summary

Processes that are associated with random phenomenon caused by noise and uncertainties are defined as stochastic or random processes. In a realistic situation, the state is often not directly obtainable, rather it is usually inferred or estimated based on the system outputs measured by instruments (sensors) along with the support of a dynamic model representing the system. The general idea of the filtering and estimation problem is to establish the best estimate for the true variable of a system from an incomplete and noisy set of observations of the system. This chapter describes the basic concepts concerning to stochastic variables, noise, probability, probability distributions, random processes, stochastic representation of general dynamic models, filtering, prediction, and state estimation.

References

[1] M.R. Spiegel, J.J. Schiller, R.A. Srinivasan, Schaum's Outline of Probability and Statistics, McGraw-Hill Education, New York, NY, 2013.

[2] A.H. Jazwinski, Stochastic Processes and Filtering Theory, Science Engineering. Academic Press., New York, 1970.

[3] J.L Doob, Stochastic Processes, John Wiley, New York, 1953.

[4] R.L. Stratonovich, On the theory of optimal nonlinear filtration of random functions, Theory Prob. Appl. 4 (1959) 223−225.

[5] R.E. Kalman, A new approach to linear filtering and prediction problems, J. Basic Eng. 82 (1) (1960) 35−45.

[6] R.E. Kalman, R.S. Bucy, New results in linear filtering and prediction theory, J. Basic Eng. 83 (1) (1961) 95−108.

[7] A. Gelb, Applied Optimal Estimation, MIT Press, Cambridge, MA, 1974.

[8] B. Anderson, J. Moore, Optimal Filtering, Prentice-Hall, Englewood Cliffs, 1979.

[9] J.B. Rawlings, B.R. Bakshi, Particle filtering and moving horizon estimation, Comput. Chem. Eng. 30 (10−12) (2006) 1529−1541.

[10] R. Kandepu, B. Foss, L. Imsland, Applying the unscented Kalman filter for nonlinear state estimation, J. Process. Control. 18 (7−8) (2008) 753−768.

[11] E. Bølviken, P.J. Acklam, N. Christophersen, J.-M. Størdal, Monte Carlo filters for non-linear state estimation, Automatica 37 (2) (2001) 177−183.

[12] A.K. Gyorgy, A. Kelemen, L. David, Unscented Kalman filters and particle filter methods for nonlinear state estimation, Procedia Technol. 12 (2014) 65−74.

[13] V. Andrieu, Exponential convergence of nonlinear Luenberger observers, in: 49th IEEE Conference on Decision and Control (CDC), 2010, pp. 2163−2168: IEEE.

[14] X. Xia, M. Zeitz, On nonlinear continuous observers, Int. J. Control. 66 (6) (1997) 943−954.

[15] A.R. Phelps, On constructing nonlinear observers, SIAM J. Control Optim. 29 (3) (1991) 516−534.

[16] M. Zeitz, The extended Luenberger observer for nonlinear systems, Syst. Control Lett. 9 (2) (1987) 149−156.

[17] D. Bestle, M. Zeitz, Canonical form observer design for non-linear time-variable systems, Int. J. Control 38 (2) (1983) 419−431.

[18] S. Rua, R.E. Vasquez, N. Crasta, C.A. Zuluaga, Observability analysis and observer design for a nonlinear three tank system: theory and experiments, Sensors 11 (214) (2020) 1−15.

[19] A. Chakrabarthy, E. Fridman, S.H. Zak, G.T. Buzzard, State and unknown input observers for nonlinear systems with delayed measurements, Automatica 95 (2018) 246−253.

Chapter 3

Linear filtering and observation techniques

3.1 Introduction

Many processes in engineering, science, economics, and other areas are described by state space representation. The state space description provides the dynamics of a system as a set of coupled ordinary differential equations in a set of internal variables known as state variables, together with a set of algebraic equations that combine the state variables into output variables. State space representation is important for mechanistic model based optimal state estimation, which has immense use in process systems engineering strategies such as process monitoring, fault diagnosis, control, and online optimization. State estimation in linear systems is performed using techniques based on optimal linear systems theory [1−4]. Linear system state estimation of this chapter sets a stage to nonlinear state estimation approaches, which are being elaborated in lateral chapters of this book. This chapter describes the contents such as the mathematical representation of a system and its associated variables in state space domain, continuous time and discrete time representation of linear dynamic systems, concepts of observability and controllability of linear systems, different continuous and discrete forms of state estimation algorithms, and their applications to linear systems.

3.2 Representation of a system and associated variables

Primarily, the states, inputs and outputs of a dynamic system are defined. Then, the continuous time and discrete time representation of time invariant and time varying linear systems are described.

3.2.1 Definition of system state

The concept of the state of a dynamic system refers to a minimum set of variables, known as state variables, that fully describe the system and its response to any given set of inputs. This definition asserts that the mathematical description of a system in terms of a minimum set of variables, $x_i(t)$, $i = 1,\ldots, n$, together with knowledge of these state variables at an initial time t_0 and the system inputs for time $t \geq t_0$ are sufficient to predict the future system state and outputs for all time $t > t_0$. This definition states that the dynamic behavior of a state-determined system is completely characterized by the response of the set of n variables $x_i(t)$, where the number n is defined as the order of the system.

3.2.2 System variables

The model representing the dynamics of a system utilizes three types of variables, namely, the input variables $u(t)$, the output variables $y(t)$, and the state variables $x(t)$. As defined in Section 3.2.1, the state is the minimum amount of information necessary at the current time to uniquely determine the dynamic behavior of the system at all future times given the inputs and parameters. The input is an independent variable analogous to the forcing function a system. Parameters consist of the physical properties used in the description of the system. An output is a variable that can be measured directly. The output may be a state variable or some function of the state.

The state variable formulation of a multivariable system and the vector notation for the system variables are given as follows:

$$u(t) = \begin{bmatrix} u_1(t) \\ u_2(t) \\ . \\ . \\ u_m(t) \end{bmatrix}, \quad y(t) = \begin{bmatrix} y_1(t) \\ y_2(t) \\ . \\ . \\ y_p(t) \end{bmatrix}, \quad x(t) = \begin{bmatrix} x_1(t) \\ x_2(t) \\ . \\ . \\ x_n(t) \end{bmatrix} \tag{3.1}$$

Optimal State Estimation for Process Monitoring, Fault Diagnosis and Control. DOI: https://doi.org/10.1016/B978-0-323-85878-6.00011-7

The n-state variables describing the behavior of a system represent the state vector or state variable vector. The n-dimensional space whose coordinate axes consist of the x_1 axis, x_2 axis, ..., x_n axis is called state space. Any state can be represented by a point in the state space.

3.2.3 State space representation of linear systems

State space representation is a mathematical model of a physical system expressed as a function of input, output, and state variables related by first-order differential equations or difference equations. The state of the system can be represented as a vector within that space.

An important property of the linear state space model is that all system variables may be represented by a linear combination of the state variables x_i and the system inputs u_i by the ordinary differential equations of the following form:

$$
\begin{aligned}
\dot{x}_1 &= a_{11}x_1 + a_{12}x_2 + ... + a_{1n}x_n & + b_{11}u_1 + b_{12}u_2 + ... + b_{1m}u_m \\
\dot{x}_2 &= a_{21}x_1 + a_{22}x_2 + ... + a_{2n}x_n & + b_{21}u_1 + b_{22}u_2 + ... + b_{2m}u_m \\
&\qquad\qquad\vdots \\
\dot{x}_n &= a_{n1}x_1 + a_{n2}x_2 + ... + a_{nn}x_n & + b_{n1}u_1 + b_{n2}u_2 + ... + b_{nm}u_m
\end{aligned}
\tag{3.2}
$$

The coefficients a_{ij} and b_{ij} are constants. The above equations can be written in a compact form as

$$\dot{x}(t) = Ax(t) + Bu(t) \tag{3.3}$$

where $x(t)$ is $n \times 1$ state vector, $u(t)$ is $m \times 1$ input vector.

In Eq. (3.3), A is $n \times n$ coefficient matrix defined by

$$
A = \begin{bmatrix}
a_{11} & a_{12} & \cdots & a_{1n} \\
a_{21} & a_{22} & \cdots & a_{2n} \\
 & & \cdot & \\
 & & \cdot & \\
a_{n1} & a_{n2} & \cdots & a_{nn}
\end{bmatrix}
$$

B is an $n \times m$ input matrix defined by

$$
B = \begin{bmatrix}
b_{11} & b_{12} & \cdots & b_{1m} \\
b_{21} & b_{22} & \cdots & b_{2m} \\
 & & \cdot & \\
 & & \cdot & \\
b_{n1} & b_{n2} & \cdots & b_{nn}
\end{bmatrix}
$$

3.2.4 Output equations

The system output variables at time t are linear combinations of the state variables x_i and input variables u_i and they are expressed by the algebraic equations of the following form:

$$
\begin{aligned}
y_1(t) &= c_{11}x_1 + c_{12}x_2 + ... + c_{1n}x_n & + d_{11}u_1 + d_{12}u_2 + ... + d_{1m}u_m \\
y_2(t) &= c_{21}x_1 + c_{22}x_2 + ... + c_{2n}x_n & + d_{21}u_1 + d_{22}u_2 + ... + d_{2m}u_m \\
&\qquad\qquad\vdots \\
y_p(t) &= c_{p1}x_1 + c_{p2}x_2 + ... + c_{pn}x_n & + d_{p1}u_1 + d_{p2}u_2 + ... + d_{pm}u_m
\end{aligned}
\tag{3.4}
$$

where $y(t)$ is $p \times 1$ output vector. The compact form of the above equations can be expressed as

$$y(t) = Cx(t) + Du(t) \tag{3.5}$$

where C is $p \times n$ output coefficient matrix which is defined by

$$
C = \begin{bmatrix}
c_{11} & c_{12} & \cdots & c_{1n} \\
c_{21} & c_{22} & \cdots & c_{2n} \\
& & \cdot & \\
& & \cdot & \\
& & \cdot & \\
c_{p1} & c_{p2} & \cdots & c_{pn}
\end{bmatrix}
$$

D is $p \times m$ input coefficient matrix which is defined by

$$
D = \begin{bmatrix}
d_{11} & d_{12} & \cdots & d_{1m} \\
d_{21} & d_{22} & \cdots & d_{2m} \\
& & \cdot & \\
& & \cdot & \\
& & \cdot & \\
d_{p1} & d_{p2} & \cdots & d_{pm}
\end{bmatrix}
$$

If the system output variables at time t are linear combinations of the state variables alone, the output equation in Eq. (3.5) becomes

$$
y(t) = Cx(t) \tag{3.6}
$$

3.2.5   Discrete time representation of linear systems

Discrete time processes arise in either of two ways. Firstly, a sequence of events takes place naturally in discrete steps. In this process, the length of each time step may be a fixed value, and the time variable moves in discrete jumps. Discrete time processes may also arise from sampling a continuous process at discrete times. In this case, the observation times may not be equally spaced. Most linear systems in the real world are described with continuous time dynamics of the type shown in Eq. (3.3). However, state estimation and control algorithms are mostly implemented in digital electronics. This often requires a transformation of continuous time dynamics to discrete time dynamics. This section discusses how a continuous time linear system can be transformed into a discrete time linear system.

The solution of a continuous time linear system in Eq. (3.3) is given by

$$
x(t) = e^{A(t-t_0)}x(t_0) + \int_{t_0}^{t} e^{A(t-\tau)}Bu(\tau)d\tau \tag{3.7}
$$

Let $t = t_k$ be some discrete time point, and t_{k-1} be the previous discrete time point. Assume that $A(\tau)$, $B(\tau)$, and $u(\tau)$ are approximately constant in the interval of integration. Then the solution at $x(t_k)$ is obtained as

$$
x(t_k) = e^{A(t_k-t_{k-1})}x(t_{k-1}) + \int_{t_{k-1}}^{t_k} e^{A(t_k-\tau)}Bu(\tau)d\tau \tag{3.8}
$$

By defining $\Delta t = t_k - t_{k-1}$, $\alpha = \tau - t_{k-1}$ and substituting for τ, the above equation is expressed as

$$
\begin{aligned}
x(t_k) &= e^{A\Delta t}x(t_{k-1}) + \int_0^{\Delta t} e^{A(\Delta t-\alpha)}Bu(t_{k-1})d\alpha \\
&= e^{A\Delta t}x(t_{k-1}) + e^{A\Delta t}\int_0^{\Delta t} e^{-A\alpha}Bu(t_{k-1})d\alpha \\
&= Fx_{k-1} + Gu_{k-1}
\end{aligned} \tag{3.9}
$$

where F and G in the above equation are defined by

$$
F = e^{A\Delta t}
$$

$$
G = F\int_0^{\Delta t} e^{-A\tau}d\tau B
$$

The expression G can be further simplified as follows.

The term $\int_0^{\Delta t} e^{-A\tau} d\tau$ in the expression can be further simplified as follows:

$$\int_0^{\Delta t} e^{-A\tau} d\tau = \int_0^{\Delta t} \sum_{j=0}^{\alpha} \frac{(-A\tau)^j}{j!} d\tau$$

$$= \int_0^{\Delta t} [I - A\tau + A^2\tau^2/2! + A^3\tau^3/3! - \cdots] d\tau$$

$$= \left[I\tau - A\tau^2/2! + A^2\tau^3/3! - \cdots \right]_0^{\Delta t}$$

$$= \left[I\Delta\tau - A(\Delta\tau)^2/2! + A^2(\Delta\tau)^3/3! - \cdots \right]$$

$$= \left[A\Delta\tau - (A\Delta\tau)^2/2! + (A\Delta\tau)^3/3! - \cdots \right] A^{-1}$$

$$= \left[I - e^{-A\Delta t} \right] A^{-1}$$

Thus the conversion from continuous time system matrices A and B to discrete time system matrices F and G can be summarized as

$$F = e^{A\Delta t}$$

$$G = F\left[I - e^{-A\Delta t} \right] A^{-1} B$$

Eq. (3.9) is a linear approximation to the continuous time dynamics given in Eq. (3.3). Note that this system defines x_k only at the discrete time points $\{t_k\}$. This expression does not say anything about what happens to the continuous time signal $x(t)$ in between the discrete time points.

3.2.6 Time variant representation of linear systems

The linear system models presented by Eqs. (3.3) and (3.6) are referred to as time invariant as the system matrices are independent of time. For continuous time linear systems in which the system matrices are functions of time, the time variant representation of Eqs. (3.3) and (3.6) are given by

$$\dot{x}(t) = A(t)\, x(t) + B(t)\, u(t) \tag{3.11}$$

$$y(t) = C(t)\, x(t) \tag{3.12}$$

For discrete time linear system with time dependent coefficient matrices, the model is represented by

$$x_k = F_{k-1} x_{k-1} + G_{k-1} u_{k-1}$$

$$y_k = C_k x_k$$

3.3 Concepts of observability and controllability

The concepts of observability and controllability introduced by Kalman in 1960 were subsequently analyzed for linear systems [5–7]. The study of observability is closely related to observer design, a technique used to construct a dynamic estimator, which produces estimates of the system state variables using information about the system inputs and outputs. The design of full order and reduced order observers, along with other estimators, are presented in later sections of this chapter. The study of controllability is also important in the design of an estimator based state feedback control system. In this section, the concepts of observability and controllability of linear systems are discussed.

3.3.1 Observability of a continuous system

A system is said to be completely observable if every state $x(t)$ can be completely identified by measurements of the output $y(t)$ over a finite time interval. Consider an nth order linear invariant system defined by Eqs. (3.3) and (3.6).

To test the observability of a linear time invariant system, the observability matrix Q_o is defined as

$$Q_o = \begin{bmatrix} C \\ CA \\ CA^2 \\ \vdots \\ CA^{n-1} \end{bmatrix} \tag{3.14}$$

The system defined by Eqs. (3.3) and (3.6) is completely observable if and only if Q_o has full rank, that is, rank $(Q_o) = n$. This condition implies that the system with the pair (A, C) is observable.

3.3.2   Observability of a discrete system

Consider a linear, time invariant, and discrete time system in the form

$$x(k + 1) = A_d x(k) + B_d u(k)$$

$$y(k) = C_d x(k) \tag{3.15}$$

where k refers the discrete point, A_d, B_d, and C_d are constant matrices with appropriate dimensions. For this system, the observability matrix is defined as

$$Q_{od} = \begin{bmatrix} C_d \\ C_d A_d \\ C_d A_d^2 \\ \vdots \\ C_d A_d^{n-1} \end{bmatrix} \tag{3.16}$$

The system defined by Eq. (3.15) is completely observable if and only if Q_{od} has full rank, that is, rank $(Q_{od}) = n$. This condition implies that the system with the pair (A_d, C_d) is observable.

3.3.3   Controllability of a continuous system

Consider an nth order linear time invariant continuous system defined by Eqs. (3.3) and (3.6). Controllability is a property of the coupling between the input and the state and thus involves the matrices A and B. A system is said to be completely state controllable, if it is possible to transfer the system from any initial state $x(t_0)$ to any other desired state $x(t)$ in specified finite time by a control vector $u(t)$.

The system defined by Eqs. (3.3) and (3.6) is completely controllable if and only if the rank of the composite matrix

$$Q_c = \begin{bmatrix} B \vdots AB \vdots ... \vdots A^{n-1}B \end{bmatrix} \tag{3.17}$$

contains n linearly independent row or column vectors, that is, rank $(Q_c) = n$. Since this criterion involves the matrices A and B, we say the pair (A, B) is controllable.

3.3.4   Controllability of a discrete system

Consider a linear invariant system in Eq. (3.15). The system controllability is defined as its ability to transfer the system from any initial state $x(0) = x_0$ to any desired state $x(k) = x_f$ in a finite time. The system defined by Eq. (3.15) is completely controllable if and only if the rank of the composite matrix

$$Q_{cd} = \begin{bmatrix} B_d \vdots A_d B_d \vdots ... \vdots A_d^{n-1} B_d \end{bmatrix} \tag{3.18}$$

contains n linearly independent row or column vectors, that is, rank $(Q_{cd}) = n$.

Example 3.1: Determine the observability and controllability of a continuous time system represented by:

$$\dot{x}(t) = \begin{bmatrix} 0 & 1 & 1 \\ -2 & -3 & 0 \\ 0 & 2 & -3 \end{bmatrix} x(t) + \begin{bmatrix} 0 \\ 1 \\ 0 \end{bmatrix} u(t)$$

$$y(t) = \begin{bmatrix} 1 & 0 & 0 \end{bmatrix} x(t)$$

Solution:

The system is of the form:

$$\dot{x}(t) = Ax(t) + Bu(t)$$
$$y(t) = Cx(t)$$

The system coefficient matrices are

$$A = \begin{bmatrix} 0 & 1 & 1 \\ -2 & -3 & 0 \\ 0 & 2 & -3 \end{bmatrix}, \ B = \begin{bmatrix} 0 \\ 1 \\ 0 \end{bmatrix}, \ C = \begin{bmatrix} 1 & 0 & 0 \end{bmatrix}$$

Test for observability:

$$Q_o = \begin{bmatrix} C \\ CA \\ CA^2 \end{bmatrix} = \begin{bmatrix} 1 & 0 & 0 \\ 0 & 1 & 1 \\ -2 & -1 & -3 \end{bmatrix}$$

$$|Q_o| = -2$$

The system is completely state observable as the rank condition is satisfied.

Test for controllability:

$$Q_c = \begin{bmatrix} B \vdots AB \vdots A^2B \end{bmatrix}$$
$$= \begin{bmatrix} 0 & 1 & -1 \\ 1 & -3 & 7 \\ 0 & 2 & -12 \end{bmatrix}$$

$$|Q_c| = 10$$

The system is completely state controllable as the rank condition is satisfied.

Example 3.2: Determine the observability and controllability of a discrete time system given by:

$$x(k+1) = \begin{bmatrix} -3 & -2 & 0 \\ 2 & 1 & 0 \\ 0 & -1 & -1 \end{bmatrix} x(k) + \begin{bmatrix} 1 \\ 0 \\ 1 \end{bmatrix} u(k)$$

$$y(k) = \begin{bmatrix} 1 & 0 & 1 \end{bmatrix} x(k)$$

Solution:

The system is of the form:

$$x(k+1) = A_d x(k) + B_d u(k)$$
$$y(k) = C_d x(k)$$

The system coefficient matrices are

$$A_d = \begin{bmatrix} -3 & -2 & 0 \\ 2 & 1 & 0 \\ 0 & -1 & -1 \end{bmatrix}, \ B_d = \begin{bmatrix} 1 \\ 0 \\ 1 \end{bmatrix}, \ C_d = \begin{bmatrix} 1 & 0 & 1 \end{bmatrix}$$

Test for observability:

$$Q_{od} = \begin{bmatrix} C_d \\ C_d A_d \\ C_d A_d^2 \end{bmatrix} = \begin{bmatrix} 1 & 0 & 1 \\ -3 & -3 & -1 \\ 3 & 4 & 1 \end{bmatrix}$$

$$|Q_o| = -2$$

The system is completely state observable.
Test for controllability:

$$Q_{cd} = \begin{bmatrix} B_d \vdots A_d B_d \vdots A_d^2 B_d \end{bmatrix}$$

$$= \begin{bmatrix} 1 & -3 & 5 \\ 0 & 2 & -4 \\ 1 & -1 & -1 \end{bmatrix}$$

$$|Q_{cd}| = -4$$

The system is completely state controllable.

3.4   Recursive weighted least squares estimator

Recursive least squares (RLS) is an algorithm that recursively computes the model parameters in order to minimize the weighted least squares cost function based on the input data [8,9]. The iterative, recursive algorithm adaptively modifies the vector of parameters of the presented model using the input data obtained from the process. More often, we obtain measurements sequentially and want to update our estimate with each new measurement. This section shows how to recursively compute the weighted least squares estimate for each new measurement condition.

Suppose $x = (x_1, x_2, \ldots, x_n)^T$ is a constant but unknown vector, and $y = (y_1, y_2, \ldots, y_l)^T$ is an l-element noisy measurement vector. Our task is to find the "best" estimate $\tilde{x}$ of x. We obtain measurements y sequentially and want to update our estimate $\tilde{x}$ with each new measurement. We consider that each y_i is a linear combination of x_j, $1 \le j \le n$, with the addition of some measurement noise ν_i. Thus, we are working with the following linear system,

$$y = Hx + \nu \tag{3.19}$$

where $\nu = (\nu_1, \nu_2, \ldots, \nu_l)^T$, and H is an $l \times n$ matrix.
The expanded form Eq. (3.19) in vector-matrix notation is represented as

$$\begin{bmatrix} y_1 \\ y_2 \\ \vdots \\ y_l \end{bmatrix} = \begin{bmatrix} H_{11} & \cdots & H_{1n} \\ \vdots & & \vdots \\ \vdots & & \vdots \\ H_{l1} & \cdots & H_{ln} \end{bmatrix} \begin{bmatrix} x_1 \\ x_2 \\ \vdots \\ x_n \end{bmatrix} + \begin{bmatrix} v_1 \\ v_2 \\ \vdots \\ v_l \end{bmatrix} \tag{3.20}$$

For kth instant, Eq. (3.19) is written as

$$y_k = H_k x_k + \nu_k \tag{3.21}$$

Suppose we have $\hat{x}_{k-1}$ as an estimate at $k-1$ instant and the prediction measurement corresponding to $\hat{x}_{k-1}$ is denoted as

$$\hat{y}_k = H_k \hat{x}_{k-1} + v_k \tag{3.22}$$

In the above equation, H_k and v_k are assumed to be the same as in Eq. (3.21).
We want to update the estimate at kth instant from its previous estimate based on the new measurement y_k. For this purpose, we consider the following recursive equation,

$$\begin{aligned} \hat{x}_k &= \hat{x}_{k-1} + K_k \left(y_k - \hat{y}_k \right) \\ &= \hat{x}_{k-1} + K_k (y_k - H_k \hat{x}_{k-1}) \end{aligned} \tag{3.23}$$

Here H_k is an $m \times n$ matrix, K_k is an $n \times m$ matrix referred to as the estimator gain matrix and $(y_k - H_k \hat{x}_{k-1})$ is the correction term. Thus the new estimate $\hat{x}_k$ is obtained by modifying the previous estimate $\hat{x}_{k-1}$ with a correction term multiplied by the gain factor.

The current estimation error is the difference between the actual and estimated states,

$$\varepsilon_k = x_k - \hat{x}_k \tag{3.24}$$

$$\begin{aligned} &= x_k - \hat{x}_{k-1} - K_k (y_k - H_k \hat{x}_{k-1}) \\ &= \varepsilon_{k-1} - K_k (y_k - H_k \hat{x}_{k-1}) \\ &= \varepsilon_{k-1} - K_k (H_k x_k + v_k - H_k \hat{x}_{k-1}) \\ &= \varepsilon_{k-1} - K_k H_k (x_k - \hat{x}_{k-1}) - K_k v_k \\ &= \varepsilon_{k-1} - K_k H_k \varepsilon_{k-1} - K_k v_k \end{aligned}$$

$$= (I - K_k H_k)\varepsilon_{k-1} - K_k v_k \tag{3.25}$$

where I is $n \times n$ identity matrix. The mean of this error is

$$E(\varepsilon_k) = (I - K_k H_k)E(\varepsilon_{k-1}) - K_k E(v_k) \tag{3.26}$$

If $E(\nu_k) = 0$ and $E(\varepsilon_{k-1}) = 0$, then $E(\varepsilon_k) = 0$. Thus if the measurement noise ν_k has zero mean for all k, and the initial estimate of x is set equal to its expected value, then $\hat{x}_k = x_k$ for all k.

This shows that, on average, the estimate $\hat{x}$ will be equal to the true value x.

Each new estimate in Eq. (3.23) requires the computation of its associated error covariance matrix P_k. The minimization of the aggregated variance of the estimation errors at time k can be used to obtain the estimation error covariance matrix P_k. Thus

$$J_k = E(\|x_k - \hat{x}_k\|^2) \tag{3.27}$$

$$= E(\varepsilon_k \varepsilon_k^T) \tag{3.28}$$

$$= E(Tr(\varepsilon_k \varepsilon_k^T))$$

$$= Tr(P_k) \tag{3.29}$$

where Tr is the trace operator.

The $n \times n$ estimation error covariance matrix P_k is obtained by the substitution of Eq. (3.26) into Eq. (3.28):

$$P_k = E(\varepsilon_k \varepsilon_k^T)$$

$$= [(I - K_k H_k)E(\varepsilon_{k-1}) - K_k E(v_k)]\left[\left((I - K_k H_k)E(\varepsilon_{k-1}) - K_k E(v_k)\right)^T\right]$$

$$= (I - K_k H_k)E(\varepsilon_{k-1}\varepsilon_{k-1}^T)(I - K_k H_k)^T - K_k E(v_k \varepsilon_{k-1}^T)(I - K_k H_k)^T - (I - K_k H_k)E(\varepsilon_{k-1} v_k^T)K_k^T + K_k E(v_k v_k^T)K_k^T \tag{3.30}$$

The estimation error ε_{k-1} at time $k - 1$ is independent of the measurement noise ν_k at time k, which implies that

$$\begin{aligned} E(v_k \varepsilon_{k-1}^T) = E(v_k)E(\varepsilon_{k-1}^T) = 0 \\ E(\varepsilon_{k-1} v_k^T) = E(\varepsilon_{k-1})E(v_k^T) = 0 \end{aligned} \tag{3.31}$$

The error covariance matrix P_{k-1} is thus represented by

$$P_{k-1} = E(\varepsilon_{k-1}\varepsilon_{k-1}^T) \tag{3.32}$$

Also the covariance of v_k is expressed as the $m \times m$ measurement covariance matrix R_k:

$$R_k = E(v_k v_k^T) \tag{3.33}$$

By using the terms in Eqs. (3.31)−(3.33), the expression P_k in Eq. (3.30) becomes

$$P_k = (I - K_k H_k)P_{k-1}(I - K_k H_k)^T + K_k R_k K_k^T \tag{3.34}$$

The P_k in Eq. (3.34) is the covariance matrix of the estimation error, which is positive and can be recursively computed for each kth time estimate.

The recursive formula in Eq. (3.23) requires the determination of the optimal gain K_k at time k. This can be computed by the substitution of Eq. (3.34) into Eq. (3.29) and taking its partial derivative with respective K_k and equating it to zero:

$$\begin{aligned} J_k = Tr(P_k) = Tr((I - K_k H_k)P_{k-1}(I - K_k H_k)^T + K_k R_k K_k^T) \\ = Tr(P_{k-1} - K_k H_k P_{k-1} - P_{k-1}K_k^T H_k^T + K_k(H_k P_{k-1} H_k^T)K_k^T) + Tr(K_k R_k K_k^T) \end{aligned} \tag{3.35}$$

$$\begin{aligned} \frac{\partial J_k}{\partial K_k} = \frac{\partial}{\partial K_k}(Tr(P_{k-1} - K_k H_k P_{k-1} - P_{k-1}K_k^T H_k^T + K_k(H_k P_{k-1} H_k^T)K_k^T)) \\ + \frac{\partial}{\partial K_k}(Tr(K_k R_k K_k^T)) \end{aligned} \tag{3.36}$$

$$\begin{aligned} = -2P_{k-1}H_k^T + 2K_k H_k P_{k-1} H_k^T + 2K_k R_k \\ = -2P_{k-1}H_k^T + 2K_k(H_k P_{k-1} H_k^T + R_k) \end{aligned}$$

By setting the partial derivative of Eq. (3.36) to be zero, we have

$$K_k = \frac{P_{k-1}H_k{}^T}{(H_k P_{k-1} H_k{}^T + R_k)} \tag{3.37}$$

We substitute K_k into Eq. (3.34) to further simplify the P_k matrix. For this we write

$$S_k = H_k P_{k-1} H_k{}^T + R_k \tag{3.38}$$

Eq. (3.37), thus becomes

$$K_k = P_{k-1} H_k{}^T S_k^{-1} \tag{3.39}$$

On substituting the above K_k into Eq. (3.34),

$$P_k = \left(I - P_{k-1}H_k^T S_k^{-1} H_k\right) P_{k-1} \left(I - P_{k-1}H_k{}^T S_k^{-1} H_k\right)^T + P_{k-1}H_k{}^T S_k^{-1} R_k S_k^{-1} H_k P_{k-1}$$

$$= \left(P_{k-1} - P_{k-1}H_k^T S_k^{-1} H_k P_{k-1}\right)\left(I - P_{k-1}H_k{}^T S_k^{-1} H_k\right)^T + P_{k-1}H_k{}^T S_k^{-1} R_k S_k^{-1} H_k P_{k-1}$$

$$= P_{k-1} - P_{k-1}H_k^T S_k^{-1} H_k P_{k-1} - \left(P_{k-1} - P_{k-1}H_k^T S_k^{-1} H_k P_{k-1}\right)\left(P_{k-1}H_k{}^T S_k^{-1} H_k\right)^T + P_{k-1}H_k{}^T S_k^{-1} R_k S_k^{-1} H_k P_{k-1}$$

$$= P_{k-1} - P_{k-1}H_k^T S_k^{-1} H_k P_{k-1} - P_{k-1}H_k{}^T S_k^{-1} H_k P_{k-1} + P_{k-1}H_k^T S_k^{-1} H_k P_{k-1} H_k{}^T S_k^{-1} H_k P_{k-1} + P_{k-1}H_k{}^T S_k^{-1} R_k S_k^{-1} H_k P_{k-1}$$

$$= P_{k-1} - P_{k-1}H_k^T S_k^{-1} H_k P_{k-1} - P_{k-1}H_k{}^T S_k^{-1} H_k P_{k-1} + P_{k-1}H_k^T S_k^{-1} S_k^{-1} H_k P_{k-1} \left(H_k P_{k-1} H_k{}^T + R_k\right) \tag{3.40}$$

On replacing $(H_k P_{k-1} H_k{}^T + R_k)$ by S_k, the above equation becomes

$$P_k = P_{k-1} - P_{k-1}H_k^T S_k^{-1} H_k P_{k-1} - P_{k-1}H_k{}^T S_k^{-1} H_k P_{k-1} + P_{k-1}H_k^T S_k^{-1} S_k S_k^{-1} H_k P_{k-1}$$

$$= P_{k-1} - 2P_{k-1}H_k^T S_k^{-1} H_k P_{k-1} + P_{k-1}H_k^T S_k^{-1} H_k P_{k-1}$$

$$= P_{k-1} - P_{k-1}H_k^T S_k^{-1} H_k P_{k-1} \tag{3.41}$$

$$= P_{k-1} - K_k H_k P_{k-1} \tag{3.42}$$

$$= (I - K_k H_k) P_{k-1} \tag{3.43}$$

An alternative expression for covariance matrix can also be derived as follows. For this, we substitute S_k in Eq. (3.38) into Eq. (3.41), thus

$$P_k = P_{k-1} - P_{k-1}H_k^T (H_k P_{k-1} H_k{}^T + R_k)^{-1} H_k P_{k-1} \tag{3.44}$$

By taking inverse in both sides of Eq. (3.44),

$$(P_k)^{-1} = (\overbrace{P_{k-1}}^{A} - \overbrace{P_{k-1}H_k^T}^{B} \overbrace{(H_k P_{k-1} H_k{}^T + R_k}^{D})^{-1} \overbrace{H_k P_{k-1}}^{C})^{-1} \tag{3.45}$$

By matrix inversion lemma, suppose we have

$$(A - BD^{-1}C)^{-1} = A^{-1} + A^{-1}B(D - CA^{-1}B)^{-1} CA^{-1} \tag{3.46}$$

In accordance with the above analogy,

$$P_k{}^{-1} = P_{k-1}^{-1} + P_{k-1}^{-1} P_{k-1} H_k^T ((H_k P_{k-1} H_k{}^T + R_k) - H_k P_{k-1} P_{k-1}^{-1}(P_{k-1}H_k^T))^{-1} H_k P_{k-1} P_{k-1}^{-1}$$

$$= P_{k-1}^{-1} + H_k^T R_k^{-1} H_k$$

The alternative expression for covariance matrix is thus given as

$$P_k = (P_{k-1}^{-1} + H_k^T R_k^{-1} H_k)^{-1} \tag{3.47}$$

This expression is more complicated than the expression given by Eq. (3.43) as it involves three matrix inversions.

3.4.1 Estimation procedure

The procedure is summarized as follows.

1. Initialize the estimator with following state and covariances:

$$\hat{x}_0 = E(x_0)$$

$$P_0 = E((x - x_0)(x - x_0)^T)$$

In the case of perfect prior knowledge of the state, $P_0 = 0$, and in the case of no prior knowledge about the state, $P_0 = \infty I$.

2. Assume the measurement y_k as given by the Eq. (3.21):

$$y_k = H_k x_k + v_k$$

where the noise v_k has zero mean and covariance R_k. The measurement noise at each time step k is independent. On assuming white measurement noise,

$$E(v_i v_j)^T = \begin{cases} 0, & \text{if } i \neq j \\ R_j, & \text{if } i = j \end{cases}$$

3. Update the estimate $\hat{x}_k$ and the covariance of the estimation error sequentially according to the equations given by Eqs. (3.37), (3.23), and (3.43):

$$K_k = P_{k-1} H_k^T (H_k P_{k-1} H_k^T + R_k)^{-1}$$

$$\hat{x}_k = \hat{x}_{k-1} + K_k(y_k - H_k \hat{x}_{k-1})$$

$$P_k = (I - K_k H_k) P_{k-1}$$

4. Iterate the equations corresponding to the gain, covariance, and state in Step 3 recursively for each new measurement condition.

Example 3.3: Suppose we wish to estimate the state x of a simple process using the noisy measurement y of the process instrument. Assume the noisy measurement has a variance $E(v^2) = \sigma^2$. At kth sampling instant, the measurement is assumed to have the following equation

$$y_k = H_k x + v_k$$

Apply the RLS to iteratively update the estimate x.

Solution:
The process has a single state, x, which is measured as y. Thus H is a scalar quantity, $H = 1$. At kth sampling instant, the measurement is given by

$$y_k = x + v_k$$

The measurement noise covariance is given by

$$R_k = E(v_k^2) = \sigma^2$$

Assume R_k to be constant. Thus $R_{k=R}$.
Consider the initial estimate as

$$\hat{x}_0 = E(x_0)$$

If x_0 is initial state and its estimate is $\hat{x}_0$, the uncertainty in the initial estimate is given by its covariance

$$P_0 = E(x_0 - \hat{x}_0)^2$$

Assume the covariance P_0 be a constant.

$$k = 1:$$

At the first sampling instant of $k = 1$, we update the estimate and its covariance using the measurement y_1 according to the equations given in Step 3 of the above estimation procedure.

$$K_1 = \frac{P_0}{P_0 + R}$$

$$\hat{x}_1 = \hat{x}_0 + \frac{P_0}{P_0 + R}(y_1 - \hat{x}_0)$$

$$P_1 = (P_0 - K_1 P_0)$$

$$= \frac{P_0 R}{P_0 + R}$$

$k = 2$:

At $k = 2$, the measurement is obtained as y_2.

$$K_2 = P_1 (P_1 + R)^{-1}$$

$$= \frac{P_1}{P_1 + R}$$

$$= \frac{P_0}{2P_0 + R}$$

$$\hat{x}_2 = \hat{x}_1 + \frac{P_1}{P_1 + R}(y_2 - \hat{x}_1)$$

$$= \hat{x}_1 + \frac{P_0}{2P_0 + R}(y_2 - \hat{x}_1)$$

$$P_2 = P_1 - K_2 P_1$$

$$= \frac{P_1 R}{P_1 + R}$$

$$= \frac{P_0 R}{2P_0 + R}$$

$k = k$:

$$K_k = \frac{P_{k-1}}{P_{k-1} + R}$$

$$= \frac{P_0}{kP_0 + R}$$

$$\hat{x}_k = \hat{x}_{k-1} + \frac{P_0}{kP_0 + R}(y_k - \hat{x}_{k-1})$$

$$P_k = \frac{P_0 R}{kP_0 + R}$$

3.5   Luenberger observer for state estimation

In many physical systems, full state information is required for various applications concerning improving plant performance. However, it may not be possible to measure all the states due to certain technical and economical reasons. It is also required to reconstruct the state variables for changes in their dynamics even if they are available as system measurements. A state observer is a device that provides an estimate of the internal state of a given real system based on the measurements of the input and output of the real system. The theory of observers was introduced with the work of Luenberger [10−12]; thus these observers are called Luenberger observers. Observers are designed as full order observers or reduced order observers. The full order observer design builds an observer that provides the state variables which has the same dimension as the original system. Reduced order observer design exploits the knowledge of some state variables available through system measurement so that the observer is constructed only for estimating state variables that are not directly obtainable from the system measurements. An observer can be implemented either in discrete form or in continuous form depending on the requirement of a system. It is usually implemented through a computer and provides the basis for many practical applications. We first describe the design of a full order state observer in its discrete and continuous form followed by a reduced order observer. We also present observer based state feedback controllers for both discrete and continuous systems.

3.5.1 Discrete form of observer design

Consider the state and measurements of a true linear, time-invariant discrete-time system represented by

$$x(k + 1) = Ax(k) + Bu(k) \tag{3.48}$$

$$y(k) = Cx(k)$$

where $x(k)$ is the system state and $u(k)$ is the system input at instants k.

If $\hat{x}(k)$ is the estimated state, the state and measurement equations of the above process are given by

$$\hat{x}(k + 1) = A\hat{x}(k) + Bu(k) \tag{3.49}$$
$$\hat{y}(k) = C\hat{x}(k)$$

The above equations can be used in conjunction with the system measurements to derive the state observer. We correct the estimation Eq. (3.49) with feedback from estimation error $y(k) - \hat{y}(k)$:

$$\hat{x}(k + 1) = A\hat{x}(k) + Bu(k) + L[y(k) - \hat{y}(k)]$$

$$= A\hat{x}(k) + Bu(k) + L[y(k) - C\hat{x}(k)]$$

$$= (A - LC)\hat{x}(k) + Bu(k) + Ly(k) \tag{3.50}$$

where L is the observer gain. The observer is said asymptotically stable if the observed error, $e(k) = x(k) - \hat{x}(k)$ converges to zero when $k \to \infty$.

3.5.1.1 Observer gain matrix

The observer error is

$$e(k) = x(k) - \hat{x}(k)$$
$$e(k + 1) = x(k + 1) - \hat{x}(k + 1) \tag{3.51}$$

Substitution of Eqs. (3.48) and (3.50) into Eq. (3.51) gives

$$e(k + 1) = A[x(k) - \hat{x}(k)] - L[y(k) - \hat{y}(k)]$$

$$= A[x(k) - \hat{x}(k)] - LC[x(k) - \hat{x}(k)]$$

$$= A - LC[x(k) - \hat{x}(k)] \tag{3.52}$$

$$= (A - LC)e(k)$$

The observer error $e \to 0$ when $k \to \infty$. The design of the state observer is based on the observability property of the system. If the observer gain L is chosen such that the feedback matrix $A - LC$ is asymptotically stable, then the estimation error $e(t)$ will decay to zero for any initial condition $e(t_0)$. This stabilization requirement can be achieved if the pair (A, C) is observable. The rate of convergence of observer error to zero can be arbitrarily chosen by designing the L appropriately. An observer gain L is selected by computing the eigenvalues of $(A - LC)$ using the polynomial of the following determinant:

$$|\lambda I - A + LC| = 0 \tag{3.53}$$

The general scheme of Luenberger for state estimation in linear system is shown in Fig. 3.1.

3.5.2 Discrete observer based state feedback control

In the control system design, state feedback requires the measurement of all the state variables with time. If all the states (x) are available as measurements from the plant, they can be fed back to the plant input (u) through a constant feedback gain matrix (K). Thus the implementation of state feedback control requires access to all the states of the

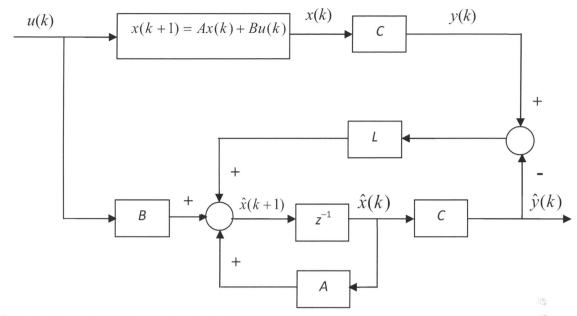

FIGURE 3.1 Luenberger observer for state estimation in a linear system.

plant. In most practical situations, the availability of all the state variables for the measurement is not possible, and for such cases, a state observer is used to estimate the unmeasured states.

For perfect control of a linear state feedback system, the control input for the system in Eq. (3.48) becomes $u(k) = -Kx(k)$. The closed loop form of the state feedback control system is thus represented by

$$x(k + 1) = (A - BK)x(k) \tag{3.54}$$

The eigenvalues of the matrix $A - BK$ represent the closed-loop poles of the system. For the design of state feedback controllers, pole placement is used to place the closed-loop poles of the system in predetermined locations.

For the plant described by Eq. (3.48), the state observer with state feedback can be formed with the following equations:

$$\begin{aligned} \hat{x}(k + 1) &= A\hat{x}(k) - BK\hat{x}(k) + L[y(k) - \hat{y}(k)] \\ &= (A - BK)\hat{x}(k) + L[y(k) - \hat{y}(k)] \end{aligned} \tag{3.55}$$

Since $y(k) = Cx(k)$, $\hat{y}(k) = C\hat{x}(k)$, $u(k) = -K\hat{x}(k)$, the state observer with state feedback control is given by

$$\begin{aligned} \hat{x}(k + 1) &= (A - LC)\hat{x}(k) - BK\hat{x}(k) + Ly(k) \\ &= (A - LC - BK)\hat{x}(k) + Ly(k) \end{aligned} \tag{3.56}$$

The augmented form of the system and observer with state feedback is represented by

$$\begin{bmatrix} x(k + 1) \\ \hat{x}(k + 1) \end{bmatrix} = \begin{bmatrix} A & -BK \\ LC & A - LC - BK \end{bmatrix} \begin{bmatrix} x(k) \\ \hat{x}(k) \end{bmatrix} \tag{3.57}$$

Since $u(k) = -K\hat{x}(k)$ and $x(k) - \hat{x}(k) = e(k)$, the augmented form of the system and observer error is represented by

$$\begin{bmatrix} x(k + 1) \\ e(k + 1) \end{bmatrix} = \begin{bmatrix} A - BK & -BK \\ 0 & A - LC \end{bmatrix} \begin{bmatrix} x(k) \\ e(k) \end{bmatrix} \tag{3.58}$$

The observer-based state feedback control scheme for linear system is shown in Fig. 3.2.

3.5.2.1 Separation principle for designing K and L matrices

The set of eigenvalues of the complete observer based control system is the union of the eigenvalues of the state feedback system and the eigenvalues of the observer. The dimension of the augmented system matrix in Eqs. (3.57) and (3.58) is of $R^{2n \times 2n}$. The $2n$ eigenvalues of the augmented matrix are equal to the individual eigenvalues of $A - BK$ and $A - LC$. This shows that the design of state feedback control law, that is, $A - BK$, can be separated from the design of

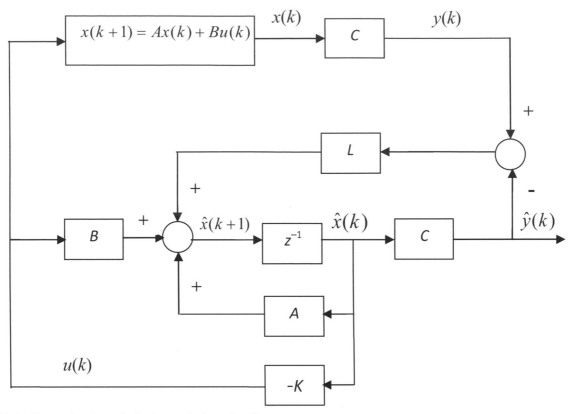

FIGURE 3.2 Observer based state feedback control scheme for a linear system.

the state observer, that is, $A - LC$. We can independently design the system using the system state feedback gain K, and independently design the observer using the observer state feedback gain L. This type of design is referred to the separation principle. Accordingly, first design the state feedback control law $u = -Kx$, by choosing the gain K by pole placement design. Next, design the observer to estimate the state x by choosing appropriate observer gain L. Design of state observers and observer based state feedback controllers have been widely discussed in the literature [13–18].

3.5.2.2 Design procedure

1. Test for the observability and controllability of the given system.
2. Determine the characteristic equation of the state feedback control system $\left| \lambda I - (A - BK) \right| = 0$
3. Assign the poles for the closed loop system and obtain the desired characteristic equation.
4. Find the elements of the gain matrix K such that the characteristic polynomial in step 2 matches the desired characteristic polynomial in Step 3. This establishes the control law $u(k) = -Kx(k)$ with known elements of K.
5. Determine the desired characteristic polynomial and apply any pole placement technique to find the elements of the observer gain matrix L such that the matrix $(A - LC)$ is asymptotically stable.
6. The equation in Eq. (3.50) with the known K and L matrices represents the observer based state feedback control system.

Example 3.4: Design the Luenberger observer for state estimation in discrete time linear system given in Example 3.2. Consider the desired pole locations for the observer are -3.5, -4.0, and -4.5, respectively.

Solution:
The system is represented by

$$x(k + 1) = \begin{bmatrix} -3 & -2 & 0 \\ 2 & 1 & 0 \\ 0 & -1 & -1 \end{bmatrix} x(k) + \begin{bmatrix} 1 \\ 0 \\ 1 \end{bmatrix} u(k)$$

$$y(k) = \begin{bmatrix} 1 & 0 & 1 \end{bmatrix} x(k)$$

The system coefficient matrices are

$$A = \begin{bmatrix} -3 & -2 & 0 \\ 2 & 1 & 0 \\ 0 & -1 & -1 \end{bmatrix}, \ B = \begin{bmatrix} 1 \\ 0 \\ 1 \end{bmatrix}, \ C = \begin{bmatrix} 1 & 0 & 1 \end{bmatrix}$$

The observability of the system is tested in Example 3.2.

Let $L = \begin{bmatrix} l_1 \\ l_2 \\ l_3 \end{bmatrix}$ be the elements of the observer gain.

The characteristic polynomial of the observer in Eq. (3.53) is

$$|\lambda I - A + LC| = 0$$

Thus,

$$\left| \begin{bmatrix} \lambda & 0 & 0 \\ 0 & \lambda & 0 \\ 0 & 0 & \lambda \end{bmatrix} - \begin{bmatrix} -3 & -2 & 0 \\ 2 & 1 & 0 \\ 0 & -1 & -1 \end{bmatrix} + \begin{bmatrix} l_1 \\ l_2 \\ l_3 \end{bmatrix} \begin{bmatrix} 1 & 0 & 1 \end{bmatrix} \right| = 0$$

$$= \left| \begin{bmatrix} \lambda & 0 & 0 \\ 0 & \lambda & 0 \\ 0 & 0 & \lambda \end{bmatrix} - \begin{bmatrix} -3 & -2 & 0 \\ 2 & 1 & 0 \\ 0 & -1 & -1 \end{bmatrix} + \begin{bmatrix} l_1 & 0 & l_1 \\ l_2 & 0 & l_2 \\ l_3 & 0 & l_3 \end{bmatrix} \right| = 0$$

$$= \left| \begin{bmatrix} \lambda + 3 + l_1 & 2 & l_1 \\ -2 + l_2 & \lambda - 1 & l_2 \\ l_3 & 1 & \lambda + 1 + l_3 \end{bmatrix} \right| = 0$$

The characteristic equation of this determinant is

$$\lambda^3 + (l_1 + 3)\lambda^2 + (-3l_2 - l_1 l_3 + 3)\lambda + (-3l_1 - 5l_2 + 4l_3 + l_1 l_3 + 1) = 0 \quad \text{(i)}$$

Since the desired poles of the observer are -3.5, -4.0, and -4.5, the characteristic equation corresponding to the poles is

$$(\lambda + 3.5)(\lambda + 4)(\lambda + 4.5) = 0$$

that is,
$$(\lambda^3 + 12\lambda^2 + 47.75\lambda + 63) = 0 \quad \text{(ii)}$$

Comparing Eqs. (i) and (ii) or applying any eigenvalue assignment technique, we obtain $l_1 = -12.625$, $l_2 = -0.5$, and $l_3 = 21.625$.

$$\text{Thus} \ \ L = \begin{bmatrix} -12.625 \\ -0.5 \\ 21.625 \end{bmatrix}$$

The state observer in Eq. (3.56) is obtained as

$$\hat{x}(k+1) = (A - LC)\hat{x}(k) + Bu(k) + Ly(k)$$

$$= \begin{bmatrix} 9.625 & -2 & -12.625 \\ 2.5 & 1 & 0.5 \\ -21.625 & -1 & -22.625 \end{bmatrix} \hat{x}(k) + \begin{bmatrix} 1 \\ 0 \\ 1 \end{bmatrix} u(k) + \begin{bmatrix} -12.625 \\ -0.5 \\ 21.625 \end{bmatrix} y(k)$$

Example 3.5: Design the Luenberger observer and observer based state feedback control for the discrete time linear system represented by

$$x(k+1) = \begin{bmatrix} 1.5 & 3.5 \\ 2 & 3 \end{bmatrix} x(k) + \begin{bmatrix} 1 \\ 0 \end{bmatrix} u(k)$$

$$y(k) = \begin{bmatrix} 0.5 & 1 \end{bmatrix} x(k)$$

Consider the desired poles for the observer are -1.5 and -2.5, and the desired poles for state feedback control are -1 and -2.

Solution:

The system is characterized by

$$A = \begin{bmatrix} 1.5 & 3.5 \\ 2 & 3 \end{bmatrix}, \quad B = \begin{bmatrix} 1 \\ 0 \end{bmatrix}, \quad C = \begin{bmatrix} 0.5 & 1 \end{bmatrix}$$

Test for controllability:

$$Q_c = [B \quad AB] = \begin{bmatrix} 1 & 1.5 \\ 0 & 2 \end{bmatrix}$$

$$|Q_c| = 2$$

The system is completely state controllable.

State feedback control:

To design a linear state feedback control, let the poles to compute the controller gain matrix are $K = [k_1 \quad k_2]$

The characteristic polynomial of the system in Eq. (3.54) is

$$|\lambda I - A + BK| = 0$$

Thus

$$\left| \begin{bmatrix} \lambda & 0 \\ 0 & \lambda \end{bmatrix} - \begin{bmatrix} 1.5 & 3.5 \\ 2 & 3 \end{bmatrix} + \begin{bmatrix} 1 \\ 0 \end{bmatrix} [k_1 \quad k_2] \right| = 0$$

$$\left| \begin{bmatrix} \lambda & 0 \\ 0 & \lambda \end{bmatrix} - \begin{bmatrix} 1.5 & 3.5 \\ 2 & 3 \end{bmatrix} + \begin{bmatrix} k_1 & k_2 \\ 0 & 0 \end{bmatrix} \right| = 0$$

$$= \left| \begin{bmatrix} \lambda - 1.5 + k_1 & -3.5 + k_2 \\ -2 & \lambda - 3 \end{bmatrix} \right| = 0$$

The characteristic equation of this determinant is

$$\lambda^2 + (k_1 - 4.5)\lambda + (-3k_1 + + 2k_2 - 2.5) = 0 \quad \text{(i)}$$

Since the desired poles of state feedback control are -1 and -2, the characteristic equation corresponding to this poles is

$$(\lambda + 1)(\lambda + 2) = 0$$

that is,

$$(\lambda^2 + 3\lambda + 2) = 0 \quad \text{(ii)}$$

Comparing Eqs. (i) and (ii) or applying any pole placement technique, we have $k_1 = 7.5$ and $k_2 = 13.5$

Thus, we have

$$K = [7.5 \quad 13.5]$$

The control input is

$$u = -Kx = -7.5x_1 - 13.5x_2$$

State observer:

Test for observability:

$$Q_o = \begin{bmatrix} C \\ CA \end{bmatrix} = \begin{bmatrix} 0.5 & 1 \\ 2.75 & 4.75 \end{bmatrix}$$

$$|Q_o| = -0.375$$

The system is completely state observable.

Let $L = \begin{bmatrix} l_1 \\ l_2 \end{bmatrix}$ be the elements of the observer gain.

The characteristic polynomial of the observer in Eq. (3.53) is

$$|\lambda I - A + LC| = 0$$

Thus

$$\left\| \begin{bmatrix} \lambda & 0 \\ 0 & \lambda \end{bmatrix} - \begin{bmatrix} 1.5 & 3.5 \\ 2 & 3 \end{bmatrix} + \begin{bmatrix} l_1 \\ l_2 \end{bmatrix} \begin{bmatrix} 0.5 & 1 \end{bmatrix} \right\| = 0$$

$$\left\| \begin{bmatrix} \lambda & 0 \\ 0 & \lambda \end{bmatrix} - \begin{bmatrix} 1.5 & 3.5 \\ 2 & 3 \end{bmatrix} + \begin{bmatrix} 0.5l_1 & l_1 \\ 0.5l_2 & l_2 \end{bmatrix} \right\| = 0$$

$$= \left\| \begin{bmatrix} \lambda - 1.5 + 0.5l_1 & -3.5 + l_1 \\ -2 + 0.5l_2 & \lambda - 3 + l_2 \end{bmatrix} \right\| = 0$$

The characteristic equation of this determinant is

$$\lambda^2 + (0.5l_1 + l_2 - 4.5)\lambda + (0.5l_1 + 0.25l_2 - 2.5) = 0 \qquad \text{(iii)}$$

Since the desired poles of the observer are -1.5 and -2.5, the characteristic equation corresponding to this poles is

$$(\lambda + 1.5)(\lambda + 2.5) = 0$$

that is,

$$(\lambda^2 + 4\lambda + 3.75) = 0 \qquad \text{(iv)}$$

Comparing Eqs. (iii) and (iv) or applying any eigenvalue assignment technique, we obtain $l_1 = 11$ and $l_2 = 3$.

Thus $L = \begin{bmatrix} 11 \\ 3 \end{bmatrix}$

The state observer is

$$\hat{x}(k+1) = (A - LC)\hat{x}(k) + Bu(k) + Ly(k)$$

$$= \begin{bmatrix} -4 & -7.5 \\ -0.5 & 0 \end{bmatrix} \hat{x}(k) + \begin{bmatrix} 1 \\ 0 \end{bmatrix} u(k) + \begin{bmatrix} 11 \\ 3 \end{bmatrix} y(k)$$

The observer based state feedback control is given by

$$\dot{x}(k+1) = (A - LC - BK)\hat{x}(k) + Ly(k)$$

$$= \begin{bmatrix} -11.5 & -21 \\ -0.5 & 0 \end{bmatrix} \hat{x}(k) + \begin{bmatrix} 11 \\ 3 \end{bmatrix} y(k)$$

3.5.3   Continuous form of observer

The state and output of a continuous time linear system are represented by Eqs. (3.3) and (3.6).

The state observer for this linear system is given by

$$\dot{\hat{x}}(t) = A\hat{x}(t) + Bu(t) + L[y(t) - \hat{y}(t)]\hat{y}(t) = C\hat{x}(t) \qquad (3.59)$$

Observer error is defined as

$$e = x - \hat{x} \qquad (3.60)$$

Differentiating the above equation,

$$\dot{e} = \dot{x} - \dot{\hat{x}} \qquad (3.61)$$

Substituting Eqs. (3.3) and (3.59) into Eq. (3.61), we have

$$\begin{aligned} \dot{e} &= Ax + Bu - [A\hat{x} + Bu + L(y - \hat{y})] \\ &= A(x - \hat{x}) - L(y - \hat{y}) \\ &= A(x - \hat{x}) - LC(x - \hat{x}) \\ &= A - LC(x - \hat{x}) \\ &= (A - LC)e \end{aligned} \qquad (3.62)$$

Solution of Eq. (3.61) is given by

$$e(t) = e^{A - LC} e(0) \qquad (3.63)$$

The observer error $e \to 0$ when $t \to \infty$. The eigenvalues of the matrix $A - LC$ can be made arbitrary by the appropriate choice of the observer gain L when the pair $[A, C]$ is observable. The given system is observable if and only if the $n \times n$ matrix, $[C^T A^T C^T \dots (A T)^{(n-1)} C^T]$ is of rank n.

3.5.4 Continuous observer based state feedback control

In state feedback control, all the states (x) are available as measurements from the plant, and they are fed back to the plant input (u) through a constant feedback gain matrix (K). Thus for a linear perfect state feedback control, the control input for the system in Eq. (3.3) becomes $u(t) = -Kx(t)$. The state feedback control scheme of a continuous system when all the states are accessible for measurement is shown in Fig. 3.3.

With the substitution of $u(t) = -Kx(t)$ in Eq. (3.3), the system with state feedback is represented by

$$\dot{x}(t) = (A - BK)x(t) \tag{3.64}$$

The solution of this equation is

$$x(t) = e^{(A-BK)t} x(0) \tag{3.65}$$

The stability and response characteristics of the system are determined by the eigenvalues of the matrix $(A - BK)$. The state feedback control gain K is selected by computing the eigenvalues of $(A - BK)$ using the polynomial of the following determinant:

$$|\lambda I - A + BK| = 0 \tag{3.66}$$

As $y(t) = Cx(t)$, $\hat{y}(t) = C\hat{x}(t)$, $u(t) = -K\hat{x}(t)$, the state observer in Eq. (3.59) with state feedback control is given by

$$\dot{\hat{x}}(t) = (A - LC)\hat{x}(t) - BK\hat{x}(t) + Ly(t) \tag{3.67}$$

The augmented form of the system and observer with state feedback is similar to the discrete form in Eq. (3.57) as given by

$$\begin{bmatrix} \dot{x} \\ \dot{\hat{x}} \end{bmatrix} = \begin{bmatrix} A & -BK \\ LC & A - LC - BK \end{bmatrix} \begin{bmatrix} x \\ \hat{x} \end{bmatrix} \tag{3.68}$$

The augmented form of the system and observer error is similar to the discrete form in Eq. (3.58) as given by

$$\begin{bmatrix} \dot{x} \\ \dot{e} \end{bmatrix} = \begin{bmatrix} A - BK & -BK \\ 0 & A - LC \end{bmatrix} \begin{bmatrix} x \\ e \end{bmatrix} \tag{3.69}$$

The observer-based state feedback control scheme for continuous time linear system is shown in Fig. 3.4.

3.5.4.1 Design of observer gain and state feedback gain matrices

As discussed in the separation principal of the above section, first design the state feedback control law $u = -Kx$, by choosing the gain K by pole placement design. Next, design the observer to estimate the state x by choosing appropriate observer gain L. Use the eigenvalue assignment technique to choose L so that the eigenvalues of $A - LC$ are at the desired locations. An observer gain L is designed by computing the eigenvalues of $(A - LC)$ using the polynomial of the determinant as $|\lambda I - A + LC| = 0$. Different methods for pole placement in state feedback control systems are discussed in the literature [19–22].

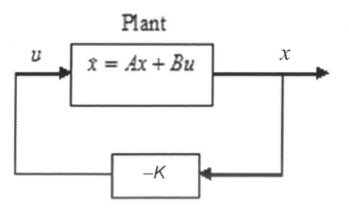

FIGURE 3.3 State feedback control scheme when all the states are accessible as measurements.

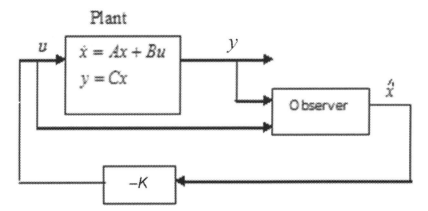

FIGURE 3.4 Observer-based state feedback control scheme with a continuous observer.

Example 3.6: Design an observer and observer based state feedback control scheme for the continuous time linear control system given in Example 3.1. Consider the desired poles to compute the state feedback gain are -2, -3, and -4, and the desired poles to compute the observer gain are -3, -4, and -5, respectively.

Solution:

The system in Example 3.1 is of the form:

$$\dot{x}(t) = Ax(t) + Bu(t)$$
$$y(t) = Cx(t)$$

The system coefficient matrices are

$$A = \begin{bmatrix} 0 & 1 & 1 \\ -2 & -3 & 0 \\ 0 & 2 & -3 \end{bmatrix}, \ B = \begin{bmatrix} 0 \\ 1 \\ 0 \end{bmatrix}, \ C = \begin{bmatrix} 1 & 0 & 0 \end{bmatrix}$$

State feedback control:

The controllability of the system is tested via solution in Example 3.1.

To design a linear state feedback control, let the poles of the controller gain matrix as

$$K = \begin{bmatrix} k_1 & k_2 & k_3 \end{bmatrix}$$

The characteristic polynomial of the system in Eq. (3.66) is

$$|\lambda I - A + BK| = 0$$

Thus

$$\left| \begin{bmatrix} \lambda & 0 & 0 \\ 0 & \lambda & 0 \\ 0 & 0 & \lambda \end{bmatrix} - \begin{bmatrix} 0 & 1 & 1 \\ -2 & -3 & 0 \\ 0 & 2 & -3 \end{bmatrix} + \begin{bmatrix} 0 \\ 1 \\ 0 \end{bmatrix} \begin{bmatrix} k_1 & k_2 & k_3 \end{bmatrix} \right| = 0$$

$$= \left| \begin{bmatrix} \lambda & 0 & 0 \\ 0 & \lambda & 0 \\ 0 & 0 & \lambda \end{bmatrix} - \begin{bmatrix} 0 & 1 & 1 \\ -2 & -3 & 0 \\ 0 & 2 & -3 \end{bmatrix} + \begin{bmatrix} 0 & 0 & 0 \\ k_1 & k_2 & k_3 \\ 0 & 0 & 0 \end{bmatrix} \right| = 0$$

$$= \left| \begin{bmatrix} \lambda & -1 & -1 \\ 2 + k_1 & \lambda + 3 + k_2 & k_3 \\ 0 & -2 & \lambda + 3 \end{bmatrix} \right| = 0$$

The characteristic equation of this determinant is

$$\lambda^3 + (k_2 + 6)\lambda^2 + (k_1 + 3k_2 + 2k_3 + 11)\lambda + (5k_1 + 10) = 0 \quad \text{(i)}$$

Since the desired poles of the closed-loop control are -2, -3 and -4, the characteristic equation corresponding to this poles is

$$(\lambda + 2)(\lambda + 3)(\lambda + 4) = 0$$

that is,
$$(\lambda^3 + 9\lambda^2 + 26\lambda + 24) = 0 \quad \text{(ii)}$$

Comparing Eqs. (i) and (ii) or applying any pole placement technique, we have

$$k_1 = 2.8, \quad k_2 = 3.0, \quad k_3 = 1.6$$

Thus, we have

$$K = [2.8 \quad 3.0 \quad 1.6]$$

The control input is

$$u = -Kx = -2.8x_1 - 3.0x_2 - 1.6x_3$$

State observer:

The observability of the system is verified in Example 3.1.

Let $L = \begin{bmatrix} l_1 \\ l_2 \\ l_3 \end{bmatrix}$ be the observer gain.

The characteristic polynomial of the observer in Eq. (3.59) is

$$|\lambda I - A + LC| = 0$$

Thus

$$\left| \begin{bmatrix} \lambda & 0 & 0 \\ 0 & \lambda & 0 \\ 0 & 0 & \lambda \end{bmatrix} - \begin{bmatrix} 0 & 1 & 1 \\ -2 & -3 & 0 \\ 0 & 2 & -3 \end{bmatrix} + \begin{bmatrix} l_1 \\ l_2 \\ l_3 \end{bmatrix} \begin{bmatrix} 1 & 0 & 0 \end{bmatrix} \right| = 0$$

$$= \left| \begin{bmatrix} \lambda & 0 & 0 \\ 0 & \lambda & 0 \\ 0 & 0 & \lambda \end{bmatrix} - \begin{bmatrix} 0 & 1 & 1 \\ -2 & -3 & 0 \\ 0 & 2 & -3 \end{bmatrix} + \begin{bmatrix} l_1 & 0 & 0 \\ l_2 & 0 & 0 \\ l_3 & 0 & 0 \end{bmatrix} \right| = 0$$

$$= \left| \begin{bmatrix} \lambda + l_1 & -1 & -1 \\ 2 + l_2 & \lambda + 3 & 0 \\ l_3 & -2 & \lambda + 3 \end{bmatrix} \right| = 0$$

The characteristic equation of this determinant is

$$\lambda^3 + (l_1 + 6)\lambda^2 + (6l_1 + l_2 + l_3 + 11)\lambda + (9l_1 + 5l_2 + 3l_3 + 10) = 0 \quad \text{(iii)}$$

Since the desired poles of the observer are -3, -4, and -5, the characteristic equation corresponding to the poles is

$$(\lambda + 3)(\lambda + 4)(\lambda + 5) = 0$$

that is,
$$(\lambda^3 + 12\lambda^2 + 47\lambda + 60) = 0 \quad \text{(iv)}$$

Comparing Eqs. (iii) and (iv) or applying any eigenvalue assignment technique, we obtain $l_1 = 6.0$, $l_2 = -2.0$, $l_3 = 2.0$

Thus $L = \begin{bmatrix} 6.0 \\ -2.0 \\ 2.0 \end{bmatrix}$

The state observer is

$$\dot{\hat{x}}(t) = (A - LC)\hat{x}(t) + Bu(t) + Ly(t) = \begin{bmatrix} -6 & 1 & 1 \\ 0 & -3 & 0 \\ -2 & 2 & 3 \end{bmatrix} \hat{x}(t) + \begin{bmatrix} 0 \\ 1 \\ 0 \end{bmatrix} u(t) + \begin{bmatrix} 6 \\ -2 \\ 2 \end{bmatrix} y(t)$$

The observer based state feedback control is given by

$$\tilde{x}(t) = (A - LC - BK)\hat{x}(t) + Ly(t) = \begin{bmatrix} -6 & 1 & 1 \\ -2.8 & -6 & -1.6 \\ -2 & 2 & -3 \end{bmatrix} \hat{x}(t) + \begin{bmatrix} 6 \\ -2 \\ 2 \end{bmatrix} y(t)$$

3.6 Reduced order Luenberger observer for state estimation

The observer design discussed in the previous subsection reconstructs the full state vector, including the measurable states, which are the outputs. The observer design can be simplified if we build an observer that provides the estimate of unmeasured states. Because of this reason, the reduced order observer, which is different from full order observers, was introduced, to estimate only immeasurable states of the system.

We describe how to design a continuous form of a reduced order observer to estimate the unmeasured state using the knowledge of the measured state. Consider the system represented by Eqs. (3.3) and (3.6). Assume that the system is observable and the observability matrix has full rank. If n is the order of the system and m is the number of observations, then the order of the reduced order observer is $n - m$.

The system defined in Eqs. (3.3) and (3.6) can be expressed in terms of measurable and immeasurable states as

$$\begin{aligned}
\dot{x}_1 &= A_{11}x_1 + A_{12}x_2 + B_1 u \\
\dot{x}_2 &= A_{21}x_1 + A_{22}x_2 + B_2 u \\
y &= x_1
\end{aligned} \tag{3.70}$$

where x_1 is the measurable state, x_2 is the immeasurable state and y is the measured output. The system can be partitioned into matrices as

$$\begin{bmatrix} \dot{x}_1 \\ \dot{x}_2 \end{bmatrix} = \begin{bmatrix} A_{11} & A_{12} \\ A_{21} & A_{22} \end{bmatrix} \begin{bmatrix} x_1 \\ x_2 \end{bmatrix} + \begin{bmatrix} B_1 \\ B_2 \end{bmatrix} u \tag{3.71}$$

$$y = \begin{bmatrix} 1 & 0 \end{bmatrix} \begin{bmatrix} x_1 \\ x_2 \end{bmatrix}$$

The dynamics of the system is written as

$$\begin{aligned}
\dot{y} &= \dot{x}_1 = A_{11}y + A_{12}x_2 + B_1 u \\
\dot{x}_2 &= A_{21}y + A_{22}x_2 + B_2 u
\end{aligned} \tag{3.72}$$

If (A, C) in Eqs. (3.3) and (3.6) is an observable pair, then the pair (A_{22}, A_{12}) is also observable. The second equation in Eq. (3.72) can be used to estimate the unobservable state x_2.

Eq. (3.72) can be rearranged as

$$\begin{aligned}
A_{12}x_2 &= \dot{y} - A_{11}y - B_1 u \\
\dot{x}_2 &= A_{21}y + A_{22}x_2 + B_2 u
\end{aligned} \tag{3.73}$$

We define the observer as

$$\dot{\hat{x}}_2 = A_{22}\hat{x}_2 + B_2 u + A_{21}y + L(\dot{y} - \dot{\hat{y}}) = A_{22}\hat{x}_2 + B_2 u + A_{21}y + L(\dot{y} - A_{11}y - B_1 u - A_{12}\hat{x}_2) \tag{3.74}$$

where L is the observer gain. There is no error in the estimation of x_1, i.e., $e_1 = x_1 - \hat{x}_1 = 0$. Thus, the estimation error dynamics are given by

$$e_2 = x_2 - \hat{x}_2 \tag{3.75}$$

It is necessary to ensure the convergence of e_2 to be zero.
Thus

$$\dot{e}_2 = \dot{x}_2 - \dot{\hat{x}}_2 \tag{3.76}$$

$$= A_{21}y + A_{22}x_2 + B_2 u - \left[A_{22}\hat{x}_2 + B_2 u + A_{21}y + L(\dot{y} - \dot{\hat{y}}) \right]$$

$$= A_{22}x_2 - \left[A_{22}\hat{x}_2 + L(\underbrace{\dot{y} - A_{11}y - B_1u}_{A_{12}x_2} - A_{12}\hat{x}_2) \right]$$

$$= (A_{22} - LA_{12})x_2 - (A_{22} - LA_{12})\hat{x}_2$$

$$= (A_{22} - LA_{12})(x_2 - \hat{x}_2)$$

$$= (A_{22} - LA_{12})e_2 \tag{3.77}$$

The derivative of the output $\dot{y}$ required for the observer equation can be avoided from the observer design equation. The gain matrix L in reduced order observer is chosen by assigning the eigenvalues of $(A_{22} - LA_{12})$ such that they lie in the left half-plane. The elements of the L matrix can be computed using the polynomial of the following determinant:

$$|\lambda I - A_{22} + LA_{12}| = 0 \tag{3.78}$$

The selection of L matrix can also be accomplished by any pole placement technique.
From Eq. (3.74), we have

$$\dot{\hat{x}}_2 = (A_{22} - LA_{12})\hat{x}_2 + (B_2 - LB_1)u + (A_{21} - LA_{11})y + L\dot{y} \tag{3.79}$$

Define

$$z = \hat{x}_2 - Ly \tag{3.80}$$

Thus

$$\dot{z} = \dot{\hat{x}}_2 - L\dot{y} \tag{3.81}$$

$$= (A_{22} - LA_{12})\underbrace{\hat{x}_2}_{z + Ly} + (B_2 - LB_1)u + (A_{21} - LA_{11})y + L\dot{y} - L\dot{y}$$

$$= (A_{22} - LA_{12})z + (B_2 - LB_1)u + (A_{21} - LA_{11})y + (A_{22} - LA_{12})Ly$$

This equation does not depend on $\dot{y}$
Thus the reduced order observer equation can be written as

$$\dot{z} = (A_{22} - LA_{12})z + (B_2 - LB_1)u + (A_{21} - LA_{11})y + (A_{22} - LA_{12})Ly \tag{3.82}$$

and

$$\hat{x}_2 = z + Ly \tag{3.83}$$

The discrete form of reduced order observer can be designed using the discrete model and discrete measurements given by Eq. (3.48) and adapting the above procedure as used for continuous form of reduced order observer design.

Example 3.7: Design the reduced order Luenberger observer for state estimation in continuous time linear system described by the following equations:

$$\dot{x}(t) = \begin{bmatrix} -1 & 1 & 3 \\ 2 & 1 & 2 \\ 1 & 2 & -3 \end{bmatrix} x(t) + \begin{bmatrix} 0 \\ 0 \\ 1 \end{bmatrix} u(t)$$

$$y(t) = \begin{bmatrix} 1 & 0 & 0 \end{bmatrix} x(t)$$

Consider the desired poles to compute the observer gain are -2 and -3.

Solution:
The system is of the form

$$\dot{x} = Ax + Bu$$

$$y = Cx$$

where

$$A = \begin{bmatrix} -1 & 1 & 3 \\ 2 & 1 & 2 \\ 1 & 2 & -3 \end{bmatrix}, \quad B = \begin{bmatrix} 0 \\ 0 \\ 1 \end{bmatrix}, \quad C = \begin{bmatrix} 1 & 0 & 0 \end{bmatrix}$$

Observability of the system is found as

$$Q_o = \begin{bmatrix} C \\ CA \\ CA^2 \end{bmatrix} = \begin{bmatrix} 1 & 0 & 0 \\ -1 & 1 & 3 \\ 6 & 6 & -10 \end{bmatrix}$$

$$|Q_o| = -28$$

The rank of Q_o is 3. The given system is observable.
The matrices are partitioned as

$$A = \begin{bmatrix} A_{11} & \vdots & A_{12} \\ \cdots & \vdots & \cdots \\ A_{21} & \vdots & A_{22} \end{bmatrix} = \begin{bmatrix} -1 & \vdots & 1 & 3 \\ \cdots & \vdots & \cdots & \cdots \\ 2 & \vdots & 1 & 2 \\ 1 & \vdots & 2 & -3 \end{bmatrix}; \quad B = \begin{bmatrix} B_1 \\ \cdots \\ B_2 \end{bmatrix} = \begin{bmatrix} 0 \\ \cdots \\ 0 \\ 1 \end{bmatrix}; \quad C = [C_1 \vdots C_2] = [1 \vdots 0 \ 0]$$

As the system (A, C) is observable, the pair (A_{22}, A_{12}) is also observable.
The observability of the reduced system is found as

$$Q_o = \mathrm{rank} \begin{bmatrix} A_{12} \\ A_{12}A_{22} \end{bmatrix} = \mathrm{rank} \begin{bmatrix} 1 & 3 \\ 3 & -7 \end{bmatrix} = 2$$

Thus the reduced order system is observable.
The reduced order observer in Eq. (3.82) is

$$\dot{z} = (A_{22} - LA_{12})z + (B_2 - LB_1)u + (A_{21} - LA_{11})y + (A_{22} - LA_{12})Ly$$

To compute the observer gain, let $L = \begin{bmatrix} l_1 \\ l_2 \end{bmatrix}$.

L is selected such that it satisfies the following characteristic polynomial defined by Eq. (3.78):

$$|\lambda I - A_{22} + LA_{12}| = 0$$

Thus

$$\left| \begin{bmatrix} \lambda & 0 \\ 0 & \lambda \end{bmatrix} - \begin{bmatrix} 1 & 2 \\ 2 & -3 \end{bmatrix} + \begin{bmatrix} l_1 \\ l_2 \end{bmatrix} \begin{bmatrix} 1 & 3 \end{bmatrix} \right| = 0$$

$$= \left| \begin{bmatrix} \lambda & 0 \\ 0 & \lambda \end{bmatrix} - \begin{bmatrix} 1 & 2 \\ 2 & -3 \end{bmatrix} + \begin{bmatrix} l_1 & 3l_1 \\ l_2 & 3l_2 \end{bmatrix} \right| = 0$$

$$= \left| \begin{bmatrix} \lambda + l_1 - 1 & -2 + 3l_1 \\ -2 + l_2 & \lambda + 3l_2 + 3 \end{bmatrix} \right| = 0$$

The characteristic equation of this determinant is

$$\lambda^2 + (l_1 + 3l_2 + 2)\lambda + (9l_1 - l_2 - 7) = 0 \quad \text{(i)}$$

Since the desired poles of the observer are -2 and -3, the characteristic equation corresponding to this poles is

$$(\lambda + 2)(\lambda + 3) = 0$$

that is,

$$(\lambda^2 + 5\lambda + 6) = 0 \quad \text{(ii)}$$

Comparing Eqs. (i) and (ii) or applying any eigenvalue assignment technique, we obtain $l_1 = 1.5$ and $l_2 = 0.5$.

Thus $L = \begin{bmatrix} 1.5 \\ 0.5 \end{bmatrix}$

The reduced order state observer in Eq. (3.82) is obtained as:

$$\dot{z} = \left\{ \begin{bmatrix} 1 & 2 \\ 2 & -3 \end{bmatrix} - \begin{bmatrix} 1.5 \\ 0.5 \end{bmatrix} \begin{bmatrix} 1 & 3 \end{bmatrix} \right\} z + \left\{ \begin{bmatrix} 0 \\ 1 \end{bmatrix} - \begin{bmatrix} 1.5 \\ 0.5 \end{bmatrix} [0] \right\} u + \left\{ \begin{bmatrix} 2 \\ 1 \end{bmatrix} - \begin{bmatrix} 1.5 \\ 0.5 \end{bmatrix} [-1] \right\} y +$$

$$\left\{ \begin{bmatrix} 1 & 2 \\ 2 & -3 \end{bmatrix} - \begin{bmatrix} 1.5 \\ 0.5 \end{bmatrix} \begin{bmatrix} 1 & 3 \end{bmatrix} \right\} \begin{bmatrix} 1.5 \\ 0.5 \end{bmatrix} y$$

$$= \begin{bmatrix} -0.5 & -2.5 \\ 1.5 & -4.5 \end{bmatrix} z + \begin{bmatrix} 5.5 \\ 1.5 \end{bmatrix} y + \begin{bmatrix} 0 \\ 1 \end{bmatrix} u$$

Thus

$$\hat{x}_2 = z + Ly$$

3.7 Kalman filter for state estimation

Kalman filter is used to provide optimal estimates of unmeasured and measured states for varying linear systems in the presence of noise by combining known measurements with the mathematical model of a process. The Kalman filter equations provide an extremely convenient procedure for digital computer implementation. It was primarily developed by R.E. Kalman in 1960 [5] for the recursive solution of a discrete data linear filtering problem. The Kalman filter has two distinct recursive steps, prediction and correction. The prediction step uses the state estimate from the previous time step to produce an estimate of the state at the current time step. This predicted state estimate is also known as a priori state estimate. This step does not include observation information from the current time step. In the correction step, the current a priori prediction is combined with current observation information to update the state estimate. The Kalman filter has been applied for state estimation in discrete as well as continuous time systems. More extensive references on the Kalman filter are reported elsewhere [1,3,23−26].

3.7.1 Discrete Kalman filter

Consider the state estimation of a process governed by a linear stochastic difference equation:

$$x_k = A_k x_{k-1} + B_k u_k + w_k \tag{3.84}$$

The observation of the process is assumed to occur at discrete points in time in accordance with the linear relationship:

$$y_k = H_k x_k + v_k \tag{3.85}$$

where $x_k = (n \times 1)$ process state vector at time t_k, $u_{k=d}$ dimensional input vector, $A_k = (n \times n)$ matrix relating x_k to x_{k+1}, $B_k = (n \times d)$ dimensional input coefficient matrix, $w_k = (n \times 1)$ vector assumed to be white noise sequence, $y_k = (m \times 1)$ vector measurement at time t_k, $H_k = (m \times n)$ matrix relating the measurement and state vector at time t_k, and $v_k = (m \times 1)$ vector of measurement error assumed to be white noise sequence.

The discrete Kalman filter algorithm is implemented in two recursive steps, a prediction step and an updated step. The prediction step uses the state estimate from the previous time step to produce an estimate of the state at the current time step. In the updated step, the current a priori prediction is combined with current observation information to refine the state estimate. This improved estimate is termed as the *posteriori* state estimate. By starting with an initial estimate $\hat{x}_0$ and its covariance matrix P_0, the Kalman filter prediction and correction equations for recursive state estimation are given as follows.

3.7.1.1 Prediction equations

The predicted state estimate and its covariance matrix are computed by:

$$\begin{aligned} \hat{x}_{k/k-1} &= A\hat{x}_{k-1/k-1} + Bu_{k-1} \\ P_{k/k-1} &= AP_{k-1/k-1}A^T + Q_k \end{aligned} \tag{3.86}$$

3.7.1.2 Correction equations

The measurements available at discrete time k are used to obtain the update estimates $\hat{x}_{k/k}$ and its covariance matrix $P_{k/k}$ by using the following equations

$$\hat{x}_{k/k} = \hat{x}_{k/k-1} + K_k \left[y_k - H_k \hat{x}_{k/k-1} \right]$$

$$P_{k/k} = P_{k/k-1} - K_k H_k P_{k/k-1} \tag{3.87}$$

$$K_k = P_{k/k-1} H_k^T \left[H_k P_{k/k-1} H_k^T + R_k \right]^{-1}$$

The recursive initial conditions for state estimate and covariance are

$$\hat{x}_{k-1/k-1} = \hat{x}_{k/k}$$
$$P_{k-1/k-1} = P_{k/k} \tag{3.88}$$

The initial values of $\hat{x}_{0/0} = 0$ and $P_{0/0} = 0$ reflect the distribution of the initial state and covariance values. The initial system state has a known mean and covariance matrix as given by

$$\hat{x}_{0/0} = E[x_0]$$

$$E\left[\left(\hat{x}_{0/0} - x_0 \right) \left(\hat{x}_{0/0} - x_0 \right)^T \right] = P_0 \tag{3.89}$$

The covariance matrices of state, process, and measurement noises are given by:

$$E\left[(x_k - \hat{x}_k)(x_k - \hat{x}_k)^T \right] = P_k$$

$$E\left[w_k w_i^T \right] = \begin{cases} Q_k, \; i = k \\ 0, \; i \neq k \end{cases}$$

$$E\left[w_k v_k^T \right] = 0$$

$$E\left[v_i v_k^T \right] = R_k \tag{3.90}$$

The recursive computation loop for discrete Kalman filter is shown in Fig. 3.5.

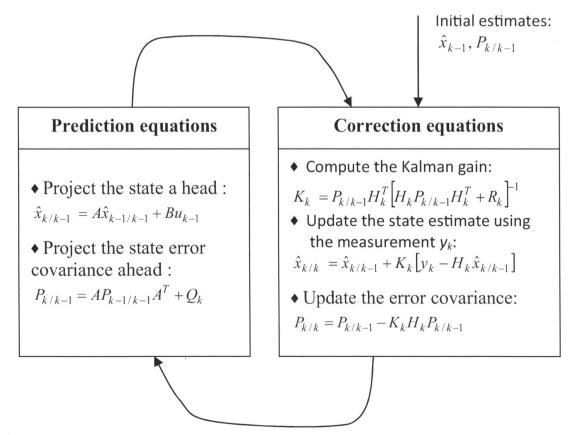

FIGURE 3.5 Computation scheme for discrete Kalman filter.

The measurements used in the Kalman filter can be actual process measurements or simulated measurements. In order to obtain the simulation measurements, the model equations are solved in the simulation program to provide true values for the states. These state variables are then corrupted with random Gaussian noise to obtain the simulated measurements. The measurements thus obtained are used as inputs to the Kalman filter. Fig. 3.6 shows the block diagram of how either the simulated or the actual process measurements can be used as inputs to the Kalman filter.

3.7.2 Filter covariance matrices

The filter covariance matrices P_0, Q, R, and the measurement matrix H_k are treated as the design parameters of the Kalman filter. The initial state of the process x_0 is also unknown. But we have some knowledge about x_0 and its expected value of covariance P_0. It is assumed that x_0 is independent of v and w. The relation for the initial state covariance matrix is given by Eq. (3.89).

Process noise w and measurement noise v reflect the uncertainty in the process and measurements. The noises w and v are not actually implemented in Kalman filter calculations, since they are assumed to be random errors with zero mean. In Kalman filter computation, these noises are expressed in terms of process and measurement noise covariance matrices **Q** and **R**. Process noise w is a random noise with zero mean which represents modeling uncertainties and disturbances. It is expressed by its covariance matrix $Q = E(ww^T)$. The n states and the random disturbance vector w of a process is represented by,

$$x = \begin{bmatrix} x_1 \\ x_2 \\ \vdots \\ x_n \end{bmatrix} ; \quad w = \begin{bmatrix} w_1 \\ w_2 \\ \vdots \\ w_n \end{bmatrix} \tag{3.91}$$

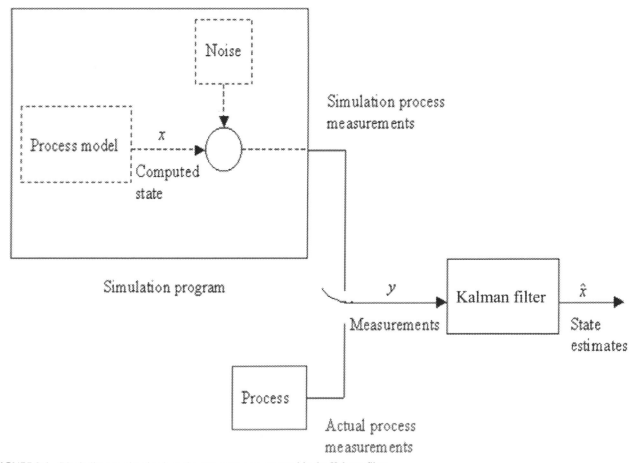

FIGURE 3.6 Block diagram showing how the measurements are used in the Kalman filter.

For n states, the process noise covariance matrix Q is expressed as

$$Q = \begin{bmatrix} E[w_1^2] & E[w_1 w_2] & \cdots & E[w_1 w_n] \\ E[w_1 w_2] & E[w_2^2] & \cdots & E[w_2 w_n] \\ \vdots & & \ddots & \\ E[w_1 w_n] & \cdots & & E[w_n^2] \end{bmatrix} \tag{3.92}$$

If Q is the auto-covariance of w, its standard assumption is that

$$Q = \begin{bmatrix} Q_{11} & 0 & \cdots & 0 \\ 0 & Q_{22} & \cdots & 0 \\ \vdots & & \ddots & \\ 0 & \cdots & & Q_{nn} \end{bmatrix} \tag{3.93}$$

where Q_{ii} is the variance of w_i.

If v is a random measurement noise vector, it is represented by

$$v = \begin{bmatrix} v_1 \\ v_2 \\ \vdots \\ v_m \end{bmatrix} \tag{3.94}$$

In Kalman filter computation, measurement noise v is expressed by its covariance matrix, $R = E(vv^T)$. For m measuremts, the measurement noise covariance matrix R is expressed by

$$R = \begin{bmatrix} E[v_1^2] & E[v_1 v_2] & \cdots & E[v_1 v_m] \\ E[v_1 v_2] & E[v_2^2] & \cdots & E[v_2 v_m] \\ \vdots & & \ddots & \\ E[v_1 v_m] & \cdots & & E[v_m^2] \end{bmatrix}$$

$$= \begin{bmatrix} \sigma_1^2 & \sigma_{12} & \cdots & \sigma_{1m} \\ \sigma_{12} & \sigma_2^2 & \cdots & \sigma_{2m} \\ \vdots & & \ddots & \\ \sigma_{1m} & \cdots & & \sigma_m^2 \end{bmatrix} \tag{3.95}$$

If R is the auto-covariance of v, its standard assumption is that

$$R = \begin{bmatrix} R_{11} & 0 & \cdots & 0 \\ 0 & R_{22} & \cdots & 0 \\ \vdots & & \ddots & \\ 0 & \cdots & & R_{mm} \end{bmatrix} \tag{3.96}$$

where R_{ii} is the variance of v_i. It is assumed that v and w are mutually uncorrelated, so that $[v\, w^T] = 0$.

The selection of the assumed covariance matrices P_0, Q, and R can have a significant effect on the estimation performance of a Kalman filter. The selection of P_0 is coupled with the assumed initial state and affects the initial convergence of the filter. In many situations, the effect of P_0 is not significant, and for simplicity, it is often arbitrarily initialized to an identity matrix. The effects of Q and R are more significant, and they affect the overall performance of the filter. The H matrix is used to convert the states into outputs. In simple cases, H is just used to select certain states, which are measured when other states are not. The recursive computation of the covariance P_k and the gain K_k are given by Eq. (3.87) based on their derivation in Section 3.4.

3.7.2.1 Evaluation of Q_k matrix

Consider a first order process in Laplace domain in two states x_1 and x_2 as described by

$$x_1(s) = F_1(s)w(s)$$

$$x_2(s) = F_2(s)w(s) \tag{3.97}$$

The process noise covariance matrix Q_k is evaluated by expected correlations between the variables x_1 and x_2. These correlations, in turn, are expressed in terms of mean square values. Thus, the Q_k matrix for the above process is represented by

$$Q_k = \begin{bmatrix} E(x_1, x_1) & E(x_1, x_2) \\ E(x_1, x_2) & E(x_2, x_2) \end{bmatrix} \tag{3.98}$$

The elements in Eq. (3.98) are evaluated as

$$E[x_i, x_k] = \int_0^{\Delta t} \int_0^{\Delta t} f_i(u) f_k(v) \delta(u, v) du dv \tag{3.99}$$

The f_i and f_k in Eq. (3.99) are obtained from the transfer functions in Eq. (3.97) as follows:

$$\begin{aligned} F_1(s) &= x_1(s)/w(s) \\ F_2(s) &= x_2(s)/w(s) \end{aligned} \tag{3.100}$$

Assume α and β are the roots of the above first order process transfer functions. Then the time domain representation of the transfer functions are given by

$$\begin{aligned} f_1(t) &= L^{-1}[F_1(s)] = e^{-\alpha t} \\ f_2(t) &= L^{-1}[F_2(s)] = e^{-\beta t} \end{aligned} \tag{3.101}$$

The elements of the Q_k matrix are thus evaluated as

$$\begin{aligned} E[x_1, x_1] &= \int_0^{\Delta t} \int_0^{\Delta t} f_1(u) f_1(v) \delta(u, v) du dv \\ &= \int_0^{\Delta t} \int_0^{\Delta t} e^{-\alpha u} e^{-\alpha v} \delta(u, v) du dv \\ E[x_1, x_2] &= \int_0^{\Delta t} \int_0^{\Delta t} f_1(u) f_2(v) \delta(u, v) du dv \\ &= \int_0^{\Delta t} \int_0^{\Delta t} e^{-\alpha u} e^{-\beta v} \delta(u, v) du dv \\ E[x_2, x_2] &= \int_0^{\Delta t} \int_0^{\Delta t} f_2(u) f_2(v) \delta(u, v) du dv \\ &= \int_0^{\Delta t} \int_0^{\Delta t} e^{-\beta u} e^{-\beta v} \delta(u, v) du dv \end{aligned}$$

Example 3.8: Consider the Wiener process represented by

$$\dot{x} = w(t)$$

The process has Gaussian statistics and is assumed to be zero at $t = 0$. A sequence of independent measurements of the process are assumed to be 1.5 and 2.5, corresponding to the sampling instants of 1 and 2 min, respectively. Let the initial process state has a zero covariance, and the standard deviation of the measurement error be 0.5. Compute the filter design parameters and apply the discrete Kalman filter for state estimation.

Solution:

In order to apply the discrete Kalman filter, the process needs to be expressed in algebraic representation. Considering the process is of the form,

$$\dot{x} = Fx + w$$

Its state transition (state coefficient) matrix is computed as

$$A = L^{-1}[(sI - F)^{-1}] = 1$$

The Laplace representation of the process is

$$x(s) = \frac{1}{s} w(s)$$

Assuming the process is of the form,

$$X(s) = G(s) w(s)$$

we have $G(s) = 1/s$ and $g(t) = 1$.

The process noise covariance Q is evaluated as

$$Q = E[x,x] = \int_0^{\Delta t} \int_0^{\Delta t} g(u)g(v)\delta(u,v)dudv$$
$$= E[x,x] = \int_0^1 \int_0^1 1.1.\delta(u,v)dudv$$
$$= 1$$

As there is one state and it is measured, the measurement coefficient matrix is

$$H = 1$$

The variance of the measurement error is

$$R = \left(\frac{1}{2}\right)^2 = \frac{1}{4}$$

The initial state and covariance conditions are

$$x_{0/0} = 0, \quad \hat{x}_{0/0} = 0, \quad P_0 = 0$$

The process and observation models are given by

$$x_k = Ax_{k-1}$$

with

$$y_k = Hx_k$$
$$A = 1; \quad H = 1$$
$$k = 1:$$

Prediction:

$$\hat{x}_{1/0} = A\hat{x}_{0/0}$$
$$= 0$$

$$P_{1/0} = AP_{0/0}A^T + Q$$
$$= 0 + 1$$
$$= 1$$

Correction:

$$K_1 = P_{1/0}H^T\left[HP_{1/0}H^T + R_1\right]^{-1}$$
$$= 1\{1 + 0.25\}^{-1}$$
$$= 0.8$$

$$\hat{x}_{1/1} = \hat{x}_{1/0} + K_1\left[y_1 - H\hat{x}_{1/0}\right]$$
$$= 0 + 0.8[1.5 - 0]$$
$$= 1.2$$

$$P_{1/1} = P_{1/0} - K_1HP_{1/0}$$
$$= 1 - 0.8$$
$$= 0.2$$

$k = 2:$
Prediction:

$$\hat{x}_{2/1} = A\hat{x}_{1/1}$$
$$= 1.2$$

$$P_{2/1} = AP_{1/1}A^T + Q$$
$$= 0.2 + 1$$
$$= 1.2$$

Correction:

$$K_2 = P_{2/1}H^T\left[HP_{2/1}H^T + R_2\right]^{-1}$$
$$= 1.2\{1.2 + 0.25\}^{-1}$$
$$= 0.8275$$

$$\hat{x}_{2/2} = \hat{x}_{2/1} + K_2\left[y_2 - H\hat{x}_{2/1}\right]$$
$$= 1.2 + 0.8275[2.5 - 1.2]$$
$$= 2.2757$$

$$P_{2/2} = P_{2/1} - K_2HP_{2/1}$$
$$= 1.2 - 0.8275 \times 1.2$$
$$= 0.207$$

Example 3.9: Apply linear discrete Kalman filter to determine the optimal estimates of the process defined by

$$x_k = \begin{bmatrix} 1 & 0.3 \\ 0.35 & 1 \end{bmatrix}x_{k-1} + \begin{bmatrix} 1 \\ 0 \end{bmatrix}u_k$$

The process is stationary with the process noise

$$Q_k = \begin{bmatrix} 1 & 0 \\ 0 & 1 \end{bmatrix}$$

The covariance matrix of the initial state vector is given by

$$P_0 = \begin{bmatrix} 2 & 0 \\ 0 & 2 \end{bmatrix}$$

The measurement noise covariance is given by

$$R_k = 1.5 + (-1)^k$$

The initial state estimate is considered as

$$x_0 = \begin{bmatrix} 30 \\ 40 \end{bmatrix}$$

Consider the measurement data of 35 and 38 corresponding to the first state available at the discrete sampling instant of $k = 1$ and $k = 2$, respectively. The input u_k is assumed to be 3.
Solution:
The initial state condition is

$$\hat{x}_{0/0} = \begin{bmatrix} 30 \\ 40 \end{bmatrix}$$

The input condition is

$$u = u_k = 3$$

The measurement condition is $y = 35$ at $k = 1$ and $y = 38$ at $k = 2$.
The discrete process and observation models in Eqs. (3.84) and (3.85) are represented by

$$x_k = Ax_{k-1} + Bu_k$$

with

$$y_k = H_k x_k$$

$$A = \begin{bmatrix} 1 & 0.3 \\ 0.35 & 1 \end{bmatrix}; \quad B = \begin{bmatrix} 1 \\ 0 \end{bmatrix}; \quad H_k = H = \begin{bmatrix} 1 & 0 \end{bmatrix}; \quad u_k = u = 3$$

$k = 1$:

Prediction:

$$\hat{x}_{1/0} = A\hat{x}_{0/0} + Bu$$

$$= \begin{bmatrix} 1 & 0.3 \\ 0.35 & 1 \end{bmatrix} \begin{bmatrix} 30 \\ 40 \end{bmatrix} + \begin{bmatrix} 1 \\ 0 \end{bmatrix} [3]$$

$$= \begin{bmatrix} 45 \\ 50.5 \end{bmatrix}$$

$$P_{1/0} = AP_{0/0}A^T + Q$$

$$= \begin{bmatrix} 1 & 0.3 \\ 0.35 & 1 \end{bmatrix} \begin{bmatrix} 2 & 0 \\ 0 & 2 \end{bmatrix} \begin{bmatrix} 1 & 0.35 \\ 0.3 & 1 \end{bmatrix} + \begin{bmatrix} 1 & 0 \\ 0 & 1 \end{bmatrix}$$

$$= \begin{bmatrix} 3.18 & 1.30 \\ 1.30 & 3.24 \end{bmatrix}$$

Correction:

$$K_1 = P_{1/0}H^T \left[HP_{1/0}H^T + R_1 \right]^{-1}$$

$$= \begin{bmatrix} 3.18 & 1.30 \\ 1.30 & 3.24 \end{bmatrix} \begin{bmatrix} 1 \\ 0 \end{bmatrix} \left\{ \begin{bmatrix} 1 & 0 \end{bmatrix} \begin{bmatrix} 3.18 & 1.30 \\ 1.30 & 3.24 \end{bmatrix} \begin{bmatrix} 1 \\ 0 \end{bmatrix} + [0.5] \right\}^{-1}$$

$$= \begin{bmatrix} 0.8641 \\ 0.3532 \end{bmatrix}$$

$$\hat{x}_{1/1} = \hat{x}_{1/0} + K_1 \left[y_1 - H\hat{x}_{1/0} \right]$$

$$= \begin{bmatrix} 45 \\ 50.5 \end{bmatrix} + \begin{bmatrix} 0.8641 \\ 0.3532 \end{bmatrix} \left[35 - \begin{bmatrix} 1 & 0 \end{bmatrix} \right] \begin{bmatrix} 45 \\ 50.5 \end{bmatrix}$$

$$= \begin{bmatrix} 36.355 \\ 46.968 \end{bmatrix}$$

$$P_{1/1} = P_{1/0} - K_1 HP_{1/0}$$

$$= \begin{bmatrix} 3.18 & 1.30 \\ 1.30 & 3.24 \end{bmatrix} - \begin{bmatrix} 0.8641 \\ 0.3532 \end{bmatrix} \begin{bmatrix} 1 & 0 \end{bmatrix} \begin{bmatrix} 3.18 & 1.30 \\ 1.30 & 3.24 \end{bmatrix}$$

$$= \begin{bmatrix} 0.4322 & 0.1767 \\ 0.1767 & 2.7859 \end{bmatrix}$$

$k = 2$:

Prediction:

$$\hat{x}_{2/1} = A\hat{x}_{1/1} + Bu$$

$$= \begin{bmatrix} 1 & 0.3 \\ 0.35 & 1 \end{bmatrix} \begin{bmatrix} 36.355 \\ 46.968 \end{bmatrix} + \begin{bmatrix} 1 \\ 0 \end{bmatrix} [3]$$

$$= \begin{bmatrix} 53.4454 \\ 59.6922 \end{bmatrix}$$

$$P_{2/1} = AP_{1/1}A^T + Q$$

$$= \begin{bmatrix} 1 & 0.3 \\ 0.35 & 1 \end{bmatrix} \begin{bmatrix} 0.4322 & 0.1767 \\ 0.1767 & 2.7859 \end{bmatrix} \begin{bmatrix} 1 & 0.35 \\ 0.3 & 1 \end{bmatrix} + \begin{bmatrix} 1 & 0 \\ 0 & 1 \end{bmatrix}$$

$$= \begin{bmatrix} 1.7889 & 1.1833 \\ 1.1833 & 3.9624 \end{bmatrix}$$

Correction:

$$K_2 = P_{2/1}H^T \left[HP_{2/1}H^T + R_2 \right]^{-1}$$

$$= \begin{bmatrix} 1.7889 & 1.1833 \\ 1.1833 & 3.9624 \end{bmatrix} \begin{bmatrix} 1 \\ 0 \end{bmatrix} \left\{ \begin{bmatrix} 1 & 0 \end{bmatrix} \begin{bmatrix} 1.7889 & 1.1833 \\ 1.1833 & 3.9624 \end{bmatrix} \begin{bmatrix} 1 \\ 0 \end{bmatrix} + [2.5] \right\}^{-1}$$

$$= \begin{bmatrix} 0.4170 \\ 0.2758 \end{bmatrix}$$

$$\hat{x}_{2/2} = \hat{x}_{2/1} + K_2 \left[y_2 - H\hat{x}_{2/1} \right]$$

$$= \begin{bmatrix} 53.4454 \\ 59.6922 \end{bmatrix} + \begin{bmatrix} 0.4170 \\ 0.2758 \end{bmatrix} \left[38 - \begin{bmatrix} 1 & 0 \end{bmatrix} \right] \begin{bmatrix} 53.4454 \\ 59.6922 \end{bmatrix}$$

$$= \begin{bmatrix} 47.0047 \\ 55.4324 \end{bmatrix}$$

$$P_{2/2} = P_{2/1} - K_2 HP_{2/1}$$

$$= \begin{bmatrix} 1.7889 & 1.1833 \\ 1.1833 & 3.9624 \end{bmatrix} - \begin{bmatrix} 0.4170 \\ 0.2758 \end{bmatrix} \begin{bmatrix} 1 & 0 \end{bmatrix} \begin{bmatrix} 1.7889 & 1.1833 \\ 1.1833 & 3.9624 \end{bmatrix}$$

$$= \begin{bmatrix} 1.0430 & 0.6899 \\ 0.6899 & 3.6391 \end{bmatrix}$$

3.7.3 Continuous Kalman filter

The continuous time linear process can be represented in the following form:

$$\dot{x}(t) = Ax(t) + Bu(t) + w(t), x(0) = x_0 \tag{3.102}$$

The measurements of the process are considered at discrete points in time with the following relationship:

$$y(t_k) = H_k x(t_k) + v(t_k) \tag{3.103}$$

where $x = n$ dimensional state vector, $u = d$ dimensional input vector, $y = m$ dimensional observation vector, $w = n$ dimensional process noise vector, $v = m$ dimensional Gaussian noise vector, $A = (n \times n)$ dimensional model coefficient matrix, $B = (n \times d)$ dimensional input coefficient matrix, and $t_k =$ discrete time.

The covariance matrices for the initial state $x(0)$, process noise $w(t)$, and observation noise $v(t_k)$ are given by the following relations:

$$P_0 = E\left[(x_0 - x(0))(x_0 - x(0))^T\right]$$
$$Q(t) = E\left[w(t)w^T(t)\right] \qquad (3.104)$$
$$R(t_k) = E\left[v(t_k)v^T(t_k)\right]$$

where P_0 = initial state noise covariance matrix, $Q(t)$ = process noise covariance matrix, and $R(t_k)$ = observation noise covariance matrix.

The Kalman filter algorithm consists of two recursive steps. In the first step, the process model is used to propagate the initial state estimates to that time at which the first measurement is available. In the second step, the propagated model estimates are combined with the measurements to provide updated or corrected estimates. By starting with an initial estimate $\hat{x}_0$ and its covariance matrix P_0, the Kalman filter prediction and correction equations for recursive state estimation are given as follows.

3.7.3.1 Prediction equations

Predicted state and its covariance matrix are computed by

$$\dot{\hat{x}}\left(t/t_{k-1}\right) = A\hat{x}\left(t/t_{k-1}\right) + Bu\left(t/t_{k-1}\right) \dot{P}\left(t/t_{k-1}\right) = AP\left(t/t_{k-1}\right) + P\left(t/t_{k-1}\right)A^T + Q(t) \qquad (3.105)$$

3.7.3.2 Correction equations

The propagated estimate $\hat{x}(t)$ and its associated covariance matrix $P(t)$ at time t_k can be denoted by $\hat{x}\left(t_k/t_{k-1}\right)$ and $P\left(t_k/t_{k-1}\right)$. The measurements at time t_k are used to obtain the update estimates $\hat{x}\left(t_k/t_k\right)$ and its covariance matrix $P\left(t_k/t_k\right)$ by using the following equations:

$$\hat{x}\left(t_k/t_k\right) = \hat{x}\left(t_k/t_{k-1}\right) + K(t_k)\left[y(t_k) - H_k\hat{x}\left(t_k/t_{k-1}\right)\right]$$
$$P\left(t_k/t_k\right) = P\left(t_k/t_{k-1}\right) - K(t_k)H_kP\left(t_k/t_{k-1}\right) \qquad (3.106)$$
$$K(t_k) = P\left(t_k/t_{k-1}\right)H_k^T(t_k)\left[H_kP\left(t_k/t_{k-1}\right)H_k^T + R_k\right]^{-1}$$

The recursive initial conditions for state estimate and covariance are

$$\hat{x}\left(t/t_{k-1}\right) = \hat{x}\left(t_k/t_k\right)$$
$$P\left(t/t_{k-1}\right) = P\left(t_k/t_k\right) \qquad (3.107)$$

where P is state covariance matrix, K is Kalman gain matrix, H_k is measurement matrix, T_k is time index, and T is transpose matrix. The filter parameters P_0, Q, and R can be used as known values, or they can be treated as design parameters. The recursive computation of continuous Kalman filter is shown in Fig. 3.7.

Example 3.10: A continuous linear plant is described by

$$\begin{bmatrix} \dot{x}_1 \\ \dot{x}_2 \end{bmatrix} = \begin{bmatrix} 1 & 1 \\ -2 & -1 \end{bmatrix} \begin{bmatrix} x_1 \\ x_2 \end{bmatrix} + \begin{bmatrix} 0 \\ 1 \end{bmatrix} u$$

The measurement equation is given by

$$y = \begin{bmatrix} 1 & 0 \end{bmatrix} \begin{bmatrix} x_1 \\ x_2 \end{bmatrix}$$

Assume the initial state to be

$$\begin{bmatrix} x_1(0) \\ x_2(0) \end{bmatrix} = \begin{bmatrix} 1 \\ 1 \end{bmatrix}$$

Consider the value of u to be unity. The diagonal elements of the initial state covariance matrix P_0 and process noise covariance matrix Q are considered as the squared values of 1% state deviations from the initial condition, and their off-diagonal

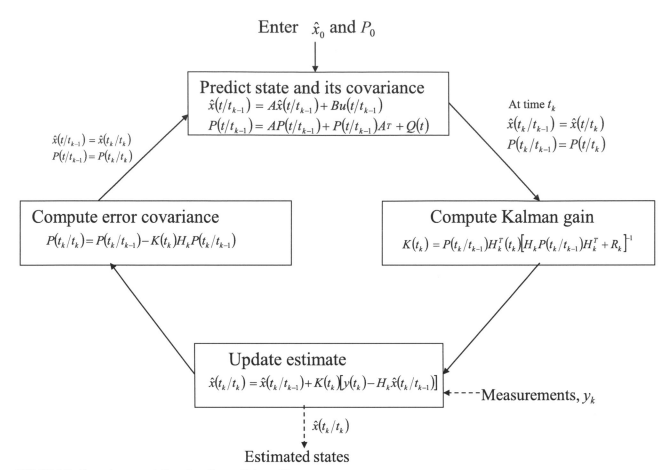

FIGURE 3.7   Recursive computation of continuous Kalman filter.

elements are assumed to be zero. The measurement noise covariance matrix R is taken as a squared value of 2% deviation from the initial measurement condition. Apply continuous linear Kalman filter to estimate the measured and unmeasured state vector using the measurement data of 1.0 for the first state available at the discrete sampling instant of $t = 0.1$ min.

Solution:

The initial state covariance matrix is formed as its diagonal elements with squared values of 1% deviation from the initial state condition:

$$P_0 = \begin{bmatrix} (0.01)^2 & 0 \\ 0 & (0.01)^2 \end{bmatrix} = \begin{bmatrix} 10^{-4} & 0 \\ 0 & 10^{-4} \end{bmatrix}$$

The process noise covariance matrix is formed as its diagonal elements with squared values of 1% deviation from the initial state condition:

$$Q = \begin{bmatrix} 10^{-4} & 0 \\ 0 & 10^{-4} \end{bmatrix}$$

The measurement coefficient matrix:

$$H_k = \begin{bmatrix} 1 & 0 \end{bmatrix}$$

The measurement noise covariance matrix is considered as a squared value of 2% deviation from the initial measurement condition.

$$R_k = 4 \times 10^{-4}$$

The input condition:

$$u = 1.0$$

The measurement condition:

$$y(0.1) = 1.0; \quad \Delta t = 0.1$$

The model is of the form:

$$\dot{x}(t) = Ax(t) + Bu(t)$$
$$y(t_k) = H_k x(t_k)$$

For convenience, the time step in prediction and the step in correction are considered the same.
$t = 0$:

$$\begin{bmatrix} \hat{x}_1(0) \\ \hat{x}_2(0) \end{bmatrix} = \begin{bmatrix} 1 \\ 1 \end{bmatrix}$$

$t = 0.1, k = 1$:
Prediction:
By simple Euler integration

$$\begin{bmatrix} \hat{x}_1 \\ \hat{x}_2 \end{bmatrix}_{(1/0)} = \begin{bmatrix} \hat{x}_1 \\ \hat{x}_2 \end{bmatrix}_{(0/0)} + \Delta t \begin{bmatrix} x_1 + x_2 \\ -2x_1 - x_2 + u \end{bmatrix}_{(0/0)}$$

$$= \begin{bmatrix} 1 \\ 1 \end{bmatrix} + 0.1 \begin{bmatrix} 2 \\ -2 \end{bmatrix}$$

$$= \begin{bmatrix} 1.2 \\ 0.8 \end{bmatrix}$$

$$P_{(1/0)} = P_{(0/0)} + \Delta t \left\{ [AP + PA^T]_{(0/0)} + Q \right\}$$

$$= \begin{bmatrix} 10^{-4} & 0 \\ 0 & 10^{-4} \end{bmatrix} + 0.1 \left\{ \begin{bmatrix} 1 & 1 \\ -2 & -1 \end{bmatrix} \begin{bmatrix} 10^{-4} & 0 \\ 0 & 10^{-4} \end{bmatrix} + \begin{bmatrix} 10^{-4} & 0 \\ 0 & 10^{-4} \end{bmatrix} \begin{bmatrix} 1 & -2 \\ 1 & -1 \end{bmatrix} + \begin{bmatrix} 10^{-4} & 0 \\ 0 & 10^{-4} \end{bmatrix} \right\}$$

$$= \begin{bmatrix} 1.3 \times 10^{-4} & -0.1 \times 10^{-4} \\ -0.1 \times 10^{-4} & 0.9 \times 10^{-4} \end{bmatrix}$$

Correction:

$$K_1 = P_{(1/0)} H_1^T \left[H_1 P_{(1/0)} H_1^T + R_1 \right]^{-1}$$

$$= \begin{bmatrix} 1.3 \times 10^{-4} & -0.1 \times 10^{-4} \\ -0.1 \times 10^{-4} & 0.9 \times 10^{-4} \end{bmatrix} \begin{bmatrix} 1 \\ 0 \end{bmatrix} \left\{ \begin{bmatrix} 1 & 0 \end{bmatrix} \begin{bmatrix} 1.3 \times 10^{-4} & -0.1 \times 10^{-4} \\ -0.1 \times 10^{-4} & 0.9 \times 10^{-4} \end{bmatrix} \begin{bmatrix} 1 \\ 0 \end{bmatrix} + [4.0 \times 10^{-4}] \right\}^{-1}$$

$$= \begin{bmatrix} 1.3 \times 10^{-4} \\ -0.1 \times 10^{-4} \end{bmatrix} \left[\frac{1}{5.3 \times 10^{-4}} \right]$$

$$= \begin{bmatrix} 0.2453 \\ -0.0189 \end{bmatrix}$$

$$\hat{x}_{1/1} = \hat{x}_{1/0} + K_1 \left[y_1 - H_1 \hat{x}_{1/0} \right]$$

$$= \begin{bmatrix} 1.2 \\ 0.8 \end{bmatrix} + \begin{bmatrix} 0.2453 \\ -0.0189 \end{bmatrix} [1.0 - 1.2]$$

$$= \begin{bmatrix} 1.2 \\ 0.8 \end{bmatrix} + \begin{bmatrix} -0.04906 \\ 0.00378 \end{bmatrix}$$

$$= \begin{bmatrix} 1.1509 \\ 0.8038 \end{bmatrix}$$

$$P_{1/1} = P_{1/0} - K_1 H_1 P_{1/0}$$

$$= \begin{bmatrix} 1.3 \times 10^{-4} & -0.1 \times 10^{-4} \\ -0.1 \times 10^{-4} & 0.9 \times 10^{-4} \end{bmatrix} - \begin{bmatrix} 0.2453 \\ -0.0189 \end{bmatrix} \begin{bmatrix} 1 & 0 \end{bmatrix} \begin{bmatrix} 1.3 \times 10^{-4} & -0.1 \times 10^{-4} \\ -0.1 \times 10^{-4} & 0.9 \times 10^{-4} \end{bmatrix}$$

$$= \begin{bmatrix} 0.9811 \times 10^{-4} & -0.0754 \times 10^{-4} \\ -0.0754 \times 10^{-4} & 0.8981 \times 10^{-4} \end{bmatrix}$$

3.8 State estimation applications of linear filtering and observation techniques

Kalman filtering and observer techniques are useful for a variety of applications. Applications of Kalman filters include trajectory estimation, state estimation for control or diagnosis, control of vehicles such as aircraft and spacecraft, communication systems, data merging, signal processing, navigation, health monitoring, econometrics, weather forecasting, and pollution estimation. Applications of observers include fault detection, medical imaging, control, and tracking problems.

3.9 Summary

This chapter describes the topics such as mathematical representation of a system and its associated variables in state space domain, continuous time and discrete time representation of linear dynamic systems, concepts of observability and controllability of linear systems, recursive least state estimation, different continuous and discrete form of state estimation algorithms and their applications to linear systems. The state estimation algorithms described include discrete and continuous forms of full order observers, reduced order observer, observer-based state feedback controllers, discrete and continuous forms of Kalman filter along with the computation procedures of filter design parameters. The application of most of these methods are illustrated in terms of different examples concerning to linear systems.

References

[1] P.B. Liebelt, An Introduction to Optimal Estimation, Addison-Wesley, Reading, MA, 1967.
[2] A. Gelb (Ed.), Applied Optimal Estimation, M.I.T. Press, Cambridge, MA, 1974.
[3] T. Kailath, Linear Systems, Prentice-Hall, Englewood Cliffs, NJ, 1980.
[4] J.S. Meditch, Stochastic Optimal Linear Estimation, and Control, McGraw-Hill, New York, NY, 1969.
[5] R.E. Kalman, A new approach to linear filtering and prediction problems, Trans. ASME − J. Basic Eng. 82 (1) (1960) 35−45.
[6] E. Kreindler, P.E. Sarachik, On the concepts of controllability and observability of linear systems, IEEE Trans. Automat. Control 9 (2) (1964) 129−136.
[7] L.M. Silverman, H.E. Meadows, Controllability and observability in time variable linear systems, SIAM J. Control 5 (1) (1967) 64−73.
[8] D. Simon, Optimal State Estimations, John Wiley & Sons, Inc, Hoboken, NJ, 2006.
[9] S.S. Haykin, Adaptive Filter Theory, Prentice-Hall, Upper Saddle River, NJ, 2002.
[10] D. Luenberger, Observing the state of a linear system, IEEE Trans. Mil. Electron. 8 (2) (1964) 74−80.
[11] D. Luenberger, Observers for multivariable systems, IEEE Trans. Autom. Control, 11 (2) (1966) 190−197.
[12] D. Luenberger, An introduction to observers, IEEE Trans. Autom. Control 16 (6) (1971) 596−602.
[13] P.J. Antsaklis, A.N. Michel, Linear Systems, Birkhäuser, Boston, MA, 2006.
[14] J. O'Reilly, Observers for Linear Systems, Academic Press, New York, NY, 1983.

[15] W.A. Wolovich, Linear Multivariable Systems, Springer-Verlag, New York, NY, 1974.

[16] C.T. Chen, Linear System Theory and Design, third ed., Oxford University Press, New York, NY, 1999.

[17] A.I. Astrovskii, I.V. Gaishun, State estimators for linear time-varying observation systems, Differential Equations 55 (2019) 363−373.

[18] D.C. Huong, V.T. Huynh, H. Trinh, Integral outputs-based robust state observers design for time-delay systems, SIAM Journal on Control and Optimization 57 (3) (2019) 2214−2239.

[19] K. Ogata, Modern Control Engineering, Prentice-Hall, India, 2006.

[20] L.A. Zadeh, C.A. Desoer, Linear System Theory: The State Space Approach, Springer Verlag, 2008.

[21] Arie Nakhmani, Modern Control: State-Space Analysis and Design Methods, McGraw Hill, 2020.

[22] T.H. Abdelaziz, Pole placement for single-input linear system by proportional-derivative state feedback, Journal of Dynamic Systems, Measurement, and Control 137 (4) (2015) 041015.

[23] A. Babiarz, A. Czornik, E. Makarov, M. Niezabitowski, S. Popova, Pole placement theorem for discrete time-varying linear systems, SIAM Journal on Control and Optimization 55 (2) (2017) 671−692.

[24] A.H. Jazwinski, Stochastic Processes and Filtering Theory, Academic Press, New York, NY, 1970.

[25] T.P. McGarty, Stochastic Systems, and State Estimation, John Wiley & Sons, Inc, New York, NY, 1974.

[26] F.L. Lewis, Optimal Estimation with an Introductory to Stochastic Control Theory, John Wiley & Sons, Inc, New York, NY, 1986.

Chapter 4

Mechanistic model-based nonlinear filtering and observation techniques for optimal state/parameter estimation

4.1 Introduction

In the previous chapter, various linear filtering and observation techniques have been discussed for optimal state estimation in linear systems. However, most physical systems are inherently nonlinear in nature. Nonlinear dynamical systems can present a variety of behaviors depending on the values of the physical parameters and their intrinsic features. Efficient operation of estimator-based nonlinear systems requires the need for reliable and accurate methods of state estimation. This chapter presents various mechanistic model-based nonlinear filtering and observation techniques for optimal state estimation of nonlinear systems. The topics described include nonlinear systems and their mathematical representation, observability of nonlinear systems, different continuous, and discrete forms of state estimation algorithms along with the examples of applications.

4.2 General nonlinear system and system models

The state estimator uses the mathematical model of the process in conjunction with the known process measurements to provide the unmeasured process states that capture the fast-changing nonlinear dynamics of the process. The design of a mechanistic model-based state estimator requires the knowledge of the nonlinear dynamic system and its mathematical description. The mechanistic models are also called the first principle models. In practice, most physical processes exhibit nonlinear behavior. Though certain processes exhibit mildly nonlinear dynamic behavior, they also present significant nonlinear behavior when the operating conditions change widely. The dynamic behavior of nonlinear processes can be better described by nonlinear models. The dynamic model developed based on the inherent process knowledge and physical observations made on the actual process can replicate the real process. The first principle models that encapsulate substantial amount of process knowledge can provide a better platform to build estimator based strategies for online optimization, process monitoring, fault diagnosis, and control. Thus first principle models of nonlinear processes become an essential component in the design of model-based state estimators.

In general, the mathematical model formulation is represented in the form:

$$\text{Dependent variables} = f(\text{independent variables, parameters, forcing functions})$$

The dependent variables reflect the behavior of the status of the system in terms of certain basic process variables. The independent variables are usually dimensions such as time and space, along with the system behavior is determined. The parameters are reflective of system properties, and the forcing functions are external influences or disturbances acting on the system. This functional relation also involves inputs that manipulate the process, and such inputs are called manipulating variables.

4.2.1 Mechanistic/first principle models

Mechanistic or first principle models are dynamic models built based on fundamental physical and chemical laws such as mass and energy balances, thermodynamics, and chemical reaction kinetics. These models can be classified into lumped parameter models and distributed parameter models.

Optimal State Estimation for Process Monitoring, Fault Diagnosis and Control. DOI: https://doi.org/10.1016/B978-0-323-85878-6.00006-3

4.2.1.1 Lumped parameter models

These are dynamic models in which the basic process variables vary with respect to one space coordinate, and they are described by differential equations. An example of such a model is a continuous stirred tank reactor with a first order exothermic reaction as shown in Fig. 4.1.

By assuming constant liquid level, constant density, and perfect mixing, such model is expressed by

$$\text{Accumulation} = \text{Input} - \text{Output} + \text{Generation due to reaction}$$

$$V\frac{dC_A}{dt} = F(C_{Ao} - C_A) - VkC_A \tag{4.1}$$

where V is the volume of reaction mixture, F is feed flow rate, k is reaction rate constant, C_{A0} is initial feed concentration of component A, and C_A is the concentration of component A in reaction mixture.

4.2.1.2 Distributed parameter models

These are the dynamic models in which the basic process variables vary with respect to more than one space coordinate, and they are described by partial differential equations. An example of such a model can be heat conduction through a long thin rod which is insulated in all directions except in x-direction, as shown in Fig. 4.2.

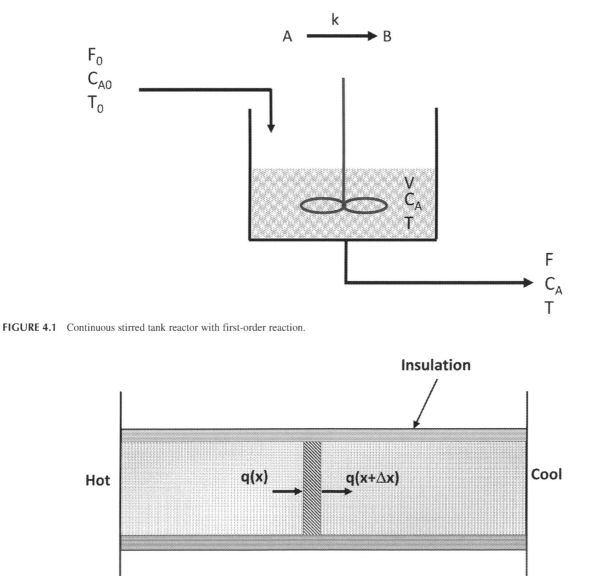

FIGURE 4.1 Continuous stirred tank reactor with first-order reaction.

FIGURE 4.2 Heat conduction through a long thin rod.

Heat balance over a unit time period Δt through the distance Δx is given by

$$\text{Heat input} - \text{Heat output} = \text{Heat accumulation}$$

$$[q(x) - q(x + \Delta x)]\Delta t = \rho c_p \Delta x \Delta T \tag{4.2}$$

$$\frac{q(x) - q(x + \Delta x)}{\Delta x} = \rho c_p \frac{\Delta T}{\Delta t}$$

$$Lim_{\Delta x \to 0} \frac{q(x) - q(x + \Delta x)}{\Delta x} = Lim_{\Delta t \to 0} \rho c_p \frac{\Delta T}{\Delta t}$$

$$\frac{-\partial q}{\partial x} = \rho c_p \frac{\partial T}{\partial t} \tag{4.3}$$

where ρ is the density, c_p is the specific heat and T is the temperature of the material. ΔT is the change in temperature over the distance Δx.

4.2.2   General representation of dynamic models

A nonlinear time varying dynamic system can be described by an n-component vector of ordinary differential equations representing its state dynamics, and an m-component vector of algebraic equations representing its output/observation. The dynamic equation of the system is given by the ordinary differential equation of the form,

$$\frac{dx(t)}{dt} = \dot{x}(t) = f[x(t), u(t), \theta(t), w(t), t] \tag{4.4}$$

where x is state vector, u is input vector, θ is parameter vector and w is disturbance/noise. The algebraic equation portraying the observable outputs is represented by

$$y(t) = h[x(t), u(t), \theta(t), v(t), t] \tag{4.5}$$

where v is observation noise. If random disturbances, measurement errors, and parameter uncertainties are omitted, the above equations become

$$\dot{x}(t) = f[x(t), u(t), t] \tag{4.6}$$

$$y(t) = h[x(t), t]$$

If there is no explicit dependence of equation coefficients on time, these equations further become time invariant as expressed by

$$\dot{x}(t) = f[x(t), u(t), w(t)] \tag{4.7}$$

$$y(t) = h[x(t)]$$

4.3   Observability of nonlinear systems

Observability is a necessary condition for state estimation. A system is said to be observable if its state at any instant can be determined by measuring its outputs over a finite time interval. Mathematically, a system is said to be observable, if for a given input function, the output map uniquely determines the state of the system at any instant.

Let us consider a continuous time nonlinear system, whose dynamics is described by the following general equation,

$$\dot{x} = f(x, u) \quad f:[t_0, t_1] \times \Omega \subset \to E^n \tag{4.8}$$

$$y = h(x) \quad h:[t_0, t_1] \times \Omega \subset \to E^m$$

where x, y, and u are vectors of state, output, and input, respectively. The f and h are nonlinear functions defined in the vector fields E^n and E^m, respectively, and Ω is the set in the field. The n and m refer the dimension of states and measurements, respectively. It is assumed that the state x is not available for direct measurement. The initial state $x(t_0)$ is unknown and belongs to the set $\Omega_0 \subseteq \Omega$. The observability problem consists of determining whether there exist relations binding the state variables $x(t)$ to outputs $y(t)$ and the time derivatives of $y(t)$, thus locally defining them uniquely in terms of measurable quantities. If no such relations exist, the initial state of the system cannot be deduced from observing its output behavior. In this section, we present observability of continuous nonlinear systems.

4.3.1   Local observability

Observability of a nonlinear system is determined locally about a given state or equilibrium point [1−3]. The system is locally observable when it is distinguishable at a point x_0 if there exists a neighborhood of x_0 such that in this neighborhood,

$$x_0 \neq x_1 \Rightarrow y(x_0) \neq y(x_1) \tag{4.9}$$

Local observability of a nonlinear system in Eq. (4.8) can be found using the Lie derivatives of h with respect to f when u is constant as defined by

$$
\begin{aligned}
y &= h(x) = L_f^0 h(x) \\
\dot{y} &= \frac{d}{dt} h(x) = \frac{\partial h(x)}{\partial x} \dot{x} = \frac{\partial h(x)}{\partial x} f(x, u) := Lfh(x) \\
\ddot{y} &= \frac{\partial L_f h(x)}{\partial x} \dot{x} = \frac{\partial L_f h(x)}{\partial x} f(x, u) := L_f^2 h(x) \\
&\vdots \\
y^{n-1} &= \frac{\partial L_f^{n-1} h(x)}{\partial x} \dot{x} = \frac{\partial L_f^{n-1} h(x)}{\partial x} f(x, u) := L_f^{n-1} h(x)
\end{aligned}
\tag{4.10}
$$

Let $\phi(x)$ denote the set of all finite linear combinations of the Lie derivatives as given by

$$
\phi(x) = \begin{bmatrix} y \\ \dot{y} \\ \ddot{y} \\ \vdots \\ y^{n-1} \end{bmatrix} = \begin{bmatrix} L_f^0 h(x) \\ L_f h(x) \\ L_f^2 h(x) \\ \vdots \\ L_f^{n-1} h(x) \end{bmatrix}
\tag{4.11}
$$

The first derivative of $\phi(x)$ about x_0 at constant u is evaluated as

$$O = J(\phi) = \frac{\partial \phi(x, u)}{\partial x} \Big|_{x=x_0} \tag{4.12}$$

The system satisfies local observability if $J(\phi)$ has rank n.

A simple observability test can be applied to nonlinear systems by linearizing the system in Eq. (4.8) about x_0 by using the Taylor series expansion as given by

$$
\begin{aligned}
\dot{x} &= f(x_0, u) + \frac{\partial f}{\partial x} \Big|_{x=x_0} (x - x_0) + \cdots + \frac{\partial f}{\partial u} \Big|_{x=x_0} (u - u_0) + \cdots \\
y &= h(x_0) + \frac{\partial h}{\partial x} \Big|_{x=x_0} (x - x_0) + \cdots
\end{aligned}
\tag{4.13}
$$

The linearized system matrices can be obtained from the first order terms as

$$A = \frac{\partial f}{\partial x}; \quad B = \frac{\partial f}{\partial u}; \quad C = \frac{\partial h}{\partial x} \tag{4.14}$$

The standard state space representation of the linearized system is of the form:

$$
\begin{aligned}
\delta \dot{x} &= A \delta x + B \delta u \\
\delta y &= C \delta x
\end{aligned}
\tag{4.15}
$$

A necessary and sufficient condition for the system in Eq. (4.15) to be observable is that the $n \times n$ observability matrix:

$$
W_o = \begin{bmatrix} C \\ CA \\ CA^2 \\ \vdots \\ CA^{n-1} \end{bmatrix}
\tag{4.16}
$$

has rank n. If the rank of $W_0 < n$, the system is unobservable. Thus an approximation to the nonlinear observability matrix defined in Eq. (4.16) can be determined using the linear system observability techniques. Observability of this linearized system may not imply the local observability of the original nonlinear system.

Example 1

Find the local observability of the system given by

$$\dot{x}_1 = x_1^2 + e^{x_2}$$
$$\dot{x}_2 = x_1^2$$
$$y = x_1$$

Solution

$$n = 2$$

Let $y = h(x)$

$$y = L_f^0 h(x) = x_1$$
$$\dot{y} = L_f h(x) = \frac{\partial h(x)}{\partial x} f(x)$$

$$= \begin{bmatrix} 1 & 0 \end{bmatrix} \begin{bmatrix} x_1^2 + e^{x_2} \\ x_1^2 \end{bmatrix}$$
$$= x_1^2 + e^{x_2}$$

$$\phi(x) = \begin{bmatrix} y \\ \dot{y} \end{bmatrix} = \begin{bmatrix} L_f^0 h(x) \\ L_f h(x) \end{bmatrix} = \begin{bmatrix} x_1 \\ x_1^2 + e^{x_2} \end{bmatrix}$$

$$J(\phi) = \begin{bmatrix} 1 & 0 \\ 2x_1 & e^{x_2} \end{bmatrix}$$

$J(\phi)$ has rank 2. Hence the system is observable from the output y.

Example 2

Consider the synthesis of cyclopentenol from cyclopentadien [4] that follows the reaction sequence as the van de Vusse reaction [5]. The reaction scheme has the form

$$A \rightarrow B \rightarrow C$$
$$2A \rightarrow D$$

In the above reaction, A is cyclopentadien, B is cyclcopentenol, C is cyclopentandiol, and D is dicyclopentadien.

On Assuming constant volume, isothermal conditions, and mass action kinetics, the mass balances for A and B can be written as:

$$\dot{c}_A = \frac{q}{V_R}(c_{A0} - C_A) - k_1 C_A - k_3 C_A^2$$

$$\dot{c}_B = -\frac{q}{V_R} c_B + k_1 C_A - k_2 C_B$$

where C_A and C_B are the concentrations of A and B, respectively, C_{A0} is the inlet concentration of A, q is the inlet flow rate, V_R is the reactor volume, and k_1, k_2, and k_3 are reaction rate constants. On considering the dilution rate (q/V_R) as the manipulated variable (u) and the desired output (y) is the concentration of $B(c_B)$ [6], the nonlinear state space equations can be written as:

$$\dot{x}_1 = -k_1 x_1 - k_3 x_1^2 - u(x_{10} - x_1)$$
$$\dot{x}_2 = k_1 x_1 - k_2 x_2 + u(-x_2)$$
$$y = x_2$$

where $x_1 = c_A$, $x_2 = c_B$, $x_{10} = c_{A0}$ and u is considered to be constant. Determine the local obsrtvability for the system.

Solution

$$n = 2$$

$$y = L_f^0 h(x) = x_2$$

$$\dot{y} = \dot{x}_2 = L_f h(x) = \frac{\partial h(x)}{\partial x} f(x)$$

$$= \begin{bmatrix} 0 & 1 \end{bmatrix} \begin{bmatrix} -k_1 x_1 - k_3 x_1^2 - u(x_{10} - x_1) \\ k_1 x_1 - k_2 x_2 + u(-x_2) \end{bmatrix}$$

$$= k_1 x_1 - k_2 x_2 + u(-x_2)$$

$$\phi(x) = \begin{bmatrix} y \\ \dot{y} \end{bmatrix} = \begin{bmatrix} L_f^0 h(x) \\ L_f h(x) \end{bmatrix} = \begin{bmatrix} x_2 \\ k_1 x_1 - k_2 x_2 + u(-x_2) \end{bmatrix}$$

$$J(\phi) = \begin{bmatrix} 0 & 1 \\ k_1 & -k_2 - u \end{bmatrix}$$

$J(\phi)$ has rank 2. The system is locally observable from the output y.

4.3.2 Global observability

The nonlinear system given by Eq. (4.8) is *globally observable* for all initial conditions, x_0, then its observability can be determined uniquely from $y(t)$ and $u(t)$ in the whole domain of $x \in X$ and $u \in U$. We suppose that the nth order derivatives of f and h exist for every $x \in \Omega$ and every $t \in [t_0, t_1]$, and that $y(t)$ is smooth so that we can approximate $y(t)$ by a truncated Taylor series as follows:

$$y(t) = y(t_0) + \dot{y}(t_0)(t - t_0) + \frac{\ddot{y}(t_0)}{2!}(t - t_0)^2 + \cdots + \frac{y^{n-1}(t_0)}{(n-1)!}(t - t_0)^{n-1} + \frac{y^n(t^*)}{n!}(t - t_0)^n \tag{4.17}$$

where $t^* \in (t_0, t)$. We define

$$y(t_0) = h(x(t_0), \ t_0) = h_0(x(t_0), \ t_0)$$

$$\dot{y}(t_0) = \frac{\partial h_0}{\partial t}(x(t_0), \ t_0) + \left(\frac{\partial h_0}{\partial x(t_0)}(x(t_0), \ t_0) \right) f(x(t_0), \ t_0) = h_1(x(t_0), \ t_0)$$

$$\vdots$$

$$y^{(n-1)}(t_0) = \frac{\partial h_{n-2}}{\partial t}(x(t_0), \ t_0) + \left(\frac{\partial h_{n-2}}{\partial x(t_0)}(x(t_0), \ t_0) \right) f(x(t_0), \ t_0) = h_{n-1}(x(t_0), \ t_0)$$

These equations can be expressed as a nonlinear map given by

$$z = H(x(t_0)) \tag{4.18}$$

where

$$z = \begin{bmatrix} y(t_0) \\ \dot{y}(t_0) \\ \vdots \\ y^{n-1}(t_0) \end{bmatrix}; \quad H(x(t_0)) = \begin{bmatrix} h_0(x(t_0), t_0) \\ h_1(x(t_0), t_0) \\ \vdots \\ h_{n-1}(x(t_0), t_0) \end{bmatrix}$$

This nonlinear map is called the observability map of the system.

The system described by Eq. (4.8) is completely observable in Ω_0 over the time interval $[t_0, t_1]$ if there exists a one-to-one correspondence between the set Ω_0 of initial states and the set of trajectories of the observed output $y(t)$ for $t \in [t_0, t_1]$. Thus if the observability map H is one-to-one form Ω_0 to $H(\Omega_0)$, then by the data set z, the initial state $x(t_0)$ of the system can be uniquely determined. From the above definition, the system is completely observable. The univalence (one-to-one mapping) of H form Ω_0 to $H(\Omega_0)$ is only a sufficient condition of observability for nonlinear continuous time systems.

4.3.2.1 Ratio condition

The mth leading principle minor of the Jacobian matrix $J(x)$ of $H(x)$, denoted by Δ_m, is defined as the determinant of the matrix $J(x)$. If there exists a constant $\varepsilon > 0$ such that the absolute values of the leading principle minors Δ_1, Δ_2, ..., Δ_n of $J(x)$ can be defined by a condition:

$$|\Delta_1| \ge \varepsilon, \quad \frac{|\Delta_2|}{|\Delta_1|} \ge \varepsilon, \ldots, \frac{|\Delta_n|}{|\Delta_{n-1}|} \ge \varepsilon \tag{4.19}$$

This condition satisfies the sufficient condition for nonlinear observability. The ratio condition defined using the leading principle minors of the Jacobian matrix $J(x)$ of $H(x)$ and the conditions for complete observability of the nonlinear system can be referred elsewhere [7]. Such conditions state that the observability mapping of the system must be differentiable, and the Jacobian matrix of the observability mapping satisfy the ratio condition uniformly. If the ratio conditions are satisfied, then H has one-to-one mapping from Ω_0 to $H(\Omega_0)$. We note that the ratio condition is only a sufficient condition for H to be a one-to-one mapping.

Apart from ratio condition, we also define a strongly positive semidefinite condition as a sufficient condition for global observability. This condition states that if there exists a constant $n \times km$ matrix A such that the square matrix $AJ(x)$ satisfies: (1) det $AJ(x) > 0$ for all $x \in \Omega$ and (2) $AJ(x) + (AJ(x))^T$ has nonnegative principle minors for all $x \in \Omega$, then H has one to one mapping from Ω to $H(\Omega)$. The notation for n, k, and m is given below. The conditions (1) and (2) are called strongly positive semidefinite conditions.

The conditions for complete observability of the nonlinear system in the set Ω_0 of initial states over the interval $[t_0, t_1]$ are defined as follows [8]:

1. $km \ge n$, where k refers kth derivatives of $f(.)$ and $h(.)$, n denotes the number of state variables, and m refers the number of outputs.
2. The observability mapping $H(x)$ of the system is continuously differentiable.
3. Ω is an open-ended subset.
4. There exists an $n \times km$ matrix A such that $AJ(x)$ satisfies the strongly positive semidefinite condition for all $x \in \Omega$. Here $J(x)$ is the Jacobian matrix of $H(x)$.

Example 3

Find the global observability of the system given by

$$\dot{x}_1 = 2x_2$$
$$\dot{x}_2 = -2x_1 + e^{x_2}$$
$$y = x_2$$

Solution

The observability matrix $H(x)$ is given by

$$H(x) = \begin{bmatrix} x_2 \\ -2x_1 + e^{x_2} \end{bmatrix}$$

The observability matrix $J(x)$ is given by

$$J(x) = \begin{bmatrix} 0 & 1 \\ -2 & e^{x_2} \end{bmatrix}$$

Let A be the identity matrix, that is, $A = I$.

It can be seen that $J(x)$ satisfies the strongly positive semidefinite condition. Hence the system is observable.

4.4 Extended Kalman filter

Kalman filter equations are used to provide the state estimate and its covariance with time for linear systems. Most systems of practical interest, however, are nonlinear and for such systems extended Kalman filter (EKF) is a useful state estimation method. EKF is a combined state estimation and parameter identification method which is used to solve nonlinear problems by using the noise corrupted measurements of the process. By this method, estimation is carried out by linearizing the nonlinear equations around the current estimate and applying the Kalman filter to the linearized equations. The EKF has been widely reported in literature [9−13].

4.4.1 Process representation for state estimation

The general process representation for state estimation by EKF is described as follows. For convenience, the time varying nonlinear dynamic process model in Eq. (4.6) for state estimation by EKF is represented by

$$\dot{x}(t) = f(x(t), t) + w(t), \quad x(0) = x_0 \tag{4.20}$$

where $x(t)$ is n dimensional state vector, f is a nonlinear function of state $x(t)$ and $w(t)$ is an additive Gaussian noise with zero mean.

The nonlinear measurement model in Eq. (4.6) with observation noise can be expressed as

$$y(t_k) = h(x(t_k)) + v(t_k) \tag{4.21}$$

where h is a nonlinear function of state $x(t_k)$. The process and measurement noises w and v are uncorrelated, zero-mean white-noise processes with known covariance matrices. The initial state x_0 is a Gaussian random vector with mean $\bar{x}_0$. The expected values of noise covariance matrices for the initial state x_0, process noise $w(t)$ and observation noise $v(t_k)$ are given by the following relations:

$$\begin{aligned}
E\left[x_0 - \bar{x}_0)(x_0 - \bar{x}_0)^T\right] &= P_0 \\
E\left[w(t)w^T(t)\right] &= Q(t) \\
E\left[v(t_k)v^T(t_k)\right] &= R(t_k)
\end{aligned} \tag{4.22}$$

where P_0 is initial state covariance matrix, $Q(t)$ is process noise covariance matrix and $R(t_k)$ is observation noise covariance matrix.

4.4.2 Process representation for state and parameter estimation

By considering time varying parameters in the process model in Eq. (4.20), the expressions for states and parameters are given by:

$$\dot{x}(t) = f_x(x(t), \quad \theta(t), \quad t) + w_x(t) \quad x(0) = x_0 \tag{4.23}$$

$$\dot{\theta}(t) = g_\theta(x(t), \quad \theta(t), \quad t) + w_\theta(t) \quad \theta(0) = \theta_0$$

where f_x and g_θ are nonlinear functions of states x, parameters θ and also input u. The w_x and w_θ are process noise with covariance matrices Q_x and Q_θ. The nonlinear observation model in Eq. (4.21) with the inclusion of parameters can be expressed as

$$y(t_k) = h_k(x(t_k), \quad \theta(t_k), \quad t) + v(t_k) \tag{4.24}$$

where h is a nonlinear function of states and parameters, and v is observation noise with zero mean. The states $x(t)$ and parameters $\theta(t)$ of Eq. (4.23) can be estimated by EKF using the known process measurements.

4.4.3 Extended Kalman filter for continuous time nonlinear systems

An EKF extends the linear Kalman filter theory to nonlinear systems [11]. The EKF is computed in two steps. The first is a prediction step, which is used to extrapolate the previous best estimates, and the second is a correction step by which the updated estimates are formed. Since prediction is based on process model, continuous prediction and discrete correction is employed for estimation. Here, we describe the EKF algorithm for state estimation by considering the process and measurement models represented by Eqs. (4.20) and (4.21).

4.4.3.1 Prediction equations

By starting with an initial estimate x_0 and its covariance P_0 at time zero and no measurements are taken between t_{k-1} and t_k, the propagating expressions for the state estimate and its covariance from t_{k-1} to t_k are,

$$\dot{\hat{x}}(t/t_{k-1}) = f(\hat{x}(t/t_{k-1}), t) \tag{4.25}$$

$$\dot{P}(t/t_{k-1}) = F(\hat{x}(t/t_{k-1}), \ t)P(t/t_{k-1}) + P(t/t_{k-1})F^T(\hat{x}(t/t_{k-1}), \ t) + Q(t) \tag{4.26}$$

where $F(\hat{x}(t/t_{k-1}),\ t)$ is the state transition matrix whose i jth element is given by

$$F(\hat{x}(t/t_{k-1}),t) = \frac{\partial f_i(x(t),t)}{\partial x_j(t)}\bigg|_{x(t)=\hat{x}(t/t_{k-1})} \tag{4.27}$$

The solution of the propagated estimate $\hat{x}(t/t_{k-1})$ and its covariance $P(t/t_{k-1})$ at time t_k are denoted by $\hat{x}(t_k/t_{k-1})$ and $P(t_k/t_{k-1})$. By using measurements at time t_k, the update estimate $\hat{x}(t_k/t_k)$ and its covariance $P(t_k/t_k)$ are computed.

4.4.3.2 Correction equations

The equations to obtain the corrected estimates are given by

$$\hat{x}(t_k/t_k) = \hat{x}(t_k/t_{k-1}) + K(t_k)\big[y(t_k) - h\big(\hat{x}(t_k/t_{k-1})\big)\big] \tag{4.28}$$

$$P(t_k/t_k) = (I - K(t_k)H(x(t_k)))P(t_k/t_{k-1}) \tag{4.29}$$

$$K(t_k) = P(t_k/t_{k-1})H^T(x(t_k))(H(x(t_k))P(t_k/t_{k-1})H^T(x(t_k))+R)^{-1} \tag{4.30}$$

where

$$H(x(t_k)) = \frac{\partial h_i(x(t_k))}{\partial x(t_k)}\bigg|_{x(t_k)=\hat{x}(t_k/t_{k-1})} \tag{4.31}$$

The recursive initial conditions for state and covariance are,

$$\hat{x}(t/t_{k-1}) = \hat{x}(t_k/t_k)$$

$$P(t/t_{k-1}) = P(t_k/t_k) \tag{4.32}$$

The recursive computation of the EKF is shown in Fig. 4.3.

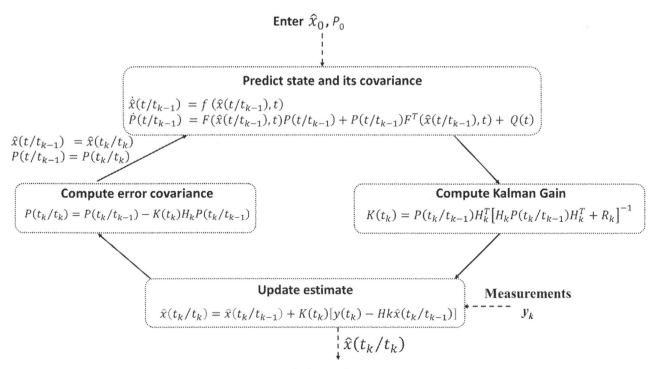

FIGURE 4.3   Recursive computation of extended Kalman filter.

4.4.4 Extended Kalman filter for discrete time nonlinear systems

Discrete time EKF is used for state estimation in discrete time nonlinear systems. The process and measurement models of a discrete time stochastic process are represented by the following equations:

$$x_{k+1} = f(x_k, \ u_k) + w_k \tag{4.33}$$
$$y_{k+1} = h(x_k) + v_k$$

where f and h are nonlinear functions of states and inputs, and u_k is the control vector. The w_k and v_k are the process and observation noises which are both assumed to be zero mean multivariate Gaussian noises. The covariances of w_k and v_k are Q_k and R_k, which are expressed as follows:

$$Q_k = E\left[w_k w_k^T\right]$$
$$R_k = E\left[v_k v_k^T\right] \tag{4.34}$$

4.4.4.1 Prediction equations

By starting with an initial estimate x_0 and its covariance P_0 at time zero, the prediction equations for the state estimate and its covariance are given by

$$\hat{x}_{k+1/k} = f(\hat{x}_{k/k}, u_k) \tag{4.35}$$
$$P_{k+1/k} = F_k P_{k/k} F_k^T + Q_k$$

where F_k is the state transition matrix whose element is given by

$$F_k = \frac{\partial f}{\partial x}\Big|_{\hat{x}_{k/k}, u_k} \tag{4.36}$$

Thus at each time step, the Jacobian is evaluated by linearizing the nonlinear process function around the current predicted states. By using measurements at time t_k, the update estimate $\hat{x}(t_k/t_k)$ and its covariance $P(t_k/t_k)$ are computed.

4.4.4.2 Correction equations

The equations to obtain corrected estimates are given by

$$K_k = P_{k+1/k} H_k^T (H_k P_{k+1/k} H_k^T + R_k)^{-1}$$
$$\hat{x}_{k+1/k+1} = \hat{x}_{k+1/k} + K_k(y_k - h(\hat{x}_{k+1/k})) \tag{4.37}$$
$$P_{k+1/k+1} = (I - K_k H_k) P_{k+1/k}$$

where the measurement Jacobian is evaluated s

$$H_k = \frac{\partial h}{\partial x}\Big|_{\hat{x}_{k+1/k}} \tag{4.38}$$

The recursive computation of the discrete time EKF can be carried out as shown in Fig. 4.4.

4.4.5 Emphasis on covariance matrices of extended Kalman filter

Initial state guess ($\hat{x}_0$), initial state noise covariance matrix (P_0), process noise covariance matrix (Q), and observation noise covariance matrix (R) can significantly inflence the performance of EKF. The uncertainty in initial state estimates cannot be measured experimentally. If the initial state estimate is not close to the actual initial state, the converge of filter to the correct state becomes slow. If the initial state covariance matrix is chosen too small along with the improper initial guess, the filter gain becomes small, and the estimator relies more on model predictions. Further, the impact of measurements on the estimator also becomes less. The importance of appropriately choosing the initial state estimate and initial state covariance matrix have been discussed in literature [14–16]. If the process noise covariance matrix is chosen smaller, the filter will rely on the model excessively and will not use the process measurements effectively to correct the states. This can lead to poor

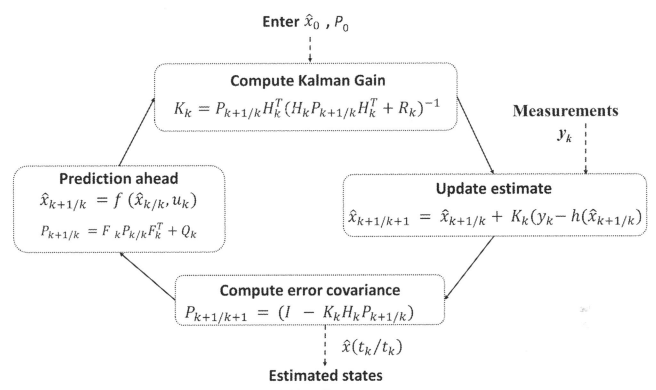

FIGURE 4.4 Recursive computation of discrete time extended Kalman filter.

performance of the estimator. If the process noise covariance matrix is chosen higher than the desired, the state estimates will be noisy and would lead to increased values of the state covariance matrix [14,17]. The selection of measurement noise covariance matrix is more or less straightforward as it can be directly derived from the measuring instruments. Thus choosing the appropriate values of tuning parameters is very important for achieving the effective performance of EKF.

The derivation of equations for state covariance matrix and Kalman gain are presented in Section 3.4 of Chapter 3, Linear Filtering and Observation Techniques.

Example 4

Consider a simple example of a nonlinear system described by the following equations,

$$\dot{x}_1 = x_1^2 + e^{x_2}$$
$$\dot{x}_2 = x_1^2$$

with the measurement equation

$$y = x_1$$

The initial state is assumed to be $[1.0 \; 1.0]^T$.

Assume the measurements are corrupted with a random Gaussian noise within the range of 0.2 to -0.2. The diagonal elements of the initial state covariance matrix P_0 and process noise covariance matrix Q are considered as the squared values of 1% deviations from initial state condition, and their off diagonal elements are assumed to be zero. The measurement noise covariance matrix R is taken as squared value of 2% deviation from initial measurement condition. Apply continuous form of EKF to estimate the state vector using the measurement data of 1.15 corresponding to the first state available at the discrete sampling instant of $t = 0.1$ min.

Solution

The system:

$$f(x) = \begin{bmatrix} f(x_1) \\ f(x_2) \end{bmatrix} = \begin{bmatrix} x_1^2 + e^{x_2} \\ x_1^2 \end{bmatrix}$$

The measurement equation:

$$y = h(x) = x_1$$

The initial state condition:

$$\begin{bmatrix} x_1(0) \\ x_2(0) \end{bmatrix} = \begin{bmatrix} 1.0 \\ 1.0 \end{bmatrix}$$

The measurement condition:

$$y(0.1) = 1.15; \quad \Delta t = 0.1$$

The diagonal elements of initial state covariance matrix are obtaind as squared values of 1% deviation from initial state condition:

$$P_0 = \begin{bmatrix} (0.01)^2 & 0 \\ 0 & (0.01)^2 \end{bmatrix} = \begin{bmatrix} 10^{-4} & 0 \\ 0 & 10^{-4} \end{bmatrix}$$

The diagonal elements of process noise covariance matrix are obtained as squared values of 1% deviation from initial state condition:

$$Q = \begin{bmatrix} 10^{-4} & 0 \\ 0 & 10^{-4} \end{bmatrix}$$

The measurement coefficient matrix:

$$H_k = \begin{bmatrix} 1 & 0 \end{bmatrix}$$

The measurement noise covariance matrix is considered as squared value of 2% deviation from initial measurement condition:

$$R_k = 4 \times 10^{-4}$$

The state and measurement models are of the form given in Eqs. (4.20) and (4.21), respectively.

For convenience, the continuous time step size in prediction and discrete time step size in correction are assumed to be the same. The initial state x_0 is considered same as the initial state estimate $\hat{x}_0$.

Observability: The nonlinear system is observable, as shown in Example 1.

$t = 0$:

$$\begin{bmatrix} \hat{x}_1(0) \\ \hat{x}_2(0) \end{bmatrix} = \begin{bmatrix} \hat{x}_1 \\ \hat{x}_2 \end{bmatrix}_{(0,0)} = \begin{bmatrix} 1 \\ 1 \end{bmatrix}$$

The measurement matrix is

$$H = \begin{bmatrix} 1 & 0 \end{bmatrix}$$

$$t = 0.1, \quad k = 1:$$

Prediction [Eqs. (4.25) and (4.26)]:
By simple Euler integration

$$\begin{bmatrix} \hat{x}_1 \\ \hat{x}_2 \end{bmatrix}_{(1/0)} = \begin{bmatrix} \hat{x}_1 \\ \hat{x}_2 \end{bmatrix}_{(0/0)} + \Delta t \begin{bmatrix} x_1^2 + e^{x_2} \\ x_1^2 \end{bmatrix}_{(0/0)}$$

$$= \begin{bmatrix} 1 \\ 1 \end{bmatrix} + 0.1 \begin{bmatrix} 3.718 \\ 1 \end{bmatrix}$$

$$= \begin{bmatrix} 1.3718 \\ 1.1 \end{bmatrix}$$

The state transition matrix as in Eq. (4.27):

$$F = \begin{bmatrix} 2x_1 & e^{x_2} \\ 2x_1 & 0 \end{bmatrix}$$

$$F|_{(0,0)} = \begin{bmatrix} 2 & 2.718 \\ 2 & 0 \end{bmatrix}$$

$$P_{(1/0)} = P_{(0/0)} + \Delta t \left\{ [FP + PF^T]_{(0/0)} + Q \right\}$$

$$= \begin{bmatrix} 10^{-4} & 0 \\ 0 & 10^{-4} \end{bmatrix} + 0.1 \left\{ \begin{bmatrix} 2 & 2.718 \\ 2 & 0 \end{bmatrix} \begin{bmatrix} 10^{-4} & 0 \\ 0 & 10^{-4} \end{bmatrix} + \begin{bmatrix} 10^{-4} & 0 \\ 0 & 10^{-4} \end{bmatrix} \begin{bmatrix} 2 & 2 \\ 2.718 & 0 \end{bmatrix} + \begin{bmatrix} 10^{-4} & 0 \\ 0 & 10^{-4} \end{bmatrix} \right\}$$

$$= \begin{bmatrix} 1.5 \times 10^{-4} & 0.4718 \times 10^{-4} \\ 0.4718 \times 10^{-4} & 1.1 \times 10^{-4} \end{bmatrix}$$

Correction [Eqs. (4.28)−(4.30)]:

$$K_1 = P_{(1/0)}H^T \left[HP_{(1/0)}H^T + R \right]^{-1}$$

$$= \begin{bmatrix} 1.5 \times 10^{-4} & 0.4718 \times 10^{-4} \\ 0.4718 \times 10^{-4} & 1.1 \times 10^{-4} \end{bmatrix} \begin{bmatrix} 1 \\ 0 \end{bmatrix} \left\{ \begin{bmatrix} 1 & 0 \end{bmatrix} \begin{bmatrix} 1.5 \times 10^{-4} & 0.4718 \times 10^{-4} \\ 0.4718 \times 10^{-4} & 1.1 \times 10^{-4} \end{bmatrix} \begin{bmatrix} 1 \\ 0 \end{bmatrix} + [4.0 \times 10^{-4}] \right\}^{-1}$$

$$= \begin{bmatrix} 1.5 \times 10^{-4} \\ 0.4718 \times 10^{-4} \end{bmatrix} \begin{bmatrix} \dfrac{1}{5.5 \times 10^{-4}} \end{bmatrix}$$

$$= \begin{bmatrix} 0.2727 \\ 0.0857 \end{bmatrix}$$

Assume the measurement value $y_1 = 1.15$ is corrupted with random Gaussian noise of random number of -0.1. Thus the value of y_1 becomes $= 1.05$

$$\hat{x}_{1/1} = \hat{x}_{1/0} + K_1 \left[y_1 - H\hat{x}_{1/0} \right]$$

$$= \begin{bmatrix} 1.3718 \\ 1.10 \end{bmatrix} + \begin{bmatrix} 0.2727 \\ 0.0857 \end{bmatrix} [1.05 - 1.3718]$$

$$= \begin{bmatrix} 1.3718 \\ 1.10 \end{bmatrix} + \begin{bmatrix} -0.0877 \\ -0.0275 \end{bmatrix}$$

$$= \begin{bmatrix} 1.2841 \\ 1.0725 \end{bmatrix}$$

$$P_{1/1} = P_{1/0} - K_1 H P_{1/0}$$

$$= \begin{bmatrix} 1.5 \times 10^{-4} & 0.4718 \times 10^{-4} \\ 0.4718 \times 10^{-4} & 1.1 \times 10^{-4} \end{bmatrix} - \begin{bmatrix} 0.2727 \\ 0.0857 \end{bmatrix} \begin{bmatrix} 1 & 0 \end{bmatrix} \begin{bmatrix} 1.5 \times 10^{-4} & 0.4718 \times 10^{-4} \\ 0.4718 \times 10^{-4} & 1.1 \times 10^{-4} \end{bmatrix}$$

$$= \begin{bmatrix} 1.091 \times 10^{-4} & 0.3432 \times 10^{-4} \\ 0.3432 \times 10^{-4} & 1.0596 \times 10^{-4} \end{bmatrix}$$

The estimated state becomes the initial condition for the next recursive step.

Example 5

Consider the discrete time nonlinear model of the van der Pol oscillator described by

$$\begin{bmatrix} x_{1,k+1} \\ x_{2,k+1} \end{bmatrix} = \begin{bmatrix} x_{1,k} + \tau x_{2,k} \\ x_{2,k} + \tau(\mu(0.5 - x_{1,k}^2)x_{2,k}) + \tau(-x_{1,k} + 0.5\cos(1.1k\tau)) \end{bmatrix}$$

where the sampling time $\tau = 0.05$ and $\mu = 1$. The output is given by

$$y_k = x_{1,k} + v_k$$

Assume v_k is random noise in the range of -0.1 to 0.1.

Consider the initial state covariance matrix as

$$P_0 = \begin{bmatrix} 10^{-4} & 0 \\ 0 & 10^{-4} \end{bmatrix}$$

Assume the process noise covariance matrix as:

$$Q_k = \begin{bmatrix} 10^{-4} & 0 \\ 0 & 10^{-4} \end{bmatrix}$$

and the measurement noise covariance matrix as

$$R_k = 4 \times 10^{-4}$$

Consider the initial actual state as $x0 = [0\ 0.5]^r$, and the initial state estimate as

$$\hat{x}_0 = \begin{bmatrix} 0.1 & 0.3 \end{bmatrix}.$$

Apply discrete form of EKF to estimate the state vector for one iteration.

Solution

The state and measurement models are of the form in Eq. (4.33). For convenience, the discrete time steps in prediction and correction are assumed to be the same.

The system model in example can be written as:

$$\begin{bmatrix} x_{1,k} \\ x_{2,k} \end{bmatrix} = \begin{bmatrix} x_{1,k-1} + \tau x_{2,k-1} \\ x_{2,k-1} + \tau(\mu(0.5 - x_{1,k-1}^2)x_{2,k-1}) + \tau(-x_{1,k-1} + 0.5\cos(1.1k\tau)) \end{bmatrix}$$

where time step $\tau = 0.05$ and $\mu = 1$.

The state transition matrix [Eq. (4.36)]:

$$F_k = \begin{bmatrix} 1 & \tau \\ -2\mu\tau x_{1,k-1}x_{2,k-1} - \tau & 1 + \mu\tau(0.5 - x_{1,k-1}^2) \end{bmatrix}$$

The measurement coefficient matrix:

$$H_k = \begin{bmatrix} 1 & 0 \end{bmatrix}$$

The initial state condition:

$$\begin{bmatrix} \hat{x}_{1,0} \\ \hat{x}_{2,0} \end{bmatrix} = \begin{bmatrix} 0.1 \\ 0.3 \end{bmatrix}$$

$$k = 1:$$

Prediction [Eq. (4.35)]:

The prediction state :

$$\begin{bmatrix} \hat{x}_{1,1} \\ \hat{x}_{2,1} \end{bmatrix} = \begin{bmatrix} \hat{x}_{1,0} + \tau\hat{x}_{2,0} \\ x_{2,0} + \tau(\mu(0.5 - x_{1,0}^2)x_{2,0}) + \tau(-x_{1,0} + 0.5\cos(1.1\tau)) \end{bmatrix}$$
$$= \begin{bmatrix} 0.1150 \\ 0.3272 \end{bmatrix}$$

$$F_k = \begin{bmatrix} 1 & \tau \\ -2\mu\tau\hat{x}_{1,0}\hat{x}_{2,0} - \tau & 1 + \mu\tau(0.5 - \hat{x}_{1,0}^2) \end{bmatrix}$$

$$F_k = \begin{bmatrix} 1.0 & 0.05 \\ -0.053 & 1.024 \end{bmatrix}$$

The prediction covariance matrix:

$$P_{k/k-1} = F_{k-1}P_{k-1/k-1}F_{k-1}^T + Q_{k-1}$$

$$P_{1/0} = \begin{bmatrix} 1.0 & 0.05 \\ -0.053 & 1.024 \end{bmatrix} \begin{bmatrix} 10^{-4} & 0 \\ 0 & 10^{-4} \end{bmatrix} \begin{bmatrix} 1.0 & -0.053 \\ 0.05 & 1.024 \end{bmatrix} + \begin{bmatrix} 10^{-4} & 0 \\ 0 & 10^{-4} \end{bmatrix}$$

$$= \begin{bmatrix} 2.0025 \times 10^{-4} & -0.0018 \times 10^{-4} \\ -0.0018 \times 10^{-4} & 2.1042 \times 10^{-4} \end{bmatrix}$$

Correction [Eq. (4.37)]:

$$K_1 = P_{1/0}H^T\left[HP_{1/0}H^T + R_1\right]^{-1}$$

$$= \begin{bmatrix} 2.0025 \times 10^{-4} & -0.0018 \times 10^{-4} \\ -0.0018 \times 10^{-4} & 2.1042 \times 10^{-4} \end{bmatrix} \begin{bmatrix} 1 \\ 0 \end{bmatrix} \left\{ \begin{bmatrix} 1 & 0 \end{bmatrix} \begin{bmatrix} 2.0025 \times 10^{-4} & -0.0018 \times 10^{-4} \\ -0.0018 \times 10^{-4} & 2.1042 \times 10^{-4} \end{bmatrix} \begin{bmatrix} 1 \\ 0 \end{bmatrix} + [4 \times 10^{-4}] \right\}^{-1}$$

$$= \begin{bmatrix} 0.3336 \\ -0.0003 \end{bmatrix}$$

Since $x_{1,1}$ obtained from state model based on actual initial condition is 0.025, assuming a random noise of 0.005, we have $y_1 = 0.03$ at $k = 1$.

Thus

$$\begin{bmatrix} \hat{x}_{1,2} \\ \hat{x}_{2,2} \end{bmatrix} = \begin{bmatrix} \hat{x}_{1,1} \\ \hat{x}_{2,1} \end{bmatrix} + K_1\left[y_1 - H\begin{bmatrix} \hat{x}_{1,1} \\ \hat{x}_{2,1} \end{bmatrix}\right]$$

$$= \begin{bmatrix} 0.1150 \\ 0.3272 \end{bmatrix} + \begin{bmatrix} 0.3336 \\ -0.0003 \end{bmatrix} \left[0.03 - \begin{bmatrix} 1 & 0 \end{bmatrix}\begin{bmatrix} 0.1150 \\ 0.3272 \end{bmatrix}\right]$$

$$= \begin{bmatrix} 0.0866 \\ 0.3272 \end{bmatrix}$$

$$P_{1/1} = P_{1/0} - K_1HP_{1/0}$$

$$= \begin{bmatrix} 2.0025 \times 10^{-4} & -0.0018 \times 10^{-4} \\ -0.0018 \times 10^{-4} & 2.1042 \times 10^{-4} \end{bmatrix} - \begin{bmatrix} 0.3336 \\ -0.0003 \end{bmatrix}\begin{bmatrix} 1 & 0 \end{bmatrix}\begin{bmatrix} 2.0025 \times 10^{-4} & -0.0018 \times 10^{-4} \\ -0.0018 \times 10^{-4} & 2.1042 \times 10^{-4} \end{bmatrix}$$

$$= \begin{bmatrix} 1.3345 \times 10^{-4} & -0.0012 \times 10^{-4} \\ -0.0012 \times 10^{-4} & 2.1042 \times 10^{-4} \end{bmatrix}$$

The estimated state becomes the initial condition for the next recursive step.

4.5 Steady state extended Kalman filter

The EKF uses the linearized model of the nonlinear system. It uses the partial derivatives of the nonlinear state equations and the nonlinear measurement equations. These partial derivatives are evaluated for each state estimate. Computation of partial derivatives of the nonlinear system and measurement equations for each estimate along with the Kalman gain matrix is computationally intensive. The computational effort associated with the matrix inversion increases considerably when the EKF is used for state estimation in high dimensional systems. In order to overcome this problem, it may be possible to compute the gain matrix in advance by choosing discrete points in the system's state space about which the nonlinear system is linearized. In the so-called steady state Kalman filter, the matrices K_k and P_k are computed in advance, and they remain constant. The steady state Kalman filter greatly reduces the computational burden of EKF. However, the steady state Kalman filter will not provide effective performance unless the initial guesses for state and covariance are properly selected. But if only steady state performance is important, the simplicity of this method could be attractive.

Here, we describe how steady state extended Kalman filter (SSEKF) is used for state estimation in continuous time nonlinear systems. The use of SSEKF is based on a precomputed Kalman gain matrix. By starting with an initial estimate x_0 and its covariance P_0 at time 0, the EKF equations, Eqs. (4.26), (4.29), and (4.30) are iteratively updated to converge to a constant state transition matrix. The iterative convergence results the gain matrix $K(t_k)$ to be a steady state gain matrix. The covariance matrix is also iteratively updated along with the gain matrix. The SSEKF with the constant gain and state covariance matrices is used for state estimation with the recursive computation of propagated and updated states. The propagated states are computed through Eq. (4.25). The solution of the propagated estimate

$\hat{x}(t/t_{k-1})$ at time t_k is denoted by $\hat{x}(t_k/t_{k-1})$. By using the measurements at time t_k, the update estimate $\hat{x}(t_k/t_k)$ is computed through Eq. (4.28). The computation of SSEKF is shown in Fig. 4.5.

4.6 Two-level extended Kalman filter

The process representation for state and parameter estimation by two-level extended Kalman filter (TLEKF) is described in Section 4.4.2. By this method, states are estimated separately in the first level by a state extended Kalman filter, and in the second level, uncertain process parameters are identified separately by a parameter extended Kalman filter. States and parameters are exchanged between the estimators for each new value of measurement. Here we describe the TLEKF for state and parameter estimation in continuous time nonlinear systems [18].

4.6.1 State estimation filter

Starting with an initial estimate, $\hat{x}_0$ and its covariance, P_{x0}, the correct estimate, $\hat{x}(t_k/t_k)$ and its associate covariance, $P_x(t_k/t_k)$ at time t_k are computed by the following equations:

$$K_x(t_k) = P_x(t_k/t_k)H_x^T(t_k)R^{-1} \tag{4.39}$$

$$P_x(t_k/t_k)^{-1} = P_x(t_k/t_{k-1})^{-1} + H_x^T(t_k)R^{-1}H_x(t_k)$$

$$\hat{x}(t_k/t_k) = \hat{x}(t_k/t_{k-1}) + K_x(t_k)\left\{y(t_k) - h_k\left[\hat{x}(t_k/t_{k-1}), \hat{\theta}(t_k/t_{k-1})\right]\right\}$$

The propagated expressions for the estimate and its covariance from t_k to t_{k+1} are:

$$\hat{x}(t/t_k) = f_x\left[\hat{x}(t/t_k), \hat{\theta}(t/t_k), t\right]$$

$$P_x(t/t_k) = F_x\left[\hat{x}(t/t_k)\right]P_x(t/t_k) + P_x(t/t_k)F_x^T\left[\hat{x}(t/t_k)\right] + Q_x \tag{4.40}$$

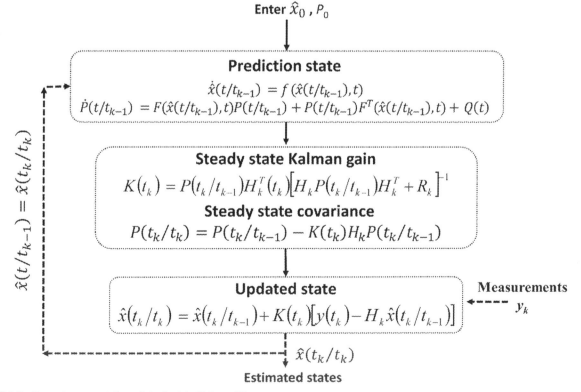

FIGURE 4.5 Recursive computation of steady state Kalman filter.

where

$$F_x\left[\hat{x}(t/t_k)\right] = \frac{\partial f_x}{\partial x}\Big|_{\substack{x=\hat{x}(t/t_k)\\ \theta=\hat{\theta}(t/t_k)}}$$

$$H_x(t_k) = \frac{\partial h_k}{\partial x}\Big|_{\substack{x=\hat{x}(t_k/t_{k-1})\\ \theta=\hat{\theta}(t_k/t_{k-1})}} \tag{4.41}$$

The propagated expressions at time t_{k+1}, $\hat{x}(t_{k+1}/t_k)$ and $P_x(t_{k+1}/t_k)$ form the recursive initial conditions for correction.

4.6.2 Parameter identification filter

Starting with the parameter estimate, $\hat{\theta}_0$ and its covariance, $P_{\theta 0}$, the correct estimate $\hat{\theta}(t_k/t_k)$ and its covariance $P_\theta(t_k/t_k)$, at time t_k, are computed by the following equations:

$$P_\theta(t_k/t_k)^{-1} = P_\theta(t_k/t_{k-1})^{-1} + H_\theta^T(t_k)R^{-1}H_\theta(t_k)$$

$$K_\theta(t_k) = P_\theta(t_k/t_k)H_\theta^T(t_k)R^{-1} \tag{4.42}$$

$$\hat{\theta}(t_k/t_k) = \hat{\theta}(t_k/t_{k-1}) + K_\theta(t_k)\left\{y(t_k) - h_k\left[\hat{x}(t_k/t_{k-1}), \hat{\theta}(t_k/t_{k-1})\right]\right\}$$

The propagating expressions for the estimate and its covariance from t_k to t_{k+1} are

$$\hat{\theta}(t/t_k) = p_\theta\left[\hat{x}(t/t_k), \hat{\theta}(t/t_k), \ t\right]$$

$$P_\theta(t/t_k) = G_\theta\left[\hat{\theta}(t/t_k)\right]P_\theta(t/t_k) + P_\theta(t/t_k)G_\theta^T\left[\hat{\theta}(t/t_k)\right] + Q_\theta \tag{4.43}$$

where

$$G_\theta\left[\theta(t/t_k)\right] = \frac{\partial p_\theta}{\partial \theta}\Big|_{\substack{x=\hat{x}(t/t_k)\\ \theta=\hat{\theta}(t/t_k)}}$$

$$H_\theta(t_k) = \frac{dh_k}{d\theta}\Big|_{\theta=\hat{\theta}(t_k/t_{k-1})}$$

$$= \frac{\partial h_k}{\partial \theta} + \frac{\partial h_k}{\partial \hat{x}(t_k/t_{k-1})} \cdot \frac{d\hat{x}(t_k/t_{k-1})}{d\theta}\Big|_{\theta=\hat{\theta}(t_k/t_{k-1})} \tag{4.44}$$

The propagated expressions at time t_{k+1}, $\hat{\theta}(t_{k+1}/t_k)$ and $P_\theta(t_{k+1}/t_k)$ form the recursive initial conditions for correction.

The initial covariance matrices P_{x0} and $P_{\theta 0}$ are used to reflect errors in the initial states and parameters. The process and observation noise covariance matrices Q_x, Q_θ, and R are used to reflect uncertainty in the process model and measurements. Matrices P_{x0}, $P_{\theta 0}$, Q_x, Q_θ, and R are generally selected as design parameters. The state and parameter estimation by this method is shown in Fig. 4.6.

The performance index J of the TLEKF is expressed by:

$$J = \sum_{k=1}^N \left\{(x(0)-\hat{x}(0))^T P_{x0}^{-1}(x(0)-\hat{x}(0)) + \left(\theta(0)-\hat{\theta}(0)\right)^T P_{\theta 0}^{-1}\left(\theta(0)-\hat{\theta}(0)\right)\right\}$$

$$+ \left\{\left(y(t_k) - h_k(\hat{x}(t_k)), \hat{\theta}(t_k)\right)R^{-1}(y(t_k)) - h_k(\hat{x}(t_k), \ \hat{\theta}(t_k))\right\}$$

$$+ \left\{(x(t_k)-\hat{x}(t_k))^T Q_x^{-1}(x(t_k)-\hat{x}(t_k)) + \left(\theta(t_k)-\hat{\theta}(t_k)\right)^T Q_\theta^{-1}\left(\theta(t_k)-\hat{\theta}(t_k)\right)\right\} \tag{4.45}$$

4.7 Adaptive fading extended Kalman filter

A discrete version of fading Kalman filter has been presented by Xia et al. [19]. Here we present a continuous version of adaptive fading extended Kalman filter (AFEKF) for state estimation in continuous nonlinear systems [20]. The

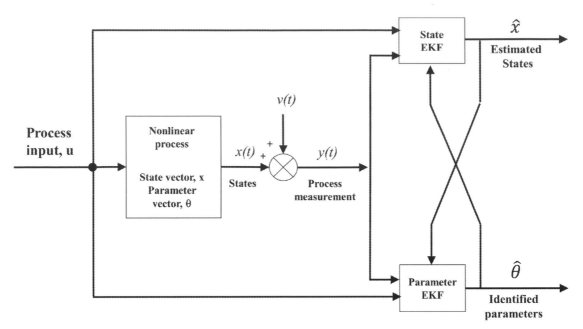

FIGURE 4.6 Two-level extended Kalman filter.

recursive computation of EKF Eqs. (4.25)–(4.32) are used for continuous state estimation by AFEKF with the modification of covariance matrix in Eq. (4.26). The computational procedure of this algorithm is similar to that of EKF except the covariance of the prediction equations. The covariance matrix in Eq. (4.26) is introduced with a forgetting factor, λ to form an adaptive covariance matrix, which is defined by:

$$\dot{P}(t/t_{k-1}) = \lambda(t/t_k)\left\{F(\hat{x}(t/t_{k-1}),t)P(t/t_{k-1}) + P(t/t_{k-1})F^T\hat{x}(t/t_{k-1}),\ t\right\} \\ + G(t/t_k)Q(t)G^T(t/t_k) \tag{4.46}$$

The λ is computed according to the following equation:

$$\lambda(t/t_k) = \max\left\{1,\ \ \text{trace}\left[N(t/t_k)\right]/\text{trace}\left[M(t/t_k)\right]\right\} \tag{4.47}$$

where

$$M(t/t_k) = H(x(t/t_k))\{F(\hat{x}(t/t_{k-1}),\ t)P(t/t_{k-1}) + P(t/t_{k-1})F^T(\hat{x}(t/t_{k-1}),\ t)\}H^T(x(t/t_k)),$$

$$N(t/t_k) = C_0(t/t_k) - H(x(t/t_k))G(t/t_{k-1})Q(t/t_{k-1})G^T(t/t_{k-1})H^T(x(t/t_k)) - R(t/t_k) \tag{4.48}$$

The $C_0(t/t_k)$ is the covariance of the error residual defined by the following equation:

$$C_0(t/t_k) = H(x(t/t_k))P(t/t_{k-1})H^T(x(t/t_k)) + R(t/t_k) \tag{4.49}$$

The AFEKF with the introduction of adaptive covariance in EKF is used for state estimation in continuous time nonlinear systems.

4.8 Unscented Kalman filter

In EKF, the process and measurement functions are linearized around the current predicted state to compute the Jacobians. When the process and observation functions are highly nonlinear, the EKF may not provide effective performance. This is due to the propagation of mean through severe nonlinearity. The unscented Kalman filter (UKF) is a recursive estimator that addresses some of the approximation issues of EKF [21]. Unlike the EKF, the UKF does not approximate the nonlinear process and observation models but uses the true nonlinear models as such. The unscented transformation (UT) is a method for calculating the statistics of a random variable which undergoes a nonlinear transformation and builds on the principle that is easier to approximate a probability distribution than an arbitrary non-linear function [22,23]. By using the unscented transform to compute the mean and covariance, the UKF avoids the need of Jacobians in the algorithm. The probability distribution is approximated by a set of deterministic points which captures the mean and covariance of the distribution. These points, called sigma points, are then processed through the nonlinear model of the system, producing a set of propagated sigma points. By choosing appropriate weights, the weighted average and weighted outer product of the transformed points provide an estimate of the mean and covariance of the transformed distribution.

We describe the UKF algorithm for state estimation in discrete time nonlinear process in which the process and measurement models are represented by Eq. (4.33). The covariance matrices Q_k and R_k, reflecting the process and observation noises are given by Eq. (4.34). Consider the propagation of a random variable x (dimension $n = L$) distributed according to a Gaussian with mean x and covariance P. The aim is to obtain a Gaussian approximation of the distribution over a nonlinear function, $y = f(x)$. The unscented transform performs this approximation by selecting the so-called sigma points χ from the Gaussian estimate and passing these points through f. Assume x has mean $\hat{x}$ and covariance P_x. Specifically, $2L + 1$ sigma points are chosen, and the computational procedure is given below.

The n-dimensional random variable x_{k-1} with mean $\hat{x}_{k-1}$ and covariance P_{k-1} is approximated by $2L + 1$ sigma points selected from the following equations:

$$\chi_{i,k-1} = \hat{x}_{k-1}, \quad i = 0$$

$$\chi_{i,k-1} = \hat{x}_{k-1} + \left(\sqrt{(L+\lambda)\hat{P}_{k-1}} \right)_i, \quad i = 1, \ldots, L \tag{4.50}$$

$$\chi_{i,k-1} = \hat{x}_{k-1} - \left(\sqrt{(L+\lambda)\hat{P}_{k-1}} \right)_i, \quad i = L+1, \ldots, 2L$$

where $(\sqrt{(L+\lambda)P_{k-1}})_i$ is the ith column of the matrix square root of $(L + \lambda)P_{k-1}$, which is symmetric and positive definite, and λ is define by

$$\lambda = \alpha^2(L + \kappa) - L$$

where α is a scaling parameter which determines the spread of the sigma points, κ is a secondary scaling parameter and L is the dimension of the state.

The weights for the states and covariance matrices are defined by

$$W_0^m = \lambda/(L + \lambda) \tag{4.51}$$

$$W_0^c = \lambda/(L + \lambda) + (1 - \alpha^2 + \beta)$$

$$W_i^m = W_i^c = 1/[2(L + \lambda)], \quad i = 1, \ldots, 2L$$

where β is a tuning parameter which is used to incorporate prior knowledge of the distribution. For Gaussian distribution, β is set as 2. The transformed state samples are computed by propagating the sigma points through a nonlinear process function:

$$\chi_{i,k/k-1} = f(\chi_{i,k-1}), \quad i = 0, \cdots, 2L \tag{4.52}$$

The transformed measurement samples are computed by propagating the sigma points through a nonlinear observation function:

$$Y_{i,k/k-1} = h(\chi_{i,k-1}), \quad i = 0, \cdots, 2L \tag{4.53}$$

Prediction step: This step involves the computation of mean $\hat{x}_{k/k-1}$ and covariance $P_{k/k-1}$ of the states using the transformed sigma points,

$$\hat{x}_{k/k-1} = \sum_{i=0}^{2L} W_i^m \chi_{i,k/k-1} \tag{4.54}$$

$$P_{k/k-1} = \sum_{i=0}^{2L} W_i^c (\chi_{i,k/k-1} - \hat{x}_{k/k-1})(\chi_{i,k/k-1} - \hat{x}_{k/k-1})^T + Q_k$$

where Q_k is the process noise covariance matrix. The transformed sigma points are used to compute the prediction measurement $\hat{y}_{k/k-1}$, the measurement covariance P_{y_k/y_k}, and the cross covariance of state-observations P_{x_k/y_k},

$$\hat{y}_{k/k-1} = \sum_{i=0}^{2L} W_i^m Y_{i,k/k-1}$$

$$P_{y_k y_k} = \sum_{i=0}^{2L} W_i^c (Y_{i,k/k-1} - \hat{y}_{k/k-1})(Y_{i,k/k-1} - \hat{y}_{k/k-1})^T + R_k \qquad (4.55)$$

$$P_{x_k y_k} = \sum_{i=0}^{2L} W_i^c (\chi_{i,k/k-1} - \hat{x}_{k/k-1})(Y_{i,k/k-1} - \hat{y}_{k/k-1})^T$$

where R is the observation noise covariance matrix.

Update step: By using the measurements y_k at time t_k, the corrected estimate $\hat{x}_{k/k}$ and its covariance $P_{k/k}$ are computed along with the gain matrix K_k. The equations to obtain corrected estimates are given by

$$K_k = P_{x_k y_k} P_{y_k y_k}^{-1}$$

$$\hat{x}_{k/k} = \hat{x}_{k/k-1} + K_k(y_k - \hat{y}_{k/k-1}) \qquad (4.56)$$

$$P_{k/k} = P_{k/k-1} - K_k P_{y_k y_k} K_k^T$$

$\hat{x}_{k/k}$ and $P_{k/k}$ form the recursive initial conditions.

The parameters used in UKF are Q_k, R_k, α, β, and κ. The choice for α of $-3 < \alpha \leq 1$ and β of 2 may be considered. κ is chosen to be small value. A proper choice of these parameters can influence the performance of UKF. The UKF estimates are said to be accurate up to third order statistics for Gaussian distributions and up to second order statistics for non-Gaussian distributions [20]. The true mean and covariance estimate of the UKF are more accurate than those obtained through the EKF. In addition, UKF does not require computation of the Jacobians, thereby reducing the computational complexity of the problem.

4.9   Square root unscented Kalman filter

The most computationally expensive operation in the UKF corresponds to calculating the new set of sigma points at each time update. This requires taking a matrix square-root of the state covariance matrix P given by $SS^T = P$. In the SRUKF implementation, S will be propagated directly, avoiding the need to refactorize at each time step. The square-root algorithm guaranties the positive semidefiniteness of the state covariance, P and improves the stability of numerical calculation [24]. The square-root form of the UKF makes use of three linear algebra techniques, namely, QR decomposition, Cholesky factor updating, and efficient least squares [25,26].

We describe the quare root unscented Kalman filter (SRUKF) algorithm for state estimation in discrete time nonlinear process in which the process and measurement models are represented by Eq. (4.33). The variances Q_k and R_k, reflecting the process and observation noises are given by Eq. (4.34).

The steps of the SRUKF algorithm are as follows:

1. The initial state $\hat{x}_0$ and Cholesky factor S_0 of the covariance are expressed as

$$\hat{x}_0 = E(x_0)$$

$$S_0 = Chol\left[E\left\{(x_0 - \hat{x}_0)(x_0 - \hat{x}_0)^T\right\}\right] \qquad (4.57)$$

2. $2L + 1$ sigma points are obtained by UT as given by

$$\chi_{i,k-1} = \hat{x}_{k-1}, \quad i = 0$$

$$\chi_{i,k-1} = \hat{x}_{k-1} + \sqrt{(L+\lambda)}S_i, \quad i = 1, \ldots, L \qquad (4.58)$$

$$\chi_{i,k-1} = \hat{x}_{k-1} - \sqrt{(L+\lambda)}S_i, \quad i = L+1, \ldots, 2L$$

The weights for the states and covariance matrices are defined by

$$W_0^m = \lambda/(L + \lambda)$$
$$W_0^c = \lambda/(L + \lambda) + (1 - \alpha^2 + \beta)$$
$$W_i^m = W_i^c = 1/[2(L + \lambda)], \quad i = 1, \ldots, 2L \tag{4.59}$$

where L represents the number of state variables, S_i is the ith column of the covariance Cholesky decomposition factor. W^m and W^c denote the weights of the sigma sample points and the sample point covariance. β is a constant. For Gaussian distributions, $\beta = 2$ is the appropriate value. λ is a scaling factor and is mathematically expressed as

$$\lambda = \alpha^2(L + \kappa) - L \tag{4.60}$$

where α is a scaling parameter which determines the spread of the sigma points, κ is a secondary scaling parameter. The sigma points are expressed in the form of a matrix as

$$\chi_{i,k/k} = \left[\hat{\chi}_{k/k}, \; \hat{\chi}_{k/k} + \sqrt{(L + \lambda)}S_k, \; \hat{\chi}_{k/k} - \sqrt{(L + \lambda)}S_k \right], \quad i = 0, \cdots, 2L \tag{4.61}$$

3. The sigma points are first brought into the prediction function as given by

$$\chi_{i,k+1/k} = f(\chi_{i,k/k}, u_k), \quad i = 0, \cdots, 2L \tag{4.62}$$

Then the mean and Cholesky factor of the covariance of sigma points are calculated as follows:

$$\hat{x}_{k+1/k} = \sum_{i=0}^{2L} W_i^m \chi_{i,k+1/k} \tag{4.63}$$

$$S_{xk}^- = qr\left\{ \left[\sqrt{W_i^{c,i=1:2L}}(\chi_{i,k+1/k}^{i=1:2L} - \hat{x}_{k+1/k}), \quad \sqrt{Q_k} \right] \right\}$$

where qr refers QR decomposition of the compound matrix. The Cholesky update of the above matrix is represented by

$$S_{xk} = Cholupdate\left\{ \left[S_{xk}^-, \; (\chi_{0,k+1/k} - \hat{x}_{k+1/k}), \quad W_0^c \right] \right\} \tag{4.64}$$

4. The sigma points are brought into the measurement function to obtain $2L + 1$ predicted values,

$$Y_{i,k+1/k} = h(\chi_{i,k+1/k}, u_k), \quad i = 0, \cdots, 2L \tag{4.65}$$

The mean and Cholesky factor of the covariance of sigma points are calculated as follows:

$$\hat{y}_{k+1/k} = \sum_{i=0}^{2L} W_i^m Y_{i,k+1/k}$$

$$S_{yk}^- = qr\left\{ \left[\sqrt{W_i^{c,i=1:2L}}(Y_{i,k+1/k}^{i=1:2L} - \hat{y}_{k+1/k}), \quad \sqrt{R_k} \right] \right\} \tag{4.66}$$

The Cholesky update of the above matrix is represented by

$$S_{yk} = Cholupdate\left\{ \left[S_{yk}^-, \; (Y_{0,k+1/k} - \hat{y}_{k+1/k}), \quad W_0^c \right] \right\} \tag{4.67}$$

The cross-covariance relation between state and observation is represented by

$$P_{x_k y_k} = \sum_{i=0}^{2L} W_i^c (\chi_{i,k+1/k} - \hat{x}_{k+1/k})(Y_{i,k+1/k} - \hat{y}_{k+1/k})^T \tag{4.68}$$

5. The gain is calculated by

$$K_k = P_{x_k y_k}(S_{y_k} S_{y_k}^T)^{-1} \tag{4.69}$$

6. The system state and Cholesky factor of the covariance updates are as follows:

$$\hat{x}_{k+1/k+1} = \hat{x}_{k+1/k} + K_k(y_{k+1} - \hat{y}_{k+1/k})$$

$$U_k = K_k S_{y_k}$$

$$S_k = Cholupdate\ (S_{xk},\ U_k,\ -1)$$

(4.70)

We brief how SRUKF is used for state estimation in nonlinear systems. The SRUKF is initialized by calculating the matrix square-root of the state covariance once using Cholesky factorization as given in Eq. (4.57). The propagated and updated Cholesky factor is then used in subsequent iterations to directly evaluate the sigma points. In Eq. (4.63), the time-update of the Cholesky factor, S_{xk}, is calculated using QR decomposition of the compound matrix containing the weighted propagated sigma points and the matrix square-root of the additive process noise covariance. The Cholesky update of S_{xk} is then performed in Eq. (4.64) along with the sigma points and the zero weight. These two steps replace the time update of the covariance P in UKF. The same two-step approach is applied to the calculation of QR decomposition, S_{yk}, and Cholesky update of the observation error covariance as in Eqs. (4.66) and (4.67). The updated S_{xk} and S_{yk} along with the cross covariance, $P_{x_k y_k}$ is used to compute the gain matrix K_k as in Eq. (4.69). Since S_{yk} is square and triangular, we can find K_k directly by using back substitution operation without the need for a matrix inversion. Finally, the posterior measurement update of the Cholesky factor of the state covariance, S_k in Eq. (4.70) is calculated by applying sequential Cholesky down dates to S_{xk}. The down date vectors are the columns of $U_k = K_k S_{y_k}$.

Since QR-decomposition and Cholesky factorization tend to control better the round off errors and there are no matrix inversions, the SRUKF has better numerical properties, and it also guarantees positive semidefiniteness of the underlying state covariance [27].

4.10 Ensemble Kalman filter

The ensemble Kalman filter (EnKF) proposed by Evensen [28] is an approximate version or extension of the Kalman filter. It uses ensembles of random samples to obtain error statistics of state variable or parameter. Error statistics can be propagated with a dynamic model through simply implementing ensemble simulations of the model. The EnKF has several variants, and it has been discussed in different review papers [29−31]. Consider a discrete time nonlinear system in which the process and measurement models are represented by Eq. (4.33). The variances Q_k and R_k, reflecting the process and observation noises are given by Eq. (4.34). We describe the EnKF in the following steps.

The starting point for ensemble filter is choosing a set of sample points, that is, an ensemble of state estimates, that captures the initial probability distribution of the state. These sample points are then propagated through the true nonlinear system, and the probability density function (PDF) of the actual state is approximated by the ensemble of the estimates. The number of ensembles required in the EnKF is heuristic.

An initial ensemble is generated by using a mean value of the initial augmented state vector $\overline{X}_0^{fi}$ and a corresponding covariance matrix Q_0. The mean value of the initial ensemble should be a good estimate of the true initial state. The members of the ensemble are generated randomly according to a Gaussian distribution. The first step is the forecast step which represents the error statistics of the state estimates. The initial ensemble is defined as

$$X_0^f = \overline{X}_0^{fi} + w_0^i$$

(4.71)

$$X_0^f = [x_0^{f_1}, \cdots, x_0^{f_N}]$$

where the first term in RHS of Eq. (4.71) represents the mean of the initial state vector and the second term is random Gaussian numbers. We assume that at time k, we have an ensemble of N forecasted state estimates with random sample errors. We denote this ensemble as $X_k^f \in \mathfrak{R}^{n \times N}$, as expressed by

$$X_k^f = [x_k^{f_1}, \cdots, x_k^{f_N}]$$

(4.72)

where $x_k^{f_i}$ refers ith forecast ensemble member. The ensemble mean $\overline{x}_k^f \in \mathfrak{R}^{n \times N}$ is defined by

$$\overline{x}_k^f = \frac{1}{N} \sum_{i=1}^{N} x_k^{f_i}$$

(4.73)

We define the ensemble error matrix $E^f_{x_k} \in R^{n \times N}$ around the ensemble mean by

$$E^f_{x_k} = \left[x^{f_1}_k - \bar{x}^f_k, \cdots, x^{f_N}_k - \bar{x}^f_k \right] \tag{4.74}$$

The ensemble output matrix $E^f_{y_k} \in R^{p \times N}$ is given by

$$E^f_{y_k} = \left[y^{f_1}_k - \bar{y}^f_k, \cdots, y^{f_N}_k - \bar{y}^f_k \right] \tag{4.75}$$

We then approximate $P^f_{x_k}$, $P^f_{xy_k}$, and $P^f_{y_k}$ as follows.

$$P^f_{x_k} = \frac{1}{N-1} E^f_{x_k} (E^f_{x_k})^T$$

$$P^f_{xy_k} = \frac{1}{N-1} E^f_{x_k} (E^f_{y_k})^T \tag{4.76}$$

$$P^f_{y_k} = \frac{1}{N-1} E^f_{y_k} (E^f_{y_k})^T$$

Thus we interpret the forecast ensemble mean as the best forecast estimate of the state, and the spread of the ensemble members around the mean as the error between the best estimate and the actual state.

The second step is the analysis step which performs the analysis of the state estimates. The approximations of the error covariances are used to determine the Kalman filter gain K_k as given by

$$K_k = P^f_{xy_k} (P^f_{y_k})^{-1} \tag{4.77}$$

The updated state estimate in the analysis step is obtained as

$$x^{a_i}_k = x^{f_i}_k + K_k [y^i_k - h(x^{f_i}_k)], \quad i = 1, \cdots, N \tag{4.78}$$

The perturbed observations y^i_k are given by

$$y^i_k = y_k + v^i_k \tag{4.79}$$

where v^i_k is a zero-mean random variable with a normal distribution and covariance R_k. The analysis of updated state error covariance matrix P^a_k is approximated by

$$P^a_k = \frac{1}{N-1} E^a_k (E^a_k)^T \tag{4.80}$$

where E^a_k is defined as

$$E^a_k = [x^{a_1}_k - \bar{x}^a_k, \cdots, x^{a_N}_k - \bar{x}^a_k] \tag{4.81}$$

$$\bar{x}^a_k = \frac{1}{N} \sum_{i=1}^N x^{a_i}_k \tag{4.82}$$

The last step is the prediction of error statistics in the forecast step:

$$x^{f_i}_{k+1} = f(x^{a_i}_k, u_k) + w^i_k \tag{4.83}$$

where the values w^i_k are sampled from a normal distribution with average zero and covariance Q_k.

Unlike the EKF, the evaluation of the filter gain K_k in the EnKF does not involve an approximation of the nonlinearity $f(x, u)$ and $h(x)$. Hence, the computational burden of evaluating the Jacobians of $f(x, u)$ and $h(x)$ is absent in the EnKF.

4.11 Particle filter

Particle filtering is based on recursive Bayesian filtering with Monte Carlo simulations. The method approximates the Bayesian posterior PDF with a set of randomly chosen weighted samples. Each sample of the state vector is referred to as a particle. Random samples (particles) obtained from a distribution are advanced through the system equations to give prior particles. The measurement information is combined with this prior distribution of particles to generate the posterior distribution. A sufficiently large number of particles guarantee almost sure convergence to the true probability distribution function [32,33].

A state space model is defined by the following state and measurement functions:

$$x_k = f_k(x_{k-1}, w_{k-1})$$
$$y_k = h_k(x_k, v_k) \tag{4.84}$$

where k is the discrete time index, and f_k is a nonlinear function describing the evolution of the state with identically distributed process noise, w_{k-1}. The h_k is a nonlinear function mapping the state space to the measurements with independent and identically distributed noise v_k. Let $y_{1:k} = (y_i, i = 1, \ldots, k)$ be the measurement sequence. On assuming the known prior distribution $p(x_0)$, the posterior probability can be obtained sequentially by prediction and updated as follows:

$$p(x_k | y_{1:k-1}) = \int p(xx_k | x_{k-1}) p(x_{k-1} | y_{1:k-1}) dx_{k-1} \tag{4.85}$$

$$p(x_k | y_{1:k}) = \frac{p(y_{1:k} | x_k) p(x_k | x_{k-1})}{p(y_k | y_{1:k-1})} \tag{4.86}$$

The above equations are the optimal solution from a Bayesian perspective to the nonlinear state estimation problem. However, certain approximation for the posterior density calculation is required as it cannot, in general, be determined analytically. Particle filters (PFs) approximate the posterior probability by a set of support points (particles) x_k^i, $i = 1, \ldots, N$, with associated weights w_k^i:

$$p(x_k | y_{1:k}) = \sum_{i=1}^{N} w_k^i \delta(x_k - x_k^i) \tag{4.87}$$

where δ is a Dirac delta measure, N is number of samples, and w_k^i, $i = 1, \cdots, N$ are normalized random weights,

$$\sum_{i=1}^{N} w_k^i = 1 \tag{4.88}$$

The normalized weights w_k^i are chosen based on the principle of importance sampling. The sequential importance sampling (SIS) is a basic framework for PFs [34]. The recursive propagation of the weights w_k^i and support points x_k^i with the reception of sequential measurements is referred to as the SIS. The main idea is to represent the required posterior density by a set of random samples with associated weights and to compute estimates based on these samples and weights,

$$w_k^i \propto w_{k-1}^i \cdot \frac{p(y_k | x_k^i) p(x_k^i | x_{k-1}^i)}{q(x_k^i | x_{1:k-1}^i, y_k)} \tag{4.89}$$

where $q(x_k^i | x_{1:k-1}^i, y_k)$ is the importance PDF and $\propto$ is the proportionality symbol. The choice of importance function plays a critical part in the performance of PF algorithm. Such a function can be approximated by

$$q(x_k^i | x_{1:k-1}^i, y_k) = \frac{p(y_k | x_k, x_{k-1}^i) p(x_k | x_{k-1}^i)}{q(y_k | x_{k-1}^i)} \tag{4.90}$$

A main problem with the SIS algorithm is the degeneracy phenomena, where just one particle will have nonnegligible weight after few iterations. This implies that a large computational effort is needed to update the particles whose contribution to the approximation is almost zero. One measure of the degeneracy is the estimated value of the effective sample size:

$$N_{eff} = \sum_{i=1}^{N} \left(\left(w_k^i \right)^2 \right)^{-1} \tag{4.91}$$

The value of N_{eff} lies in the interval $1 \leq N_{eff} \leq N$. If N_{eff} is so small, it indicates degeneracy. Resampling is a strategy to overcome the degeneracy of samples in SIS and is a crucial step in particle filtering algorithms when N_{eff} falls below a predetermined threshold [35]. The basic idea of resampling is to eliminate particles that have small weights and to concentrate on particles with large weights.

The PF algorithm for state estimation is presented below:

1. *Initialization* ($k = 0$): Set the initial state vector $\{x_0^i, \ i = 1, \cdots, N\}$ and $\{w_0^i, \ i = 1, \cdots, N\}$.
   Assume that all the particles are equally probable at the start of the algorithm.
2. *Update the weights*: $w_k^i = w_{k-1}^i \cdot p(y_k|x_k^i)$, with normal PDF and normalize the weights

$$w_k^i = \frac{w_k^i}{\sum w_k^i}.$$

3. *Resampling*: apply the chosen resampling method to the set of particles $\{x_k^i, \ i = 1, \cdots, N\}$ and their $\{w_k^i, \ i = 1, \cdots, N\}$ to obtain a new set of particles and set of weights.
4. *Compute the estimate*: $\tilde{x}_k = \sum_{j=1}^{N} w_k^j x_k^j$
5. Set $k = k + 1$, go to step 2 and iterate the algorithm until end of the simulation time interval.

Fig. 4.7 shows the PF computation with resampling.

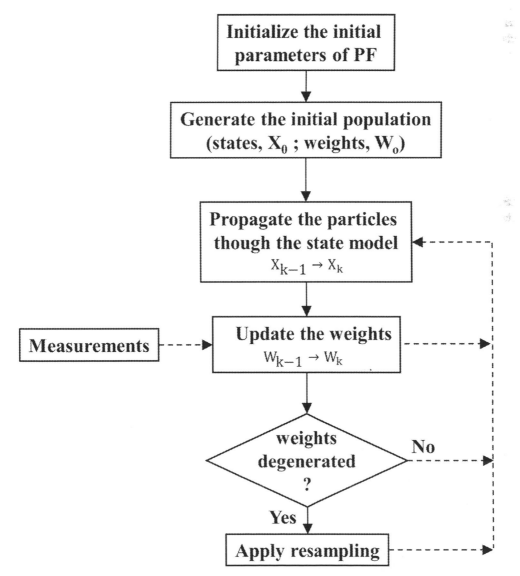

FIGURE 4.7   Particle filter with resampling.

4.12 Reduced order Luenberger observer

Luenberger observer presented in Section 3.5 of Chapter 3, Linear Filtering and Observation Techniques, can be designed correctly only if the system is described by linear differential equations without any system uncertainties. However, most real systems include system uncertainties arising due to time variation of parameters, nonlinearities, and noise of the system. The state estimator designed neglecting these uncertainties leads to incorrect estimates. Hence, an observer for nonlinear systems must be designed by taking care of such uncertainties. Luenberger observer can be applied to nonlinear systems by using the following design procedure [36].

Consider a stochastic system with unknown parameters as described by the following state and measurement equations:

$$\dot{x}(t) = Ax(t) + Bu(t) + Df(t, \theta) + Ew(t), \quad x(0) = x_0 \tag{4.92}$$

$$y(t) = Cx(t) + v(t)$$

where $x(t)$ is n dimensional state vector, $u(t)$ is d dimensional input vector, $y(t)$ is m dimensional output vector, $w(t)$ is d dimensional process noise vector, $v(t)$ is m dimensional observation noise vector and, θ is p dimensional unknown parameter, $f(t,\theta)$ is l dimensional vector with elements split into a set of nonlinear functions and another set of linear and nonlinear functions with unknown coefficients, D is coefficient matrix of the vector $f(t, \theta)$, C is process output coefficient matrix and A, B, and E are constant coefficient matrices in the linear part of the system model with appropriate dimensions.

The vector function $f(t, \theta)$ in Eq. (4.92) is specified as

$$f(t, \theta) = [\theta_1 f_1(t), \quad \theta_2 f_2(t), \cdots, \quad \theta_p f_p(t) ; f_{p+1}(t), \cdots, f_l(t)] \tag{4.93}$$

where θ_i, $i = 1, 2, \ldots, p$ are the coefficients of the vector θ. These coefficients represent the parameters that may change in the system. The $f_i(t)$, $i = 1, 2, \ldots, p$, may be linear and/or nonlinear functions of $x(t)$ and $u(t)$, but $f_i(t)$, $i = p + 1, \ldots, l$, are strictly nonlinear functions of $x(t)$ and $u(t)$. The nonlinear terms and the terms associated with unknown parameters are incorporated in the term $Df(t, \theta)$ of Eq. (4.92).

The observer is described by the following equations:

$$\dot{z}(t) = Fz(t) + Ky(t) + MBu(t), \quad z(0) = z_0 \tag{4.94}$$

$$\hat{x}(t) = Gz(t) + H_b y(t) \tag{4.95}$$

where $z(t)$ is $(n-m)$ dimensional state vector of the observer, $\hat{x}(t)$ is n dimensional estimated state vector, $u(t)$ is d dimensional input vector of the system and $y(t)$ is m dimensional output vector of the system. The F, K, M, G, and H_b are constant observer coefficient matrices, where F is $(n-m) \times (n-m)$ dimensional matrix, K is $(n-m) \times m$ dimensional matrix, M is $(n-m) \times n$ dimensional matrix, G is $n \times (n-m)$ dimensional matrix and H_b is $n \times m$ dimensional matrix. The state estimator gives an estimate of $\hat{x}(t)$ which tends to $x(t)$ for $t > t_s$, where t_s is the settling time.

4.12.1 Determination of observer coefficient matrices

The existence conditions for the observer coefficient matrices F, K, M, G, and H_b, and the evaluation of these elements is further illustrated. A matrix W is defined and the matrices V and G are obtained by the following relation

$$\begin{bmatrix} C_0 \\ \hline W \end{bmatrix}^{-1} = [V : G] \tag{4.96}$$

where C_0 is an $m \times n$ dimensional observation matrix, W is $(n-m) \times n$ dimensional matrix whose elements can be arbitrarily assigned, and with $\left[C_0^T : W^T\right]^T \neq 0$, V is $n \times m$ dimensional matrix, and G is $n \times (n-m)$ dimensional matrix. The matrices V and G are automatically determined by Eq. (4.96) if W is specified. The following relations define the observer coefficient matrices:

$$\begin{aligned} L &= (WD)(C_0 D)^{\#} + K_0 (I_m - (C_0 D)(C_0 D)^{\#}) \\ M &= W - LC_0 \\ H &= V + GL \\ F &= MAG \\ K &= MAH \end{aligned} \tag{4.97}$$

where K_0 is $(n-m) \times m$ dimensional matrix, whose elements can be arbitrarily assigned. The matrix $(C_0 D)^{\#}$ is generalized inverse defined by

$$(C_0D)^{\#} = [(C_0D)^T(C_0D)]^{-1}(C_0D)^T \tag{4.98}$$

The matrix F is rearranged with the substitution of L and M relations as

$$F = (WAG - WD(C_0D)^{\#}C_0AG) - K_0[(I_m - (C_0D)(C_0D)^{\#}C_0AG)] \\ = F_0 - K_0H_0 \tag{4.99}$$

The equation for the error is given by

$$\dot{e}(t) = Fe(t) + Kv(t) - MEw(t), \quad e(0) = e_0 \tag{4.100}$$

where $e(t)$ is $n-m$ dimensional error vector. If the real parts of all the eigen values of F are negative, the error term converges to zero exponentially. If the system is error free, the estimated state vector $\hat{x}(t)$ converges to the true value of $x(t)$.

The observer coefficient matrices given in Eq. (4.97) can be determined if and only if the system model satisfies the following conditions:

1. rank $C_0D = l \leq m$, where l is the rank of C_0D and m is the dimension of observation vector,
2. $F = F_0 - K_0H_0$ is to be satisfied by the proper choice of K_0, and
3. The term generalized inverse $(C_0D)^{\#}$ should be nonsingular.

If the pair (F_0, H_0) in Eq. (4.99) is completely observable, then F can made stable by assigning all the eigen values of F arbitrarily. If the number of elements of the vector $f(t,\theta)$ in Eq. (4.92) is equal to the number of output measurements, $l = m$, the $H_0 = 0$. Then the pair (F_0, H_0) is not observable, and consequently, all the eigen values of F cannot be assigned arbitrarily. F should be stable to obtain correct state estimates.

4.13 Reduced order extended Luenberger observer

The reduced order Luenberger observer discussed in the earlier section can be used for state estimation of nonlinear systems in the presence of parameter variations. An extended version of reduced order Luenberger observer can be formed by linearizing the process model around the current operating point and applying it for state estimation in nonlinear systems in the presence of parameter changes.

The nonlinear process model represented by $\dot{x}(t) = f(x, u, \theta)$ is linearized and rearranged with the separation of parameter terms as

$$\delta\dot{x}(t) = A\delta x(t) + B\delta u(t) + Df_e(t,\theta), \quad \delta x(0) = \delta x_0 \\ \delta y(t) = C\delta x(t) + v(t) \tag{4.101}$$

where the coefficient matrices are $A = \partial f / \partial x$, $B = \partial f / \partial u$, and $D = \partial f / \partial \theta$. The terms $\delta x(t)$ is the deviation state vector from the initial operating point, $\delta u(t)$ is the deviation input vector from the initial operating point, $f_e(t, \theta)$ is the vector function of parameters, and θ is p dimensional unknown parameter vector. The deviation variables $\delta x(t)$ and $\delta u(t)$ are represented by

$$\delta x(t) = x(t) - x^* \\ \delta u(t) = u(t) - u^* \tag{4.102}$$

where x^* and u^* are state and input vectors corresponding to the initial operating point. The vector function $f_e(t, \theta)$ is represented as

$$f_e(t, \theta) = [\theta_1 f_1(t), \cdots, \theta_p f_p(t)]^T \tag{4.103}$$

where θ_i, $i = 1, \cdots, p$ are the elements of the vector θ. The terms associated with the unknown parameters are incorporated in the term $Df_e(t, \theta)$.

The observer for state estimation of a process represented by the linearized state space model in Eq. (4.101) is described by the following equations:

$$\dot{z}(t) = Fz(t) + K\delta y(t) + MB\delta u(t), \quad z(0) = z_0 \\ \delta\hat{x}(t) = Gz(t) + H_b\delta y(t) \tag{4.104} \\ \hat{x}_b(t) = x^* + \delta\hat{x}(t)$$

where $z(t)$ is $n - m$ dimensional state vector of the observer, $\hat{x}_b(t)$ is n dimensional estimated state vector and $\delta y(t)$ is the output deviation vector of the process. F, K, M, G, and H_b are constant observer coefficient matrices whose dimensions are given in the earlier section 4.12. The existence conditions for the observer, the determination of observer coefficient matrices F, K, M, G, and H_b, and the conditions that the model should satisfy to determine the coefficient matrices are given in the earlier section.

4.14 Nonlinear observer

Various nonlinear observers have been reported for state estimation in nonlinear systems [37−40]. Consider the continuous time nonlinear process model and the nonlinear measurement model represented by Eqs. (4.20) and (4.21). Here we describe the nonlinear observer of Frank [41] for state estimation in nonlinear processes. The state vector x in the nonlinear process model can be expressed in terms of available state by measurement x_a, and unavailable state by measurement x_u as given by

$$x^T = (x_a, x_u)^T \tag{4.105}$$

The nonlinear process and measurement models in Eqs. (4.20) and (4.21) can be expressed in the following form:

$$\begin{aligned}
\dot{x} &= f(x_a, x_u) + w \\
y_a &= C_a(x_a) \\
y_u &= C_u(x_u) \\
x_a(0) &= x_{a0} \\
x_u(0) &= x_{u0}
\end{aligned} \tag{4.106}$$

In the above equations, x_a is $n_a \times 1$ state vector, x_u is $n_u \times 1$ state vector, y_a is $m_a \times 1$ output vector, y_u is $m_u \times 1$ output vector which is unavailable by measurement. f, C_a, and C_u are partially differentiable nonlinear functions.

The observer is defined by

$$\dot{\hat{x}} = f(\hat{x}_a, \hat{x}_u) + K_a(y_a - \hat{y}_a) + K_u(y_u - \hat{y}_u) \tag{4.107}$$

where

$$\begin{aligned}
\hat{y}_a &= C_a(\hat{x}_a) \\
\hat{y}_u &= C_u(\hat{x}_u) \\
x_a(0) &= \hat{x}_{a0} \\
x_u(0) &= \hat{x}_{u0} \\
\hat{x}^T &= (\hat{x}_a \ \hat{x}_u)^T
\end{aligned}$$

The estimated state vector $\hat{x}_u$ of $\dot{\hat{x}}$ is used to replace $\hat{y}_u$.

The estimation error equation which is linearized around the state estimates $\hat{x}_a$ and $\hat{x}_u$ is given by

$$e_a = F_a e_a + F_u e_u \tag{4.108}$$

$$e_u = x_u - \hat{x}_u \tag{4.109}$$

$$F_a = \left(\frac{df}{dx_a} - K_a \frac{dC_a}{dx_a} \right) \Big|_{\hat{x}_a, \hat{x}_u} \tag{4.110}$$

$$F_u = x \left(\frac{df}{dx_u} - K_u \frac{dC_u}{dx_u} \right) \Big|_{\hat{x}_a, \hat{x}_u} \tag{4.111}$$

The stability of the observer may be assured by an appropriate choice of the feedback matrix K_a. The matrix K_u is in general nonconstant and can be chosen as

$$K_u = \frac{df}{dx_u} \Big|_{\hat{x}_a, \hat{x}_u} \times \frac{dC_u^{\#}}{dx_u} \Big|_{\hat{x}_a, \hat{x}_u} \tag{4.112}$$

where # sign indicates pseudo inverse.

Example 6

Apply nonlinear observer described in Section 4.14 for state estimation in a nonlinear system expressed by the following equations,

$$\dot{x}_1 = x_2 - x_1(x_1^2 + x_2^2)$$

$$\dot{x}_2 = -x_1 - x_2(x_1^2 + x_2^2)$$

$$y = x_1$$

Illustrate the calculations for one-time step considering the initial state $x_0 = \begin{bmatrix} 1 & 1 \end{bmatrix}^T$, and the initial state estimate $\hat{x}_0 = \begin{bmatrix} 1.1 & 0.9 \end{bmatrix}^T$. Assume the sample time to be $\Delta t = 0.1$.

Solution

At time $t = 0$, we have

$$x_0 = \begin{bmatrix} 1 \\ 1 \end{bmatrix}$$
$$\hat{x}_0 = \begin{bmatrix} 1.1 \\ 0.9 \end{bmatrix}$$

The available state is $x_a = x_1$ and the unavailable state is $x_u = x_2$.
Thus

$$\begin{bmatrix} x_a & x_u \end{bmatrix}^T = \begin{bmatrix} x_1 & x_2 \end{bmatrix}^T$$
$$\begin{bmatrix} \hat{x}_a & \hat{x}_u \end{bmatrix}^T = \begin{bmatrix} \hat{x}_1 & \hat{x}_2 \end{bmatrix}^T$$

The observer is represented by [Eq. (4.107)]:

$$\dot{\hat{x}} = f(\hat{x}_a, \hat{x}_u) + K_a(y_a - \hat{y}_a) + K_u(y_u - \hat{y}_u)$$

where

$$f(\hat{x}_a, \hat{x}_u) = f(\hat{x}_1, \hat{x}_2) = \begin{bmatrix} \hat{x}_2 - \hat{x}_1(\hat{x}_1^2 + \hat{x}_2^2) \\ -\hat{x}_1 - \hat{x}_2(\hat{x}_1^2 + \hat{x}_2^2) \end{bmatrix}$$

$$y_a = C_a(x_a) = C_a(x_1) = x_2 - x_1(x_1^2 + x_2^2)$$
$$y_u = C_u(x_u) = C_u(x_2) = -x_1 - x_2(x_1^2 + x_2^2)$$

$$\hat{y}_a = C_a(\hat{x}_a) = C_a(\hat{x}_1) = \hat{x}_2 - \hat{x}_1(\hat{x}_1^2 + \hat{x}_2^2)$$
$$\hat{y}_u = C_u(\hat{x}_u) = C_u(\hat{x}_2) = -\hat{x}_1 - \hat{x}_2(\hat{x}_1^2 + \hat{x}_2^2)$$

K_a is chosen arbitrarily to satisfy the observer convergence.

K_u is defined by

$$K_u = \left. \frac{df}{dx_u} \right|_{\hat{x}_a, \hat{x}_u} \times \left. \frac{dC_u^{\#}}{dx_u} \right|_{\hat{x}_a, \hat{x}_u}$$

where

$$\left. \frac{df}{dx_u} \right|_{\hat{x}_a, \hat{x}_u} = \begin{bmatrix} 1 - 2\hat{x}_1\hat{x}_2 \\ -\hat{x}_1^2 - 3\hat{x}_2^2 \end{bmatrix}$$

$$\left. \frac{dC_u}{dx_u} \right|_{\hat{x}_a, \hat{x}_u} = -\hat{x}_1^2 - 3\hat{x}_2^2$$

$$\left. \frac{dC_u^{\#}}{dx_u} \right|_{\hat{x}_a, \hat{x}_u} = \frac{1}{-\hat{x}_1^2 - 3\hat{x}_2^2}$$

We evaluate the components of the observer at the initial conditions.
At $t = t_0$, we have

$$f(\hat{x}_a, \hat{x}_u) = \begin{bmatrix} -1.322 \\ -2.918 \end{bmatrix}, \quad y_a = -1, \quad y_u = -3.0, \quad \hat{y}_a = -1.322, \quad \hat{y}_u = -2.918,$$

$$K_u = \begin{bmatrix} 0.2776 \\ 1.0 \end{bmatrix} \text{ and } K_a \text{ is assigned as } K_a = \begin{bmatrix} 0.5 \\ 0.5 \end{bmatrix}.$$

The observer state equation in simple Euler's integration is expressed as

$$\hat{x}_{t+1} = \hat{x}_t + \Delta t \left[f(\hat{x}_a, \hat{x}_u) + K_a(y_a - \hat{y}_a) + K_u(y_u - \hat{y}_u) \right]$$

We substitute the numerical quantities in the above equation.
At $t = 0.1$, we have

$$\hat{x}_1 = \hat{x}_0 + \Delta t \left[f(\hat{x}_a, \hat{x}_u) + K_a(y_a - \hat{y}_a) + K_u(y_u - \hat{y}_u) \right]$$

$$= \begin{bmatrix} 1.1 \\ 0.9 \end{bmatrix} + 0.1 \left\{ \begin{bmatrix} -1.322 \\ -2.918 \end{bmatrix} + \begin{bmatrix} 0.5 \\ 0.5 \end{bmatrix} 0.322 + \begin{bmatrix} 0.2776 \\ 1.0 \end{bmatrix} - 0.082 \right\}$$

$$= \begin{bmatrix} 0.9817 \\ 0.6161 \end{bmatrix}$$

The estimated state will form the initial condition for the next recursive step.

4.15 State estimation applications of nonlinear filtering and observation techniques

Nonlinear filtering, estimation, and observation techniques have numerous applications in almost all fields of engineering and science. These techniques are abundantly used in fields concerning to electronics, communications, aerospace, navigation, neuroscience and related ones. The application areas include machine learning, sensorless control, diagnosis, fault-tolerant control, signal processing, data merging, robotic, vision, sensor fusion, control, instrumentation, target tracking, surveillance, guidance, obstacle avoidance trajectory estimation, state and parameter estimation, and several others. These techniques have also wide range of applications in chemical, biochemical, environmental, civil, mechanical, and related fields. The application areas in these fields include soft sensing, process monitoring, fault detection, product quality monitoring, weather forecasting, pollution monitoring, data reconciliation, system identification, nonlinear and model-based control, and others. These techniques have also various applications in economics, finance, political science, and many others.

4.16 Summary

This chapter presents various mechanistic model-based nonlinear filtering and observation techniques for optimal state estimation of nonlinear systems. The nonlinear system and system models that constitute a basis for nonlinear state estimation are described. The observability criteria of nonlinear systems for optimal state estimation is illustrated. The description and implementation procedures of various nonlinear algorithms for optimal state and parameter estimation are presented. These methods include EKF, SSEKF, TLEKF, AFEKF, UKF, SRUKF, EnKF, PF, and different nonlinear observers. These methods have vast applications in almost all fields of engineering and science.

References

[1] R. Hermann, A.J. Krener, Nonlinear controllability and observability, IEEE Trans. Autom. Control 22 (5) (1977) 728–740.

[2] A. Isidori, Nonlinear Control Systems – An Introduction, Springer-Verlag, New York, NY, 1989.

[3] H. Sussmann, V. Jurdjevic, Controllability of nonlinear systems, J. Diff. Eqs. 12 (1) (1972) 95–116.

[4] S. Engell, K.U. Klatt, Gain scheduling of a non-minimum phase CSTR, in: Proceedings of the European Control Conference, pp. 2323–2328, 1993.

[5] J.G. van de Vusse, Plug flow reactor vs tank reactor, Chem. Eng. Sci. 19 (12) (1964) 994–997.

[6] C. Kravaris, P. Daoutidis, Nonlinear state feedback control of second ordern on minimum phase systems, Comput. Chem. Eng. 14 (4/5) (1990) 439–449.

[7] T. Fujisawa, E.S. Kuh, Some results on experience in uniqueness of solutions in nonlinear networks197 IEEE Trans. Circuit Theory CT-18 (5) (1971) 501–506.

[8]   S.R. Kou, D.L. Elliott, T.J. Torn, Observability of nonlinear systems, Inf. Control 22 (1) (1973) 89—99.

[9]   C.H. Wells, Application of modern estimation and identification techniques to chemical processes, AIChE J. 17 (4) (1971) 966—973.

[10] A. Gelb, Applied Optimal Estimation, MIT, Cambridge, 1974.

[11] A.H. Jazwinski, Stochastic Processes and Filtering. Mathematics in Science and Engineering, Academic Press, New York, NY, 1970.

[12] P.S. Maybeck, Stochastic Models, Estimation, and Control. Mathematics in Science and Engineering, Academic Press, New York, NY, 1979.

[13] D. Simon, Optimal State Estimation, John Wiley & Sons, Hoboken, NJ, 2006.

[14] J. Valappil, C. Georgakis, Systematic estimation of state noise statistics for extended Kalman filters, AIChE J. 46 (2) (2000) 292—398.

[15] P. Vachhani, S. Narasimhan, R. Rengaswamy, Recursive state estimation in nonlinear processes, in: Proceedings of the American Control Conference, Boston, 2004.

[16] J. Prakash, S.C. Patwardhan, S.L. Shah, Constrained nonlinear state estimation using ensemble Kalman filters, Ind. Eng. Chem. Res. 49 (5) (2010) 2242—2253.

[17] N.P.G. Salau, J.O. Trierweiler, A.R. Secchi, Numerical pitfalls by state covariance computation, Comput. Aided Chem. Eng. 27 (2009) 1215—1220.

[18] C. Venkateswarlu, K. Gangiah, M. Bhagavantha Rao, Two-level methods for incipient fault diagnosis in nonlinear chemical processes, Comput. Chem. Eng. 16 (5) (1992) 463—476.

[19] Q. Xia, M. Rao, Y. Ying, X. Shen, Adaptive fading Kalman filter with application, Automatica 30 (8) (1994) 1333—1338.

[20] C. Venkateswarlu, S. Avantika, Optimal state estimation of multicomponent batch distillation, Chem. Eng. Sci. 56 (20) (2001) 5771—5786.

[21] S.J. Julier, J.K. Uhlmann, A new extension of the Kalmanfilter to nonlinear systems, in: Proceedings of the AeroSense: Eleventh International Symposium on Aerospace/Defense Sensing, Simulation and Control, pp.182—193, 1997.

[22] S. Julier, J. Uhlmann, A general method for approximating nonlinear transformations of probability distributions, University of Oxford, November 1996.

[23] S. Julier, J. Uhlmann, H.F. Durant-Whyte, A new method for the nonlinear transformation of means and covariances in filters and estimators, IEEE Trans. Autom. Control 45 (3) (2000) 477—482.

[24] R.V.D. Merwe, E.A. Wan, The square-root unscented Kalman filter for state and parameter-estimation. In: Acoustics, Speech, and Signal Processing, pp. 3461—3464, 2002.

[25] W. Press, S. Teukolsky, W. Vetterling, B. Flannery, Numerical Recipes in C: The Art of Scientific Computing, second (ed.), Cambridge University Press, New York, NY, 1992.

[26] G. Stewart, Matrix Algorithms, Basic Decompositions, vol. I, SIAM, Philadelphia, 1998.

[27] R. van der Merwe, E. Wan, The square-root unscented Kalman filter for state and parameter-estimation, in: Proceedings of the IEEE International Conference on Acoustics (Cat. No. 01CH37221), 2001.

[28] G. Evensen, Sequential data assimilation with a nonlinear quasi-geostrophic model using Monte-Carlo methods to forecast error statistics, J. Geophys. Res. 99 (C5) (1994) 10143—10162.

[29] G. Evensen, The Ensemble Kalman Filter: theoretical formulation and practical implementation, Ocean Dyn. 53 (4) (2003) 343—367.

[30] A.C. Lorenc, The potential of the ensemble Kalman filter for NWP, a comparison with 4D-Var, Q. J. R. Meteorol. Soc. 129 (595) (2003) 3183—3203.

[31] P.L. Houtekamer, H.L. Mitchell, Ensemble kalman filtering, Q. J. R. Meteorol. Soc. 131 (613) (2005) 3269—3289.

[32] A. Manya, Particle filter and extended Kalman filter for nonlinear estimation: a comparative study. <http://users.isr.ist.utl.pt/~pjcro/cadeiras/dsfps0708/SEMS/PF_EKF_TermPaper.pdf>, 2008.

[33] A. Doucet, A. Johansen, A tutorial on particle filtering and smoothing. <http://www.cs.ubc.ca/~arnaud/doucet_johansen_tutorialPF.pdf>, 2008.

[34] C. Robert, G. Casella, Monte Carlo Statistical Methods, Springer, New York, NY, 1999.

[35] A. Doucet, S. Godsill, C. Andrieu, On sequential Monte Carlo sampling methods for Bayesian filtering, Statist. Comput. 10 (3) (2000) 197—208.

[36] K. Watanabe, D.M. Himmelblau, Fault diagnosis in nonlinear chemical processes, AIChE J. 29 (2) (1983) 243—260.

[37] J.M. Ali, N.H. Hoang, M.A. Hussain, D. Dochain, Review and classification of recent observers applied in chemical process systems, Computers & Chemical Engineering 76 (2015) 27—41.

[38] T. Menard, E. Moulay, W. Perruquetti, Fixed-time observer with simple gains for uncertain systems, Automatica 81 (2017) 438—446.

[39] B. Yi, R. Ortega, W. Zhang, On state observers for nonlinear systems: A new design and a unifying framework, IEEE Transactions on Automatic Control 64 (3) (2018) 1193—1200.

[40] S. Lisci, M. Grosso, S. Tronci, A geometric observer-assisted approach to tailor state estimation in a bioreactor for ethanol production, Processes 8 (4) (2020) 480.

[41] P.M. Frank, Advanced fault detection and isolation schemes using nonlinear and robust observers, in: Proceedings of the IFAC Tenth Triennial World Congress, Munich, pp. 63—68, 1987.

Chapter 5

Data-driven modeling techniques for state estimation

5.1 Introduction

A data-driven model is based on the analysis of the data about a specific system. The main concept of data-driven model is to find relationships between the system input and output variables without explicit knowledge of the physical behavior of the system. Data-driven models are not the first principles, but are derived directly from experimental data or other physical observations of the system. Unlike first principles models, data-driven models make no attempt to model the internal features of the system. Instead, they focus on matching the input−output behavior to observational data. Data-driven modeling assumes the availability of a considerable and sufficient amount of data describing the underlying system.

Currently, the data-driven models are becoming more and more common in several applications. The rapid increase in the availability of data of physical systems has stimulated the development of many data-driven methods for modeling and prediction. These include the conventional statistical methods such as linear regression, autoregressive moving average, and autoregressive integrated moving average models. Recently, the presence of massive data in different fields and real-world applications has encouraged the use of advanced methods based on data mining, machine learning, and artificial intelligence approaches. Data-driven model development generally does not require to fundamentally understand the underlying process. Data-driven models have various advantages with wide applications in tasks related to classification, pattern recognition, and associative & predictive analysis in different fields. This chapter describes various data-driven methods and algorithms, including principal component analysis (PCA), projection to latent structures (PLS), nonlinear iterative partial least squares (NIPALS), artificial neural networks (ANN), and radial basis function networks (RBFN). These tools form a basis for developing state estimators to various applications concerning process systems engineering which will be discussed in later chapters of this book.

5.2 Principal component analysis

PCA is a multivariate statistical data reduction method, which has become more popular with many advantages. The name of PCA comes from the *principal axes* of an ellipsoid representing the new coordinate axes. The PCA represents high dimensional process data in a reduced dimension and finds components (latent variables) that are efficient for describing the data. PCA has been used for various multivariate data analysis techniques such as process monitoring, quality control, sensor, and process fault diagnosis. More comprehensive description of PCA can be referred to elsewhere [1−3].

5.2.1 Basic principles

The idea behind PCA is to reduce the dimension of a dataset while retaining as much of its information as possible. Industrial process data often involves many variables, which are highly correlated. PCA transforms a number of correlated variables into a smaller number of uncorrelated variables called *principal components* (PCs). The first PC accounts for as much of the variability in the data as possible, and each succeeding component accounts for as much of the remaining variability as possible. All PCs are orthogonal. PCA can be described as a method to project a high-dimensional measurement space onto a space with significantly fewer dimensions. By finding a new orthogonal basis for the data set and by using a reduced number of base vectors, a dimension reduction is obtained in PCA. The original variables are projected onto the PCs as scores.

Optimal State Estimation for Process Monitoring, Fault Diagnosis and Control. DOI: https://doi.org/10.1016/B978-0-323-85878-6.00010-5

5.2.2 Geometric interpretation of PCA

The PCA model divides the monitoring space into two orthogonal subspaces, called principal component subspace and the residual subspace (RS), as shown in Fig. 5.1. Out of these two subspaces, normal data variation will be captured by the PC subspace, and the abnormal variation and noise may be captured by the RS. Every data vector is decomposed into two orthogonal vectors by projecting on to the two subspaces.

In general, a normalized data matrix X of n samples (rows) and m variables (columns) can be decomposed as follows:

$$
\begin{aligned}
X &= \hat{X} + \tilde{X} \\
\tilde{X} &= \tilde{T}\,\tilde{X}^T \\
\hat{X} &= TP^T
\end{aligned}
\tag{5.1}
$$

where $\hat{X}$ and $\tilde{X}$ represent the modeled and unmodeled variations of the data X. The T and P are the scores and loading matrices for the data.

PCA finds a new orthogonal basis for the data set. The base vectors of the new basis are ordered in descending order to explain the variance from the original data. When a data set is subjected to PCA, it is observed that only the first few PCs are associated with systematic variation in the data, and the remaining PCs are associated with "noise." The noise here refers to uncontrolled experimental and instrumental variations arising from random processes. PCA models are formed by retaining only the PCs that are descriptive of systematic variation in the data.

5.2.3 Eigen structure for data matrix

The mathematical technique used in PCA is called Eigen analysis. The basic underlying mechanism in PCA is developing an Eigen structure for the data matrix. We solve for the Eigen values and Eigen vectors of a square symmetric matrix. PCs are nothing but the Eigen vectors of the covariance matrix of the data matrix. The Eigen vector associated with the largest Eigen value has the same direction as the first PC. The Eigen vector associated with the second largest Eigen value determines the direction of the second PC. The sum of the Eigen values equals the trace of the square

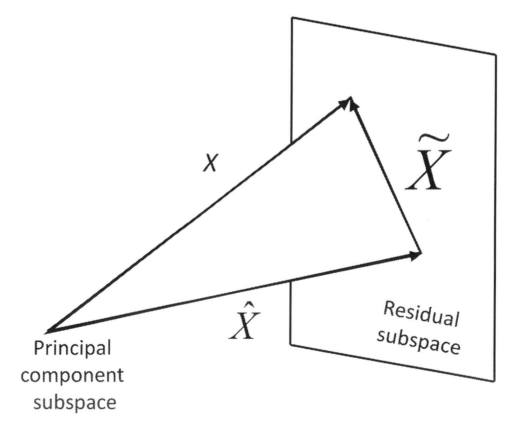

FIGURE 5.1 Principal component analysis monitoring space.

matrix, and the maximum number of Eigen vectors equals the number of rows (or columns) of this matrix. Thus, establishing the Eigen structure is an important task in process monitoring by PCA.

The Eigen values, also called characteristic or latent roots of a square matrix A are the solutions of the determinantal equation

$$|A - \lambda I| = 0 \tag{5.2}$$

A nonzero vector V with the property that $AV = \lambda V$ is called an Eigen vector of A corresponding to the Eigen value λ, which is represented by

$$(A - \lambda I)V = 0 \tag{5.3}$$

The significance of the above equation is that the product of the matrices $(A - \lambda I)$ and the nonzero matrix V is the zero vector,

The roots of the characteristic equation in Eq. (5.2) are the Eigen values of A. Eigen vectors can be calculated using Eq. (5.3). If A is symmetric, the Eigen vectors x_i and x_j that correspond with the Eigen values λ_i and λ_j for $i \neq j$ are orthogonal. Several methods for Eigen vector calculation can be found in any standard engineering mathematics books. More algorithmic procedures for the calculation of Eigen values and Eigen vectors of square matrices and nonsquare matrices can be found in the book, *Numerical Recipes in C* [4] or any other numerical mathematics books.

5.2.4   PCA model establishment

The PCA model is established by the Eigen value decomposition of the covariance matrix of the original data matrix. The PCA assumes linearity among the process variables and results a low dimensional model by extracting the process information stored in a large amount of measurement data. The data matrix X is of $m \times n$ dimension with rows as samples and the columns as variables. To compare the variables with different amplitudes and variability, data are normally mean centered and scaled prior to analysis. Mean centering is done by calculating the mean of each variable from all of the data and then subtracting it from every observation data of that variable. Mean centering has the effect of enhancing the subtle differences between the data. The mean centered data matrix is assumed such that its all columns have zero mean. The covariance of the data matrix is defined as

$$\text{Covariance } (X) = \left(X^T X\right) / (m - 1) \tag{5.4}$$

Thus

$$\text{Covariance } (X)P_i = \lambda_i P_i \tag{5.5}$$

where λ_i is the Eigen value associated with the Eigen vector P_i. In the PCA model, X is decomposed into the sum of the product of n pairs of vectors [2,5]. Each pair consists of a $n \times 1$ vector called the *loadings*, P_i, and a $m \times 1$ vector called the *scores*, T_i. The loadings denote the Eigen vectors. The loading vectors are orthonormal and represent the directions with maximum variability. The score vectors are the coordinates of the data points projected on the PCs. Thus X is expressed as

$$X = T_1 P_1{}^T + T_2 P_2{}^T + \ldots + T_n P_n{}^T + E \tag{5.6}$$

$$X = TP^T + E \tag{5.7}$$

where E is the residual matrix representing the information not captured by the model.

The matrix of loading vectors P forms a new orthogonal basis for the space spanned by X, and the individual P_i are the Eigen vectors of the covariance matrix of the mean centered data. The loadings vectors P_i are the PCs, which are linear combinations of the original variables that together account for large fractions of the variance in the original matrix. In general, the first few PCs describe the maximum amount of variance in the data, thereby facilitates the dimensionality reduction. Each of the scores vectors T_i is simply the projection of X onto the new basis vector P_i

$$T_i = X P_i \tag{5.8}$$

When PCA is performed on a data set, it is often found that only the first few PCs are associated with systematic variation in the data, and the remaining PCs are associated with "noise." PCA models are formed by retaining only the PCs that are descriptive of systematic variation in the data. Determination of the proper number of PCs can be done by several techniques [6,7].

5.3 Projection to latent structures

Statistical modeling techniques attempt to understand the relationship between predictor and response variables. PLS is a PCA-related statistical regression technique that simultaneously extracts information from both input and output data. The PLS has the potential to summarize the underlying structures of the process data as a linear combination of the latent variables and to relate the predictor and response variables by means of regression models. A detailed description of the PLS algorithm and its mathematical formulation can be found in the literature [8,9].

5.3.1 The regression problem

The regression problem is how to model one or several dependent variables/responses, Y, by means of a set of predictor variables, X. In the PLS method, we divide the variables (columns) into two blocks denoted as X and Y. Both X and Y must be available when building the model, but later, when using the model, only X is required. The major use of PLS is to estimate the variables in Y using variables in X. Specifically, PLS regression searches for a set of latent vectors that performs a simultaneous decomposition of X and Y with the constraint that these components explain the covariance between X and Y. PLS has been widely used to extract the relevant information from the process data representing the measured variables and quality variables [10−12].

5.3.2 Scaling and centering the data

The effectiveness of PLS depends on the scaling of the data. The training data matrices in X and Y blocks are of dimensions $m \times n_x$, and $m \times n_y$, respectively, where n_x refers the input variables and n_y refers to the output variables. The standard multivariate approach is to scale each variable to unit variance by dividing them by their standard deviations, and center them by subtracting their averages. This is also referred to as *auto-scaling*. This gives each variable (column) the same weight. For ease of interpretation and for numerical stability, it is advised to *center* the data before the analysis. This can be done either before or after scaling by subtracting the averages from all variables both in X and Y.

5.3.3 PLS model

PLS is conceptually similar to PCA, in which inferential model development is based on two data matrices X and Y associated with the process measurements and the quality measurements, respectively. PLS extracts the latent variables from the respective data matrices. Each pair of latent variables accounts for a certain amount of variability in the input and output data sets. Usually, most of the variance of the two data blocks can be accounted for by the first few latent variables, whilst the higher order latent variables are typically associated with the random noise in the data. The appropriate number of latent variables required to describe the latent structures is generally identified by means of cross validation [13,14].

The PLS algorithm decomposes the input and output data matrices X and Y of dimensions $m \times n_x$ and $m \times n_y$, respectively, into two lower dimensional score matrices T and U of dimensions $n_x \times k$ and $n_y \times k$, respectively. The score matrices T and U represent the projection of the original matrices X and Y onto the latent variable space along with two residual matrices E and F of dimensions $m_x \times n_x$ and $m_y \times n_y$ that contain part of X and Y that left of the regression:

$$Y = TP^T + E \tag{5.9}$$

$$X = UQ^T + F \tag{5.10}$$

The latent variables are aligned along with the k columns of the two score matrices, T ($n_x \times k$) and U ($n_y \times k$), and are ordered in such a way that the amount of variance of the original data described by each variable decreases as the number of latent variables increases. The scores (T) obtained from X data due to PLS decomposition are used separately for regression to predict Y scores. The PLS algorithm automatically predicts Y using the extracted Y-scores (U). The PLS transformation is performed so that the score vectors of each t and h latent variable are mutually related through an inner linear relationship.

$$u_i = b_i t_i + h_i \tag{5.11}$$

where b_i is the coefficient determined by minimizing the norm of the residual vector h_i.

5.4   Artificial neural networks

ANN are computer systems that mimic the operations of the human brain by mathematically modeling its neurophysiological structure [15,16]. They consist of a large number of computational units connected in a massively parallel structure. These computational units are called neurons. The neurons replace the nerve cells in the brain, and the strengths of the interconnections are represented by weights, in which the learned information is stored. The interconnections between the layers of the network and the transfer functions of the neuron processing functions represent the distributed relationships between input and output data. This unique arrangement can acquire some of the neurological processing ability of the biological brain, such as learning and drawing conclusions from experience.

5.4.1   ANN structure and its components

The widely used ANN paradigm is a multi-layered feed-forward network (MFFN) with multi-layered perceptron, mostly comprising three sequentially arranged layers of processing units. The MFFN provides a mapping between an input (x) and output (y) through a nonlinear function f as $y = f(x)$. The three layered MFFN has input, hidden, and output layers, and each layer comprises of its own nodes. All the nodes in the input layer are connected using weighted links to the hidden layer nodes, and similar links exist between the hidden and output layer nodes. The weights represent the state of the knowledge and are adjusted during the learning stage to improve the network performance. Information processing in the network is achieved through activation and inhibition of interconnections between the nodes in the layers of the network. Usually, the input and hidden layers also contain a bias node with its output as unity. The nodes in the input layer do not perform any numerical processing. All numerical processing is done by the hidden and output layer nodes. The structure of a typical multiinput–multioutput neural network is shown in Fig. 5.2.

5.4.1.1   Components of neuron

ANNs adopt a brain metaphor of information processing. They acquire intelligent learning while processing information. Neural networks consist of massively interconnected simple processing elements called neurons or nodes arranged in a layered structure. Fig. 5.3 represents processing of information by a single neuron of the network. It has four components: (1) input connections consist of neurons which receive activation from other nodes; (2) summation function

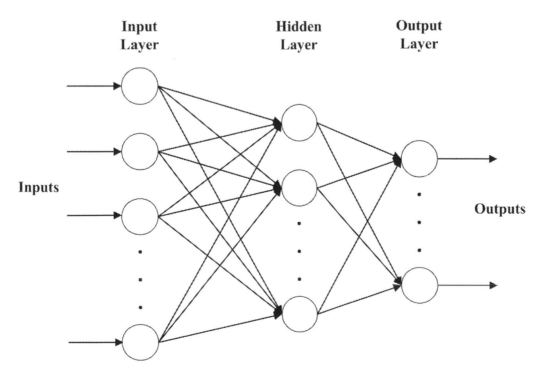

FIGURE 5.2   A typical neural network structure.

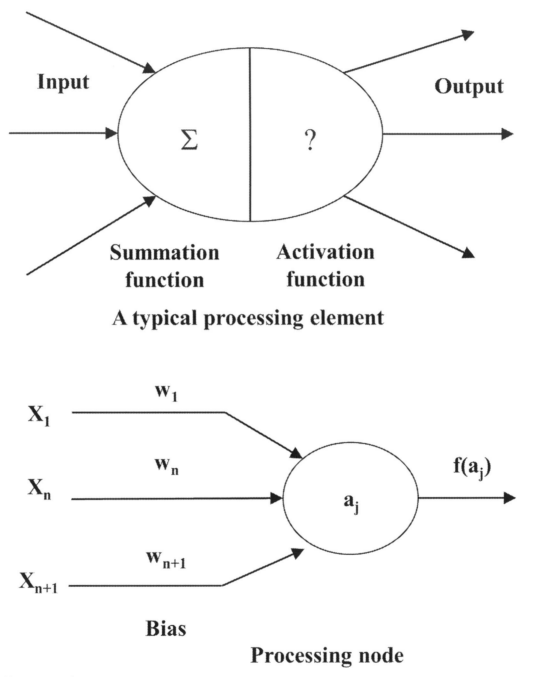

FIGURE 5.3 Neuron processing.

combines various input activations into a single activation; (3) processing function converts this summation of input activation into output activation; and (4) output activation of the neuron arrives as input activation to other units.

The jth node activation a_j is represented by

$$a_j = \sum w_{ij} x_i \tag{5.12}$$

where w_{ij} are interconnection weights and x_i are inputs to the node. The output activation of jth node is given by

$$o_j = f(a_j) = \frac{1}{1 + e^{-aj}} \tag{5.13}$$

5.4.2   Neuron processing functions

The nonlinear neuron processing function is the node activation function. Various forms of activation functions are used to compute the node output. The commonly used activation functions are sigmoid, hyperbolic tangent, and Gaussian functions. The sigmoid function can be expressed as

$$f(x) = \frac{1}{1 + e^{-x}} \tag{5.14}$$

The output of the processing node using sigmoid function is of the form shown in Fig. 5.4A. The hyperbolic tangent function is represented by

$$f(x) = \frac{e^x - e^{-x}}{e^x + e^{-x}} \tag{5.15}$$

The output of the processing node using hyperbolic tangent function is of the form shown in Fig. 5.4B. The Gaussian function is represented by

$$f(x) = e^{-x^2/2} \tag{5.16}$$

The output of the processing node using the Gaussian function is of the form shown in Fig. 5.4C.

5.4.3   Learning paradigms

ANN map input patterns to their associated output patterns by learning from examples. ANN learning paradigms include supervised learning, unsupervised learning, and reinforced learning.

5.4.3.1   Supervised learning

This type of learning is acquired by training the network with a sequence of input and output data and comparing the network predictions with the expected responses. Network interconnection weights are adjusted according to the learning algorithm and training is continued until the network provides desired responses. Given a set of example pairs (x, y), $x \in X$, $y \in Y$, NN learning finds a function f that matches the example data. Many modeling and prediction tasks fall under this paradigm.

5.4.3.2   Unsupervised learning

In this learning paradigm, the target output vector is not known for the corresponding input vector. Network interconnection weights are modified by assigning a similar input vector as the output. The network learns on its own by discovering and adapting to structural features in the input patterns. Given some data x and a cost function, the target output is determined to minimize the cost function. It involves looping connections back to the feedback layers. The network with this learning is called a self-learning network or self-organizing network. This learning enables the network to extract statistical properties of the data and grouping vectors into classes. Applications fall in the domain of statistical modeling, filtering, and clustering.

5.4.3.3   Reinforced learning

In this learning, input data x is usually not given and it is generated by agent interactions with the problem environment. The agent performs action y, and the environment generates the observation x while satisfying the performance function called a reinforced signal. This learning is performed when the knowledge required for supervised training is not sufficient. This is not a popular form of learning. This learning paradigm include games and sequential decision-making applications.

5.4.4   Learning algorithms and training procedure

ANN learning is an optimization process. Learning is generally based on a cost function, which is defined as the squared sum of the difference between the desired output and output obtained from the output layer of the network. The iterative learning makes the network to recognize patterns in the data and creates an internal model, which provides predictions for the new input conditions. For example, in a feed-forward network, inputs are fed through a hidden layer to an output layer. Each neuron forms a weighted sum of inputs from the previous layers to which it is connected, adds a threshold

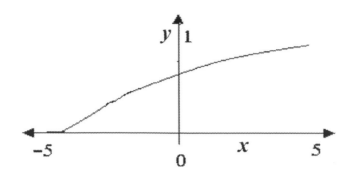

(A) Output of sigmoid function

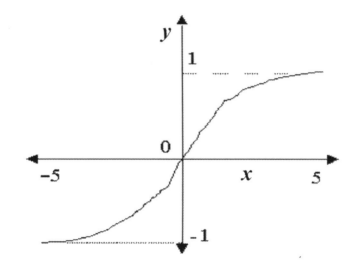

(B) Output of hyperbolic tangent

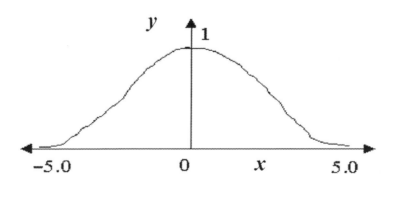

(C) Output of Gaussian function

FIGURE 5.4 Output representation of neuron processing functions.

value and produces a nonlinear function of this sum as its output. The output value serves as input to the next layer to which the neuron is connected, and the process is repeated until convergence is obtained for the actual and predicted values in the output layer. A number of learning algorithms are reported [17] to update the interconnection weights in neural networks. These algorithms include backpropagation, modified backpropagation, conjugate gradient, quasi-Newton and reinforced learning. A hierarchically self-organized learning (HSOL) algorithm [18] is reported for RBFN.

5.4.4.1 ANN training procedure

The procedure to train a typical backpropagation network is described. The problem of neural network training is to obtain a set of weights such that the prediction error defined by the difference between the network's predicted outputs and the measured outputs is minimized. The iterative training makes the network recognize patterns in the data and creates an internal model that provides predictions for the new input condition. The input to the network consists of n-dimensional vector x_p and a unit bias. Each input is multiplied by a weight w_{ij}, and the products are summed to obtain the activation state S_{pj}:

$$S_{pj} = \sum_{i=1}^{N} w_{ij} x_{pi} + w_{N+1,j} \tag{5.17}$$

The output of the hidden layer neuron O_{pj} for sigmoid function is calculated as

$$O_{pj} = f(S_{pj}) = \frac{1}{1 + e^{-S_{pj}}} \tag{5.18}$$

where f represents the differentiable and nondecreasing function. The output layer of a single hidden layer network performs the same calculations as above, except that the input vector x_p is replaced by the hidden layer output O_p and the corresponding weights w_{jk}:

$$S_{pk} = \sum_{i=1}^{M} w_{jk} O_{pi} + w_{M+1,k} \tag{5.19}$$

$$O_{pk} = y_{pk} = \frac{1}{1 + e^{-S_{pk}}} \tag{5.20}$$

Similar calculations can be extended to networks containing more than one hidden layer. A simple way of measuring the progress of learning is by defining the sum of squared error, E_p for p learning patterns. The set of training examples consists of p input−output vector pairs (x_p, d_p). Weights are initially randomized. Thereafter, weights are adjusted so as to minimize the objective function $E(w)$, defined as the mean squared error between the prediction outputs, y_{pk} and the target outputs, d_{pk} for all the input patterns:

$$E(w) = \sum_{p=1}^{p} E_p \tag{5.21}$$

where E_p is the sum of squared error with each training example,

$$E_P = \sum_{K=1}^{M} (d_{pk} - y_{pk})^2 \tag{5.22}$$

The task of E_p minimization is accomplished by training the network using a gradient descent technique such as the generalized delta rule. This is also known as the delta backpropagation rule [19]. According to this rule, the error function δ_{pk} between the hidden layer neurons to the output layer neuron k is computed as

$$\delta_{pk} = (d_{pk} - y_{pk}) f'(S_{pk}) \tag{5.23}$$

The error function δ_{pj} from input neuron to hidden neuron j can be calculated as

$$\delta_{pj} = f'(S_{pj}) \sum_{k=1}^{M} \delta_{pk} w_{jk} \tag{5.24}$$

The weight change from output to hidden layer after nth data presentation is given by

$$\Delta w_{jk}(n) = \eta \delta_{pk} O_{pk} + \alpha \Delta w_{jk}(n-1) \tag{5.25}$$

where η is the learning rate and α is the momentum factor. The updated weights are given by

$$w_{jk}(n) = w_{jk}(n-1) + \Delta w_{jk}(n) \tag{5.26}$$

The weight changes from hidden to input layer can be calculated in the same way. After the weights are updated, a new training example is randomly selected, and the procedure is repeated until the satisfactory reduction of the objective function is achieved.

5.4.5 Information processing

The processing of information by a feed-forward network is explained. Network training is an iterative procedure that begins with initializing the weight matrix randomly. The network learning process involves two types of passes: a forward pass and a reverse pass. In the forward pass, an input pattern from the example data set is applied to the input nodes at the input layer of the network. The weighted sum of the inputs is then transformed into outputs at the nodes of the hidden layer using a nonlinear activation function such as the sigmoid function. The outputs of the hidden nodes form the inputs to the output layer nodes. In the reverse pass, the specific squared error defined in Eq. (5.22) is computed and used for updating the network weights in accordance with the gradient strategy. The weight updating procedure, when repeated for all the patterns in the training set completes one iteration. For a given ANN based modeling problem, the number of nodes in the network input layer and output layer is dictated by the input—output dimensionality of the pattern being modeled. However, the number of hidden nodes is an adjustable structural parameter. If the network architecture contains more hidden units than necessary, it leads to an oversized network. To avoid overfitting of the network, the network simulations are to be conducted by systematically varying the number of hidden units. These simulations provide optimal network architecture with the smallest error magnitude for the test data.

5.4.6 ANN architectures

Based on the nature of learning and type of neuron processing functions, ANN can have different types of architectures, which are classified into backpropagation/feed-forward networks, RBFN, auto-associative networks, recurrent networks, and multi-hidden layer feed-forward networks [20].

5.4.6.1 Feed-forward network

The structure and components of this network are discussed in Section 5.4.1. The networks generally consist of three layers. Information flows in one direction from the input layer via hidden layer to the output layer. Mostly sigmoid type activation functions are used as node processing functions. Network learning is global in nature. The network is trained by backpropagation optimization algorithms in which error backpropagation is used to update the network parameters. The network provides a global approximation to empirical data. It is more suitable for function approximation. The structure of the network is shown in Fig. 5.5.

5.4.6.2 Gaussian potential network

This network consists of mostly three layers, the input, hidden, and output layers. The node in the hidden layer is the important processing step. The Gaussian function is used for node processing. The network is trained by self-learning algorithms. The network updating parameters are the interconnection weights, means, standard deviations, and correlation coefficients. The network provides effective interpolation and better smoothing of data. The network structure is shown in Fig. 5.6.

5.4.6.3 Auto-associative network

This network has identical input and output nodes. The network approximates identity mapping between network input and output. A typical network topology is shown in Fig. 5.7. The network has one or more hidden layers. The hidden layer is called a bottle-neck layer. The key feature of the auto-associative network is the data compression in the bottle-neck

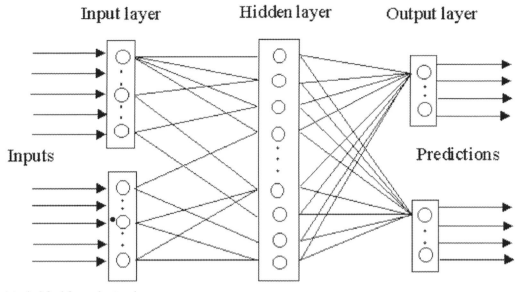

FIGURE 5.5   A typical feed-forward network.

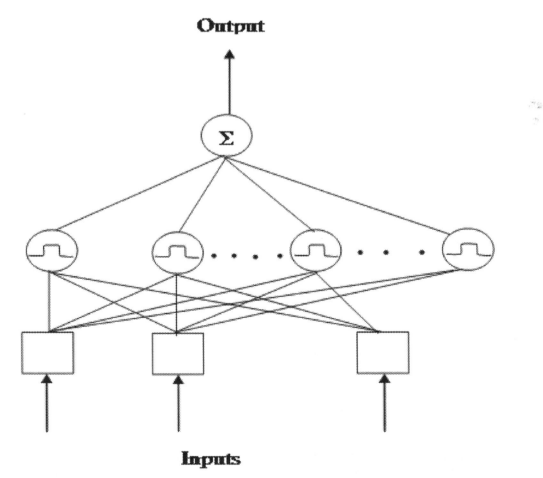

FIGURE 5.6   A typical Gaussian function network.

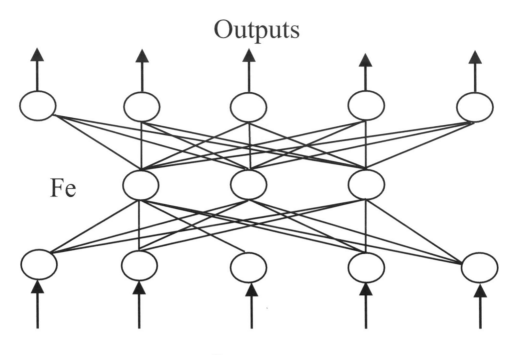

Outputs

Fe

Inputs

FIGURE 5.7 A typical auto-associative neural network.

inner layer. The bottle-neck inner layer provides the topology with powerful properties of the feature extraction. The bottle-neck inner layer on convergence provides information corresponding to significant features or signatures of the data.

5.4.6.4 Recurrent network

The structure of this network is similar to that of a feed-forward network, except that the information runs not only in the forward direction but also in a backward direction. In recurrent networks, outputs of intermediate nodes at one time step are fed back into the network as inputs at the next time step. Recurrent networks allow the network to selectively save information about the past. This network feeds the current readings as well as the stored intermediate values from the past, thus providing a historical context of dependency. Recurrent networks can be an externally recurrent type or internally recurring type. The internally recurring type network is shown in the Fig. 5.8. The dashed lines show recurrent connections with an implicit time delay. A network with external recurrent connections between the outputs and inputs is shown in Fig. 5.9. The dashed lines denote the recurrent connections with time delay but with no adjustable weights. Externally recurrent networks use values of the process outputs of the network as inputs rather than the measured process values.

5.4.6.5 Multilayer perceptron

Feed forward networks with multiple hidden layers can be used to improve the predictive performance of neural network models. This network has interconnections between the input to the first hidden layer, between hidden layers, and the last hidden layer to the output layer. The improved performance with such models is achieved at the expense of the computational requirement involved in the network training. A typical two hidden layer network configuration is shown in Fig. 5.10.

5.5 Radial basis function networks

ANN models intend to establish a static network configuration, and the network may not achieve the desired performance unless it has adequate computational units. Moreover, the backpropagation algorithm used to train the network may require longer computational time, and convergence to a minimum error solution cannot be assured. Therefore, the determination of optimal parameters of the network through its automatic configuration is very useful. RBFN is an

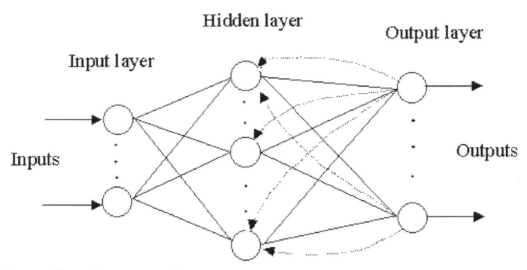

FIGURE 5.8   Structure of internally recurrent network.

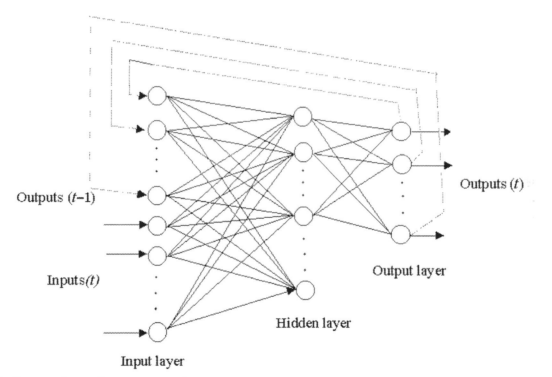

FIGURE 5.9   Structure of externally recurrent network.

alternative modeling tool to ANN. The advantage of RBFN is that it can be configured automatically while optimally updating its parameters. RBFN has been proved as an effective tool in several applications [21−23].

5.5.1   Structure of RBFN

The simple RBFN is composed of a simple structure consisting of a single input layer, single hidden layer, and single output layer. The network with a single output node is shown in Fig. 5.6. In RBFN, the nodes in the hidden layer represent radial basis functions (RBFs), which are described in the following:

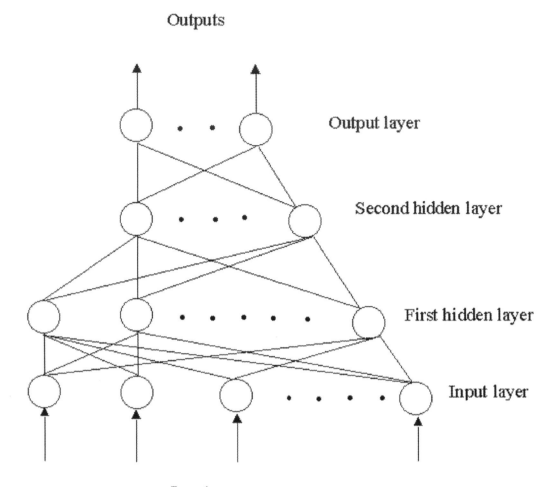

Inputs

FIGURE 5.10   A two hidden layer feed-forward network.

1. The cluster centers in the input space are made up of a center vector m_i with elements m^i_j ($j = 1$ to n).
2. A distance measure to compute the distance of the input vector I, with elements I_j ($j = 1$

   to n) from the center vector m_i as given by

$$d_i = \sqrt{\sum_{j=1}^{n} k^i_j (I_j - m^i_j)^2} \tag{5.27}$$

where k^i_j is the (i, j)th element of the shape matrix K defined as the inverse of the covariance matrix:

$$k^i_j = \frac{h^i_j}{\left(\sigma^i_j\right)^2} \tag{5.28}$$

where h^i_j is the correlation coefficient, and σ^i_j represents marginal standard deviation.

3. A Gaussian type of transfer function which transforms the Euclidian summation d_i ($i = 1$ to m), to give an output for each node as defined by

$$\varphi(d_i) = \exp \frac{\left(-d_i^2\right)}{\gamma^2} \tag{5.29}$$

where γ is a real value.

The hidden layer processes the output from the input layer using the distance measure of Eq. (5.27) and the transfer function in Eq. (5.29). The output of the network O is a weighted sum of the outputs of $\varphi(d_i)$ from the hidden layer and is given by

$$O = w_0 + \sum_{i=1}^{m} w_i \varphi(d_i) \tag{5.30}$$

5.5.2 Automatic configuration of RBFN

A HSOL algorithm [18] is proved as an efficient algorithm to automatically configure the RBFN. This algorithm automatically creates RBFs and modifies parameter vectors of the RBFN. The training algorithm is composed of two parts: the first part is concerned with the adjustment of network parameters and the second part is concerned with the recruitment of minimum necessary number of RBFs based on adjusting the accommodation boundaries of the individual RBFs.

In the HSOL algorithm, an RBF is associated with (1) the accommodation boundary defined by the input space; and (2) the class representation defined in the output space. An effective radius, r_e that defines a region of input space upon which the corresponding RBF can have an influence describes the accommodation boundary. A set of classes can be predicted at the output space, which can be pattern classes when the network is engaged in pattern classification, or signs or subranges of the output space when the network is engaged in function approximation. The accommodation boundaries are adjusted dynamically. They are initially set large for achieving rough but global training, but gradually reduced to a smaller size for fine learning. The method of reducing the size of the accommodation boundary is based on reducing the effective radius when the network performance becomes saturated with the currently available RBFs.

The progress of learning for a single output network is measured by using the root mean square error, E_{rms} for N teaching patterns s defined by

$$E_{rms} = \sqrt{\frac{2}{N} \sum_{p=1}^{N} E_p} \tag{5.31}$$

with

$$E_p = \frac{1}{2} \left(t_p - O_p(n) \right)^2 \tag{5.32}$$

where t_p represents the desired output value defined by the pth teaching pattern, O_p represents the actual output value of the pth teaching pattern, and n represents a column vector, which is a collection of all parameters associated with the output. The parameter saturation vector s is defined as:

$$s(p) = \alpha \frac{\partial E_p}{\partial n} + (1 - \alpha)s(p - 1) \tag{5.33}$$

where α is a positive constant between 0 and 1, and p represents the pth teaching pattern presented to the network. The vector s provides the weighted average of $(\partial E_p / \partial n)$ over the horizon of learning iterations. The saturation criterion ρ is defined as the inverse of $\|s\|$ as

$$\rho(p) = \begin{cases} \rho(p-1) + b \dfrac{\sqrt{d_s}}{\|s(p)\|} & \text{if } p > p_0 \\ 0 & \text{otherwise} \end{cases} \tag{5.34}$$

where d_s is the dimension of s, b is a small positive constant representing the increment rate of ρ, and p_0 is the delay factor defined as $1/\alpha$.

5.5.2.1 Updating of RBFN parameters

The network parameters are updated using the negative gradient of the error function, Eq. (5.32).

The incremental weights (Δwt) to update the weights (w_t) between the output and the ith RBF:

$$\Delta wt_{ji} = \frac{-\partial E_p}{\partial wt_{ji}} = (t_j - O_j)\varphi_i \tag{5.35}$$

The jth element of the incremental mean vector (Δmn) to update the mean vector (mn):

$$\Delta mn_j^i = \frac{-\partial E_p}{\partial mn_j^i} = \sum_{i=1}^{N} k_j^i \left(x_i - mn_j^i\right) \varphi_i \sum_{k=1}^{M} (t_k - O_k)wt_{ki} \tag{5.36}$$

The incremental marginal standard deviation ($\Delta \sigma s$) to update the standard deviation (σs):

$$\Delta \sigma s_j^i = \frac{-\partial E_p}{\partial \sigma s_j^i} = \sum_{i=1}^{N} \frac{k_j^i \left(x_i - mn_j^i\right)^2}{\sigma s_j^i} \varphi_i \sum_{k=1}^{M} (t_k - O_k)wt_{ki} \tag{5.37}$$

The incremental correlation coefficient (Δhs) to update the correlation coefficient (hs):

$$\Delta hs_{jk}^i = \frac{-\partial E_p}{\partial hs_{jk}^i} = -\frac{1}{2} \frac{\left(x_i - mn_j^i\right)\left(x_k - mn_k^i\right)}{\left(\sigma s_j^i \sigma s_k^i\right)^2} \varphi_i \sum_{k=1}^{M} (t_k - O_k)wt_{ki} \tag{5.38}$$

Thus the parameter vector of the output unit, $n = \left[wt^T, mn^T, \sigma s^T, hs^T\right]^T$ is updated as

$$n_j^{new} = n_j^{old} + \eta \Delta n_j \tag{5.39}$$

where η is the positive constant called the learning rate, and $\Delta n = \left[\Delta wt^T, \Delta mn^T, \Delta \sigma s^T, \Delta hs^T\right]^T$.

In the above equations, N and M represent the number of inputs and outputs, φ is the output of RBF, t is the target output and O is the prediction output.

5.5.3   Implementation procedure

Initially, there is no RBF assigned to the network, and the output of the network is set to zero. The HSOL algorithm automatically creates and shapes RBF's and adjusts weight vectors for the output. The algorithm is implemented according to the following procedure:

1. Set $i = 1$, $j = 0$ and $p = 0$, where i represents learning cycles, j represents the number of RBF's and p represents the number of patterns presented to the network. Set $s = 0$ and $\rho = 0$ (Eqs. (5.33) and (5.34)).
2. Invoke the ith learning cycle, where one learning cycle implies the random presentation of all the learning patterns in the pool to the network. The procedure of one learning cycle is as follows:
   a. Get the next teaching pattern, set $p = p + 1$.
   b. Apply the following kernel procedure:
      If $\left|t_p - O_p\right| > \varepsilon_m$, where t_p and O_p represent the desired and actual values of the output for the pth training pattern, and ε_m represents error margin, then do the following:
      If there is an RBF having the same class representation as that of the teaching pattern and the pth teaching pattern falls inside the hypersphere (accommodation boundary) of the RBF, apply parameter update rules, Eqs. (5.35)–(5.39).
      If there is no such RBF, then generate a new RBF
      set $j = j + 1$ and $p = 0$.
      set $s = 0$ and $\rho = 0$.
      The following parameter values are assigned to the new RBF:
      the mean vector, $m_j =$ input sample, x_p of the pth teaching pattern,
      the weight value $w_j =$ output sample, t_p of pth teaching pattern,
      the shape matrix $k = \left(\frac{1}{\sigma_0^2}\right)I$, where σ_0 is the predetermined nominal variance and the effective radius $r_{i=r_0}$,
   where r_0 is the predefined initial effective radius.
      Go to Step 2a.
      If $\left(t_p - O_p\right) \le \varepsilon_m$, then apply the parameter update rules, Eqs. (5.35)–(5.39).
      Calculate $\|s\|/\sqrt{d_s}$, based on Eq. (5.33).
      Calculate ρ based on Eq. (5.34).
      If $\|s\|/\sqrt{d_s} < \rho$, then reduce the effective radius of individual RBFs.

$$r_l^{new} = \begin{cases} r_l^{old} r_d \text{ for } l = 1, \ldots, j, & \text{if } r_i > r_L \\ r_l^{old} \text{ for } l = 1, \ldots, j, & \text{otherwise} \end{cases}$$

where r_l is the lower bound of radius and r_d is the radius decrement rate.
3. If all the teaching patterns are presented, go to the next step. Otherwise, go to Step 2a.
   a. Set $i = i + 1$.
   b. If the network shows satisfactory performance, then stop. If not, go to Step 2.

Apart from adjusting the RBFN parameters, the training method automatically generates the nodes of the hidden layer, which otherwise had to be specified a priori.

5.6 Nonlinear iterative partial least squares

NIPALS algorithm is more descriptive in computing the loadings and scores of the input and output data. The use of preprocessed normalized data of inputs and outputs is a convenient way for this method to computes the loadings and scores of the respective data. This method is very useful to treat the data of high dimensional matrices and requires lower computational effort.

5.6.1 NIPALS algorithm for PLS

In this work, scores of input and output matrices are computed for PLS using the NIPALS algorithm. A detailed description of the NIPALS algorithm and its mathematical formulation can be found elsewhere [24].

The implementation of the NIPALS algorithm is carried out using the following steps.

1. Mean center and scale X and Y.
2. Set the output scores u equal to a column of Y.
3. Compute input weights w by regressing X on u.

$$w^T = \frac{u^T \cdot X}{u^T \cdot u}$$

4. Normalize w to unit length.

$$w = \frac{w}{\|w\|}$$

5. Calculate the input scores t.

$$t = \frac{X \cdot w}{w^T \cdot w}$$

6. Compute output loadings q by regressing columns of Y on t.

$$q^T = \frac{t^T Y}{t^T t}$$

7. Normalize q to unit length.

$$q = \frac{q}{\|q\|}$$

8. Calculate new output scores u.

$$u = \frac{Y.q}{q^T q}$$

9. Calculate the input loadings p by regressing X on t.

$$p^T = \frac{t^T X}{t^T t}$$

10. Compute the inner model regression coefficients b.

$$b = \frac{t^T u}{t^T t}$$

The output scores are related to the input scores by a linear relationship:

$$u_i = b_i t_i + h_i$$

where b_i is the coefficient determined by minimizing the norm of the residual vector h_i.

5.6.2 Nonlinear PLS within PLS framework

PLS models provide the relationships between sets of observed variables by means of latent variables. The basic assumption of linear PLS is that the relation between observed data sets is linear, and the same assumption of linearity holds for modeling the relation in the projected subspace of latent variables. The projection of the observed data onto a subspace of orthogonal latent variables is found useful for the data that is highly correlated and noisy. However, in many situations a strong nonlinear relation between sets of data may exist. Although linear PLS can be used to approximate this nonlinearity, in many applications, such approximation may not be adequate, and the use of a nonlinear model is needed. Nonlinear PLS is therefore required for the analysis of nonlinear data.

The concept of nonlinear PLS modeling was introduced by Wold et al. [25]. Nonlinear PLS techniques represent nonlinear relationships existing in the process data matrices X and Y while maintaining the robust generalization property of the linear PLS approach. These techniques retain the framework of linear PLS but use a nonlinear relationship for each pair of latent variables u_i and t_i. The general relationship has the form

$$u_i = f_i(t_i) + h_i$$

where f stands for nonlinear vector function and h denotes a vector of residuals. The relation between each pair of latent variables is modeled separately. Nonlinear PLS can be modeled by using the standard NIPALS steps until the convergence and the subsequent nonlinear fitting of the inner relation between the extracted pair of the t and u vectors. Different techniques such as higher order polynomial functions, smoothing splines, ANN and RBFN are used to model the nonlinear PLS [26–33].

5.6.3 NIPALS algorithm for RBFN

The data collected from a nonlinear system can be modeled by linear regression techniques to approximate complex relationships over small regions of the predictor variables. This approach is based upon the assumption that the underlying nonlinear relationship can be locally approximated by a linear model, and hence over these small regions, the predictor variables exhibit linear behavior which can be approximated by linear regression techniques. However, in many practical situations, industrial data can exhibit nonlinear features that are required to be encapsulated within a single model. In this case, empirical modeling relies on either transforming the original variables using a nonlinear function and applying linear regression to the transformed variables, or alternatively, by adopting a more complex nonlinear regression approach. Techniques that fall within the framework of nonlinear regression include SPLINE functions, sigmoid neural networks and the RBFN [30]. A number of methods have been proposed in the literature to integrate nonlinear features within the linear PLS framework and thus provide a nonlinear algorithm. Integrating nonlinear features within the PLS algorithm requires modifications to be made to the original PLS regression algorithm. A step forward in nonlinear PLS modeling is described in the work of Wold et al. [25], where the benefits achievable by applying an updating procedure to the parameters of the PLS model are shown. Baffi et al. [29] proposed an alternative procedure for updating the PLS model parameters in combination with quadratic mapping. This algorithm is referred to NIPALS, which is integrated with the RBFN (Section 5.5 of Chapter 5) to form the NIPALS–RBFN algorithm.

The NIPALS–RBFN algorithm uses the normalized input and output data of mean zero to compute the loadings and scores. The implementation of the NIPALS algorithm is carried out using the following steps.

1. Mean center and scale X and Y.
2. Set the output scores u equal to a column of Y.
3. Compute input weights w by regressing X on u.

$$w^T = \frac{u^T \cdot X}{u^T \cdot u}$$

4. Normalize w to unit length

$$w = \frac{w}{\|w\|}$$

5. Calculate the input scores t

$$t = \frac{X.w}{w^T.w}$$

6. Train the RBF network between t and u

$$(c_j,\ \rho_j,\ w_j) \leftarrow (t,\ u)$$

where c_j is the jth center chosen for the data sample t, ρ_j center width for the jth center, and w_j denotes the weight associated with jth RBF. This notation can be matched with the notation used in Section 5.5.

7. Calculate nonlinear prediction of u

$$\hat{u} = \sum_{j=1}^{NC} w_j \exp\left(-\|\frac{c_j - t}{\rho_j^2}\|\right) + w_0$$

8. Compute output loadings q by regressing columns of Y on $\hat{u}$

$$q^T = \frac{\hat{u}^T Y}{\hat{u}^T \hat{u}}$$

9. Normalize q to unit length

$$q = \frac{q}{\|q\|}$$

10. Calculate new output scores u

$$u = \frac{Y \cdot q}{q^T \cdot q}$$

11. Compute the input weights updating parameters Δw
12. Compute new input weights w.

$$w = w + \Delta w$$

13. Normalize w to unit length

$$w = \frac{w}{\|w\|}$$

14. Calculate the new input scores t

$$t = \frac{X.w}{w^T.w}$$

15. Check convergence on t: if Yes go to 16, Else go to 6
16. Calculate the input loadings p by regressing X on t

$$p^T = \frac{t^T X}{t^T t}$$

17. Train the RBF network between t and u

$$(c_p,\ \rho_p,\ w_j) \leftarrow (t,\ u)$$

18. Calculate nonlinear prediction of u

$$\hat{u} = \sum_{j=1}^{NC} w_j \exp\left(-\|\frac{c_j - t}{\rho_j^2}\|\right) + w_0$$

19. Calculate input residual matrix

$$E = X - t.p^T$$

20. Calculate output residual matrix

$$F = Y - \hat{u}.q^T$$

21. Replace X and Y by with E and F for additional NIPALS dimensions and repeat steps $2-21$.

5.7 State estimation applications of data-driven modeling techniques

Data-driven methods have seen extensive interest over the last few years and are being successfully applied across an extraordinary range of problem domains in diverse areas such as engineering, finance, medicine, geology, physics, and biology. These methods have various advantages with wide applications in tasks related to classification, pattern recognition, associative and predictive analysis in various fields. These methods have numerous applications in the areas of optimal state estimation, process monitoring, process control, fault diagnosis, product quality monitoring, Weather forecasting, pollution monitoring, and in many fields concerning to science and technology.

5.8 Summary

Data-driven modeling framework is often more suitable to describe the real-world systems that are associated with complexities. The input−output data available from a system can be used to derive different forms of computationally efficient data-driven models for the purpose of prediction, state estimation, monitoring, and other process systems engineering applications. In this chapter, various data-driven methods and algorithms, including PCA, PLS, NIPALS, ANN, and RBFN are presented. The state estimators derived based on these methods can serve various applications concerning to process systems engineering.

References

[1] S. Wold, K. Esbensen, P. Geladi, Principal component, analysis, Chemomet. Intell. Lab. Syst. 2 (1987) 37−52.

[2] I.T. Jolliffe, Principal Component Analysis, second ed., Springer, New York, 2002.

[3] D.F. Morrison, Multivariate Statistical Methods, third ed., McGraw-Hill, New York, 1990.

[4] W.H. Press, S.A. Teukolsky, W.T. Vetterling, B.P. Flannery, Numerical Recipes in C: The Art of Scientific Computing, Cambridge University Press, Cambridge, 1988.

[5] J.E. Jackson,. A user's guide to principal components, in: V. Barnett, J. Stuart Hunter, D. G. Kendall, J. L. Teugels (Eds.), Wiley Series in Probability and Statistics, Wiley, New York, 1991.

[6] Y. Choi, J. Taylor, R. Tibshirani, Selecting the number of principal components: estimation of the true rank of a noisy matrix, Ann. Stat. 45 (2017) 2590−2617.

[7] S.J. Qin, R. Dunia, Determining the number of principal components for best reconstruction, J. Proc. Control 10 (2000) 245−250.

[8] P. Geladi, B.R. Kowalski, Partial least squares regression: a tutorial, Anal. Chim. Acta 185 (1986) 1−17.

[9] H. Wold, Partial least squares, in: S. Kotz, C. B. Read, N. Balakrishnan, B. Vidakovic, N. L. Johnson (Eds.), Encyclopedia of Statistical Sciences, 6, Wiley, New York, 1984, pp. 581−591.

[10] J.V. Kresta, J.F. MacGregor, T.E. Marlin, Multivariate statistical monitoring of process operating performance, Can. J. Chem. Eng. 69 (1991) 35−47.

[11] T. Kourti, J.F. MacGregor, Process analysis, monitoring and diagnosis using multivariate projection methods, Chemomet. Intell. Lab. Syst. 28 (1995) 3−21.

[12] M. Kano, K. Miyazaki, S. Hasebe, I. Hashimoto, Inferential control system of distillation compositions using dynamic partial least squares regression, J. Proc. Control 10 (2000) 157−166.

[13] B. Li, J. Morris, E.B. Martin, Model selection for partial least squares regression, Chemomet. Intell. Lab. Syst. 64 (2002) 79−89.

[14] H.T. Eastment, W.J. Krzanowski, Cross-validatory choice of the number of components from a principal component analysis, Technometrics 24 (1982) 73−77.

[15] F. Girosi, T. Poggio, Networks and the best approximation property, Biol. Cybern. 63 (1990) 169−179.

[16] W.P. Jones, J. Hoskins, Back propagation, BYTE Mag. 12 (1987) 155−162.

[17] M.C. Palancar, J.M. Aragon, J.S. Torrecilla, pH − control system based on artificial neural networks, Ind. Eng. Chem. Res. 37 (1998) 2729.

[18] S. Lee, R.M. Kil, A Gaussian potential function network with hierarchically self-organizing learning, Neural Netw. 4 (1991) 207.

[19] J.E. Dayhoff, Neural Network Architectures, Van Nostrand Reinhold, New York, 1990.

[20] D.R. Baughman, Y.A. Liu, Neural Networks in Bioprocessing and Chemical Engineering, Academic Press, San Diego, CA, 1995.

[21] S.B. Chen, S.A. Billings, C.F.N. Covan, P.M. Grant, Nonlinear systems identification using radial basis functions, Int. J. Syst. Sci. 21 (1990) 2513–2539.

[22] M.T. Musavi, W. Ahmed, K.H. Chan, K.B. Faris, D.M. Hummels, On the training of radial basis function classifiers, Neural Netw. 5 (1992) 595–603.

[23] R.V. Mayorga, J. Carrera, A radial basis function network approach for the computational of inverse continuous time variant functions, Int. J. Neural Syst. (2007) 149–160.

[24] H. Wold, Estimation of principal components and related models by iterative least squares, Multivariate Analysis, Academic Press, New York, 1966, pp. 391–420.

[25] S. Wold, N.K. Wold, B. Skagerberg, Nonlinear PLS modelling, Chemomet. Intell. Lab. Syst. 7 (1989) 53–65.

[26] I. Frank, A nonlinear PLS model, Chemomet. Intell. Lab. Syst. 8 (1990) 109–119.

[27] S. Qin, T. McAvoy, Non-linear PLS modelling using neural networks, Comput. Chem. Eng. 16 (1992) 379–391.

[28] D. Wilson, G. Irwin, G. Lightbody, Nonlinear PLS modeling using radial basis functions, in: American Control Conference. Albuquerque, New Mexico, 1997.

[29] G. Baffi, E. Martin, A. Morris, Nonlinear projection to latent structures revisited: the quadratic PLS algorithm, Comput. Chem. Eng. 23 (1999) 395–411.

[30] D.J.H. Wilson, G.W. Irwin, G. Lightbody, Nonlinear PLS using radial basis functions, Trans. Inst. Meas. Control 19 (1997) 211–220.

[31] K. Hazama, M. Kano, Covariance-based locally weighted partial least squares for high-performance adaptive modeling, Chemometrics and Intelligent Laboratory Systems 146 (2015) 55–62.

[32] H. Liu, C. Yang, B. Carlsson, S.J. Qin, C. Yoo, Dynamic nonlinear partial least squares modeling using Gaussian process regression, Industrial & Engineering Chemistry Research 58 (36) (2019) 16676–16686.

[33] Z. Dong, N. Ma, A novel nonlinear partial least square integrated with error-based extreme learning machine, IEEE Access 7 (2019) 59903–59912.

Chapter 6

Optimal sensor configuration methods for state estimation

6.1 Introduction

The technological advancements have made the chemical industry to undergo significant changes with the emphasis of running the plants in the most optimal manner with due consideration to the safety aspects, monitoring, and control. To meet these goals, a detailed information about the number of process variables and parameters is required. While it is possible to install additional instrumentation and sensors to monitor the plant's operation, this can be expensive in terms of the capital cost and also incurs the higher maintenance cost. In order to overcome these issues and to minimize the additional instrumentation, model-based state estimators (also called soft sensors) are found to be useful for process estimation. With the use of state estimators, it is possible to estimate many other variables and process parameters by strategically measuring key variables. However, in order to gain maximum benefit from the state estimation techniques, the sensors that provide vital information about the measured variables are to be placed at optimal locations in the process plant. Thus, selection of measurements has become an important prerequisite for successful estimation of process states. The aim of sensor configuration is to identify the minimum number of sensors that provide maximum amount of information on the states in a process. In this chapter, various sensor configuration methods that are widely used for the purpose of process state estimation are presented. The sensor configuration methods presented in this chapter will have immense use in applications and case studies related to optimal state estimation, and these are elaborated in later chapters of this book.

6.2 Brief review on sensor configuration methods

Various techniques have been reported for sensor selection in linear and nonlinear systems. The sensor selection methods reported for linear systems include sensor network design based on steady-state behavior of the systems [1], sensor location techniques for dynamic systems based on minimization of error covariance matrices of Kalman filters [2], analysis of observability matrices [3], and techniques based on minimization of measurement cost [4]. Optimal sensor location for linear or linearized systems has also been reported by various other researchers [5,6]. However, a linearized model may not represent the actual dynamics of the process over the entire region of operation.

Most processes are accurately described by nonlinear dynamic models and the sensor location techniques based on these models would be more desirable for nonlinear systems. The issue of sensor selection for state estimation in nonlinear systems has been addressed by several researchers which include the observability analysis of nonlinear systems based on geometric approach [7], use of observability functions [8], and methods for distributed systems [9,10]. Sensor selection for state estimation in nonlinear systems has also been performed by using empirical observability gramians or observability covariance matrices [11,12]. These observability covariance matrices reflect the influence of process states on the measurements and they provide a comprehensive description of the state to output behavior of the system. Certain scalar measures computed from the error covariance matrices are used to quantify the degree of observability based on which the sensors that provide maximum amount of information are ranked [13]. This chapter primarily focuses on sensor selection methods for optimal estimation of states, which serve as inferential measurements for process operational strategies.

Sensor configuration methods are broadly classified into classical methods and gramian-based methods. The gramian-based methods include both the observability and empirical observability gramians. These methods are discussed as follows.

Optimal State Estimation for Process Monitoring, Fault Diagnosis and Control. DOI: https://doi.org/10.1016/B978-0-323-85878-6.00005-1

6.3 Optimal sensor configuration: classical methods

Optimal sensor location is the most important issue of state estimation in dynamic systems. In this study, various classical methods such as sensitivity index (SI), singular value decomposition (SVD), and principal component analysis (PCA) are described for sensor placement in dynamic systems.

6.3.1 Sensitivity index

In state estimation, primary variables are the ones that are difficult to measure online, whereas the secondary variables are measured online. For example, in distillation, the product compositions represent the primary variables and tray temperatures indicate the secondary variables. Because of the intrinsic feature of the process, the secondary variables have a certain dependency on the primary variables. The main aim of sensor configuration is to identify the most suitable secondary variables that can be used as measurements to the state estimator.

The SI is used to measure the degree of sensitivity of each available secondary variable with respect to changes in each primary variable. This SI is defined as the partial derivative of each secondary variable with respect to each variable to be estimated. Tolliver and McCune [14] employed a sensitivity gain matrix for selecting the best temperature measurement location in continuous distillation for the purpose of column control. For state estimation in distillation, this SI can be employed to measure the degree of sensitivity of each measured temperature with respect to changes in state variables defining the product compositions. In batch operation, the partial derivatives of each measured variable with respect to state variables are approximated by instantaneous pseudo steady-state SI calculated at different time intervals and is expressed in the form of a gain matrix. If Y is the secondary variable (temperature) and x refers the primary variable (product composition) in a distillation column, the SI calculated for all the available process variables are collected in a gain matrix K which is defined as

$$
K(t) = \begin{bmatrix} \frac{\Delta Y_1}{\Delta x_1} \cdots \frac{\Delta Y_1}{\Delta x_i} \cdots \frac{\Delta Y_1}{\Delta x_m} \\ \vdots \vdots \vdots \\ \frac{\Delta Y_j}{\Delta x_1} \cdots \frac{\Delta Y_j}{\Delta x_i} \cdots \frac{\Delta Y_j}{\Delta x_m} \\ \vdots \vdots \vdots \\ \frac{\Delta Y_n}{\Delta x_1} \cdots \frac{\Delta Y_n}{\Delta x_i} \cdots \frac{\Delta Y_n}{\Delta x_m} \end{bmatrix}^T
\tag{6.1}
$$

Here, Y_j is the jth measured variable, x_i represents the ith state variable defining the composition, n is the number of available process measurements, and m is the number of state variables defining the compositions. Here, $\Delta x_i = x_i(t + \Delta t) - x_i(t)$ indicates the variation of the ith variable during the selected time interval Δt, and $\Delta T_{j=} T_j(t + \Delta t) - T_j(t)$ represents the variation of the J measured variable in the same period. Since the process is inherently dynamic, all the variables are time-varying during the time interval Δt. Consequently, each element $\Delta T_j/\Delta x_i$ of K is only an approximation of the corresponding partial derivative $\partial T_j/\partial x_i$. The $m \times n$ sensitivity matrix K can be determined from simulations using the process model.

The sensitivity gain matrix can be calculated for both continuous and batch processes. For continuous processes, K is time-invariant, and can be obtained by applying small perturbations in the primary variables around the steady state of the system. On the other hand, K is time-varying for batch processes. In this case, an instantaneous pseudo-steady state sensitivity matrix can be calculated at different time instants using the data. The sensitivity matrix, K can be used directly to establish instantaneous optimal sensor configuration. The element k_{ij} a measure of the sensitivity of the jth measured variable to the variation in the ith estimation variable. The magnitude of the elements of K matrix reflects the sensitivities of the secondary variables with respect to the primary variables at each time instant. At each time period, all the elements of K in their absolute form are arranged in descending order along with their corresponding measurement locations. The measured variable having the largest value of k_{ij} could be considered as the most suitable input to the estimator. Similarly, the location having the second largest value of k_{ij} is the second most appropriate input to estimator.

6.3.2 Singular value decomposition

SVD is useful in analyzing the multivariable aspects of the gain matrix K. The sensitivity information defined by the gain matrix of Eq. (6.1) can be extracted by exploiting the properties of SVD [15]. The application of this approach to the sensitivity gain matrix K leads to the identification of the measurement variables that are least interacting and most sensitive to the primary state variables. By using SVD, the matrix K can be decomposed into three unique component matrices as

$$K = U \sum V^T \tag{6.2}$$

where K is an $n \times m$ matrix, U is an orthogonal matrix, the columns of which are called the left singular vectors, V is a orthogonal matrix, the columns of which are called the right singular vectors and $\sum$ is a $n \times m$ diagonal matrix of scalar called the singular values and organized in descending order such that $\sigma_1 \geq \sigma_2 \geq \cdots \sigma_m \geq 0$. The U and K matrices are both measures of sensor sensitivity. The geometric property of U matrix provides subtle insights about the measurement sensitivity. The U vectors present the measurement sensitivities in an orthogonal coordinate system. It provides a sensor coordinate system for viewing the sensitivity of the system to change in the primary state variables. The analysis of interpretation of U vector is as follows:

$$U = [U_1 \quad U_2 \ldots.] \tag{6.3}$$

where U_1 is the first column in U matrix which indicate the most sensitive combination of the measurement. The principal component of U_1 vector is the most sensitive location for a sensor in the process. It is the most responsive to the change in primary state variables. U_2 is the second column in U matrix, representing the next most sensitive combination of measurement, which is orthogonal to the measurement represented in U_1. The principal component of U_2 vector represents the measurement sensitivity but exhibits the least possible interaction with the first sensor. The third vector is the next most responsive and so on. All the elements of U_1, U_2, ..., etc. are arranged in descending order along with its measurement location and analyzed with time.

6.3.3   Principal component analysis

PCA is one of the important data-driven modeling tool, which provides the significance and weightage of different process variables. More details about this method is described in Section 5.2 of Chapter 5, Data-driven modeling techniques for state estimation. In this section, the illustration of PCA approach for configuring the measurement sensors for state estimation is presented. The properties of PCA can be exploited in order to identify the most appropriate set of measured variables for state estimation from the information contained in the gain matrix K of Eq. (6.1). The matrix K is first scaled in such a way that each row is normalized to zero mean and unit variance:

$$\hat{k}_{ij} = \frac{k_{ij} - \overline{k}}{\sigma_i} \tag{6.4}$$

with

$$\overline{k}_i = \frac{1}{n} \sum_{j=1}^{n} k_{ij}$$

$$\sigma_i^2 = \frac{\sum_{j=1}^{n} \left(k_{ij} - \overline{k}_i \right)^2}{n-1}$$

where k_{ij} indicates an element and $\hat{k}_{ij}$ is its normalized value. $\overline{k}_i$ and σ_i are the mean and standard deviation of the ith row of K, respectively.

In the PCA analysis, the normalized gain matrix K is factored into two matrices:

$$\hat{K}(t) = TP^T \tag{6.5}$$

where $T(m \times s)$ is the score matrix and $P(n \times s)$ is the orthonormal loading matrix, whose rows are the s principal components. The loading matrix P defining the eigenvectors is computed using the covariance matrix of the normalized gain matrix, which is expressed by

$$Cov\,(\hat{K}) = \frac{\hat{K}^T \hat{K}}{m-1} \tag{6.6}$$

If the dimension $s = 1$, the loading matrix $\boldsymbol{P}$ becomes a vector, which represents the direction that is most sensitive to the primary variables, and the jth element of $\boldsymbol{P}$ can be interpreted as a measure of the contribution of the jth measured variable towards high-sensitivity direction. Thus, the first largest value of the principal components identifies the measured variable that is most sensitive to the primary state variables. The largest value of the loadings identifies the second most sensitive measurement location, and so on. The PCA transformation of the sensitivity matrix indicates the number of measurements to be identified for state estimation, because all the measured variables that correspond to largest loadings are retained while ignoring the smaller ones. Thus, the sensitivity gain matrix K calculated at each sample time t is subjected to PCA-based sensitivity analysis to identify the most sensitive measured variables at the current sampling instant. The overall optimal measurement selection for the state estimator is determined by calculating the summated PC index (SUMPC), for each measured variable:

$$\text{SUMPC}_j = \sum_{t=1}^{N_S} p_j(t) \tag{6.7}$$

where $p_j(t)$ represents the value of the principal component obtained at time t for the jth measured variable, and N_s indicates the total number of samples. The set of measured variables that have the highest SUMPC values are considered as inputs to the state estimator.

6.4 Optimal sensor configuration: gramian-based methods for linear systems

The success of state estimation in a process depends on the rigorous evaluation of the observability and appropriate selection of measurements that adequately characterize the process behavior. Inappropriate use of measurements for state estimation may lead to numerical problems such as singularity or over parameterization, which may drastically reduce the accuracy of the estimation scheme. Thus, for optimal state estimation, it has to be ascertained that the considered measurements are sufficient and the process observability criteria is satisfied.

6.4.1 Observability gramian for linear systems

State observability concerns the extent to which the states of a system influence the process outputs. Observability is a necessary condition for state estimation. The basic concepts of observability of linear systems are discussed in Section 3.3 of Chapter 3, Linear filtering and observation techniques. In this section, the observability concepts for optimal sensor configuration in linear systems are discussed.

The observability of a linear time invariant system,

$$\dot{x}(t) = Ax(t) + Bu(t)$$
$$y(t) = Cx(t) \tag{6.8}$$

is defined by the observability gramian of the form

$$W_o^{linear} = \int_0^\infty e^{A^T t} C^T C e^{At} dt \tag{6.9}$$

A system is completely observable if the observability gramian has full rank. If the gramian is observed to be rank deficient, then the system will not be observable and some of the states cannot be reconstructed from the measured data. This observability gramian W_o is interpreted as the energy generated by the outputs for perturbations in the initial states. The output energy generated by a linear system corresponds to the singular values of the observability gramian. Observability is a necessary condition for sensor placement in state estimation. However, this criterion alone may not be sufficient in plant environment. It is important that the system must have a high degree of observability for reliable estimation of unmeasurable states. Various measures based on the analysis of observability gramian have been introduced to find the degree of observability of linear systems in the context of sensor location for state estimation [3,13,16].

The observability gramian for a linear discrete time system of the form,

$$x(k + 1) = Ax(k) + Bu(k)$$
$$y(k) = Cx(k) \tag{6.10}$$

is given by

$$W_o^{linear} = \sum_{m=0}^{\alpha} (A^T)^m C^T C A^m \qquad (6.11)$$

6.4.1.1 Observability gramian metrics

Observability refers to the property of a system that allows the reconstruction of the values of the state variables given the outputs. The widely employed metrics to measure the degree of observability of a system based on the linear observability gramian are briefed as follows.

6.4.1.1.1 Observability gramian determinant

The increased degree of observability is indicated by the larger observability gramian determinant. If $\det(W_o^{linear}) = 0$, all states cannot be observed. The commonly used metric is

$$D = \left[\left(W_o^{linear} \right) \right]^{1/n} \qquad (6.12)$$

where n is the number of system states. If the determinant is zero, it is difficult to determine the states actually contributing to system observability. Because of this difficulty, this metric is less preferred to compute the degree of i.

6.4.1.1.2 Observability gramian condition number

The condition number of a matrix refers to the ratio of its largest and smallest magnitude eigenvalues. Smaller observability Gramian condition numbers correspond to improved observability. Often logarithmic scaling is used:

$$C = \log_{10} \frac{\sigma_{max}(W_o^{linear})}{\sigma_{min}(W_o^{linear})} \qquad (6.13)$$

This metric is better than the gramian determinant in assessing the degree of observability.

6.4.1.1.3 Observability gramian trace

The observability gramian trace refers to the sum of all gramian eigenvalues:

$$\mathrm{Trace}\left(W_o^{linear}\right) = \sum_{i=1}^{n} \sigma_i\left(W_o^{linear}\right) \qquad (6.14)$$

The larger the value of trace, the greater the system observability. This metric is more useful than the gramian determinant and gramian condition number in assessing the degree of observability.

6.4.2 Observability gramian-based sensor selection procedure

The following procedure is usually applied for determining optimal sensor locations for linear systems.

1. Compute the observability gramians for a variety of combinations of sensors at different locations.
2. Compute the scalar measures from the gramian and compare the degree of observability of various sensor locations.
3. Assign the sensor that corresponds to the largest value of the observability measure as a good candidate for optimal sensor location.

6.5 Optimal sensor configuration for nonlinear systems

The observability of a linear system can be determined exactly. However, the observability of a nonlinear system based on its linearized model may not be sufficient. A more rigorous measure is needed to find the degree of observability of a nonlinear system defined by the models:

$$\begin{aligned} \dot{x}(t) &= f(x(t), u(t)) \\ y(t) &= h(x(t)) \end{aligned} \qquad (6.15)$$

The observability of nonlinear systems has been discussed in Section 4.3 of Chapter 4, Mechanistic model-based nonlinear filtering and observation techniques for optimal state/parameter estimation. The observability of a nonlinear system can be computed by using the observability covariance matrices. These observability covariance matrices can be viewed as an extension of linear observability gramians to nonlinear systems.

Extensive efforts have been made to derive the conditions for observability of nonlinear systems [17–21]. However, these observability conditions are easier to implement with simple systems and are usually too complex to be interpreted for high dimensional systems [22]. The observability of systems with nonlinear dynamics can be computed by using the empirical observability covariance matrices, which can be viewed as an extension of linear observability gramians to nonlinear systems, and these are referred to empirical observability gramians [23].

6.5.1 Empirical observability gramian for nonlinear systems

The empirical observability gramian represents the output energy of a nonlinear system and this energy can be used for the observability analysis of the system over its operating region. These gramians have the advantage that they are relatively inexpensive to compute and easy to manipulate. The observability covariance matrix of a nonlinear system is defined as

$$W_o^{nonlinear} = \sum_{l=1}^{r} \sum_{m=1}^{s} \frac{1}{rsc_m^2} \int_0^\infty T_l \psi^{lm}(t) T_l^T dt \tag{6.16}$$

where

$$T^n = \left\{ T_1 T_2 \cdots T_r; \ T_i \in R^{nxn}, \ T_i^T T_i = I, \ i = 1, \ldots, r \right\}$$

$$M = \{c_1 c_2 \cdots c_s; \ c_i \in R, c_i > 0, \ i = 1, \ldots, \ s\}$$

$$E^n = \{e_1 e_2 \cdots e_n; \text{standard unit vectors in } R^n\}$$

$\psi_{ij}^{lm}(t) = (Y^{ilm}(t) - Y_{ss}^{ilm})^T (Y^{jlm}(t) - Y_{ss}^{jlm})$ The matrices T and E represent the input directions, and M is the vector which contains the information about the different excitation magnitudes. The notation r represents the number of matrices for perturbation directions, s is the number of different perturbation sizes for each direction, and n is the number of states of the full order system. Y is the measurement vector containing all the available measurements:

$$Y = [y_1, y_2, \ldots, y_{nm}]^T \tag{6.17}$$

where, nm is the number of available measurements from the system. $Y^{ilm}(t)$ is the output of the system corresponding to the initial condition defined by

$$x(0) = c_m T_l e_i + x_{ss} \tag{6.18}$$

The notation Y_{ss}^{ilm} indicate the steady-state of the output that the system will reach after this perturbation. For the computation of the empirical gramian, the parameters T, M, and E are to be chosen with appropriate dimensions. The matrix E is chosen such that all the states are selected for perturbation. This results in a $n \times n$ identity matrix where n is the number of states. The set M is the set of different magnitudes of perturbations. These are used in the computation of $x(0)$ in Eq. (6.18). Different perturbations can be selected, which should be realistic for the operating conditions. The perturbation size should excite the system such that the regular operating range is covered. Computation of the observability covariance matrix requires the perturbation of each state of the system at least once. The nonsingularity of the empirical observability gramian defined by Eq. (6.16) fulfils the observability condition of a nonlinear system [20].

The overall empirical observability gramian for a discrete time nonlinear system of the form,

$$\begin{aligned} x(k+1) &= f(x(k), u(k)) \\ y(k) &= h(x(k)) \end{aligned} \tag{6.19}$$

is expressed by

$$W_o^{nonlinear} = \sum_{l=1}^{r} \sum_{m=1}^{s} \frac{1}{rsc_m^2} \sum_{1}^{q} T_l \psi^{lm}(k) T_l^T \Delta t_k \tag{6.20}$$

In this equation Δt_k is the discrete sampling time. This gramian considers the influence of all state perturbations on all measurements. The nonsingularity condition of this gramian fulfils the observability condition of nonlinear system.

6.5.2 Empirical observability gramian metrics

Fulfilling the observability condition may not alone be sufficient for determining the sensor locations for state estimation in a nonlinear system, but the system should exhibit high degree of observability for the selected measurements. Observability is not only affected by the number of sensors but also its location. Best sensor position is the one that gives maximum signal response on the sensor for the changes in the states of the system. Various scalar quantification measures based on the analysis of empirical observability gramian computed from the nonlinear process model have been used to find the degree of observability for the purpose of sensor location for state estimation. These measures that have originally proposed for quantification of the degree of observability of linear/linearized systems for sensor placement [3,6,16] can be extended to find the degree of observability for sensor location for state estimation in nonlinear systems. The scalar quantification measures based on empirical observability gramian of a nonlinear system to find the degree of observability to locate sensors for state estimation are briefed as follows.

$\text{Max}\left\{\text{Trace}\left(W_{o\xi}^{nonlinear}\right)\right\}$: Trace of the nonlinear empirical observability gramian represents the total output energy generated by a nonlinear system, which is equivalent to the sum of the diagonal elements of the singular values,

$$\text{Trace}\left(W_{o\xi}^{nonlinear}\right) = \sum_{i=1}^{n} \sigma_i \tag{6.21}$$

This measure is interpreted as an indicator of the sum of the singular values of the matrix. The larger value corresponds to better degree of observability. Trace values of individual measurement specific gramians $W_{o\xi}^{nonlinear}$ are arranged in descending order according to their magnitudes to find the measurement ranking.

$\text{Max}\left\{\sigma_{min}\left(W_{o\xi}^{nonlinear}\right)\right\}$: This measure considers the smallest singular value of the empirical observability gramian. The degree of observability strongly depends on the smallest singular value of the gramian as it indicates the nature of the unobservability of the system. It represents the near singularity condition, which is a measure indicating how far a system is from being unobservable. The larger minimum singular value provides the better observability of the system. The minimum singular values are computed for each of the measurement specific gramian and arranged in the descending order of their magnitudes to find the measurement ranking. The measurement with larger minimum singular value is better equipped for state estimation.

$\text{Min}\left\{\text{Condition Number}\left(W_{o\xi}^{nonlinear}\right)\right\}$: The minimum and maximum singular values of each empirical observability gramian matrix are used to define the condition number as

$$\text{Condition Number}\left(W_{o\xi}^{nonlinear}\right) = \frac{\sigma_{max}}{\sigma_{min}} \tag{6.22}$$

These numbers computed for each gramian are arranged in ascending order to obtain the measurement ranking. Smaller condition number implies better degree of observability of a system.

6.5.3 Empirical observability gramian-based sensor selection procedure

The empirical observability gramian-based sensor selection procedure is summarized in the flowchart in Fig. 6.1.

6.6 Summary

Optimal sensor location is the most important issue of state estimation in dynamic systems. In this chapter, different sensor configuration methods for state estimation are discussed by classifying them under classical, gramian, and empirical gramian approaches. The methods discussed under classical approach include the SI, SVD, and PCA. In gramian approach, different observability gramian-based metrics are employed for sensor configuration of linear systems. In empirical gramian approach, different empirical observability gramian metrics computed from the empirical observability covariance matrices are employed for sensor selection in nonlinear systems. The presented sensor location methods serve as useful tools for optimal configuration of sensors for state estimation in various engineering systems.

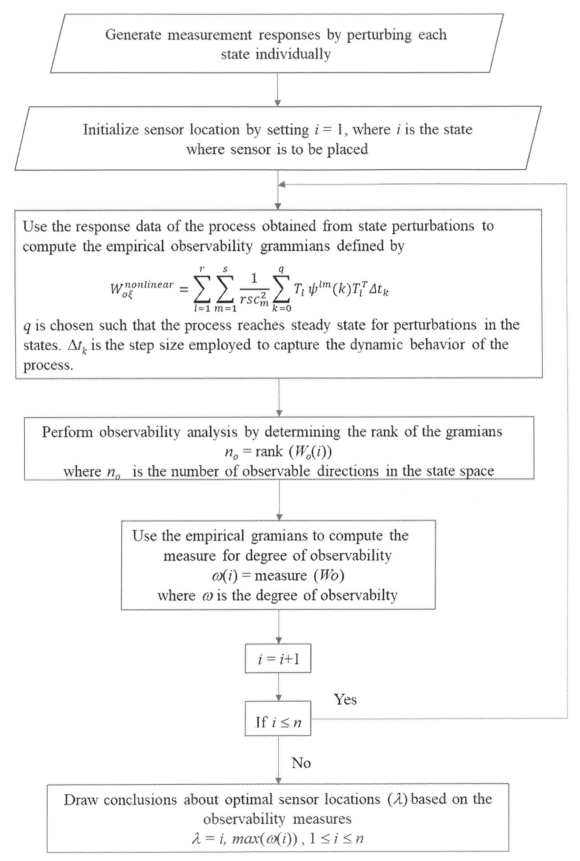

FIGURE 6.1 Flow chart for optimal sensor location.

References

[1] F. Madron, V. Veverka, Optimal selection of measurement points in a complex plant by linear methods, AIChE J. 38 (1992) 227.

[2] S. Kumar, S.H. Seinfeld, Optimal location of measurements in tubular reactors, Chem. Eng. Sci. 33 (1978) 1507–1516.

[3] D. Dochain, N. Tali-Maaner, J.P. Babary, On modeling, monitoring and control offixed bed bioreactors, Comp. Chem. Eng. 21 (1997) 1255–1266.

[4] E. Muslin, C. Benqlilou, M.J. Bagajewicz, L. Puigjaner, Instrumentation design based on optimal Kalman filtering, J. Pro. Con. 15 (2005) 629–638.

[5] K.R. Muske, A methodology for optimal sensor selection in chemical processes, in: Proc. Am. Control Conf., Anchorage, Alaska, 2002.

[6] W. Waldruff, D. Dochain, S. Bourrel, A. Magnus, On the use of observability measures for sensor location in tubular reactor, J. Process. Control 8 (1998) 497–505.

[7] T. Lopez, J. Alvarez, On the effect of estimation structure on the functioning of a nonlinear copolymer reactor estimator, J. Pro. Con. 14 (2004) 99–109.

[8] J.M.A. Scherpen, Balancing of nonlinear systems, Syst. Control Lett. 21 (1993) 143–153.

[9] A.V. Wouwer, N. Point, D. Portman, M. Remy, An approach to the selection of optimal sensor locations in distributed parameter systems, J. Pro. Con. 10 (2000) 291–300.

[10] A. Alonso, I.G. Kevrekidis, J.R. Banga, C.E. Froujakis, Optimal sensor location and reduced order observer design for distributed process system, Comp. Chem. Eng. 28 (2004) 27–35.

[11] J. Hahn, T.F. Edgar, A gramian based approach to non-linear quantification and model reduction, Ind. Eng. Chem. Res. 40 (2001) 5724–5751.

[12] S. Lall, J.E. Marsden, S. Glavaski, A subspace approach to balanced truncation for model reduction of non-linear control system, Int. J. Robust. Non Cont. 12 (2002) 519–535.

[13] F.W.J. Van den Berg, H.C.J. Hoefsloot, H.F.M. Boelens, A.K. Smilde, Selection of optimal sensor position in a tubular reactor using robust degree of observability criteria, Chem. Eng. Sci. 55 (2000) 827–837.

[14] T.L. Tolliver, L.C. McCune, Distillation control design based on steady statesimulation, ISA Trans. 17 (1978) 3–10.

[15] C.F. Moore, Selection of controlled and manipulated variables, in: W.L. Luyben (Ed.), Practical Distillation Control, Van Nostrand Reinold, New York, 1992.

[16] P.C. Muller, H.I. Weber, Analysis and optimization of certain quantities of controllability and observability for linear dynamic system, Automatica 8 (1972) 237–246.

[17] Isidori, Nonlinear Control Systems, Springer-Verlag, New York, 1995.

[18] H. Nijmeijer, A. Van der Schaft, Nonlinear Dynamical Control Systems, Springer-Verlag, New York, 1990.

[19] J.D. Stigter, D. Joubert, J. Molenaar, Observability of complex systems: Finding the gap, Scientific Reports 7 (1) (2017) 1–9.

[20] P. Lecca, A. Re, Identifying necessary and sufficient conditions for the observability of models of biochemical processes, Biophysical Chemistry 254 (2019) 106257.

[21] J. Tsinias, C. Kitsos, Observability and state estimation for a class of nonlinear systems, IEEE Transactions on Automatic Control 64 (6) (2018) 2621–2628.

[22] A.K. Singh, J.H. Hahn, State estimation for high dimensional chemical processes, Comp. Chem. Eng. 29 (2005) 2326–2334.

[23] A.K. Singh, J.H. Hahn, Determining optimal sensor locations for state and parameter estimation for stable nonlinear systems, Ind. Eng. Chem. Res. 44 (2005) 5645–5659.

Part II

Optimal state estimation for process monitoring

Chapter 7

Application of mechanistic model-based nonlinear filtering and observation techniques for optimal state estimation in multicomponent batch distillation

7.1 Introduction

On-line estimation of compositions in batch distillation is highly beneficial for its optimal operation. Batch distillation is the most important separation processes used in many chemical industries, especially those related to the production of fine and speciality chemicals. This is due to the low scale production, rapid change in market needs, and the flexibility in purifying different mixtures under a variety of operational conditions. Batch distillation has the advantage of separating a multicomponent mixture in a single column, whereas a number of interconnected columns are required to separate continuously the same mixture. In order to meet the purity specifications, the batch column has to be operated as precisely as possible. If the current compositions of the batch column are known, they can form a basis for improving the process performance through an operator decision making or for the development of an active automatic closed-loop control scheme. An automatic sensor system for composition measurement is not economical, since component analyzers like gas chromatographs involve large measurement delays as well as high investment and maintenance costs. As an alternative, component concentrations can be estimated from temperature measurements by using estimating techniques based on Kalman filter/observer. Thus, on-line estimation of compositions attains significance for efficient operation of the batch distillation column.

The purpose of on-line state estimation techniques is delivery of reliable, real-time estimates for the state variables defining a process on the basis of the available process knowledge including a dynamic process model and the incoming data from process measurement sensors. Stochastic filters address the problem of estimating process states in the presence of random process disturbances and measurement errors. Stochastic filters have been applied for process state estimation in various processes, including distillation [1–6].

In this chapter, various methods based on the Kalman filtering approach are presented for inferential estimation of compositions in multiple fraction multicomponent batch distillation [7]. Extended versions of Kalman filtering techniques, viz., extended Kalman filter (EKF), adaptive fading Kalman filter, and steady state Kalman filter, are used to derive the composition estimators. These estimators are supported by a dynamic model of batch distillation, which includes Antoine equation-based vapor–liquid equilibrium relation for bubble point calculation. This iterative bubble point calculation is used to provide column temperatures for the estimators. The ability of the composition estimators is assessed by using a performance criterion consisting of multiple performance indices. The sensitivity of the state estimators is evaluated with respect to the effect of the number of temperature measurements, measurement noise, and filter design parameters such as initial state covariance, process and observation noise covariance matrices. The design, implementation, and performance of these methods are shown for composition estimation of a hydrocarbon system involving the separation of cyclohexane/heptane/toluene mixture.

7.2 Batch distillation process and its dynamic model

In batch distillation, a liquid mixture is charged into a vessel, and heat is added to produce vapor that is fed into a rectifying column. The liquid mixture can be a fresh feed and also with any recycled slop cuts. During the initial start-up period, the column operates under total reflux condition in which vapor from the top of the column is condensed and returned to the column. The most important aspect of inferential estimation of batch distillation is the way the composition fronts move through the column

Optimal State Estimation for Process Monitoring, Fault Diagnosis and Control. DOI: https://doi.org/10.1016/B978-0-323-85878-6.00007-5

during the course of the batch cycle. The operation of batch distillation described here corresponds to a ternary system. During the column operation under total reflux condition, the concentration of the lightest component build-up on the upper trays in the column, and the concentrations of the intermediate component and heaviest component decrease in the top of the column but increases in the still pot. When the concentration of the lightest component in the distillate reaches its specified purity level, then the distillate product withdrawal is begun. During the withdrawal of the first product, there is a composition front located in the lower part of the column that separates the lightest and intermediate components. This front moves up the column as the light product is removed. When this front nears the top of the column, the distillate stream is diverted to another tank as the first slop cut. When the concentration of the intermediate component in the distillate reaches its specified purity level, the distillate is diverted to another tank in which the second product is collected. When the purity of the material in this tank drops to the specified purity level, the distillate stream is diverted into another tank, and the second slop cut is collected until the average composition of the material remaining in the still pot and on the trays in the column meets the purity specification of the heavy product. A detailed description of the operation and modeling of batch distillation column is described elsewhere [8–10]. The schematic of multicomponent batch distillation is shown in Fig. 7.1.

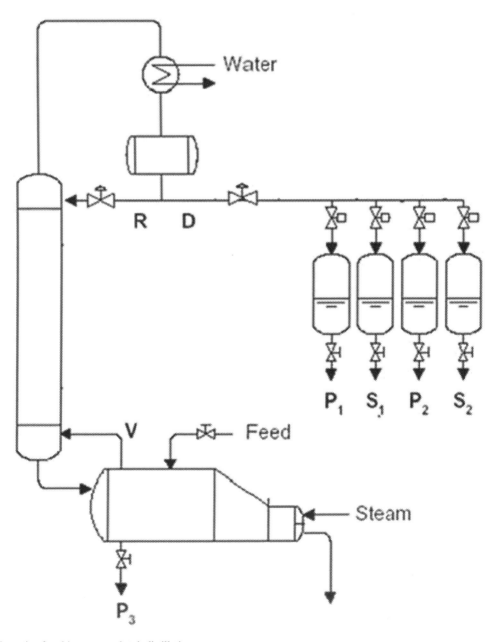

FIGURE 7.1 Schematic of multicomponent batch distillation.

In order to represent the realistic operation of an actual batch distillation column, a rigorous nonlinear model that considers the simultaneous effect of heat and mass transfer operations and fluid flow on the plates is needed. Such batch distillation model is derived from first principles involving dynamic material and components and algebraic energy equations supported by vapor—liquid equilibrium and physical properties. The multicomponent batch distillation dynamics simulator has major computation functions like vapor flow, liquid flow and tray holdup calculations, enthalpy calculations, average molecular weight and density calculations, and vapor—liquid equilibrium calculations. The equations describing the process are given by

Reboiler (subscript B):

$$\frac{dH_B}{dt} = L_1 - V_B \tag{7.1}$$

$$\frac{dx_{B,j}}{dt} = \left(L_1\left(x_{1,j} - x_{B,j}\right) - V_B\left(y_{B,j} - x_{B,j}\right)\right)/H_B, \; j = 1, \ldots, N_c \tag{7.2}$$

$$V_B = (Q_B - L_1(hl_B - hl_1) - H_B \delta hl_B)/(hv_B - hl_B) \tag{7.3}$$

Bottom tray (subscript 1):

$$\frac{dH_1}{dt} = V_B + L_2 - V_1 - L_1 \tag{7.4}$$

$$\frac{dx_{1,j}}{dt} = \left(V_B\left(y_{B,j} - x_{1,j}\right) + L_2\left(x_{2,j} - x_{1,j}\right) - V_1\left(y_{1,j} - x_{1,j}\right)\right)/H_1, \quad j = 1, \; \ldots, N_c \tag{7.5}$$

$$V_1 = (V_B(hv_B - hl_1) + L_2(hl_2 - hl_1) - H_1 \delta hl_1)/(hv_1 - hl_1) \tag{7.6}$$

$$L_1 = V_B + L_2 - V_1 - \delta H_1 \tag{7.7}$$

Intermediate trays (subscript i):

$$\frac{dH_i}{dt} = V_{i-1} + L_{i+1} - V_i - L_i \tag{7.8}$$

$$\frac{dx_{i,j}}{dt} = \left(V_{i-1}\left(y_{i-1,j} - x_{i,j}\right) + L_{i+1}\left(x_{i+1,j} - x_{i,j}\right) - V_i\left(y_{i,j} - x_{i,j}\right)\right)/H_i, \; j = 1, \ldots, N_c \tag{7.9}$$

$$V_i = (V_{i-1}(hv_{i-1} - hl_i) + L_{i+1}(hl_{i+1} - hl_i) - H_i \delta hl_i)/(hv_i - hl_i) \tag{7.10}$$

$$L_i = V_{i-1} + L_{i+1} - V_i - \delta H_i \tag{7.11}$$

Top tray (subscript Nt):

$$\frac{dH_{Nt}}{dt} = V_{Nt-1} + L_0 - V_{Nt} - L_{Nt} \tag{7.12}$$

$$\frac{dx_{Nt,j}}{dt} = \left(V_{Nt-1}(y_{Nt-1,j} - x_{Nt,j}) + L_0(x_{D,j} - x_{Nt,j}) - V_{Nt}\left(y_{Nt,j} - x_{Nt,j}\right)\right)/H_{Nt}, \; j = 1, \ldots, N_c. \tag{7.13}$$

$$V_{Nt} = (V_{Nt-1}(hv_{Nt-1} - hl_{Nt}) + L_0(hl_D - hl_{Nt}) - H_{Nt} \; \delta hl_{Nt})/(hv_{Nt} - hl_{Nt}) \tag{7.14}$$

$$L_{Nt} = L_0 + V_{Nt-1} - V_{Nt} - \delta H_{Nt} \tag{7.15}$$

Reflux drum (subscript D):

$$\frac{dH_D}{dt} = V_{Nt} - L_0 - D \tag{7.16}$$

$$\frac{dx_{D,j}}{dt} = \left(V_{Nt}\left(y_{Nt,j} - x_{D,j}\right)\right)/H_D, \; j = 1, \ldots, N_c \tag{7.17}$$

$$L_0 = V_{Nt} - D \tag{7.18}$$

In the above equations, x = liquid mole fraction, y = vapor mole fraction, V = vapor flow rate, L = liquid flow rate, H = holdup, h_v = vapor enthalpy, h_l = liquid enthalpy, N_t = top tray, and Nc = number of components. Eqs. (7.2), (7.5), (7.9), (7.13), and (7.17) describe the rate of change of compositions with time, Eqs. (7.3), (7.6), (7.10), and (7.14) describe the vapor flow rates, and Eqs. (7.7), (7.11), (7.15), and (7.18) describe liquid flow rates. In liquid and vapor rate calculations, the rate change of holdup and liquid enthalpy is approximated by a numerical differentiation procedure of low order, that replaces $\frac{dH_i}{dt}$ as δH_i, and $\frac{hl_i - hl_{i-1}}{dt}$ as δhl_i. The molar holdup variations on plates are considered by assuming volume holdup, V_i as constant, and variation of liquid density as a function of temperature, pressure, and composition.

7.3 Simplified dynamic model of batch distillation

The rigorous mathematical model of batch distillation given in Section 7.2 is simplified by assuming constant tray and reflux drum holdups, constant vapor boil-up rate, and constant vapor and liquid flow rates. The simplified model is useful for state estimation, since it is difficult to obtain the measured values of liquid and vapor flow rates and tray holdups with time in a realistic situation. The equations representing the simplified model are given as follows:

Reboiler (subscript B):

$$\frac{dx_{B,j}}{dt} = \left(R\left(x_{1,j} - x_{B,j}\right) - V_B\left(y_{B,J} - x_{B,j}\right)\right)/H_B, \quad j = 1, \ldots, N_c \tag{7.19}$$

Bottom tray (subscript 1):

$$\frac{dx_{1,j}}{dt} = \left(V\left(y_{B,j} - y_{1,j}\right) + R\left(x_{2,j} - x_{1,j}\right)\right)/H_1, \quad j = 1, \ldots, N_c \tag{7.20}$$

Intermediate trays (subscript i):

$$\frac{dx_{i,j}}{dt} = \left(V\left(y_{i-1,j} - y_{i,j}\right) + R\left(x_{i+1,j} - x_{i,j}\right)\right)/H_i, \quad j = 1, \ldots, N_c \tag{7.21}$$

Top tray (subscript Nt):

$$\frac{dx_{Nt,j}}{dt} = \left(V\left(y_{Nt-1,j} - y_{Nt,j}\right) + R\left(x_{D,j} - x_{Nt,j}\right)\right)/H_{Nt}, \quad j = 1, \ldots, N_c \tag{7.22}$$

Reflux drum (subscript D):

$$\frac{dx_{D,j}}{dt} = \left(V\left(y_{Nt,j} - x_{D,j}\right)\right)/H_D, \quad j = 1, \ldots, N_c \tag{7.23}$$

7.3.1 Equilibrium relations

The following form of vapor−liquid equilibrium relationship can be used for multicomponent mixture:

$$y_{i,j} = K_{i,j}x_{i,j} \tag{7.24}$$

where

$$K_{i,j} = f_i\left(x_{i,j}, T_i, P_i\right) \tag{7.25}$$

$K_{i,j}$ is related to the pure component vapor pressure by the relation:

$$K_{i,j} = Ps_{i,j}/P_i \tag{7.26}$$

where P_i is the column pressure or plate pressure, and $Ps_{i,j}$ is the vapor pressure of component j. Antoine equation can be used to obtain $K_{i,j}$ or $Ps_{i,j}$ and also to compute the bubble point temperature such that

$$\sum_{i=1}^{Nc} y_{i,j} = 1.0 \tag{7.27}$$

7.4 The application system

In this work, a 20-tray batch distillation column involving the separation of cyclohexane/heptane/toluene mixture [11] is considered as a simulation test bed to evaluate the methods of sensor configuration and state estimation. The column specifications and the thermodynamic coefficients for this system are given in the Table 7.1. In this system,

TABLE 7.1 Specifications.

Column specifications			
Number of stages	22		
Initial charge	2930 mol		
Reboiler heat duty	2.031×10^9 cal h^{-1}		
Pressure	761.654 mm		
Thermodynamic coefficients			
Component	a	b	c
Cyclohexane	6.85146	1206.407	223.136
n-Heptane	6.89386	1264.370	216.640
Toluene	6.95087	1342.310	219.187

cyclohexane and heptane are separated as distillate products, and toluene is separated as a product in the still pot. A vapor boil-up rate of 2690 mol, condenser holdup of 50 mol, and tray holdup of 10 mol are considered.

7.5 Measurements configuration for state estimation

Optimal measurement selection is the most important issue for state estimation in multistage dynamic systems like batch distillation. Batch distillation is an inherent dynamic system in which the interactions between mass transfer and thermodynamics dominantly influence the column dynamics, and the tray sensitivities corresponding to sensor locations change during the course of operation. This causes difficulty in selecting a priori which tray temperature measurements can be used for inferential estimation of compositions in batch distillation. Various multivariate statistical methods have been reported in literature for measurement selection either for the purpose of control in continuous distillation [12,13] or for state estimation in batch distillation [14]. In this study, a principal component analysis (PCA) is used to configure the best set of temperature measurements for model-based state estimation in batch distillation. A detailed description of PCA is given in Section 5.2 of Chapter 5, Data-driven modeling techniques for state estimation.

The data for sensor configuration is generated through numerical simulation using the rigorous model of batch distillation based on the specifications given in Table 7.1. These data include the temperature measurements of the column, and the product compositions of distillate and reboiler recorded at every sampling period during the entire operation. The Euler method with a step size of 0.001 h is used to integrate the differential equations of the system. The temperature data are obtained from Antoine equation through bubble point calculation procedure. The temperature data of every 0.001 h are corrupted with random Gaussian noise in the range of -1 to 1, and these data are considered for the scheme of measurement configuration and state estimation.

In this study, a sensitivity gain matrix $K(t)$ is formulated by using the partial derivatives of each measured variable with respect to state variables defining the product compositions:

$$
K(t) = \begin{bmatrix}
\frac{\Delta T_1}{\Delta x_1} & \cdots & \frac{\Delta T_1}{\Delta x_i} & \cdots & \frac{\Delta T_1}{\Delta x_m} \\
\vdots & & \vdots & & \vdots \\
\frac{\Delta T_j}{\Delta x_1} & \cdots & \frac{\Delta T_j}{\Delta x_i} & \cdots & \frac{\Delta T_j}{\Delta x_m} \\
\vdots & & \vdots & & \vdots \\
\frac{\Delta T_n}{\Delta x_1} & \cdots & \frac{\Delta T_n}{\Delta x_i} & \cdots & \frac{\Delta T_n}{\Delta x_m}
\end{bmatrix}^T
\tag{7.28}
$$

where T_j is the jth measured variable, x_i represents the ith state variable defining the composition, n is the number of available process measurements, and m is the number of state variables defining the compositions. Here, $\Delta x_i = x_i(t + \Delta t) - x_i(t)$ indicates the variation of the ith variable during the selected time interval Δt, and $\Delta T_j = T_j(t + \Delta t) - T_j(t)$

represents the variation of the jth measured variable in the same period. Since the process is inherently dynamic, all variables are time varying during the time interval Δt. Consequently, each element $\Delta T_j / \Delta x_i$ of K is only an approximation of the corresponding partial derivative $\partial T_j / \partial x_i$. The $m \times n$ sensitivity matrix K can be determined from simulations based on the detailed first principle process model.

PCA is a multivariate statistical technique that transforms a number of correlated variables into a smaller number of uncorrelated variables called principal components (PCs). The PCs in PCA model represent an orthogonal transformation of the data so that the variances of the data on the derived coordinates are in decreasing order of magnitude. The first few PCs represent the directions associated with high variance of observations. The last few PCs refer to directions associated with small variance. The properties of PCA can be exploited in order to identify the most appropriate set of measured variables for state estimation from the information contained in the gain matrix, K of Eq. (7.28). The matrix K is first scaled in such a way that each row is normalized to zero mean and unit variance with:

$$\hat{k}_{ij} = \frac{k_{ij} - \overline{k}_i}{\sigma_i} \tag{7.29}$$

with

$$\overline{k}_i = \frac{1}{n} \sum_{j=1}^{n} k_{ij}$$

$$\sigma_i^2 = \frac{\sum_{j=1}^{n} (k_{ij} - \overline{k}_i)^2}{n - 1}$$

where k_{ij} and $\hat{k}_{ij}$ are the elements and normalized elements of K matrix, and n is the number of elements in each row of K matrix. The $\overline{k}$ and σ_i are the mean and standard deviation of the ith row of K. In PCA, the normalized gain matrix $\hat{K}$ is factored into two matrices:

$$\hat{K}(t) = TP^T \tag{7.30}$$

where $T(m \times s)$ is the score matrix and $P(n \times s)$ is the orthonormal loading matrix, whose rows are the s PCs. The loading matrix P defining eigenvectors is computed using the covariance matrix of the normalized gain matrix, $\hat{K}$. The elements in P matrix represent the direction that is most sensitive to the primary variables. The first largest value of the PCs identifies the measured variable that is most sensitive to the primary state variables. The second largest value of the loadings identifies the second most sensitive measurement location, and so on. The sensitivity gain matrix K of Eq. (7.28) calculated at each sample time is subjected to PCA transformation to identify the most sensitive measured variables at the current sampling instant. The overall optimal measurement selection for the state estimator is determined by calculating the summated PC index (SUMPC), for each measured variable:

$$\text{SUMPC}_j = \sum_{j=1}^{N_s} p_j(t) \tag{7.31}$$

where $p_j(t)$ represents the value of the PC obtained at time t for the jth measured variable, and N_s indicates the total number of samples. The set of measured variables having the highest SUMPC values are considered as inputs to the state estimator.

The normalized gain matrix $\hat{K}$ of Eq. (7.30) is obtained by transforming the elements of the gain matrix K of Eq. (7.28). The matrix $\hat{K}$ can be further factored into a score matrix T and a loading matrix P as in Eq. (7.30). The elements of this loading vector in P are arranged in descending order along with their corresponding tray locations in order to evaluate the most sensitive measurement tray location and its change with time. Each loading of the PCA model represents the measure of sensitivity of each tray temperature. The data corresponding to various sampling instants is employed for tray sensitivity analysis using the PCA model. The results of Fig. 7.2 show the sensitive tray location intensity based on the largest elements of the principal loading vector for the data corresponding to sampling periods of 0.005 h.

The results show that the sensitive tray measurement location corresponding to the first largest element of the principle loading vector is confined to the lower or top region of the column as compared to the central region of the column. This indicates that 19th and 20th tray in the top region and reboiler and first tray in the bottom region have appeared to be more sensitive almost throughout the batch operation. The optimal measurement selection for the state estimator is also determined by calculating the SUMPC defined by Eq. (7.31). The set of measured variables having the largest

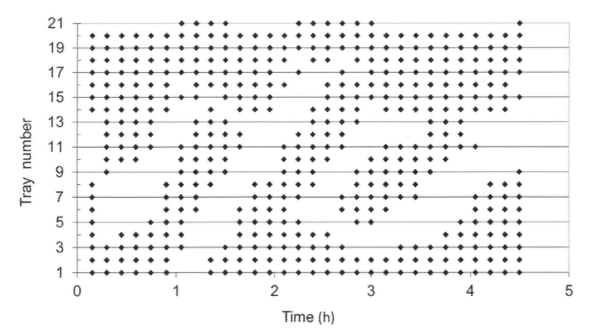

FIGURE 7.2   Tray sensitivity information using principal component analysis.

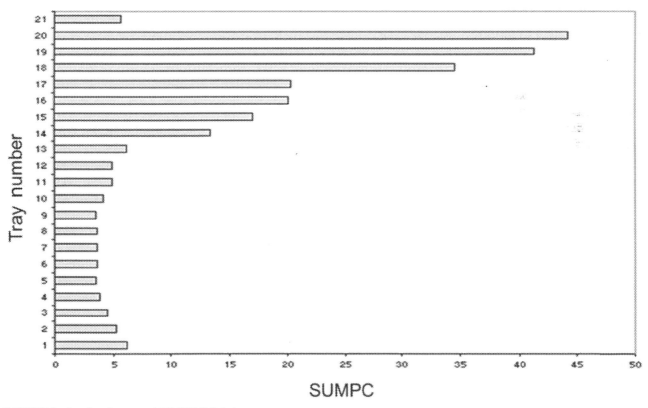

FIGURE 7.3   Results of summated PC (SUMPC) index.

SUMPC values can be considered as inputs to the state estimator. The results of SUMPC in Fig. 7.3 indicate the better suitability of 20th tray in the top region and reboiler in the bottom region for state estimation in batch distillation. Even though 19th tray is also appeared to be more sensitive, it is not considered for state estimation because considering two trays in series (19th and 20th) will hamper the estimation results.

7.6 Performance criteria

Three performance indices, namely, capacity factor (CAP), integral square error (ISE), and batch time (BT) are used to assess the performance of the state estimators used for multicomponent batch distillation.

Capacity factor: The CAP of the batch column is determined as the total specified products produced divided by the total time of the batch. A 30 min period is assumed to be needed to empty and recharge the still pot so that the total batch time is BT + 0.5 h. If P_1, P_2, and P_3 are the products specified in a ternary system, the CAP is given by:

$$CAP = (P_1 + P_2 + P_3) / (BT + 0.5) \qquad (7.32)$$

The products produced will also be equal to the net fresh feed charged to the still pot minus the sum of the two slop cuts, S_1 and S_2:

$$P_1 + P_2 + P_3 = HBO - S_1 - S_2 \qquad (7.33)$$

The CAP increases with the increase in the amount of products.

Batch time: The BT includes the time at total reflux and the time producing products and slop cuts. BT serves as a reference to compare the actual estimator timings for the batch operation.

Integral square error: The estimated states are the component concentrations on all the trays, reboiler, and reflux drum. Computation of ISE using the time-varying composition profiles of the whole column requires more computational effort. The ISE computation based on product compositions is quite simple and can be used to represent the performance of the batch column. For a ternary distillation, the ISE is expressed as:

$$ISE = \left(x_{D,1} - \hat{x}_{D,1}\right)^2 + (x_{D,2} - \hat{x}_{D,2})^2 + (x_{B,3} - \hat{x}_{B,3})^2 \qquad (7.34)$$

7.7 Extended Kalman filter for compositions estimation

The general representation of the nonlinear batch distillation process with continuous-time nonlinear model and discrete time observation model for state estimation by EKF is given in Section 4.4.1 of Chapter 4. The recursive computation of EKF algorithm along with its implementation scheme is given in Section 4.4.3. EKF is computed in two steps. The first is a prediction step, which is used to extrapolate the previous best estimates, and the second is a correction step by which the updated estimates are formed. Since prediction is based on a process model, continuous prediction and discrete correction are employed in the estimation scheme. The mechanistic model-based composition estimation scheme for multicomponent batch distillation is shown in Fig. 7.4.

7.7.1 Design of EKF estimator

The procedure used to design composition estimators for ternary batch distillation based on its simplified model is explained in the following. The same procedure is valid for binary and other multicomponent batch distillation systems.

The state variables in batch distillation are the component concentrations in reboiler, on all trays and in reflux drum. The state vector, x in nonlinear model representation in Eq. (4.25) can be expressed for ternary batch distillation in terms of its first and second components as:

$$x = \left[x_{B,1}\ x_{1,1}\ x_{2,1}\ \cdots\ x_{Nt,1}\ x_{D,1}\ x_{B,2}\ x_{1,2}\ x_{2,2}\ \cdots\ x_{Nt,2}\ x_{D,2}\right]^T \qquad (7.35)$$

The elements f_{ij} of state transition matrix F involved in Eq. (4.26) of the EKF algorithm are defined as follows:
Reboiler:

$$f_{1,1} = \left[-R + V\left(1 - E_{B,11}\right)\right]/H_B;\ \ f_{1,2} = R/H_B;\ \ f_{1,N+1} = -\left(V/H_B\right)E_{B,12};$$

$$f_{N+1,1} = -\left(V/H_B\right)E_{B,21};\ \ f_{N+1,N+1} = \left[-R + V\left(1 - E_{B,22}\right)\right]/H_B;\ \ f_{N+1,N+2} = R/H_B$$

The f_{ij} are partial derivatives of distillation model with the 1st and 2nd components in reboiler, and N represents number of stages.
Tray i:

$$f_{i+1,i} = \left(V/H_i\right)E_{i-1,11};\ \ f_{i+1,i+1} = -\left(R + VE_{i,11}\right)/H_i;\ \ f_{i+1,i+2} = R/H_i;$$

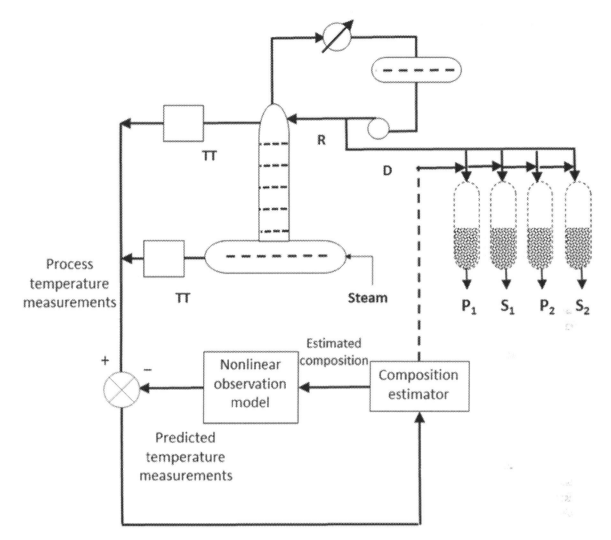

FIGURE 7.4 Mechanistic model-based composition estimation scheme for multicomponent batch distillation.

$$f_{i+1,i+N} = \left(V/H_i\right)E_{i-1,12}; \quad f_{i+1,i+N+1} = -\left(V/H_i\right)E_{i,12}; \quad f_{i+N+1,i} = \left(V/H_i\right)E_{i-1,21};$$

$$f_{i+N+1,i+1} = -\left(V/H_i\right)E_{i,21}; \quad f_{i+N+1,i+N} = \left(V/H_i\right)E_{i-1,22}; \quad f_{i+N+1,i+N+1} = -\left(R+VE_{i,22}\right)/H_i; \quad f_{i+N+1,i+N+2} = R/H_i.$$

The above $f_{i,j}$ are partial derivatives of distillation model with the 1st and 2nd components in tray i.
Reflux drum:

$$f_{N,N-1} = \left(V/H_D\right)E_{Nt,11}; \quad f_{N,N} = -\left(V/H_D\right); \quad f_{N,2N-1} = \left(V/H_D\right)E_{Nt,12};$$

$$f_{2N,N-1} = \left(V/H_D\right)E_{Nt,21}; \quad f_{2N,2N-1} = \left(V/H_D\right); \quad E_{Nt,22} \; f_{2N,2N} = -\left(V/H_D\right).$$

These $f_{i,j}$ are partial derivatives of distillation model with the 1st and 2nd components in reflux drum. In the above equations,

$$E_{i,11} = \frac{\partial y_{i,1}}{\partial x_{i,1}} = \frac{\left(\sum_{j-1}^{Nc} Ps_{i,j}x_{i,j}\right)Ps_{i,1} - Ps_{i,1}x_{i,1}(Ps_{i,1} - Ps_{i,3})}{\left(\sum_{j=1}^{Nc} Ps_{i,j}x_{i,j}\right)^2} \tag{7.36}$$

$$E_{i,12} = \frac{\partial y_{i,1}}{\partial x_{i,2}} = \frac{-Ps_{i,1}x_{i,1}(Ps_{i,2} - Ps_{i,3})}{\left(\sum_{j=1}^{Nc} Ps_{i,j}x_{i,j}\right)^2} \tag{7.37}$$

$$E_{i,21} = \frac{\partial y_{i,2}}{\partial x_{i,1}} = \frac{-Ps_{i,2}x_{i,2}(Ps_{i,1} - Ps_{i,3})}{\left(\sum_{j=1}^{Nc} Ps_{i,j}x_{i,j}\right)^2} \tag{7.38}$$

$$E_{i,22} = \frac{\partial y_{i,2}}{\partial x_{i,2}} = \frac{\left(\sum_{j-1}^{Nc} Ps_{i,j}x_{i,j}\right)Ps_{i,2} - Ps_{i,2}x_{i,2}\left(Ps_{i,2} - Ps_{i,3}\right)}{\left(\sum_{i1}^{Nc} Ps_{i,j}x_{i,j}\right)^2} \tag{7.39}$$

where

$$y_{i,j} = \frac{Ps_{i,j}x_{i,j}}{\sum_{j=1}^{Nc} Ps_{i,j}x_{i,j}} \tag{7.40}$$

and

$$\log Ps_{i,j} = a_j - b_j / \left(c_j + T_i\right) \tag{7.41}$$

The elements of measurement Jacobian matrix, H (Eq. 4.31) on the measured plate temperature are obtained from the Antoine equation as defined by

$$h_{i,1} = \frac{-b_j\left(Ps_{i,1} - Ps_{i,3}\right)}{\left(\sum_{j=1}^{Nc} Ps_{i,j}x_{i,j}\right)\left[a_j - \log\left\{\frac{Ps_{i,j}P_i}{\sum_{j=1}^{Nc} Ps_{i,j}x_{i,j}}\right\}\right]^2} \tag{7.42}$$

$$h_{i,2} = \frac{-b_j\left(Ps_{i,2} - Ps_{i,3}\right)}{\left(\sum_{j=1}^{Nc} Ps_{i,j}x_{i,j}\right)\left[a_j - \log\left\{\frac{Ps_{i,j}P_i}{\sum_{j=1}^{Nc} Ps_{i,j}x_{i,j}}\right\}\right]^2} \tag{7.43}$$

where $h_{i,1}$ and $h_{i,2}$ are partial derivatives of temperature measurement model with the 1st and 2nd components on measured plate i.

7.7.2 EKF implementation results

The implementation of EKF for state estimation in multicomponent distillation is carried out by applying it to the separation of a hydrocarbon system, cyclohexane/heptane/toluene into products and slop cuts in a 20-tray column [11]. A detailed description of the operation of the batch distillation column is given in Section 7.2. The design of EKF composition estimator based on the dynamic mathematical model of batch distillation is given in Section 7.7.1. The Euler method with a step size of 0.001 h is used to integrate the differential equations of the system. The liquid and vapor flow rates, and tray holdups are considered constant during total reflux conditions. These variables vary with time from the beginning of distillate withdrawal and with the start of the transient operation. The temperature measurements are obtained from the Antoine equation through the bubble point calculation procedure. The temperature data of every 0.001 h are corrupted with random Gaussian noise, and these measurements are used for state estimation. The random numbers in the range of -1 to 1 having a mean of 0.00096 and standard deviation of 0.5796 are used as random Gaussian noise. Three cases of measurement data are considered for state estimation. These include reboiler and Plate 10 in the first case; reboiler, Plate 8, and Plate 16 in the second case; and reboiler, Plate 5, Plate 10, Plate 15, and Plate

20 in the third case. These temperature measurements include the measurements configured PCA as well as measurements of trays that are not located in series. The initial state noise covariance matrix P_0, process noise covariance matrix, Q, and observation noise covariance matrix, R are treated as design parameters for the estimators. The elements of P_0, Q, and R are initially selected using process and measurement noise information and further tuned by computation trails such that the performance index is minimized. The diagonal elements of P_0 and Q are selected as 0.0001. The diagonal elements of R for all the cases of measurements are defined as 5.0. The simplified model of batch distillation, which supports the estimators is employed with a vapor boil-up rate of 2690 mol, condenser holdup of 50 mol, and tray holdup of 10 mol.

The component concentrations are expressed in terms of actual and estimated compositions. Actual compositions are defined as those obtained due to numerical simulation of the detailed dynamic distillation model. The estimated compositions are those predicted by the composition estimators. The initial composition of the mixture may vary from batch to batch, and also, the initial component concentration may not be the correct values of the actual mixture. Therefore, the estimator must be able to start with approximate initial conditions and converge to the actual conditions in the column. The quantities and compositions predicted by the estimators can be used to start product and slop cut withdrawals such that the operation of the batch distillation can be controlled under specified conditions.

The actual compositions and the compositions estimated by the method of EKF in the distillate and reboiler for $x_{int} = 0.340/0.335/0.325$, $\hat{x}_{int} = 0.333/0.333/0.334$, and $x_{spec} = 0.9/0.8/0.8$ for the case of two measurements are shown in Fig. 7.5. The results of the estimator are also studied by using recycled batches. The initial composition of the recycled batch is obtained as follows. When the method of EKF is applied for state estimation of a mixture with $x_{int} = 0.407/0.394/0.199$, $\hat{x}_{int} = 0.42/0.37/0.21$, and $x_{spec} = 0.85/0.91/0.92$, a first slop cut of 250 mol with 0.149/0.848/0.0068 mol fraction and a second slop cut of 165 mol with 0.0041/0.572/0.426 mol fraction are obtained. These slop cuts can be combined with the original mixture to form a new mixture with 0.382/0.429/0.189 as component mole

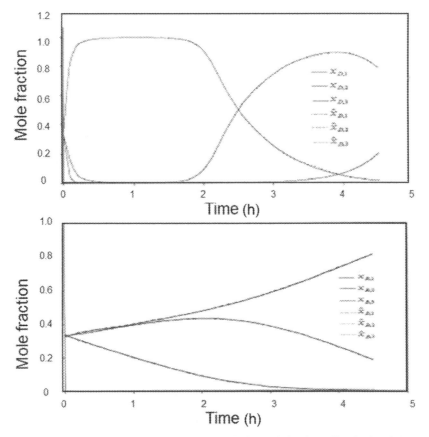

FIGURE 7.5   Actual compositions and compositions estimated by the method of extended Kalman filter in the distillate and reboiler for two measurement condition.

$(x_{int} = 0.340/0.335/0.325 ;\ \hat{x}_{int} = 0.333/0.333/0.334;\ x_{spec} = 0.9/0.8/0.8).$

fractions. Fig. 7.6 shows the actual and estimated compositions in the distillate and reboiler by the method of EKF for this recycled batch with $x_{int} = 0.382/0.429/0.189$, $\hat{x}_{int} = 0.333/0.333/0.334$, and $x_{spec} = 0.85/0.91/0.92$. The results in Fig. 7.5 show the better convergence of estimated compositions with the actual ones.

7.8 Steady state Kalman filter for compositions estimation

The SSKF is another version of the EKF used for state estimation. The recursive prediction step and correction step of this method involve a precomputed steady state transition matrix and fixed Kalman gain matrix. The description of the SSKF for nonlinear state estimation is given in Section 4.5 of Chapter 4, Mechanistic model-based nonlinear filtering and observation techniques for optimal state/parameter estimation.

7.8.1 Design of SSKF estimator

The state vector x in batch distillation is represented by Eq. (7.35). The elements of the state transition matrix F and the measurement transition matrix H used in the design of SSKF are the same as in EKF (Section 7.7.1). By starting with the initial state vector and covariance matrix, the EKF equations, Eqs. (4.26), (4.29), and (4.30) given in Chapter 4 are iteratively updated to converge to a constant state transition matrix F along with the steady state gain matrix $K(t_k)$. The SSKF with the constant gain and state covariance matrices is used for state estimation in batch distillation through recursive computation of prediction and correction equations. The propagated states $\hat{x}(t/t_{k-1})$ obtained by Eq. (4.25) are denoted at time t_k as $\hat{x}(t_k/t_{k-1})$. The updated estimate $\hat{x}(t_k/t_k)$ is computed through Eq. (4.28) by using the measurements at time t_k.

7.8.2 SSKF implementation results

The implementation procedure for SSKF is the same as the EKF procedure described in Section 7.7.2. The method of SSKF is applied for composition estimation in multicomponent batch distillation through recursive computation of prediction and correction equations with the steady state Kalman gain matrix. The cases of temperature measurements considered for SSKF are same as in the EKF estimator. The actual compositions and the compositions estimated by the method of SSKF in the distillate and reboiler for $x_{int} = 0.340/0.335/0.325$, $\hat{x}_{int} = 0.333/0.333/0.334$, and $x_{spec} = 0.9/0.8/0.8$ for the case of two measurements are shown in Fig. 7.7.

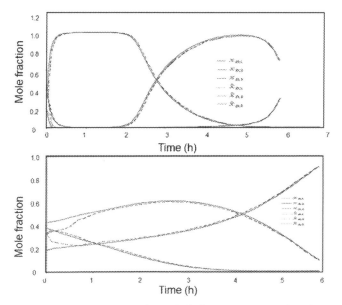

FIGURE 7.6 Actual compositions and compositions estimated by the method of extended Kalman filter in the distillate and reboiler for three measurement condition.

$$(x_{int} = 0.382/0.429/0.189\,;\ \hat{x}_{int} = 0.333/0.333/0.334;\ x_{spec} = 0.85/0.91/0.92).$$

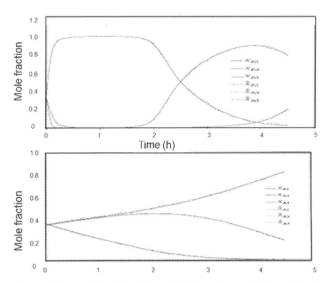

FIGURE 7.7   Actual compositions and compositions estimated by the method of steady state Kalman filter in the distillate and reboiler for two measurement condition.

$$(x_{int} = 0.340/0.335/0.325 \; ; \; \hat{x}_{int} = 0.333/0.333/0.334; \; x_{spec} = 0.9/0.8/0.8).$$

7.9   Adaptive fading extended Kalman filter for compositions estimation

The AFEKF is another version of the EKF derived for state estimation. The description of AFEKF for nonlinear state estimation is given in Section 4.7 of Chapter 4.

7.9.1   Design of AFEKF estimator

The state vector x in batch distillation is represented by Eq. (7.35). The elements of the state transition matrix F and the measurement transition matrix H used in the design of AFEKF are same as in the EKF estimator (Section 7.7.1). The recursive computation procedure of EKF represented by Eqs. (4.25)–(4.32) is used for state estimation by AFEKF. However, the covariance matrix of EKF in Eq. (4.26) is modified to form an adaptive covariance matrix by introducing a forgetting factor λ. The equations used for the computation of the adaptive covariance matrix and the forgetting factor are given by Eqs. (4.46)–(4.49) in Section 4.7 of Chapter 4.

7.9.2   AFEKF implementation results

The implementation procedure for AFEKF is the same as in EKF described in Section 7.7.2. The method of AFEKF is applied for composition estimation in multicomponent batch distillation through the recursive computation of prediction and correction equations given in Section 4.4.3 with the replacement of covariance equation in Eq. (4.26) by the adaptive covariance equation in Eq. (4.46). The cases of temperature measurements considered for AFEKF are the same as in EKF. The actual compositions and the compositions estimated by the method AFEKF in the distillate and reboiler for $x_{int} = 0.407/0.394/0.199$, $\hat{x}_{int} = 0.42/0.37/0.21$, and $x_{spec} = 0.85/0.91/0.92$ for the case of three measurements are shown in Fig. 7.8. These results show the better convergence of estimated compositions with the actual ones.

7.10   Comparative performance of composition estimators

The effectiveness of the composition estimators is assessed by using different performance indices, namely, CAP, ISE, and BT. These performance criteria are defined in Section 7.6. The performance of state estimators is investigated towards the effect of number of temperature measurements, measurement noise, and filter design parameters such as initial state covariance, process and observation noise covariance matrices. The initial random Gaussian noise v is chosen in the form of random numbers within the range of -1 to 1 having a mean of 0.00096 and standard deviation of 0.5796. The matrices P_o, Q, and R are selected such that their diagonal elements are 0.0001, 0.0001, and 5.0

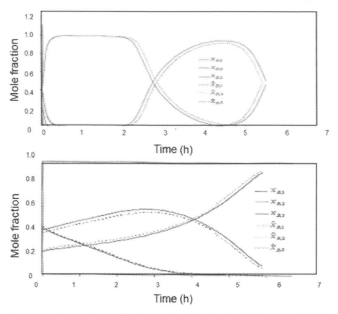

FIGURE 7.8 Actual compositions and compositions estimated by the method of adaptive fading extended Kalman filter in the distillate and reboiler for three measurement condition.

$$(x_{int} = 0.407/0.394/0.199 \; ; \; \hat{x}_{int} = 0.42/0.37/0.21; \; x_{spec} = 0.85/0.91/0.92).$$

TABLE 7.2 Comparison of performance results of state estimators.
$(x_{int} = 0.407/0.394/0.199; \; \hat{x}_{int} = 0.42/0.37/0.21; \; x_{spec} = 0.85/0.91/0.92).$

Measurements	Performance criteria	Actual	EKF	AFEKF	SSKF
2	CAP	359.655	381.703	386.274	360.644
	BT	6.066	6.068	5.910	6.080
	ISE		7.170	14.575	15.297
3	CAP	359.655	380.310	386.450	358.327
	BT	6.066	6.030	5.921	6.078
	ISE		5.457	14.782	10.336
5	CAP	359.655	380.622	386.342	348.544
	BT	6.066	6.032	5.931	6.064
	ISE		5.551	14.723	8.564

AFEKF, adaptive fading extended Kalman filter; *BT,* batch time; *CAP,* capacity factor; *EKF,* extended Kalman filter; *ISE,* integral square error; *SSKF,* steady state Kalman filter.

respectively, while keeping the off-diagonal elements to be zero. The cases of temperature measurements considered for state estimation include, reboiler and Plate 10 as two measurement case; reboiler, Plate 8, and Plate 16 as three measurement case and reboiler, Plate 5, Plate 10, Plate 15, and Plate 20 as five measurement case.

Table 7.2 shows the values of CAP, BT, and ISE for the cases of two measurements, three measurements and five measurements for the actual and estimator initial compositions, and the specified compositions shown in the table caption. The above-mentioned noise (v) and the covariance matrices (P_o, Q, and R) are used for this analysis. The results in Table 7.2, especially, the ISE values show the better performance of EKF. Table 7.3 shows the comparison of the performance of the estimators for the cases of two, three, and five measurements when measurement noise is increased five times from its initial range of -1.0 to 1.0. The actual and estimator initial compositions, and the specified compositions for this analysis are shown in the table caption. These results show that the increased noise level has

TABLE 7.3 Comparison of performance results of state estimators.
$(x_{int} = 0.340/0.335/0.325; \; \hat{x}_{int} = 0.333/0.333/0.334; \; x_{spec} = 0.90/0.80/0.80)$.

Measurements	Performance criteria	Actual	EKF		AFEKF		SSKF	
			Measurement noise		Measurement noise		Measurement noise	
			−1 to 1	−5 to 5	−1 to 1	−5 to 5	−1 to 1	−5 to 5
2	CAP	537.268	531.520	525.906	538.050	539.850	524.637	532.788
	BT	4.554	4.549	4.609	4.529	4.522	4.629	4.540
	ISE		4.711	6.043	4.527	13.157	4.614	6.802
3	CAP	537.268	531.390	525.676	536.754	537.890	528.055	534.137
	BT	4.554	4.551	4.612	4.529	4.519	4.617	4.540
	ISE		4.891	6.503	3.015	4.975	3.678	5.283
5	CAP	537.268	531.785	526.681	538.235	551.653	532.766	535.724
	BT	4.554	4.548	4.603	4.542	4.513	4.574	4.540
	ISE		4.581	4.993	5.642	7.770	3.207	4.370

AFEKF, adaptive fading extended Kalman filter; *BT*, batch time; *CAP*, capacity factor; *EKF*, extended Kalman filter; *ISE*, integral square error; *SSKF*, steady state Kalman filter.

a marginal effect on the performance of EKF and SSKF for all the measurement cases. However, AFEKF has shown inferior performance for the case of two measurements. The increase in noise level within the considered range has a minimal effect on the performance of EKF.

The effect of initial state covariance matrix P_o on the performance of estimators is investigated in terms of ISE, CAP, and BT for all the measurement cases by increasing and decreasing its value to 100 times from its base case value of 0.0001. The results indicate the degraded performance of SSKF when P_o is increased by 100 times for the case of five measurements, whereas the change in P_o has a marginal effect on the performance of SSKF for other measurement cases. Also the change in P_o is found to have nonminal influence on the performance of EKF and AFEKF for all the measurement cases. The sensitivity of the estimators is also investigated towards the effect process noise covariance matrix, Q for all the measurement cases by increasing and decreasing it 100 times from its base value of 0.0001. The results indicate that decreasing or increasing Q by 100 times has a marginal influence on the performance of EKF for all the cases of measurements. However, decreasing Q by 100 times has shown certain effect on the performance of AFEKF for the cases of two and three measurements, whereas increasing it by 100 times affected the performance of AFEKF considerably for all the measurement cases. SSKF has shown stable performance for increasing or decreasing of Q by 100 times for the cases of two and three measurements. The effect of measurement noise covariance matrix, R on the performance of the estimators is also investigated for all the measurement cases by increasing it to 10 and decreasing it to 1 from its base case value of 5.0. The results indicate the stable performance of EKF by decreasing or increasing its value within the considered range. However, decreasing its value to 1.0 has degraded the performance of AFEKF. The SSKF has shown nearly stable performance for increasing or decreasing its value within the considered range.

7.11 Summary

Batch distillation is an intrinsically dynamic process exhibiting highly interactive and nonlinear behavior. Production and separation of chemicals by batch distillation involve the collection over time of a series of product cuts of different compositions. During the course of its operation, the composition profiles can change over a wide range of values. The product withdrawal or the slop cut removal during the course of its operation is based on the desired purity specifications for the products. Despite advances in instrument analyzers, direct and instantaneous composition measurement during operation is not economical and involves measurement delays. This difficulty can overcome through on-line inference of compositions by means of estimation methods. The purpose of on-line state estimation is delivery of reliable, real-time estimates for the state variables defining a process on the basis of the available process knowledge, including a dynamic process model and the incoming data from process measurement sensors. In batch distillation, the compositions predicted by the state estimators can be used to start product and slop cut withdrawals so that the column operation is maintained to achieve desired purity specifications of the products. The state estimator can also be incorporated in on-line optimization and advanced process control schemes.

In this chapter, various stochastic model-based estimators such as EKF, AFEKF, and SSKF are presented for inferential estimation of compositions in the multicomponent batch distillation column. These state estimators are supported by a dynamic model of batch distillation that incorporates component balance equations together with thermodynamic relations, that includes bubble point temperature computation. A performance criterion with multiple performance indices is used to assess the composition estimators. The state estimators are designed and applied to estimate the component concentrations on all trays, reboiler, condenser and products of the multicomponent batch distillation involving the separation of a hydrocarbon mixture, cyclohexane/heptane/toluene. The best set of temperature measurements that are used as inputs to the state estimators are obtained by using PCA. The sensitivity of the composition estimators is investigated towards the effect of various levels of observation noise v, initial state covariance matrix P_o, process noise covariance matrix Q, and measurement noise covariance matrix R. Among these methods, the method of EKF has shown better performance for composition estimation in multiple fraction multicomponent batch distillation.

References

[1] R. Baratti, A. Bertucco, A. Da Rold, M. Morbidelli, A composition estimator for multicomponent columns — development and experimental test on ternary mixtures, Chem. Eng. Sci. 53 (1995) 3601–3612.

[2] N. Regnier, G. Defaye, L. Carlap, C. Vidal, Software sensor based control of exothermic batch reactors, Chem. Eng. Sci. 51 (1996) 5125–5136.

[3] D.R. Yang, K.S. Lee, Monitoring of a distillation column using modified extended Kalman filter and a reduced order model, Comput. Chem. Eng. 21-S (1997) 565–570.

[4] M. Soroush, State and parameter estimations and their applications in process control, Comput, Chem. Eng. 23 (1998) 229–245.

[5] R.M. Oisiovici, S.L. Cruz, State estimation of batch distillation columns using an Extended Kalman filter, Chem. Eng. Sci. 55 (2000) 4667–4680.

[6] V.M. Becerra, P.D. Roberts, G.W. Griffiths, Applying the extended Kalman filter to systems described by nonlinear differential-algebraic equations, Control. Eng. Pract. 9 (2001) 267–281.

[7] C. Venkateswarlu, S. Avantika, Optimal state estimation of multicomponent batch distillation, Chem. Eng. Sci. 56 (2001) 5771–5786.

[8] G.P. Distefano, Mathematical modeling and numerical integration of multicomponent batch distillation equations, AIChE J. 14 (1968) 190–199.

[9] H. Galindez, A.A. Fredenslund, Simulation of multicomponent batch distillation processes, Comput. Chem. Eng. 12 (1998) 281–288.

[10] W.L. Luyben, Multicomponent batch distillation, part 1. Ternary systems with slops recycle, Ind. Eng. Chem. Res. 27 (1998) 642–647.

[11] S. Farhat, M. Czernicki, L. Pibouleau, S. Domench, Optimization of multiple-fraction batch distillation by nonlinear programming, AIChE J. 36 (1990) 1349–1360.

[12] T.L. Tolliver, L.C. McCune, Distillation control design based on steady state simulation, ISA Trans. 17 (1978) 3–10.

[13] C.F. Moore, Selection of controlled and manipulated variables, in: W.L. Luyben (Ed.), Practical Distillation Control, Van Nostrand Reinold, New York, 1992.

[14] E. Zamprogna, M. Barolo, D.E. Seborg, Optimal selection of soft sensor inputs for batch distillation column using principal component analysis, J. Proc. Control. 15 (2005) 39–52.

Chapter 8

Application of mechanistic model-based nonlinear filtering and observation techniques for optimal state estimation in multicomponent reactive batch distillation with optimal sensor configuration

8.1 Introduction

Online estimation of compositions play a significant role in monitoring and operation of reactive batch distillation column. Estimation, monitoring, and control of compositions aids the process to produce fine and speciality chemicals according to the well-defined purity specifications. The transient nature of reactive batch distillation coupled with the integration of reaction kinetics and mass transfer in a single unit makes its operation more complex. In many reactive batch distillation processes, the chemical reaction rates are the dominant factor in determining the process performance. During the course of operation, the product formation and the composition profiles change with the operating conditions over a wide range of values. The product withdrawal or the slop cut removal during the course of operation is taken based on the desired purity specifications for the products. In order to meet the desired product specifications, the column has to be operated as precisely as possible. If the instantaneous compositions during the operation are known, they can form a basis for improving the process performance through an operator decision making or for the development of an active automatic closed-loop control scheme. Despite advances in online composition analyzers, direct composition measurement is not economical and involve with high investment and maintenance costs. These component analyzers also introduce measurement delays into the control loop. This difficulty can be overcome by using estimation methods which incorporate the process knowledge and the available sensor measurements to estimate the compositions.

The approach followed for designing composition estimators for reactive batch distillation is similar to that of batch distillation. More literature exists for the design of state estimators in batch distillation [1−4], which can also be adapted for state estimation in reactive batch distillation. Techniques based on extended Kalman filter (EKF), neurofuzzy inference, and neural networks have been reported for composition estimation in reactive batch distillation [5−7]. However, studies related to optimal measurement selection and online state estimation are less reported for reactive batch distillation.

This chapter presents a mechanistic model-based state estimation scheme for reactive batch distillation with optimal selection of process measurements to be used as inputs to the state estimator. Reactive batch distillation is an intrinsically dynamic process in which the compositions of all stages, the product and slop-cut compositions define the state variables. In this study, optimal temperature sensor locations for state estimation are determined by applying various methods such as sensitivity analysis, singular value decomposition (SVD) and principal component analysis (PCA) and analyzing their results. An EKF derived from the theory of stochastic filtering and observation is designed and applied for inferential estimation of compositions in reactive batch distillation. The estimator is supported by a simple model of the process that include reaction kinetics and thermodynamics. Different performance criteria are employed to assess the sensor configuration methods and the state estimator. The sensitivity of the estimator is studied with respect to the effect of changes in feed compositions, measurement noise, and EKF design parameters. The performance of the proposed scheme is evaluated through simulation by applying it for composition estimation on all trays, reboiler, reflux

Optimal State Estimation for Process Monitoring, Fault Diagnosis and Control. DOI: https://doi.org/10.1016/B978-0-323-85878-6.00002-6

drum, and products of a reactive batch distillation column in which ethyl acetate is produced through an esterification reaction between acetic acid and ethyl alcohol.

8.2 Reactive batch distillation process and its dynamic model

A reaction mixture along with catalyst is charged into a vessel and heat is added so that the reaction takes place to form products and the vapors move up the column. The reaction occurs in liquid phase in the reboiler on all trays and in the condenser. The chemical reaction and distillation proceed simultaneously and the products are collected from a reflux drum. Initially the column is under total reflux operating condition in which vapor from the top of the column is condensed and returned to the column. When the concentration of the lightest component in the distillate reaches its specified purity level, then the distillate product withdrawal is begun. Removal of the lightest distillate component shifts the chemical equilibrium further to the right and thereby improves conversion of the reactants. Thus the reaction and distillation proceed simultaneously and the distillate is collected as products and slop cuts according to the desired purity specifications for the products. This process is continued until the average composition of the material remaining in the still pot and on the trays in the column meets the purity specification of the heavy product. The schematic of reactive batch distillation column is shown in Fig. 8.1.

In order to represent realistic operation of actual reactive batch distillation column, a rigorous nonlinear model that considers simultaneous effect of chemical reaction, heat and mass transfer operations, and fluid flow on the plates is needed. Such a model is derived from first principles involving dynamic material and component balances with reaction terms, and algebraic energy equations supported by vapor—liquid equilibrium and physical properties. The derivation of the model assumes that the vapor phase holdup is negligible compared to the liquid-phase holdup, negligible chemical reactions in the vapor phase, constant column pressure during operation, ideal trays and total condensation with no subcooling in the condenser. Detailed description of the modeling of reactive batch distillation is reported elsewhere [8]. The multicomponent reactive batch distillation dynamic simulator of this study has major computation functions like vapor flow, liquid flow and tray holdup calculations, enthalpy calculations, average molecular weight and density calculations, vapor—liquid equilibrium calculations and reaction rate calculations. The equations describing the process are given as follows:

Reboiler (subscript B):

$$\frac{dM_B}{dt} = L_1 - V_B + M_B \Delta R_B \tag{8.1}$$

$$\frac{d(M_B x_{B,j})}{dt} = L_1 x_{1,j} - V_B y_{B,j} + M_B R_{B,j}, \quad j = 1, \cdots, N_c \tag{8.2}$$

$$\frac{d(M_B H_{L,B})}{dt} = L_1 H_{L,1} - V_B H_{V,B} + Q_B \tag{8.3}$$

Bottom tray (subscript 1):

$$\frac{dM_1}{dt} = V_B + L_2 - V_1 - L_1 + M_1 \Delta R_1 \tag{8.4}$$

$$\frac{dM_1 x_{1,j}}{dt} = V_B y_{B,j} + L_2 x_{2,j} - V_1 y_{1,j} - L_1 x_{1,j} + M_1 R_{1,j}, \quad j = 1, \cdots, N_c \tag{8.5}$$

$$\frac{d(M_1 H_{L,1})}{dt} = V_B H_{V,B} - V_1 h v_1 + L_2 H_{L,2} - L_1 H_{L,1} + Q_1 \tag{8.6}$$

$$L_1 = V_B + L_2 - V_1 - \delta M_1 \tag{8.7}$$

Intermediate trays (subscript i):

$$\frac{dM_i}{dt} = V_{i-1} + L_{i+1} - V_i - L_i + M_i \Delta R_i \tag{8.8}$$

$$\frac{dM_i x_{i,j}}{dt} = V_{i-1} y_{i-1,j} - V_i x_{i,j} + L_{i+1} x_{i+1,j} - L_i x_{i,j} + M_i R_{i,j}, \quad j = 1, \ldots, N_c \tag{8.9}$$

$$\frac{d(M_i H_{L,i})}{dt} = V_{i-1} H_{V,i-1} - V_i H_{V,i} + L_{i+1} H_{L,i+1} - L_i H_{L,i} + Q_i \tag{8.10}$$

$$L_i = V_{i-1} + L_{i+1} - V_i - \delta M_i \tag{8.11}$$

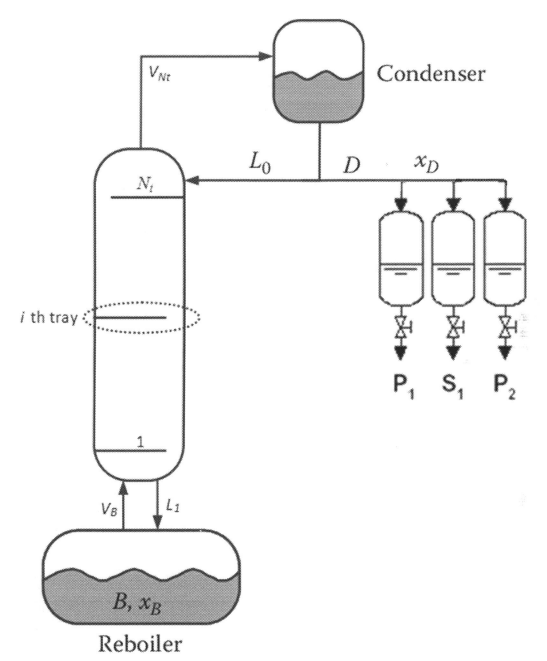

FIGURE 8.1   The schematic of reactive batch distillation column.

Top tray (subscript N_t):

$$\frac{dM_{Nt}}{dt} = V_{Nt-1} + L_0 - V_{Nt} - L_{Nt} + M_{Nt}\Delta R_{Nt} \tag{8.12}$$

$$\frac{dM_{Nt}x_{Nt,j}}{dt} = V_{Nt-1}y_{Nt-1,j} + L_0x_{D,j} - L_{Nt}x_{Nt,j} - V_{Nt}y_{Nt,j} + M_{Nt}R_{Nt,j} \quad j = 1,\ldots,N_c \tag{8.13}$$

$$\frac{d(M_{Nt}H_{L,Nt})}{dt} = V_{Nt-1}H_{V,Nt-1} - V_{Nt}H_{L,Nt} + L_0H_{L,D} - L_{Nt}H_{L,Nt} + Q_{Nt} \tag{8.14}$$

$$L_{Nt} = L_0 + V_{Nt-1} - V_{Nt} - \delta M_{Nt} \tag{8.15}$$

Reflux drum (subscript D):

$$\frac{dM_D}{dt} = V_{Nt} - L_0 - D + M_D\Delta R_D \tag{8.16}$$

$$\frac{dM_D x_{D,j}}{dt} = V_{Nt}y_{Nt,j} - Dx_{D,j} + Rx_{D,j} + M_D R_{D,j}, \quad j = 1, \ldots, N_c \tag{8.17}$$

$$L_0 = V_{Nt} - D \tag{8.18}$$

In the above equations, x = liquid mole fraction, y = vapor mole fraction, V = vapor flow rate, L = liquid flow rate, M = hold up, R = rate of reaction, H_v = vapor enthalpy, H_l = liquid enthalpy, N_t = top tray, and N_c = number of components. Eqs. (8.2), (8.5), (8.9), (8.13), and (8.17) describe the rate of change of compositions with time, Eqs. (8.3), (8.6), (8.10), and (8.14) define the enthalpy balance relations, and Eqs. (8.7), (8.11), (8.15), and (8.18) describe liquid flow rates. Vapor rates are calculated using enthalpy balance equations. In the liquid and vapor rate calculations, the rate of change of holdup and liquid enthalpy is approximated by a numerical differentiation procedure of low order, that replaces $\frac{dM_i}{dt}$ as δM_i, and $\frac{H_{L,i} - H_{L,i-1}}{dt}$ as $\delta H_{L,i}$. The molar holdup variations on plates is considered by assuming volume holdup, V_i as constant and variation of liquid density as a function of temperature, pressure, and composition.

8.2.1   Equilibrium relations

The following vapor–liquid equilibrium relationship can be employed for multicomponent mixture:

$$y_{i,j} = K_{i,j}x_{i,j} \tag{8.19}$$

where

$$K_{i,j} = f_i(x_{i,j}, \; T_i, \; P_i) \tag{8.20}$$

$K_{i,j}$ is related to the pure component vapor pressure by the relation:

$$K_{i,j} = \frac{PS_{i,j}}{P_i} \tag{8.21}$$

where P_i is the column pressure and $PS_{i,j}$ is the vapor pressure of component j. Vapor–liquid equilibrium relations can be used to obtain $K_{i,j}$ or $PS_{i,j}$ and also to compute bubble point temperature such that $\sum_{i-1}^{i-Nc} y_{i,j} = 1.0$.

8.2.2   Enthalpy relations

Vapor and liquid enthalpies are calculated using the relations:

$$\begin{aligned} H_L &= f(x, T, P) \\ H_V &= f(y, T, P) \end{aligned} \tag{8.22}$$

8.3   Simplified dynamic model of reactive batch distillation

A composition estimator designed based on a rigorous mathematical model of multicomponent reactive batch distillation is not suitable for implementation in realistic situation, since it is difficult to obtain the measured values of liquid and vapor flow rates and tray holdups with time. Therefore, a simplified model that involve the knowledge of kinetics and thermodynamics, and assume constant tray and reflux drum holdups, and constant vapor and liquid flow rates is more useful to support the estimator. The equations representing the simplified model are given as follows.

Reboiler (subscript B):

$$\frac{dx_{B,j}}{dt} = \left(R(x_{1,j} - x_{B,j}) - V_B(y_{B,j} - x_{B,j}) + M_B R_{B,j}\right)/M_B, \quad j = 1, \ldots, N_c \tag{8.23}$$

Bottom tray (subscript 1):

$$\frac{dx_{1,j}}{dt} = \left(V_B(y_{B,j} - y_{1,j}) + R(x_{2,j} - x_{1,j}) + M_1 R_{1,j}\right)/M_1, \quad j = 1, \ldots, N_c \tag{8.24}$$

Intermediate trays (subscript i):

$$\frac{dx_{i,j}}{dt} = \left(V_{i-1}(y_{i-1,j} - y_{i,j}) + R\left(x_{i+1,j} - x_{i,j}\right) + M_i R_{i,j}\right)/M_i, \quad j = 1, \ldots, N_c \tag{8.25}$$

Top tray (subscript N_t):

$$\frac{dx_{Nt,j}}{dt} = \left(V(y_{Nt-1,j} - y_{Nt,j}) + R(x_{D,j} - x_{Nt,j}) + M_{Nt} R_{Nt,j}\right)/M_{Nt}, \quad j = 1, \ldots, N_c \tag{8.26}$$

Reflux drum (subscript D):

$$\frac{dx_{D,j}}{dt} = \left(V(y_{Nt,j} - x_{D,j}) + M_D R_{D,j}\right)/M_D, \quad j = 1, \ldots, N_c \tag{8.27}$$

The vapor−liquid equilibrium relations are the same as given by Eqs. (8.19)−(8.21).

8.4   The application system

In order to design and implement the measurement selection methods as well as the state estimation scheme, a reactive batch distillation with an esterification reaction with the following reversible reaction is considered:

$$\underset{\text{Acetic acid}}{CH_3COOH(1)} + \underset{\text{Ethanol}}{C_2H_5OH(2)} \leftrightarrow \underset{\text{Ethyl acetate}}{CH_3COOC_2H_5(3)} + \underset{\text{Water}}{H_2O(4)} \tag{8.28}$$

The boiling temperatures of acetic acid, ethanol, ethyl acetate, and water are 391.1 K, 351.5 K, 350.3 K, and 373.2 K, respectively. The reaction mixture exhibits azeotropism due to the formation of ethyl acetate−ethanol, acetic acid−water, and ethyl acetate−acetic acid−water azeotropes. Reactive batch distillation can provide a means of breaking azeotropes by altering or eliminating the conditions for azeotrope formation through the combined effect of reaction and separation. Ethyl acetate is the main product which has the lowest boiling temperature in the mixture. Controlled removal of ethyl acetate by distillation shifts the chemical equilibrium further to the right and thus improves conversion of the reactants. The data and the specifications are given in Table 8.1. The amount of material charged to the column including the trays and condenser is 5 kmoles in which 4% (200 moles) is equally considered as trays holdup of 100 moles and condenser holdup of 100 moles. The trays holdup is equally divided to obtain a holdup of 12.5 moles on each tray. Initially the compositions on plates and in the reflux drum are considered to be the same as the feed compositions. Table 8.1 also shows the vapor−liquid equilibrium data [9] and the kinetic data [10]. Vapor and liquid enthalpies are calculated using data of Reid et al. [11].

The sequence of operation is briefed as follows. A feed mixture with its initial composition is introduced into the column and initially the column is operated under total reflux condition. The maximum achievable concentration of ethyl acetate under total reflux condition is 0.964 mole fraction, which imposes a limit on the achievable product purity under batch operation. Ethyl acetate and water in distillate and water in reboiler are considered as products in this operation. When the average purity of ethyl acetate satisfies its desired purity specification, it is withdrawn as a first distillate product. When the average purity of ethyl acetate drops below its specified composition, its collection is terminated and the distillate is withdrawn as a slop cut until the second product composition builds up in the distillate. When the average water purity in the distillate satisfies its desired specification, the slop cut withdrawal is terminated and water is collected as second product. During the time of second distillate product withdrawal, the composition of water in the reboiler builds up and the operation is allowed to continue until the composition of water in the reboiler is above its specified value. Thus the water in the reboiler is accumulated as a third product.

8.5   Sensor configuration for state estimation

Optimal measurement selection is the most important issue for state estimation in multistage dynamic systems like batch and reactive batch distillation. The interaction between reaction kinetics, mass transfer, and thermodynamics in reactive batch distillation leads to its complex dynamic behavior and the optimal sensor location and the state estimator design must consider these aspects for effective realization of the process dynamics. Since effective monitoring and control of the column operation is based on the estimated compositions, the measurements used as inputs to the estimator must be suitably selected in order to provide accurate estimation of compositions. All the available process measurements cannot be used as inputs to the estimator due to measurement redundancy. The distributed and dynamic nature of reactive batch distillation provide many possible choices for locating temperature sensors. Sensor selection issues such

TABLE 8.1 Column specifications for ethanol esterification process.

No. of ideal stages (including reboiler and condenser)	10
Total fresh feed (kmol)	5
Feed composition	
Acetic acid	0.45
Ethanol	0.45
Ethyl acetate	0.0
Water	0.1
Internal plate hold up (kmol)	0.0125
Condenser hold up (kmol)	0.1
Vapor boil up rate (kmolh^{-1})	2.5
Column pressure (bar)	1.013
Vapor $-$ liquid equilibrium	
Acetic acid(1) + Ethanol(2) $\leftrightarrow$ Ethyl acetate (3) + Water (4)	
$K_1 = (2.25 \times 10^{-2})\,T - 7.812,\ T > 347.6\text{K}$	
$K_1 = 0.001,\ T \leq 347.6\text{K}$	
$\log K_2 = -2.3 \times 10^3/T + 6.588$	
$\log K_3 = -2.3 \times 10^3/T + 6.742$	
$\log K_4 = -2.3 \times 10^3/T + 6.484$	
Kinetic data	
$r = k_1 C_1 C_2 - k_2 C_3 C_4$	
r is rate of reaction, (gmol L^{-1} min^{-1}) and	
C_i is concentration in (gmol L^{-1}) for ith component.	
The rate constants are	
$k_1 = 4.76 \times 10^{-4}$ (L gmol^{-1} min^{-1})	
$k_2 = 1.63 \times 10^{-4}$ (L gmol^{-1} min^{-1})	

as iterative measurement selection procedure [12], degree of freedom argument on observability criteria [1], and singularity and overparameterization problems [13] have been discussed in the context of continuous distillation and batch distillation. The application of various statistical techniques for sensor configuration for the purpose state estimation in reactive batch distillation has been reported [14]. Chapter 6.3 provides the basic description of different classical methods such as sensitivity index (SI), SVD, and PCA. Here we extend the application of these methods for determining optimal sensor locations for state estimation in reactive batch distillation.

For applying these methods for sensor configuration, the time varying data is generated through numerical simulation of the rigorous model of reactive batch distillation based on the conditions given in Table 8.1. These data include the temperature measurements of the column, and the product compositions of distillate and reboiler recorded at every sampling period during the entire operation. The data of various sampling instants including a base case sampling period of 3.6 s is employed in these methods. In order to reflect the real situation, the temperature data of every sampling period is corrupted with random Gaussian noise of zero mean within the range of −0.5 to 0.5 having a standard deviation of 0.28°C.

8.5.1 Sensor configuration using sensitivity index

The basic description of SI is given in Section 6.3.1 of Chapter 6, Optimal Sensor Configuration Methods for State Estimation. The sensitivity gain matrix represents the SI, which measures the degree of sensitivity of each available measured variable of temperature with respect to changes in state variables defining the product compositions. The partial derivatives of each measured variable with respect to state variables defining the product compositions can be approximated in the form an instantaneous pseudo steady state SI calculated at different time intervals during the operation of reactive batch column and is expressed by a gain matrix K of the following form:

$$K(t) = \begin{bmatrix} \frac{\Delta T_1}{\Delta x_1} & \cdots & \frac{\Delta T_1}{\Delta x_i} & \cdots & \frac{\Delta T_1}{\Delta x_m} \\ \vdots & & \vdots & & \vdots \\ \frac{\Delta T_j}{\Delta x_1} & \cdots & \frac{\Delta T_j}{\Delta x_i} & \cdots & \frac{\Delta T_j}{\Delta x_m} \\ \vdots & & \vdots & & \vdots \\ \frac{\Delta T_n}{\Delta x_1} & \cdots & \frac{\Delta T_n}{\Delta x_i} & \cdots & \frac{\Delta T_n}{\Delta x_m} \end{bmatrix}^T \qquad (8.29)$$

The notation for the terms of this gain matrix is clearly explained in Section 7.5 of Chapter 7, Application of Mechanistic Model-Based Nonlinear Filtering and Observation Techniques for Optimal State Estimation in Multicomponent Batch Distillation. The pseudo steady state sensitivity matrix K of Eq. (8.29) is computed based on the temperature data of eight trays and the product compositions, that is, ethyl acetate and ethanol from reflux drum, and water and acetic acid from reboiler. The reason for using the product components for sensitivity studies is due to the difficulty in defining the sensitivity matrix based on the composition information of all trays, as such information cannot be easily measured. The sensitivity matrix K of Eq. (8.29) represents the SI by directly providing the instantaneous measurement information for state estimation. Each element k_{ij} of K matrix is a measure of sensitivity of the jth measured variable to the variation in the ith estimation variable. Tray numbers are considered from bottom to top with zero representing the reboiler and eight denoting the top tray. The magnitude of the elements of K matrix reflects the temperature sensitivities with respect to compositions at each sampling instant. At each time instant, all the elements of K in their absolute form are arranged in descending order along with their corresponding measurement locations. The measured variable having the largest value of k_{ij} could be considered as the most suitable input to the estimator. Similarly, the location of the second largest value of k_{ij} could be considered the second most appropriate input to the estimator, and so on. Sensitivity analysis based on K matrix is carried out using the data of different sampling instants, that is, 3.6, 18, 54, 108, 216, and 864 s, respectively. Fig. 8.2A–C shows the sensitive tray locations with respect to first, second, and third largest elements of K matrix when the data corresponding to a sampling period of 18 s is employed. The $m \times n$ sensitivity matrix, K can be determined from simulations based on the detailed first principle process model. The results of SI calculated for different sampling periods have shown that the locations of the sensitive trays along the column vary considerably during the operation. The sensitive region is initially located at the top of the column, suddenly drops to the bottom at $t \approx 2$ h and gradually shifts towards the top of the column before dropping again to the bottom at $t \approx 24$ h. This trend suggests that no particular region or any set of trays shows consistently high sensitivity for the entire duration, and all the column trays can be considered to be the measurement locations.

8.5.2   Sensor configuration using singular value decomposition

The basic description of SVD is given in Section 6.3.2 of Chapter 6, Optimal Sensor Configuration Methods for State Estimation. The properties of SVD are exploited to extract the sensitivity information from K matrix of Eq. (8.29) to identify the best set of temperature measurements for state estimation in reactive batch distillation. By using SVD, the sensitivity matrix K of Eq. (8.29) can decomposed into three unique component matrices as

$$K^T = U \sum V^T \qquad (8.30)$$

where K^T an $n \times m$ matrix; U is an $n \times n$ orthogonal matrix, the columns of which are called the left singular vectors; V is a $m \times m$ orthogonal matrix, the columns of which are called the right singular vectors and $\sum$ is an $n \times m$ diagonal matrix of scalar called the singular values which are organized in descending order such that $\sigma_1 \geq \sigma_2 \geq \ldots \sigma_m \geq 0$. The U and K matrices are both measures of sensor sensitivity. The U vector is interpreted as:

$$U = [U_1 \; U_2 \; \cdots \; U_n] \qquad (8.31)$$

The information contained in the sensitivity gain matrix, K of Eq. (8.29) and orthogonal matrix, U of Eq. (8.30) are analyzed to obtain the temperature sensitivities with respect to compositions. The columns in U matrix are denoted by the vectors, $U_1–U_9$. The elements in the vectors of U_1, U_2, U_3, etc., are arranged in descending order along with their measurement

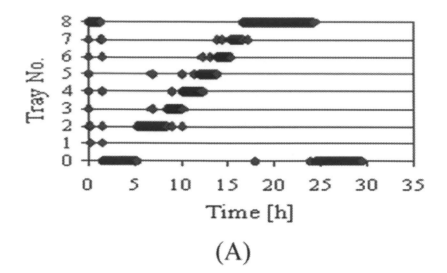

(A)

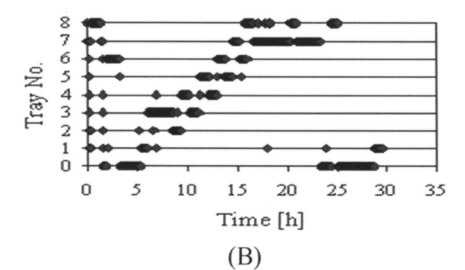

(B)

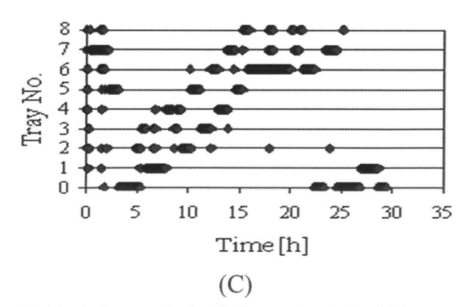

(C)

FIGURE 8.2 Tray sensitivity information for sensor configuration: (A)−(C) represent the results of sensitivity index corresponding to a sampling time of 18 s.

locations and analyzed with time. The largest element of each vector denotes the principal component which is drawn against time. The SVD analysis is carried out using the principal components of the vectors of U matrix for the data corresponding to different sampling instants, that is, 3.6, 18, 54, 108, 216, and 864 s, respectively. Fig. 8.3A–C indicates the sensitive tray locations corresponding to the principal components of U_1, U_2, and U_3 vectors when the data corresponding to a sampling period of 18 s is used. These results as well as the results of SVD corresponding to other sampling periods have shown that the location of the sensitive measurement point changes during the entire operation. The SVD results corresponding to the first sensitive measurement location indicate that the reboiler and top tray could be considered slightly more relevant, since they correspond to the most sensitive measurement point for a longer period of time compared to the other available locations. As for the second most sensitive measurement location is concerned, the SVD analysis suggests the suitability of the top two trays of the column to be the measurement locations for almost entire duration of operation. This result can be explained considering the fact that the products are collected from reflux drum, and therefore the temperature measurements obtained from the top two trays of the column are inherently very informative during the entire operation. The results concerning the third most sensitive measurement location indicate that it is difficult to interpret the sensitive tray locations. From these results, it can be concluded that at least one measurement sensor can be placed on any one of the top two trays and the remaining measurement sensors can be placed on any tray along the column.

8.5.3   Sensor configuration using principal component analysis

The basic description of PCA is given Section 6.3.3 of Chapter 6, Optimal Sensor Configuration Methods for State Estimation. The normalized gain matrix $\hat{K}$ in Eq. (6.5) of Chapter 6 is obtained by transforming the elements of the gain matrix K of Eq. (8.29) as per the normalization procedure given in Section 6.3.3. In the PCA analysis, the normalized gain matrix $\hat{K}$ is factored into two matrices:

$$\hat{K}(t) = TP^T \tag{8.32}$$

where $T(m \times s)$ is the score matrix and $P(n \times s)$ is the orthonormal loading matrix, whose rows are the s principal components. Thus the original process variable information is collectively expressed in the form of $\hat{K}$ matrix, which is of dimension (4×9) for the esterification reactive batch distillation column. In PCA approach, the decomposition of $\hat{K}$ matrix is performed in such a way that most of the original information is retained by a single principal component, which is represented by the first vector of the loading matrix, P. The elements of this loading vector are arranged in descending order along with their corresponding tray locations in order to evaluate the most sensitive measurement tray location and its change with time. Each loading of the PCA model represents the measure of sensitivity of each tray temperature. The absolute values of few of the loadings that are evaluated during the batch of reactive distillation show considerably higher magnitude than the others tentatively indicating the number of measurements to be considered for state estimation.

   The magnitude of the loadings also suggests the selection and location of sensor placement trays needed for composition estimation in reactive batch distillation. The results of Fig. 8.4 show the three most sensitive trays identified by the PCA model based on the first, second, and third largest elements of principal loading vector for the data corresponding to a sampling period of 18 s. The results of Fig. 8.4A show that at any given instant, the reboiler and top tray are the most sensitive trays to be considered for composition estimation. In contrast to the results of SI and SVD where the location of the first most sensitive tray varies along the column with time, the sensor location identified by the PCA is mostly confined to reboiler and/or top tray. The trays located in the central section of the column are not considered so important for sensor locations. The results of Fig. 8.4B show that the sensitive tray measurement location corresponding to the second largest element of the principal loading vector is mostly confined to lower or top region of the column as compared to the central region of the column. The results of Fig. 8.4C corresponding to the third most sensitive measurement location indicate that the sensor location is difficult to interpret. Similar trends are observed when the data corresponding to other sampling instants are treated for PCA transformation. These results show that the largest element of the principal loading vector provides better selection of sensitive trays. The analysis of the results thus show better identification of sensitive measurements by PCA model over SI and SVD. The PCA results are further confirmed by using the summated PC index (SUMPC) defined as follows.

   The overall optimal measurement selection for the state estimator is determined by calculating the SUMPC, for each measured variable:

$$\text{SUMPC}_j = \sum_{j=1}^{N_S} p_j(t) \tag{8.33}$$

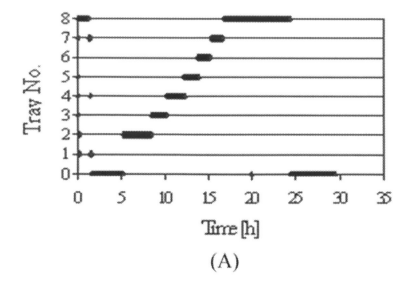

(A)

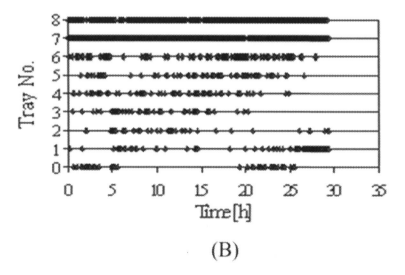

(B)

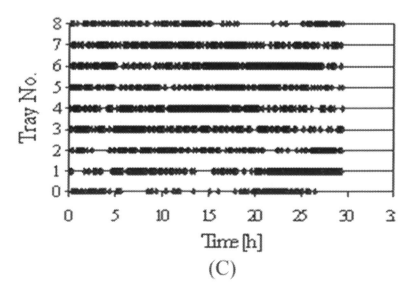

(C)

FIGURE 8.3 Tray sensitivity information for sensor configuration: (A)–(C) represent the results of singular value decomposition corresponding to a sampling time of 18 s.

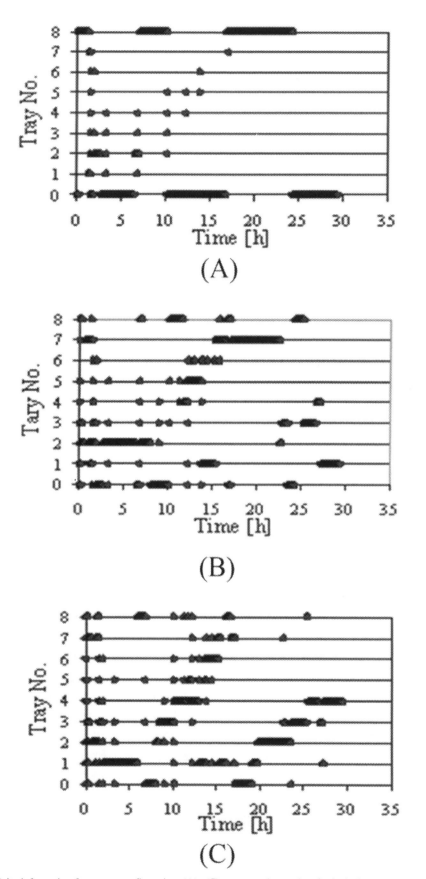

FIGURE 8.4 Tray sensitivity information for sensor configuration: (A)−(C) represent the results of principal component analysis corresponding to a sampling time of 18 s.

where $p_j(t)$ represents the value of the principal component obtained at time, t for the jth measured variable, and N_s indicates the total number of samples. The set of measured variables having the highest SUMPC values are considered as inputs to the state estimator. Fig. 8.5A—C shows the SUMPC results when the temperature data with normal and low measurement noise of different sampling instants are considered for PCA transformation. These results also ascertain the reboiler and top tray to be the most sensitive temperature measurement locations. If accurate composition estimation is not obtained by these two measurements, then the second tray can be considered as an additional temperature measurement. These results indicate the negligible influence of sampling period on the performance of PCA. The results of SUMPC are also evaluated by considering the measurement data of higher noise. Fig. 8.5D—F shows the SUMPC results when the temperature data corresponding to different sampling instants are corrupted by a random Gaussian noise of zero mean with a standard deviation of 0.5796°C. The measurement noise affects the outcome of PCA making it difficult to identify the most sensitive measurements when the temperature data of smaller sampling instants is considered. The adverse effect of measurement noise in sensor identification can be counteracted through appropriate adjustment of the sampling interval. From the results of Fig. 8.5D—F, it is observed that when noisy measurements are involved, the data corresponding to higher sampling instants provide better identification of sensitive trays. These results also indicate that when the measurements contain more noise, it is appropriate to select the data corresponding to larger sampling intervals for robust identification of sensitive trays by the PCA model. The results thus show the usefulness of PCA sensitivity analysis for measurement selection to the state estimator. The optimal set of temperature measurements selected by the PCA model can be used as inputs to the state estimator for composition estimation in reactive batch distillation.

8.6 Performance criteria

The performance indices, namely, capacity factor (CAP), batch time (BT), and mean integral squared error (MISE) are employed to assess the effect of estimator design parameters on the performance of the estimator. The CAP and BT used for reactive distillation are the same as defined for batch distillation in Section 7.6.

8.6.1 Mean integral squared error

The MISE is calculated using the estimated and actual product compositions. Actual compositions are defined as those obtained due to numerical simulation of the detailed dynamic model of reactive batch distillation. The estimated compositions are those predicted by the estimator. The computation of this measure is based on product compositions in distillate and reboiler. For the reactive distillation process of this study, this measure is expressed as

$$
\text{MISE} = \sqrt{\frac{\sum_{i=1}^{Ns} \left[(x_{D,3} - \hat{x}_{D,3})^2 + (x_{D,4} - \hat{x}_{D,4})^2 + (x_{B,4} - \hat{x}_{B,4})^2 \right]}{N_s}}
\tag{8.34}
$$

where $x_{D,3}$ $x_{D,4}$, and $x_{B,4}$ refer the actual product compositions of ethyl acetate and water in distillate, and water in reboiler, and $\hat{x}_{D,3}$, $\hat{x}_{D,4}$, and $\hat{x}_{B,4}$ refer their corresponding estimated product compositions. The N_s refers the number of samples.

8.7 Extended Kalman filter for compositions estimation

The general process representation of the nonlinear reactive batch distillation column with continuous-time nonlinear model and discrete time observation model for state estimation by EKF is given in Section 4.4.1 of Chapter 4. The recursive computation of EKF algorithm along with its implementation scheme is given in Section 4.4.3. EKF is computed in two steps. The first is a prediction step, which is used to extrapolate the previous best estimates, and the second is a correction step by which the updated estimates are formed. Since prediction is based on a process model, continuous prediction and discrete correction are employed in the estimation scheme. The mechanistic model-based composition estimation scheme for multicomponent batch distillation is shown in Fig. 8.6.

The EKF algorithm presented in the earlier section is designed and applied in conjunction with the optimal set of temperature measurements selected through the PCA model for inferential estimation of compositions in reactive batch distillation. An estimator design based on a rigorous mathematical model of multicomponent reactive batch distillation is not suitable for realistic situation, since it is difficult to obtain the measured values of liquid and vapor flow rates and tray holdups with time. Therefore a simplified model that assumes constant tray and reflux drum holdups, and constant vapor and liquid flow rates is considered to design the estimator. Wilson and Martinez [5] applied EKF for state estimation in reactive batch distillation. However, the authors have not studied the criteria of optimal measurement selection for state estimation. Zamprogna et al. [4] applied a state estimation scheme based on neural networks with optimal

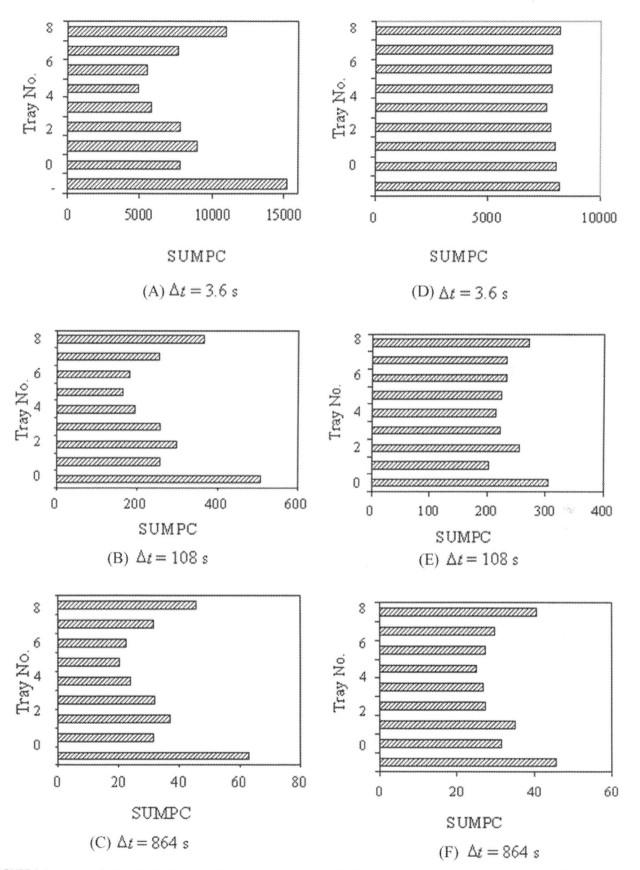

FIGURE 8.5 Results of summated PC index at different sampling instants: (A)–(C) correspond to normal measurement noise; (D)–(F) correspond to higher measurement noise.

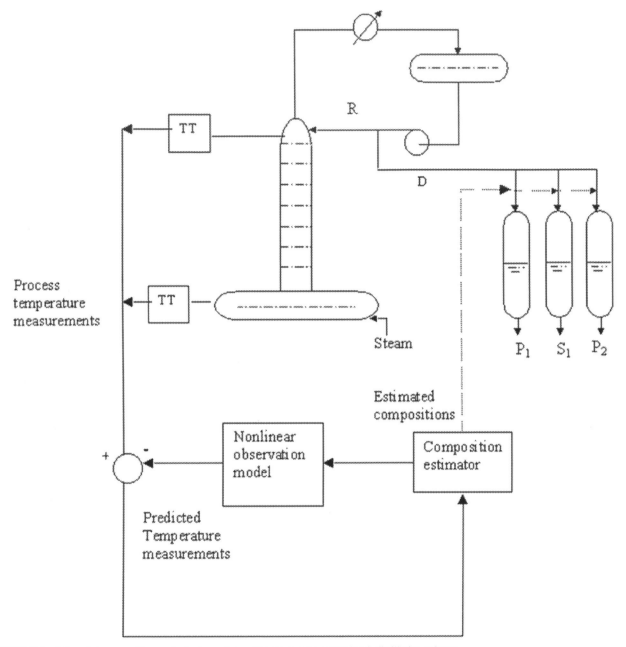

FIGURE 8.6 Inferential composition control scheme for multicomponent reactive batch distillation column.

selection of temperature measurements. However, their state estimation is concerned to batch distillation with no chemical reaction. In contrast to the above, this study explores the dual problem of optimal measurement selection and model-based state estimation for a multicomponent reactive batch distillation. The simplified model of reactive batch distillation that supports the estimator is given in Section 8.3.

8.7.1 Design of extended Kalman filter estimator

The state variables are the component concentrations in reboiler, on all trays and in reflux drum. The state vector for this case study of esterification reactive batch distillation is expressed by:

$$x = \begin{bmatrix} x_{B,2} \, x_{1,2} \ldots x_{8,2} \, x_{D,2} \, x_{B,3} \, x_{1,3} \ldots x_{8,3} \, x_{D,3} \, x_{B,4} \, x_{1,4} \ldots x_{8,4} \, x_{D,4} \end{bmatrix}^T \qquad (8.35)$$

This state vector corresponding to ethanol, ethyl acetate, and water compositions is denoted by the notation (2), (3), and (4), respectively, in Eq. (8.28). The difference in summated mole fractions of these components from unity gives acetic acid composition. The elements of the state transition matrix F involved in Eq. (4.26) of Chapter 4, are obtained by taking the partial derivatives of the component balances, Eqs. (8.23)–(8.27), with respect to each element of the state vector. The procedure for obtaining the partial derivatives is similar to that of the batch distillation given in Section 7.7.1 of Chapter 7.

The actual and prediction temperatures involved in the estimator are obtained through bubble point calculation procedure of vapor–liquid equilibrium relation. The temperature measurement on plate i is given by the expression:

$$T_i = \frac{a}{\left(b - \log \left\{ \dfrac{Ps_{i,j}}{\sum\limits_{j=1}^{Nc} Ps_{i,j} x_{i,j}} \right\} \right)} \tag{8.36}$$

where a and b are coefficients in vapor–liquid equilibrium relation (Table 8.1). The elements of the measurement Jacobean matrix, H [Eq. (4.31)] on the measurement plate are obtained by taking the partial derivatives of the measurement model, Eq. (8.36) with respect to each element of the state vector.

8.7.2   Extended Kalman filter implementation results

The EKF is designed and applied for composition estimation on all trays, reboiler, reflux drum, and products of ethyl acetate reactive batch distillation column. The simplified model that supports the estimator is used with a feed charge of 5000 moles, vapor boil up rate of 2500 moles h^{-1}, condenser holdup of 100 moles and tray holdup of 12.5 moles. The model equations are integrated by using Euler's method with a step size of 0.001 h. The temperature data of every sampling time is corrupted with a zero mean random Gaussian noise within the range of -0.5 to 0.5 having a standard deviation of 0.28°C. The temperature data with this noise is considered as the base data. The design parameters of EKF such as the initial state covariance matrix P_0, process noise covariance matrix, Q and observation noise covariance matrix, R are initially selected using the process and the measurement noise information and further tuned by computation trails such that the performance index defined by Eq. (8.34) is minimized. The diagonal elements of P_0 and Q are selected as 1×10^{-5}. The diagonal elements of R are defined as 5.0. The optimal measurement set suggested by the PCA model, that is, the reboiler and top tray temperatures are used as inputs to the EKF. In addition, various measurement combinations according to the ranking suggested by the PCA model are also considered as inputs to the EKF. The accuracy of the state estimator is assessed by using a MISE, Eq. (8.34).

To confirm the suitability of measurements selected by PCA, the state estimator is implemented by considering the same initial conditions for simulation and estimation, and evaluating the MISE values for the measurements selected according to PCA ranking. Thus, the EKF is implemented using different PCA-based measurement combinations with $x_{\text{initial}} = \hat{x}_{\text{initial}} = 0.45/0.45/0.0/0.10$, and $x_{\text{specified}} = 0.8/0.8/0.62$. The notation x_{initial} and $\hat{x}_{\text{initial}}$ refer to actual and estimator initial composition sequence of acetic acid, ethanol, ethyl acetate, and water, respectively. The notation $x_{\text{specified}}$ refers to the specified compositions of ethyl acetate and water in distillate, and water in reboiler. The MISE results shown in Fig. 8.7 support the PCA model selection by confirming the set of two temperature measurements, that is, the reboiler and top tray to be the most effective measurement set for composition estimation.

The MISE results of Fig. 8.7 evaluated using the measurement combinations based on PCA ranking indicate the unsatisfactory estimator performance for the case of a single measurement and for the cases of measurements four or more. The lower performance of estimator for the case of single measurement is due to the nonobservability condition of the process and its unsatisfactory performance in the case of four or more measurements is due to measurement redundancy. This indicates that better performance of the estimator can be realized by considering two or three measurement combinations. In order to further verify the suitability of PCA model-based measurement selection, various random combinations of two and three measurements with no PCA ranking are also subjected to estimator evaluation and the MISE results of these measurement combinations are shown in Table 8.2. The number zero in Table 8.2 corresponds to reboiler. The comparison of these results also indicate the better suitability of PCA model-based measurement selection for optimal state estimation.

In reactive batch distillation, the initial compositions of the reactants may vary from batch to batch. Therefore the estimator must be able to start with approximate initial conditions and converge to the actual

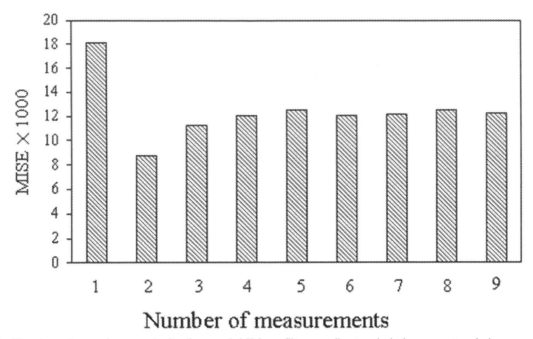

FIGURE 8.7   Mean integral squared error evaluation for extended Kalman filter according to principal component analysis measurement ranking configuration.

TABLE 8.2 Results of mean integral squared error with different measurement combinations.

Measurement combinations	Mean integral squared error (MISE)	Measurement combinations	MISE
0,8	0.00710	1,5	0.01594
1,2	0.01282	3,7	0.01687
1,4	0.01886	1,8	0.01621
1,7	0.01786	0,2,8	0.01197
2,4	0.01574	1,2,3	0.01551
2,6	0.01672	1,4,8	0.01452
2,7	0.−1798	1,3,6	0.01743
3,6	0.01870	0,3,6	0.01362
3,8	0.01237	5,7,8	0.01352
4,5	0.01576	3,5,7	0.01661
6,8	0.01068	2,5,6	0.01661
0,7	0.01193	0,4,6	0.01340
0,5	0.01596	4,6,8	0.01410

conditions in the column. The quantities and compositions predicted by the estimator can be used to start product and slop cut withdrawals while satisfying the specified conditions. Thus the performance of the estimator is studied for different initial mixture compositions and different product purity specifications. Fig. 8.8 show the responses of the actual compositions and the compositions estimated in the distillate and reboiler for $x_{\text{initial}} = 0.45/0.45/0.0/0.10$, $\hat{x}_{\text{initial}} = 0.41/0.41/0.02/0.16$ and $x_{\text{specified}} = 0.8/0.8/0.62$. The close agreement between the actual and estimated responses of Fig. 8.8 explains the state tracking ability of the model-based estimator. The state estimation scheme is also evaluated by applying it for recycled batches. The initial composition of the

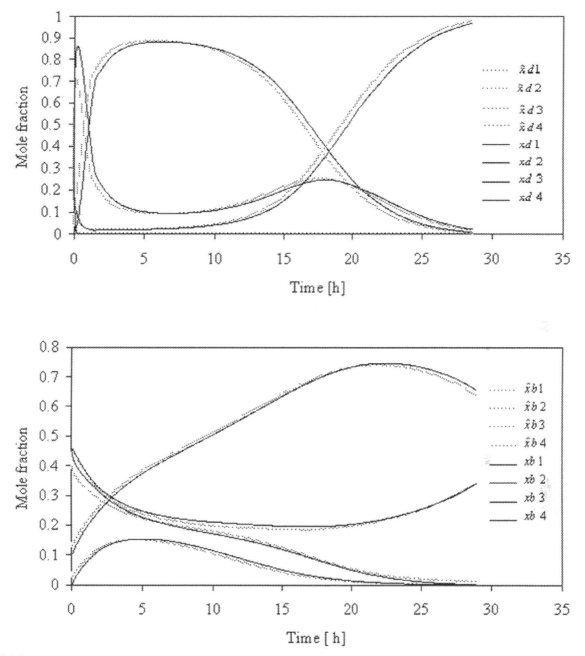

FIGURE 8.8 Actual compositions and compositions estimated by extended Kalman filter in the distillate and reboiler ($x_{\text{initial}} = 0.45/0.45/0.00/0.10$; $\hat{x}_{\text{initial}} = 0.41/0.41/0.02/0.16$).

recycled batch is obtained as follows. A fresh batch of 5000 moles with $x_{\text{initial}} = 0.45/0.45/0.0/0.10$, $\hat{x}_{\text{initial}} = 0.41/0.41/0.02/0.16$, and $x_{\text{specified}} = 0.8/0.8/0.62$ is subjected to initial total reflux operation and there upon to separate products and slop cuts so as to satisfy the desired specifications. This operation has resulted 1910 moles of ethyl acetate rich distillate product with $0.002/0.134/0.800/0.064$ as its composition and 845 moles of water rich distillate product with $0.0005/0.0540/0.0230/0.9225$ as its composition along with 847 moles of slop cut with $0.0013/0.2050/0.2624/0.5313$ as its composition. Also 1398 moles of water rich product with its composition as $0.3438/0.0014/0.0008/0.6540$ is retained in the reboiler. A case of recycled batch is then initialized by combining 847 moles of slop cut with $0.0013/0.2050/0.2624/0.5313$ as its composition and 4153 moles of fresh feed with $0.45/0.45/0.0/0.10$ as its composition to form 5000 moles of recycle mixture with its composition as $0.373/0.408/0.040/0.179$. The actual and estimated compositions for this recycled batch with $x_{\text{initial}} = 0.373/0.408/0.040/0.179$, $\hat{x}_{\text{initial}} = 0.40/0.43/0.02/0.15$ and $x_{\text{specified}} = 0.8/0.8/0.62$ are shown in Fig. 8.9. The

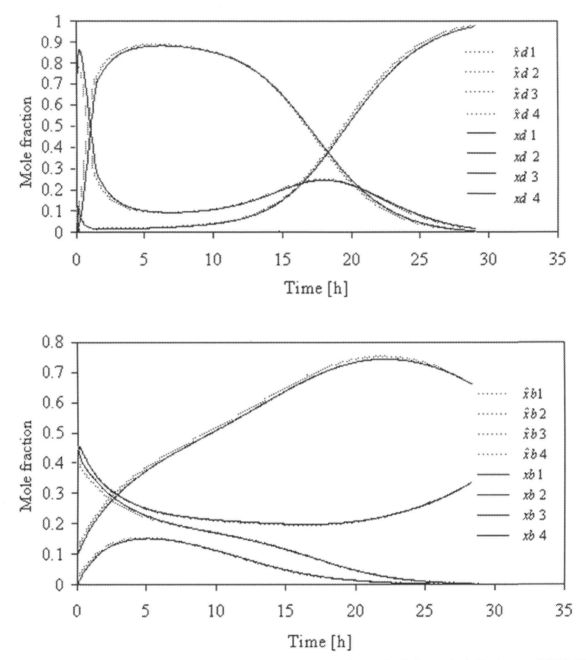

FIGURE 8.9 Actual compositions and compositions estimated by extended Kalman filter in the distillate and reboiler ($x_{\text{initial}} = 0.373/0.408/0.040/0.179$; $\hat{x}_{\text{initial}} = 0.40/0.43/0.02/0.15$).

results of this recycled batch also show good agreement between the estimated and actual compositions. These composition profiles correspond to the operation of the column that starts from the beginning of total reflux operation using specified vapor generation rate. Actually, the time taken to achieve the total reflux operation from a cold start would generate different composition profiles for the batch.

The sensitivity of the composition estimator is investigated in terms of MISE and BT for the effects of various levels of observation noise v, initial state covariance matrix P_0, process noise covariance matrix Q, and measurement noise covariance matrix R. These results are given in Table 8.3. The BT includes the time at total reflux and the time producing products and slop cuts, and it serves as a reference to compare the actual and estimator timings for the reactive batch distillation. These performance studies are evaluated by considering the initial feed compositions for the estimator and numerical simulation differently. The parameter values shown in the middle columns of

TABLE 8.3 Performance of composition estimator ($x_{initial} = 0.45/0.45/0.0/0.10$; $\hat{x}_{initial} = 0.41/0.41/0.04/0.14$; $x_{specified} = 0.80/0.80/0.62$; $N_t = 8$; $D = 125$).

(A) Effect of measurement noise

Performance criteria	Actual	Measurement noise, v	
		-0.5 to 0.5	-5 to 5
MISE		0.0071	0.0115
BT	29.860	29.030	29.350

(B) Effect of initial state covariance matrix

Performance criteria	Actual	Elements of initial state covariance matrix, P_0		
		1×10^{-7}	1×10^{-5}	1×10^{-3}
MISE		0.0069	0.0071	0.0075
BT	29.860	29.090	29.030	29.010

(C) Effect of process noise covariance matrix

Performance criteria	Actual	Elements of process noise covariance matrix, Q		
		1×10^{-7}	1×10^{-5}	1×10^{-3}
MISE		0.0065	0.0071	0.0084
BT	29.860	29.260	29.030	30.091

(D) Effect of measurement noise covariance matrix

Performance criteria	Actual	Elements of measurement noise covariance matrix, R		
		1	5	25
MISE		0.0072	0.0071	0.0089
BT	29.860	29.290	29.030	29.020

BT, batch time; MISE, mean integral squared error.

the subtables (B)−(D) of Table 8.3 correspond to those of base case design values. While studying the influence of one parameter on the estimator performance, the other parameters are considered to be those of base case design values only. Table 8.3(B) shows the performance results of the estimator when the measurement noise is increased by 10 times from the initially set range of −0.5 to 0.5. The results show marginal influence of this higher noise on the estimator BT and MISE. The results of BT and MISE in Table 8.3(B) show the negligible influence P_0 when it is decreased or increased by 100 times. The results of Table 8.3(C) show that increasing Q by 100 times has some reasonable influence on the estimator performance. The results of Table 8.2(D) show that increasing or decreasing the value of R by five times from base case has only a moderate influence on the estimator performance. These results thus show that even with considerable changes in measurement noise and filter design parameters as well as with different initial estimator compositions from the actual ones, the proposed inferential estimation scheme provides effective performance for state estimation in reactive batch distillation. The performance of the estimator is also evaluated by considering the measurements available at wider sampling instants. The discrete updating step of EKF has the flexibility to incorporate the measurements available at wider sampling instants while iterating the continuous prediction step with smaller integration time. Fig. 8.10 shows the responses of the estimated and actual compositions in distillate and reboiler for $x_{initial} = 0.40/0.40/0.05/0.15$ and $\hat{x}_{initial} = 0.36/0.44/0.02/0.18$ when the measurements of 0.005 h sampling time are corrupted by a random Guassian noise of five times higher than the normal noise. In this case, the product compositions specified for ethyl acetate in distillate and water in reboiler are 0.80 and 0.74, respectively. These results show the fast and effective convergence of the estimator even when the noisy measurements sampled at wider discrete time instants are considered for composition estimation. Thus the stochastic state estimator supported by a simplified nonlinear model that incorporate reaction kinetics and

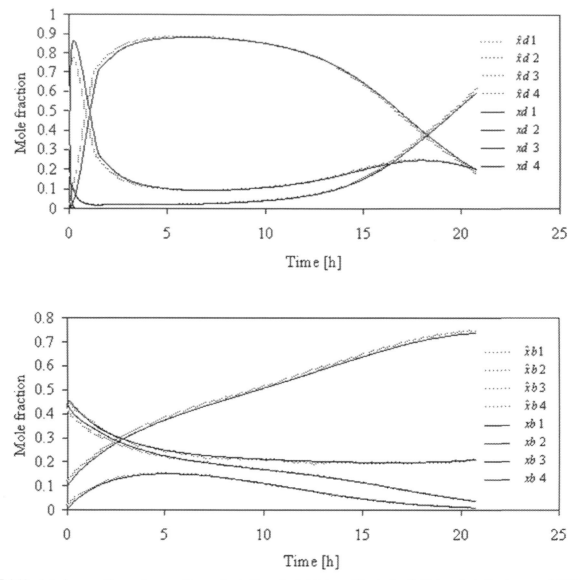

FIGURE 8.10 Actual compositions and compositions estimated by extended Kalman filter in the distillate and reboiler ($x_{initial}$ = 0.40/0.40/0.05/0.15; $\hat{x}_{initial}$ = 0.36/0.44/0.02/0.18).

vapor–liquid equilibrium along with the optimally selected measurements has shown reliable and robust performance for inferential estimation of compositions in multicomponent reactive batch distillation.

8.8 Summary

Optimal state estimation has a vital role in monitoring and operation of reactive batch distillation column. Reactive batch distillation is an intrinsically dynamic process associated with the complexities caused due to interactions between reaction kinetics, mass transfer, and thermodynamics. Optimal measurement selection and composition estimation are important tasks for effective monitoring and control of reactive batch distillation process. In this chapter, an inferential model-based estimation scheme based on EKF with optimal measurement selection is presented and applied for composition estimation in multicomponent reactive batch distillation. PCA is applied to configure the optimal set of temperature measurements required for composition estimation. The performance of the estimator is investigated toward the influence of various factors which include changes in initial batch compositions, slop cut recycle, measurement sampling instants, observation noise, and filter design parameters. The proposed model-based estimation scheme with optimal measurement selection has shown effective and robust performance for inferential estimation of compositions in reactive batch distillation column.

References

[1] E. Quintero-Marmol, W.L. Luyben, C. Georgakis, Application of an extended Luenberger observer to the control of multicomponent batch distillation, Ind. Eng. Chem. Res. 30 (1991) 1870–1880.

[2] R.M. Oisiovici, S.L. Cruz, State estimation of batch distillation columns using an extended Kalman filter, Chem. Eng. Sci. 55 (20) (2000) 4667–4680.

[3] C. Venkateswarlu, S. Avantika, Optimal state estimation of multicomponent batch distillation, Chem. Eng. Sci. 56 (2001) 5771–5786.

[4] E. Zamprogna, M. Barolo, D.E. Seborg, Estimating product composition profiles in batch distillation via partial least squares regression, Control. Eng. Pract. 12 (7) (2004) 917–929.

[5] J.A. Wilson, C.L.M. Martinez, State estimation in the control of integrated batch reaction with distillation process, Chem. Eng. Res. Design 75 (6) (1997) 603–608.

[6] A. Bahar, C. Özgen, State estimation and inferential control for a reactive batch distillation column, Eng. Appl. Artif. Intell. 23 (2) (2010) 262–270.

[7] S.M. Khazraee, A.H. Jahanmiri, Composition estimation of reactive batch distillation by using adaptive neuro-fuzzy inference system, Chin. J. Chem. Eng. 18 (4) (2010) 703–710.

[8] I.M. Mujtaba, S. Macchietto, Efficient optimization of batch distillation with chemical reaction using polynomial curve fitting techniques, Ind. Eng. Chem. Res. 36 (6) (1997) 2287–2295.

[9] J. Simandl, W.Y. Svrcek, Extension of the simultaneous solution and inside-outside algorithms to distillation with chemical reactions, Comput. Chem. Eng. 15 (5) (1991) 337–348.

[10] M.B. Bogacki, K. Alejski, J. Szymanowski, The fast method of the solution of a reacting distillation problem, Comput. Chem. Eng. 13 (9) (1989) 1081–1085.

[11] R.C. Reid, J.M. Prausnitz, T.K. Sherwood, The Properties of Gases and Liquids, third ed., McGraw-Hill, New York, NY, 1977.

[12] B. Joseph, C.B. Brosilow, Inferential control of processes. Part I: steady state analysis and design, AIChE J. 24 (3) (1978) 485–492.

[13] M. Kano, K. Miyazaki, S. Hasebe, I. Hashimoto, Inferential control system of distillation compositions using dynamic partial least squares regression, J. Process Control 10 (2–3) (2000) 157–166.

[14] C. Venkateswarlu, B. Jeevan Kumar, Composition estimation of multi-component reactive batch distillation with optimal sensor configuration, Chem. Eng. Sci. 61 (17) (2006) 5560–5574.

Chapter 9

Application of mechanistic model-based nonlinear filtering and observation techniques for optimal state estimation in complex nonlinear dynamical systems

9.1 Introduction

Nonlinear dynamical systems can exhibit different types of behavioral patterns depending on the values of the physical parameters and intrinsic features of the respective systems. These behavioral features are characterized by complex dynamic phenomena such as multiple steady state behavior, limit cycle oscillations, and chaos. The limit cycle oscillations can result when the steady state of the system loses its stability through a change in operating condition. These systems also exhibit complicated dynamic responses when the limit cycle becomes unstable. The exotic behavioral patterns displayed by the nonlinear systems may include multiperiodic and nonperiodic dynamic responses with irregular and chaotic phenomena. These unconventional dynamic phenomena exhibited by the chemical systems can be attributed to the nonlinear interaction between several quantities caused due to the perturbations in inputs and parameters that can be stored or interconnected within these systems.

Among the dynamical systems, continuous chemical reactors and continuous polymerization reactors have received special attention due to their complicated dynamics. Understanding the dynamic phenomena of these systems is important for analyzing and adapting conditions that aid in improving their operation. Several researchers have analyzed continuous stirred tank chemical reactors to characterize the phenomena of multiple steady states, oscillatory and nonlinear dynamics, periodic oscillations, limit cycles, bifurcations, and chaotic behaviour [1–11]. Several researchers have also investigated the existence of steady-state multiplicities, self sustained oscillations, oscillatory dynamics, bifurcation analysis, and chaos in continuous solution polymerization reactors [12–18].

Bifurcation and stability theories provide very useful conceptual and computational guidance in investigating complex dynamic phenomena and their evolution in chemical reaction systems. Period-doubling bifurcation, originally suggested is the most common bifurcation that causes chaos in chemical reactions. As the period-doubling continues, the period quickly approaches infinity leading to chaos. The unstable, oscillatory, and chaotic phenomenon displayed by the chemically reacting systems has desirable as well as undesirable features. The desirable feature of multiple steady state conditions is that one of the unstable steady states may correspond to a higher rate of reaction/yield, and this becomes the preferred state to operate on it to enhance the process performance. The desirable feature of chaos is that it enhances mixing and chemical reactions and provides a vibrant mechanism for transport of heat and mass. On the other hand, the intrinsic features of the reacting systems with the interactive influence of chemical or thermal energy may cause irregular dynamic behavior leading to degraded performance. In such situations, chaotic behaviour is considered undesirable and should be avoided. Chaotic processes show extreme sensitivity to initial conditions, and the process trajectories can diverge exponentially. However, chaos offers great flexibility to operate chemical systems because there are an infinite number of unstable periodic orbits (UPOs) embedded in a chaotic attractor, in which one can choose a specific UPO along with its time averaged prespecified performance defined in terms of conversion, yield, or selectivity, and the process can be stabilized to operate on the chosen UPO.

Successful control, and optimization of complex dynamic processes heavily rely on fast and accurately available real time process variable information. However, in most systems, the state variables desired by the online optimizers/

Optimal State Estimation for Process Monitoring, Fault Diagnosis and Control. DOI: https://doi.org/10.1016/B978-0-323-85878-6.00004-X

controllers cannot be easily available through measurement or available with large measurement delays. Such inaccessible or nonmeasurable state variables can be made available with the help of online state estimators. Online state estimation has several advantages for systems that are characterized by complicated dynamic behavior and the state estimator plays a crucial role in monitoring, optimization, and control of such processes. These estimated process states can serve as inferential measurements to the conventional/advanced controllers and online optimizers. The problem of model based state estimation for nonlinear systems with and without complicated dynamic phenomena has been extensively covered in literature [19−30].

The objective of this chapter is to present model based state estimators for nonlinear dynamical systems that exhibit complex dynamic phenomena such as multiple steady state behaviors, limit cycle oscillations, and chaos. In this chapter, an extended version of the Kalman filtering approach is used to design the model based state estimators for chemical systems that are associated with complex dynamic behavior so as to effectively capture their rapidly changing dynamics. The design and implementation of the state estimators are carried out by choosing two typical continuous nonlinear dynamical systems, a chemical reactor and homopolymerization reactor. The dynamic behavior of these complex dynamic systems is thoroughly analyzed by performing multiplicity, stability, and bifurcation studies prior to the design and implementation of state estimators.

9.2 Nonlinear dynamical CSTR

A nonisothermal jacketed continous stirred tank reactor (CSTR) with a first order, irreversible, exothermic, and consecutive reaction of the type $A \rightarrow B \rightarrow C$ is considered as an example of this study. The system exhibits diverse dynamic features such as multistationary, limit cycle oscillations, and even chaotic behaviour for certain parameter values [31].

9.2.1 Mathematical model

The nonlinear CSTR is described by the following dimensionless mass and energy balance equations [6]:

$$\frac{dx_1}{dt} = 1 - x_1 - Dax_1\exp\left[x_3/1 + \varepsilon_A x_3\right] + d_1 \tag{9.1}$$

$$\frac{dx_2}{dt} = -x_2 + Dax_1\exp\left[x_3/1 + \varepsilon_A x_3\right] - DaSx_2\exp\left[kx_3/1 + \varepsilon_A x_3\right] + d_2 \tag{9.2}$$

$$\frac{dx_3}{dt} = -x_3 + B\,Dax_1\exp\left[x_3/1 + \varepsilon_A x_3\right] - Da\,B\,\alpha\,Sx_2\exp\left[kx_3/1 + \varepsilon_A x_3\right] - \beta(x_3 - x_{3c}) + d_3 \tag{9.3}$$

The variables x_1 and x_2 denote the dimensionless concentrations of species A and B, respectively, and x_3 is the dimensionless temperature. The parameter x_{3c} represents the reactor coolant temperature. The load disturbances in feed compositions are denoted by d_1 and d_2, and the load disturbance in the reactor temperature is denoted by d_3. In the above equations, Da is Damkohler number, ε_A is activation number, S is the ratio of the rate constants for the series reaction, α is the ratio of heat effects for the series reaction, β is heat transfer coefficient and B is the dimensionless number signifying the adiabatic temperature rise. This reactor system exhibits multistationary behavior, oscillations, and chaos for the parameter values shown in Table 9.1.

TABLE 9.1 Characterization of steady states.

Set no.	Parameter values	x_1^s	x_2^s	x_3^s	Stability
I	$Da = 0.06$, $S = 0.0005$, $\varepsilon_A = 0$, $k = 1$, $\alpha = 0.426$, $\beta = 7.7$, $B = 55.0$	0.0378	0.9501	6.0500	Unstable
II	$Da = 0.26$, $S = 0.5$, $\varepsilon_A = 0$, $k = 1$, $\alpha = 0.426$, $\beta = 7.7$, $B = 57.77$	0.0729	0.1259	3.8900	Stable limit cycle
III	Same as Set II except $\beta = 7.9999$	0.0819	0.1391	3.7627	Chaotic

9.2.2 Stability analysis

The steady state equations of the system, Eqs. (9.1)–(9.3) with $\varepsilon_A = 0$ and $k = 1$ are given as

$$x_{1s} = \frac{1}{1 + Da\exp(x_{3s})} \tag{9.4}$$

$$x_{2s} = \frac{1 - x_{1s}}{1 + DaS\,\exp(x_{3s})} \tag{9.5}$$

$$x_{3s} = \frac{B(1 - x_{1s}) - \alpha B(1 - x_{1s} - x_{2s}) + \beta x_{3s}}{1 + \beta} \tag{9.6}$$

For fixed values of α, β, B, S, x_{3c}, the dependency of the steady state value, x_{3s}, on the Damkholer number, Da, will have certain possible shapes as shown in Fig. 9.1, where curve (A) shows the situation in which multiple steady states exists for some values of Da, while curve (B) shows the situation when only a single steady state may exist.

The Jacobian matrix for the system, Eqs. (9.1)–(9.3) is given by

$$A = \begin{bmatrix} a_{11} & a_{12} & a_{13} \\ a_{21} & a_{22} & a_{23} \\ a_{31} & a_{32} & a_{33} \end{bmatrix} \tag{9.7}$$

where a_{ij}'s are the elements obtained by taking the partial derivatives of the model, Eqs. (9.1)–(9.3), with respect to the state vector. The nature of the behavior of the system is determined from the roots of the characteristic equation,

$$|A - \lambda I| = -\lambda^3 + S_1\lambda^2 - S_2\lambda + S_3 = 0 \tag{9.8}$$

where $S_1 = a_{11} + a_{22} + a_{33}$; $S_2 = a_{11}a_{22} - a_{12}a_{21} + a_{11}a_{33} - a_{13}a_{31} + a_{22}a_{33} - a_{23}a_{32}$; $S_3 = \text{Det } A$.

S_1 and S_2 are the traces and the sum of the principle minor elements of the matrix. The Routh table for Eq. (9.8) is defined by

$$\begin{vmatrix} 1 & S_2 & 0 \\ -S_1 & -S_3 & 0 \\ \dfrac{(S_1 S_2 - S_3)}{S_1} & 0 & 0 \\ -S_3 & & \end{vmatrix} \tag{9.9}$$

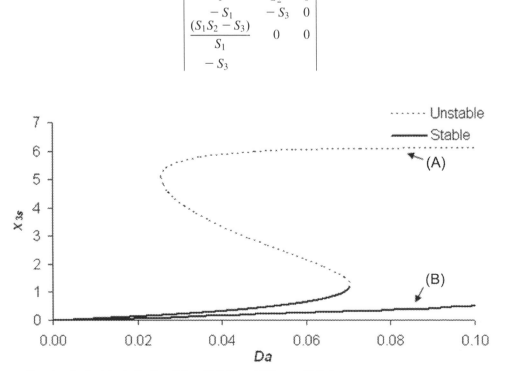

FIGURE 9.1   Locus of reactor steady states indicating: (A) multiplicity and (B) no multiplicity.

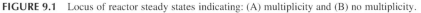

According to the Routh—Hurwitz stability criterion, the eigenvalues are the characteristic roots of Eq. (9.8) and should fall in the left half complex plane. The necessary and sufficient conditions for the steady-state to be locally asymptotically stable are:

$$
\begin{aligned}
&\text{(i) } S_1 = \text{tr } A < 0 \\
&\text{(ii) } \frac{(S_1 S_2 - S_3)}{S_1} > 0 \\
&\text{(iii) } S_3 = \text{Det } A < 0
\end{aligned}
\tag{9.10}
$$

The characteristics of the matrix A are intimately connected to the issue of steady state multiplicity. A unique steady state always satisfies the Condition (iii). If either of the S_1 or S_3 has the opposite sign, the steady state is unstable. When the eigenvalue is zero or purely imaginary, the condition for static bifurcation existing in multiple steady states is $S_3 = \text{Det } A = 0$. Fig. 9.2A and B represents the steady-state stability curves for the parameter values specified in Table 9.1 for Sets I, II, and III, respectively.

9.2.3 Bifurcation analysis

For complex systems, bifurcation arises due to perturbations in system parameters that change the stability characteristics of the steady-state solutions. Thus, stability and bifurcation impact each other. The bifurcation diagram for the system is built by varying the parameters, α, β, B, and S while keeping the others constant. These parameters completely define the state of the CSTR system. A bifurcation diagram for x_{3s} versus Da can be drawn as shown in Fig. 9.3 by choosing the process parameters as $S = 0.0005$, $\varepsilon_A = 0$, $k = 1$, $\alpha = 0.426$, $\beta = 7.7$ and allowing B to vary as $B = B$ $(1 + \Delta)$ with Δ as a perturbation in B. Here, B is the dimensionless number signifying the adiabatic temperature rise defined by $B = \frac{(-\Delta H_1) C_{Af} E_1}{\rho C_p R T_f^2}$.

The x_{3s} versus Da plot for $B = 55$ in Fig. 9.3A shows the presence of multiple steady states; however, on considering a perturbation in Δ such that $B = 30$, the x_{3s} versus Da plot in Fig. 9.3A shows the existence of a single steady state. The x_{3s} versus Da plot in Fig. 9.3B depicts how the steady state behavior of the system changes for different perturbations in B. These curves can be used to infer the influence of the parameters involved in the definition of B on the reaction system.

Similar observations can be made for the steady state behavior of the system with respect to the changes in α by defining it as $\alpha = \frac{-\Delta H_2}{-\Delta H_1}$ and drawing x_{3s} versus Da curves as shown in Fig. 9.4A. From these results, it is observed that the increase of α has caused a higher enthalpy change in $B \to C$ reaction, indicating that B is consumed faster than it is produced. As more heat is evolved in the second reaction, the conversion of B increases as temperature increases. Decrease of α is found to have the opposite effect. Similarly, x_{3s} versus Da curves shown in Fig. 9.4B can be drawn to depict the effect of β by defining it as $\beta = UA/\rho C_p V$. The increase of β is found to provide better heat transfer rates between the coolant and the reactor.

9.3 Optimal state estimation in nonlinear dynamical CSTR

The nonlinear dynamic model of the reactor system in its dimensionless form is used in conjunction with the temperature measurements to estimate the reactor species concentrations. An extended Kalman filter (EKF) is designed to obtain the instantaneous composition information from the temperature data of the system. The EKF algorithm, along with the general process representation, is given in Section 4.4 of Chapter 4, Mechanistic model-based nonlinear filtering and observation techniques for optimal state/parameter estimation. The algorithm given in Section 4.4.3 of Chapter 4 is implemented for nonlinear state estimation in the CSTR system. The schematic of inferential estimation for the reactor system is shown in Fig. 9.5.

9.3.1 State estimator design

The design of the state estimator involves the following components.

9.3.1.1 State vector

The chemical species concentrations and the temperature define the state vector as

$$
x = \begin{bmatrix} x_1 & x_2 & x_3 \end{bmatrix}^T
\tag{9.11}
$$

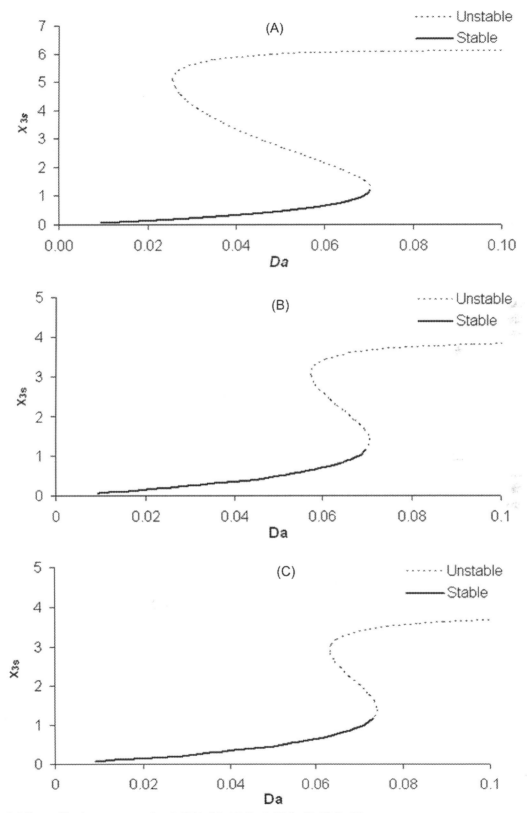

FIGURE 9.2 Stability profiles for parameter values in Table 9.1: (A) Set I; (B) Set II; (C) Set III.

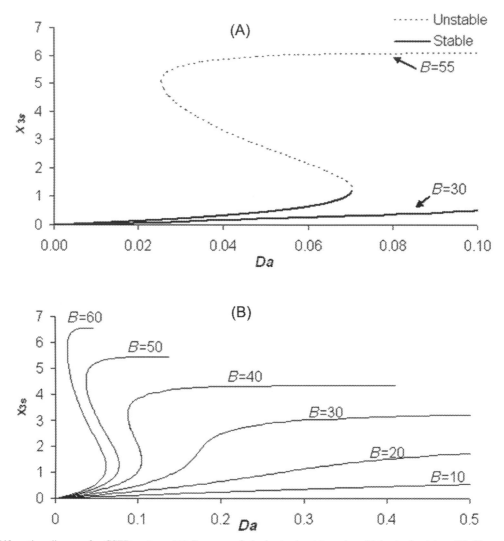

FIGURE 9.3 Bifurcation diagram for CSTR system: (A) Presence of single steady state and multiplr steady states; (B) Changes in steady state behavior of the system for different perturbations in adiabatic temperature.

9.3.1.2 State transition matrix

The elements f_{ij} of the state transition matrix, F (Eq. 4.27), are computed by taking the partial derivatives of $f(x)$ defined by Eqs. (9.1)−(9.3) with respect to the state vector:

$$F = \begin{bmatrix} f_{11} & f_{12} & f_{13} \\ f_{21} & f_{22} & f_{23} \\ f_{31} & f_{32} & f_{33} \end{bmatrix} \tag{9.12}$$

The elements of the state transition matrix in Eq. (9.12) are given as follows.

$$\begin{aligned} f_{11} &= -1 - Da\,\exp(x_3), \quad f_{12} = 0, \quad f_{13} = -Dax_1\,\exp(x_3), \quad f_{21} = Da\,\exp(x_3), \\ f_{22} &= -1 - DaS\,\exp(x_3), \quad f_{23} = Dax_1\,\exp(x_3) - DaS\,\exp(x_3), \quad f_{31} = BDa\,\exp(x_3), \\ f_{32} &= -DaB\alpha S\,\exp(x_3), \quad f_{33} = -1 + BDax_1\,\exp(x_3) - DaB\alpha Sx_2\,\exp(x_3) - \beta \end{aligned} \tag{9.13}$$

9.3.1.3 Measurement matrix

The measurement relation for temperature is

$$H = [0\ 0\ 1] \tag{9.14}$$

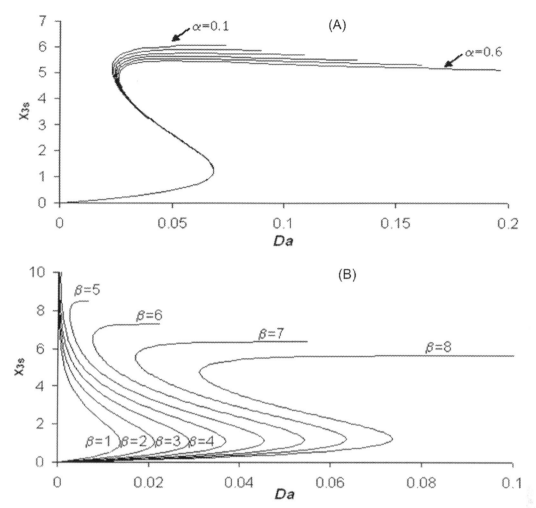

FIGURE 9.4   Steady state behavior of CSTR system: (A) Effect of α; (B) Effect of β.

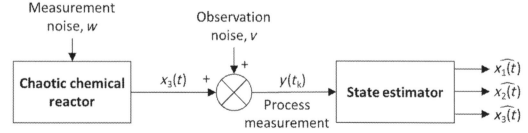

FIGURE 9.5   State estimation scheme for nonlinear dynamical CSTR system.

The temperature state equation, Eq. (9.3) in its discrete form is used as the nonlinear measurement equation, $h(x)$. The elements of the measurement transition matrix, H_x (Eq. 4.31), are computed by taking the partial derivatives of $h(x)$ with respect to the state vector

$$H_x = [h_{11} \quad h_{12} \quad h_{13}] \tag{9.15}$$

All these components are evaluated for chaotic reactor and used with the EKF estimator to obtain measured and unmeasured states of the reactor. The state estimator uses the temperature data of every sampling time as measurements and provides the estimates of temperature as well as reactor species concentrations.

9.3.2 State estimation results

The reactor system represented by Eqs. (9.1)–(9.3) exhibits multistationary behavior, oscillations, and chaos for parameter values given in Table 9.1. For parameter Set I, the system shows an unstable steady state. For parameter Sets II and III, the system exhibits limit cycle oscillations and chaotic behavior, respectively. The temperature data used for state estimation are obtained through numerical integration of model equations using Gear's method with a sampling time of 0.0001 units. The temperature measurement of every 1 s is considered for state estimation. The EKF is designed and applied in conjunction with the known temperature measurements to estimate the process states. These estimated states are the reacting species concentrations as well as temperature. The diagonal elements for each of the design parameters P_0 (initial state covariance matrix) and Q (process noise covariance matrix) of the EKF algorithm are selected as 0.000256, 0.000424, 0.53407, respectively, and the diagonal elements of the observation noise covariance are set to be 100. The performance of the EKF estimator is evaluated for different cases of process parameter values shown in Table 9.1. The results in Figs. 9.6 and 9.7 represent the actual and estimated concentration profiles, and the phase plane plots corresponding to the parameter Sets II and III of Table 9.1, respectively.

In order to represent the realistic situation, the temperature state is corrupted with random Gaussian noise having a mean zero and standard deviation of 0.25. Fig. 9.8 shows the actual and estimated concentration profiles corresponding to set III in the presence of stochastic noise in temperature obeying Gaussian distribution. In all the cases, the estimated states are in close resemblance with the actual states.

9.4 Nonlinear dynamical homopolymerization reactor

Continuous homopolymerization reactors represent one important class of dynamical systems. The homopolymerization reactions are characterized by high heat release, high viscosity, and poor heat transfer. Vinyl acetate homopolymerization has shown complex dynamic behavior for certain ranges of operating conditions due to the large heat of polymerization, gel effect, and large activation energy of the initiation step. These factors can readily produce phenomena such as multiple steady states, steady state instability, limit cycles, and chaotic behaviour [31].

The kinetic mechanism for the free-radical vinyl acetate homopolymerization reaction is as follows.

$$I \xrightarrow{K_d} 2R$$

FIGURE 9.6 Concentration profiles and phase plots for Set II: actual (x_1, x_2); estimated $(\hat{x}_1, \hat{x}_2)$.

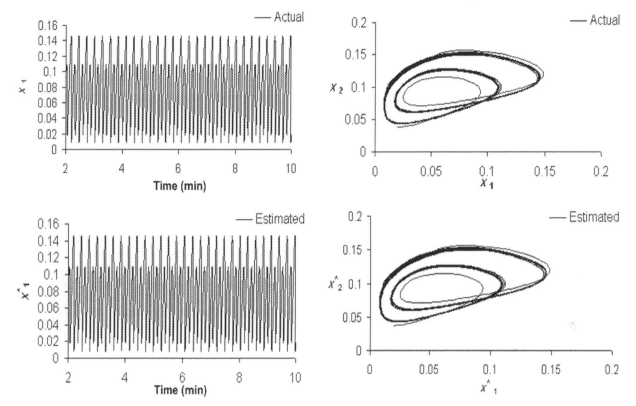

FIGURE 9.7 Concentration profiles and phase plots for Set III: actual (x_1, x_2); estimated $(\hat{x}_1, \hat{x}_2)$.

(Radical initiation)

$$R + M \xrightarrow{K_r} P_1$$

$$P_i + M \xrightarrow{K_p} P_{i+1} \text{ (Chain propagation)}$$

$$P_i + P_n \xrightarrow{K_t} \Lambda_{i+n} \text{ (Chain termination)}$$

9.4.1   Mathematical model

The mathematical model for homopolymerization reaction of vinyl acetate (VA) in tertiary butanol using Azobisisobutyronitrile (AIBN) as initiator is described by the following equations [32].

Monomer mass balance

$$\frac{dv_m}{d\tau} = \frac{\rho_{mf}v_{mf}}{\rho_m} - q_o q_i v_m - \frac{(MW)_m R_m \theta}{\rho_m} + v_m \rho_m \frac{dT}{d\tau}\frac{d(1/\rho_m)}{dT} \tag{9.16}$$

Solvent mass balance

$$\frac{dv_s}{d\tau} = \frac{\rho_{sf}v_{sf}}{\rho_s} - v_s q_o q_i + \rho_s v_s \frac{dT}{d\tau}\frac{d(1/\rho_s)}{dT} \tag{9.17}$$

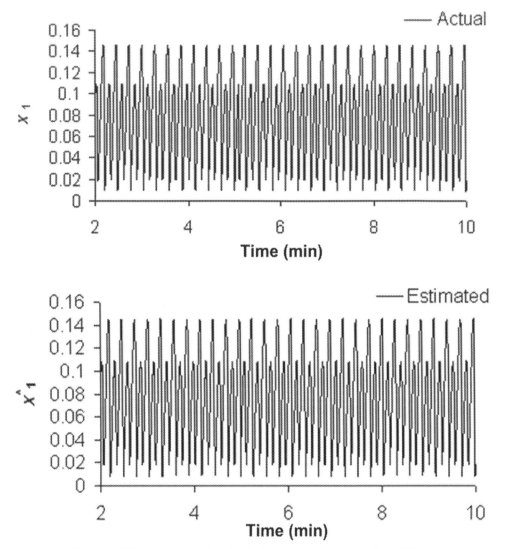

FIGURE 9.8 Concentration profile for Set III in the presence of stochastic disturbance: actual (x_1); estimated ($\hat{x}_1$).

Initiator mass balance

$$\frac{dc_i}{d\tau} = c_{if} - c_i(q_oq_i + k_d\theta) \tag{9.18}$$

Energy balance

$$\frac{dT}{d\tau} = \frac{\rho_f(T_f - T)}{\rho} + \frac{\theta}{\rho C_p}\frac{\Delta H}{(MW)_m}v_m\rho_mk_pP - \frac{\theta}{\rho C_p}\frac{UA(T - T_c)}{V} \tag{9.19}$$

In the above equations, q_oq_i is a term that considers changes in the density of the reactive mixture and is defined by

$$q_oq_i = \frac{\rho_{mf}v_{mf}}{\rho_m} + \frac{\rho_{sf}v_{sf}}{\rho_s} + \theta(MW)_m\rho_mR_m\left(\frac{\rho_m}{\rho_p} - 1\right) + \frac{dT}{d\tau}\left(\rho_sv_s\frac{d(1/\rho_s)}{dT} + \rho_mv_m\frac{d(1/\rho_m)}{dT} + \rho_pv_p\frac{d(1/\rho_p)}{dT}\right) \tag{9.20}$$

The overall density and specific heat of the mixture are

$$\begin{aligned}\rho &= \rho_mv_m + \rho_sv_s + \rho_pv_p\\C_p &= y_mC_{pm} + y_sC_{ps} + y_pC_{pp}\end{aligned} \tag{9.21}$$

Under quasi steady state assumption for the free-radical species (R and P), the normalized rate of monomer reaction is expressed as

$$Rate_m = k_p P + R_{ckr} \tag{9.22}$$

where concentration of species P can be written as

$$P = \sqrt{\frac{R_{ckr} M}{k_t}} \tag{9.23}$$

The R_{ckr} is the normalized rate of initiation as given by

$$R_{ckr} = \frac{2 f k_d c_i}{M} \tag{9.24}$$

By substituting the value of R_{ckr} from Eq. (9.24) in Eq. (9.22)

$$\begin{aligned} Rate_m &= k_p P + \frac{2 f k_d c_i}{M} \\ &= \frac{R_m (MW)_m}{\rho_m v_m} \end{aligned} \tag{9.25}$$

where R_m is the rate of monomer consumption expressed by

$$R_m = \frac{k_p P \rho_m v_m}{(MW)_m} + 2 f k_d c_i \tag{9.26}$$

The termination and propagation constants $(k_t,\ k_p)$, used in the above equations include gel effect with the form

$$\begin{aligned} k_t &= k_t^0 g_t \\ k_p &= k_p^0 g_p \end{aligned} \tag{9.27}$$

The gel effect correlations are

$$\begin{aligned} g_p &= 1 \\ g_t &= \exp(-0.4407 x_t - 6.753 x_t^2 - 0.3495 x_t^3) \end{aligned} \tag{9.28}$$

In the above equations, the k_t^0 and k_p^0 are the kinetic constants at zero polymer concentration, and g_t and g_p are the gel effect correlations that take into account the effects of increasing polymer concentration. $-\Delta H_R$ is heat of polymerization (cal gmol^{-1}), k_d is rate constant for initiator decomposition (l s^{-1}), k_p is rate constant for propagation (l gmol^{-1} s^{-1}), k_t is rate constant for termination (l gmol^{-1} s^{-1}), $(MW)_m$ is molecular weight of monomer (g gmol^{-1}), P is live radical concentration (gmol L^{-1}), $q_o q_i$ is ratio of outlet to inlet volumetric flow rate, R_m is rate of consumption of monomer (gmol L^{-1} min^{-1}), v_m is volume fraction of monomer in reactor, v_{mf} is volume fraction of monomer in feed, x_t is mole fraction of polymer and y_i refers weight fraction of respective components (monomer, solvent, and polymer). The other parameters involved in simulation are shown in Table 9.2.

9.4.2 Stability analysis

The steady state model of the system can be obtained from Eqs. (9.16)–(9.19) as follows.

Monomer mass balance

$$\frac{\rho_{mf} v_{mf}}{\rho_m} - q_o q_i v_m - \frac{(MW)_m R_m \theta}{\rho_m} = 0 \tag{9.29}$$

Solvent mass balance

$$\frac{\rho_{sf} v_{sf}}{\rho_s} - v_s q_o q_i = 0 \tag{9.30}$$

TABLE 9.2 Parameters for the homopolymerization of VA.

$\rho_m(T) = 958.4 - 1.3276(T - 273)$ (g L^{-1})

$\rho_s(T) = \left(\frac{74120}{(60.21 + 0.116\,T)}\right)$ (g L^{-1})

$\rho_p(T) = 1211 - 0.8496(T - 273)$ (g L^{-1})

$C_{pm} = 0.470$ (cal g^{-1} K^{-1})

$C_{pp} = 0.3453 + 9.55 \times 10^{-4}(T - 298)$ (cal g^{-1} K^{-1})

$C_{ps} = 0.716$ (cal g^{-1} K^{-1})

$k_p^0 = 82.212 \times 10^8 e^{-\frac{6100}{RT}}$ (L gmol^{-1} s^{-1})

$k_d = 94.8 \times 10^{15} e^{-\frac{30800}{RT}}$ (L gmol^{-1} s^{-1})

$k_t^0 = 469.392 \times 10^{10} e^{-\frac{2462}{RT}}$ (L gmol^{-1} s^{-1})

$\Delta H = 21000$ (cal gmol^{-1})

$UA = 12$ (cal min^{-1} K^{-1})

$V = 500$ (mL)

$f = 0.8$

Initiator mass balance

$$c_{if} - c_i(q_o q_i + k_d \theta) = 0 \tag{9.31}$$

Energy balance

$$\frac{\rho_f(T_f - T)}{\rho} + \frac{\theta}{\rho C_p}\frac{\Delta H}{(MW)_m} v_m \rho_m k_p P - \frac{\theta}{\rho C_p}\frac{UA(T - T_c)}{V} = 0 \tag{9.32}$$

In the above equations, the term $q_0 q_i$ is further defined by

$$q_0 q_i = \frac{\rho_{mf} v_{mf}}{\rho_m} + \frac{\rho_{sf} v_{sf}}{\rho_s} + \theta(MW)_m \rho_m R_m \left(\frac{\rho_m}{\rho_p} - 1\right) \tag{9.33}$$

9.4.3   Steady state solution

The steady state solution of the homopolymerization reactor model is not as simple as the chemical reactor system. Since a direct solution is difficult due to the interdependency of variables, an indirect approach is employed to obtain the steady solution for the variables v_s, v_m, c_i, and T, as illustrated below.

Step 1: An initial value of temperature is guessed, and the temperature dependent parameters in Table 9.2 are solved.

Step 2: An initial value of v_s is assumed.

Step 3: Using the values in Steps 1 and 2, $q_o q_i$ is found from Eq. (9.30):

$$q_0 q_i = \frac{\rho_{sf} v_{sf}}{\rho_s v_s} \tag{9.34}$$

On rearranging Eq. (9.29),

$$\frac{\rho_{mf} v_{mf}}{\rho_m} = q_o q_i v_m + \frac{\theta(MW)_m R_m}{\rho_m} \tag{9.35}$$

On substituting q_o, q_i and $\frac{\rho_{mf} v_{mf}}{\rho_m}$ in Eq. (9.33) and rearranging, we get

$$\theta = \frac{q_o q_i v_p \rho_p}{(MW)_m R_m} \tag{9.36}$$

Substitute θ and $q_o q_i$ in Eq. (9.33) to obtain v_p as

$$v_p = \frac{\left(q_0 q_i - \frac{\rho_{mf} v_{mf}}{\rho_m} - \frac{\rho_{sf} v_{sf}}{\rho_s}\right)}{q_0 q_i \rho_p \left(\frac{1}{\rho_p} - \frac{1}{\rho_m}\right)} \tag{9.37}$$

From monomer mass balance, Eq. (9.29) we get

$$\frac{\left(\frac{\rho_{mf} v_{mf}}{\rho_m} - q_0 q_i v_m\right)\rho_m}{(MW)_m} = \theta R_m \tag{9.38}$$

From Eqs. (9.23), (9.24), and (9.26), we have

$$R_m = \frac{k_p P \rho_m v_m}{(MW)_m} + P^2 k_t \tag{9.39}$$

Eq. (9.32) can be put in a simplified notation:

$$G_1 + G_2 \theta P - G_3 \theta = 0 \tag{9.40}$$

where

$$G_1 = \rho_f (T_f - T) C_p; \quad G_2 = \frac{\Delta H v_m k_p \rho_m}{(MW)_m}; \quad G_3 = UAV(T - T_c)$$

From Eq. (9.40), we can evaluate the residence time (θ) as

$$\theta = \frac{G_1}{G_3 - G_2 P} \tag{9.41}$$

Substituting Eqs. (9.39) and (9.41) in Eq. (9.38) and rearranging, we get

$$P^2 G_1 k_t + P\left(\frac{G_1 k_p \rho_m v_m}{(MW)_m} + G_2 \theta R_m\right) - G_3 \theta R_m = 0 \tag{9.42}$$

Eq. (9.42) is a quadratic expression where P is the molar concentration of free radical having two roots. By choosing the positive roots of P, the values of c_i and θ can be evaluated from Eqs. (9.23), (9.24), and (9.41). The c_i can also be evaluated from the initial mass balance steady-state Eq. (9.31) and compared with that evaluated from Eqs. (9.23) and (9.24). If both these c_i's are equal, then the corresponding temperature and volume fraction of the solvent (v_s) are chosen, and the volume fraction of monomer v_m is computed from the relation $v_m = 1 - (v_s + v_p)$. Thus, the values of θ and T can be established. If both the c_i's are not equal, then the procedure is repeated from Step 2 with a small increment in v_s.

Step 4: The procedure is repeated from Step 1 for all the temperatures in the given range with incremental guesses in temperature.

The temperature versus residence time curve in Fig. 9.9, which is drawn for the fixed set of operating conditions, $T_c=318$K, $T_f=315$K, $v_{mf}=0.3$, and $c_{if}=0.03203$ gmol L^{-1}, shows the multiple steady-state behavior of the system. To determine the stability at each point of the curve, the local stability of the system can be verified by forming the Jacobian matrix and evaluating the stability of the system based on the Routh–Hurwitz stability criterion.

The Jacobian matrix, A for this system of equations is represented as:

$$A = \begin{bmatrix} a_{11} & a_{12} & a_{13} & a_{14} \\ a_{21} & a_{22} & a_{23} & a_{24} \\ a_{31} & a_{32} & a_{33} & a_{34} \\ a_{41} & a_{42} & a_{43} & a_{44} \end{bmatrix} \tag{9.43}$$

where a_{ij}'s are the elements obtained by taking the partial derivatives of the model Eqs. (9.16)–(9.19), with respect to state vector. The nature of the behavior of the system is determined from the roots of the fourth-order characteristic equation,

$$|A - \lambda I| = \lambda^4 - S_1 \lambda^3 + S_2 \lambda^2 - S_3 \lambda + S_4 = 0 \tag{9.44}$$

For the sake of notation simplicity, the elements of A are denoted as:

$$A = \begin{bmatrix} a & b & c & d \\ e & f & g & h \\ i & j & k & l \\ m & n & o & p \end{bmatrix} \tag{9.45}$$

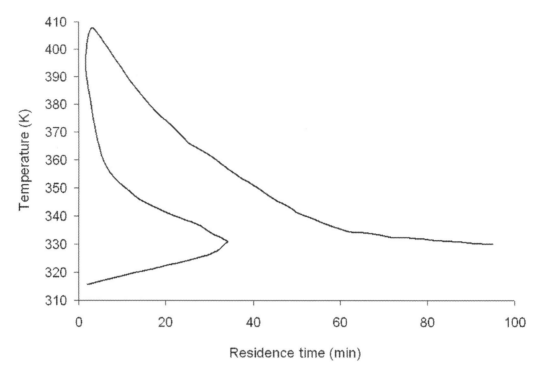

FIGURE 9.9   Multiplicity curve for homopolymerization reaction.

The characteristic equation has the form:

$$|A - \lambda I| =$$
$$\lambda^4 - \lambda^3(a + k + p + f) + \lambda^2(ak + ap + af + kp + fk + fp - ol - be - ci - dm - gj - hn) +$$
$$\lambda(- akp - afk - afp + aol - fkp + fol + bek + bep - bgi - bhm - cej + cfi + cip - cml - den + dfm - dio + dmk + agj + ahn + gjp - gnl - hjo + hnk) +$$
$$(afkp - afol - bekp + beol + bgip - bglm - bhio + bhmk + cejp - cenl - cfip + cfml + chin - chjm - dejo + denk + dfio - dfmk - dgin + dgmj - agjp + agnl + ahjo - ahnk) = 0$$

(9.46)

Eq. (9.46) is in the form of Eq. (9.44). The Routh table for Eq. (9.44) is defined by

$$\begin{vmatrix} 1 & S_2 & S_4 & 0 \\ -S_1 & -S_3 & 0 & 0 \\ C & S_4 & 0 & 0 \\ D & 0 & 0 & 0 \\ S_4 \end{vmatrix}$$

(9.47)

where S_i's are the sums of the principal minors of A. It can be observed that S_1 represents the trace of A and S_4 denotes the determinant of A. In the above table,

$$C = \frac{(S_1 S_2 - S_3)}{S_1} \quad \text{and} \quad D = \frac{C(- S_3) + S_1 S_4}{C}$$

(9.48)

According to the Routh−Hurwitz stability criterion, the conditions for the system to be stable are:

(i) $S_1 = \operatorname{tr} A < 0$

(ii) $\dfrac{(S_1 S_2 - S_3)}{S_1} > 0$

(iii) $\dfrac{(S_1 S_2 - S_3)}{S_1}(- S_3) + S_1 S_4 > 0$

(iv) $S_4 = \operatorname{Det} A > 0$

(9.49)

Conditions (ii) and (iii) are satisfied if and only if $S_2 > 0$ and $S_3 < 0$. The characteristics of the matrix A are intimately connected with the question of steady-state multiplicity. A unique steady-state always satisfies the Condition (iv). If either of the S_1 or S_4 has the opposite sign, the steady-state is unstable. When the eigenvalue is zero or purely imaginary, the condition for static bifurcation existing in multiple steady states is $S_4 = \mathrm{Det}\, A = 0$.

9.4.4 Bifurcation analysis

For this fourth-order system, the conditions for Hopf bifurcation to limit cycles at imaginary eigenvalues ($\pm i\omega$) can also be determined from the Routh–Hurwitz criterion when the Condition (iii) is equal to zero, that is,

$$S_1 S_2 S_3 - (S_3)^2 - S_1{}^2 S_4 = 0 \qquad (9.50)$$

Hopf bifurcation points are those points where a pair of complex eigenvalues cross the imaginary axis, changing the stability characteristics of the steady-state solutions. Periodic solutions are expected to evolve from these special bifurcation points. For a particular set of operating conditions, $T_c{=}318\mathrm{K}$, $T_f{=}315\mathrm{K}$, $v_{mf}{=}0.3$, and $c_{if}{=}0.03203$ gmol L^{-1}, the bifurcation diagram in Fig. 9.10 shows the multiple steady states, limit cycle, and bifurcation points as well as stable and unstable zones that are generated with respect to the variation in residence time. For the same set of conditions, the transition from stable to unstable zone (beginning at $\theta = 24$) occurs causing oscillatory behavior that eventually leads to period-doubling bifurcation cascade and chaos as depicted in Fig. 9.11. The magnification results of Fig. 9.11A are further explained in Fig. 9.11B and C, which indicates the sequence of periodic windows of periods 2, 4, 8 and so on leading to a chaotic region. It can be observed that at residence time $\theta = 27.0$ min, the system exhibits limit cycles with sustained period oscillations. The sustained oscillatory behavior of the response in the time domain is shown in Fig. 9.12A, and the corresponding phase plane plot is shown in Fig. 9.12B. The dynamic characteristics of the responses are further explained along with the state estimation results in the later sections.

9.5 Optimal state estimation in nonlinear dynamical homopolymerization reactor

The nonlinear dynamic model of the homopolymerization reactor is used in conjunction with the temperature measurements to estimate the reactor species concentrations. An EKF is designed to obtain the instantaneous composition information from the temperature data of the system. The EKF algorithm and the general process representation are given in

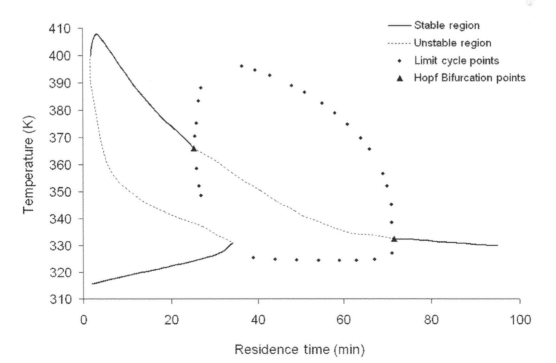

FIGURE 9.10 Period-doubling bifurcation diagram.

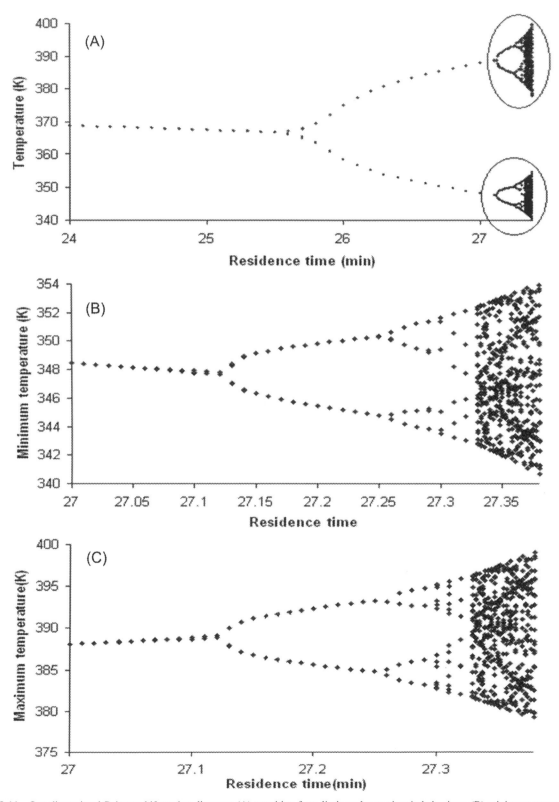

FIGURE 9.11 One dimensional Poincare bifurcation diagram: (A) transition from limit cycles to chaotic behaviour; (B) minimum temperature; (C) maximum temperature.

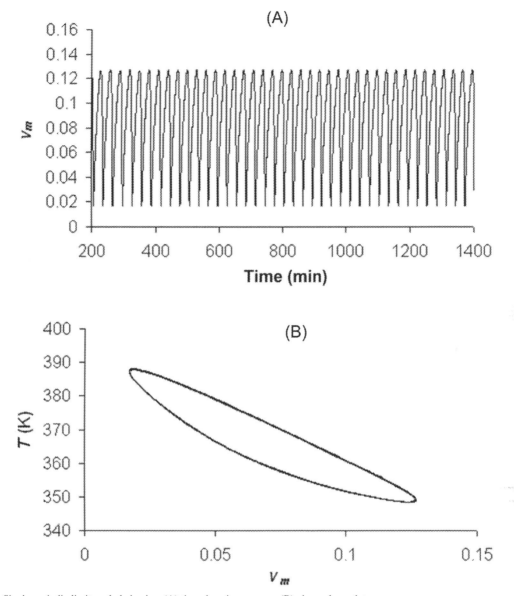

FIGURE 9.12   Single periodic limit cycle behavior: (A) time-domain response; (B) phase plane plot.

Section 4.4 of Chapter 4, Mechanistic model-based nonlinear filtering and observation techniques for optimal state/parameter estimation. The algorithm given in Section 4.4.3 is implemented for nonlinear state estimation in a homopolymerization reactor.

9.5.1   State estimator design

The nonlinear dynamic model of the VA homopolymerization system is used in conjunction with the temperature measurements to estimate the volume fraction of monomer, solvent, the concentration of initiator, and reactor temperature. The design of a soft sensor involves the following components.

9.5.1.1   State vector

The volume fraction of monomer, solvent, the concentration of initiator, and reactor temperature define the state vector as

$$x = \begin{bmatrix} v_m & v_s & c_i & T \end{bmatrix}^T \tag{9.51}$$

9.5.1.2 State transition matrix

The elements f_{ij} of the state transition matrix, F in Eq. (4.27), are computed by taking the partial derivatives of $f(x)$ defined by Eqs. (9.16)−(9.19) with respect to the state vector:

$$F = \begin{bmatrix} f_{11} & f_{12} & f_{13} & f_{14} \\ f_{21} & f_{22} & f_{23} & f_{24} \\ f_{31} & f_{32} & f_{33} & f_{34} \\ f_{41} & f_{42} & f_{43} & f_{44} \end{bmatrix} \tag{9.52}$$

The elements of the state transition matrix are given as follows:

$$f_{11} = -\frac{\left(\frac{d(q_o q_i)}{dv_m} v_m + (q_o q_i)\right)}{\theta} - \frac{MWm}{\rho_m}\frac{dR_m}{dv_m} + \frac{1}{\rho_m}\left(-\frac{d\rho_m}{dT}\right)\left(v_m \frac{d}{dv_m}\left(\frac{dT}{dt}\right) + \frac{dT}{dt}\right)$$

$$f_{12} = -\frac{v_m}{\theta}\frac{d(q_o q_i)}{dv_s} + \frac{\left(-\frac{d\rho_m}{dT}\right)}{\rho_m}\left(v_m \frac{d}{dv_s}\left(\frac{dT}{dt}\right)\right),$$

$$f_{13} = -\frac{v_m}{\theta}\frac{d(q_o q_i)}{dc_i} + \frac{\left(-\frac{d\rho_m}{dT}\right)}{\rho_m}\left(v_m \frac{d}{dc_i}\left(\frac{dT}{dt}\right)\right) - \frac{MWm}{\rho_m}\frac{dR_m}{dc_i}$$

$$f_{14} = -\frac{\varphi_m}{\theta}\frac{\rho_{mf}}{\rho_m^2}\frac{d\rho_m}{dT} - \frac{v_m}{\theta}\frac{d(q_o q_i)}{dT} + \frac{\left(-\frac{d\rho_m}{dT}\right)}{\rho_m}\frac{d}{dT}\left(\frac{1}{\rho_m}\frac{dT}{dt}\right)v_m - \frac{MWm}{\rho_m}\frac{d}{dv_m}\left(\frac{R_m}{\rho_m}\right)$$

$$f_{21} = -\frac{v_s}{\theta}\frac{d(q_o q_i)}{dv_m} - \frac{v_s}{\rho_s}\frac{d\rho_s}{dT}\frac{d}{dv_m}\left(\frac{dT}{dt}\right), \quad f_{22} = -\frac{\left(\frac{d(q_o q_i)}{dv_s} v_s + (q_o q_i)\right)}{\theta} - \frac{\frac{d\rho_s}{dT}}{\rho_s}\left(v_s \frac{d}{dv_s}\left(\frac{dT}{dt}\right) + \frac{dT}{dt}\right)$$

$$f_{23} = -\frac{v_s}{\theta}\frac{d(q_o q_i)}{dc_i} - \frac{v_s}{\rho_s}\frac{d\rho_s}{dT}\frac{d}{dc_i}\left[\frac{dT}{dt}\right],$$

$$f_{24} = -\frac{\varphi_s}{\theta}\frac{\rho_{sf}}{\rho_s^2}\frac{d\rho_s}{dT} - \frac{v_s}{\theta}\frac{d(q_o q_i)}{dT} - v_s\frac{d}{dT}\left(\frac{1}{\rho_s}\frac{d\rho_s}{dT}\frac{dT}{dt}\right),$$

$$f_{31} = -\frac{c_i}{\theta}\frac{d(q_o q_i)}{dv_m}, \quad f_{32} = -\frac{c_i}{\theta}\frac{d(q_o q_i)}{dv_s}, \quad f_{33} = \frac{-1}{\theta}\left(\frac{d(q_o q_i)}{dc_i}c_i + (q_o q_i)\right) - k_d,$$

$$f_{34} = -\frac{c_i}{\theta}\frac{d(q_o q_i)}{dT} - c_i\frac{dk_d}{dT}, \quad f_{41} = \frac{d}{dv_m}(dT/dt), \quad f_{42} = \frac{d}{dv_s}(dT/dt), \quad f_{43} = \frac{d}{dc_i}(dT/dt),$$

$$f_{44} = \frac{d}{dT}(dT/dt)$$

9.5.1.3 Measurement matrix

The measurement relation for temperature is

$$H = \begin{bmatrix} 0 & 0 & 0 & 1 \end{bmatrix} \tag{9.53}$$

The temperature state equation, Eq. (9.19) in its discrete form, is used as the nonlinear measurement equation, $h(x)$. The elements of the measurement transition matrix, H_x in Eq. (4.31) are computed by taking the partial derivatives of $h(x)$ with respect to the state vector

$$H_x = \begin{bmatrix} h_{11} & h_{12} & h_{13} & h_{14} \end{bmatrix} \tag{9.54}$$

All these components are evaluated for the homopolymerization reactor and used with the EKF estimator to obtain measured and unmeasured states of the reactor. The soft sensor uses the temperature data of every sampling time as measurements and provides the estimates of temperature as well as reactor species concentrations.

9.5.2   State estimation results

The mathematical model of the homopolymerization reactor is solved by numerical integration using Gear's method with a sampling time of 0.0001 units. The diagonal elements for each of the design parameters P_0 (initial *state covariance matrix*) and Q (process noise covariance matrix) of the EKF algorithm are selected as 0.00157, 0.00745, 0.00345, and 3.476 respectively, and the diagonal elements of the observation noise covariance, R are set to be 100. From the results of Fig. 9.12, it can be observed that for the given set of operating conditions (T_c=318K, T_f=315K, v_{mf}=0.3, c_{if}=0.03203 gmol L^{-1}), at residence time $\theta = 27.0$ min, the system exhibits limit cycle behavior with sustained period oscillations. As θ is increased by 0.2 min from 27.0 min, there exist two minimum and maximum temperatures showing double periodic limit cycles. The actual and estimated double periodic oscillatory limit cycles and the phase plane plots that are generated at $\theta = 27.2$ min are shown in Fig. 9.13. With the further increase of θ by another 0.08 min, the system exhibits period-doubling characteristics with four minimum temperatures and four maximum temperatures. This actual and estimated sustained quadruple periodic oscillatory behavior in temperature responses and the corresponding phase plane plots for the same operating conditions at $\theta = 27.28$ min are shown in Fig. 9.14. As θ is increased further, there exist multiple minimum temperatures and multiple maximum temperatures. This leads to huge temperature oscillations leading to chaotic behavior. The actual and estimated chaotic responses in temperature and the corresponding phase plane plots for the same operating conditions at $\theta = 27.37$ min are shown in Fig. 9.15. These results confirm the effectiveness of the method of EKF as a state estimator for polymerization reactor.

These results show that increasing or decreasing the P_o and Q by two to three times has only a mild influence on the performance of the polymerization reactors. It is also observed that the performance of the polymerization reactor is more affected by the measurement covariance matrix, R, when it is doubled to the set condition.

9.6   Summary

Nonlinear dynamical systems can exhibit different types of behavioral patterns depending on the values of the physical parameters and intrinsic features of the respective systems. Among the dynamical systems, continuous chemical reactors, and continuous polymerization reactors have received special attention due to their complicated dynamics. These

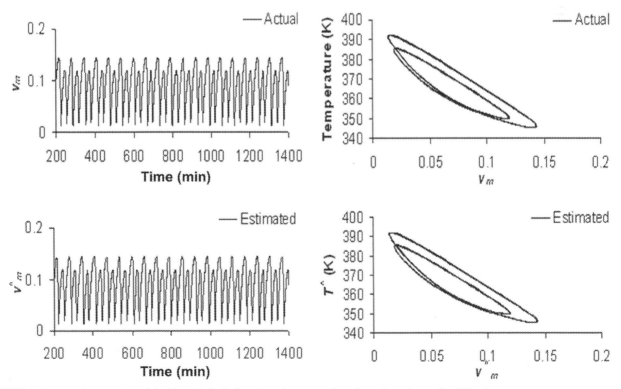

FIGURE 9.13   Actual and estimated double periodic limit cycle and corresponding phase plane plots at $\theta = 27.2$ min.

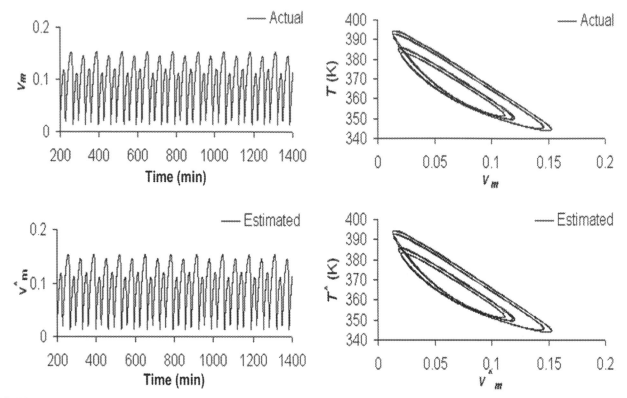

FIGURE 9.14 Actual and estimated quadruple periodic limit cycle responses and corresponding phase plane plots at $\theta = 27.28$ min.

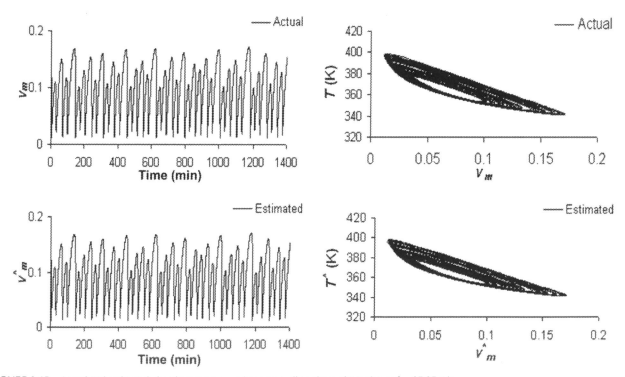

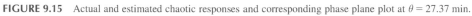

FIGURE 9.15 Actual and estimated chaotic responses and corresponding phase plane plot at $\theta = 27.37$ min.

nonlinear dynamical systems exhibit complex dynamic phenomena such as multiple steady-state behaviors, limit cycle oscillations, and chaos. In this chapter, an extended version of the Kalman filtering approach is used to design model based state estimators to capture the rapidly changing dynamics of the nonlinear dynamical systems that present complicated dynamic behavior. The dynamic steady state of these systems is thoroughly analyzed by performing multiplicity, stability, and bifurcation studies prior to design of state estimators. The design and implementation of the state estimator are studied by considering two typical continuous nonisothermal nonlinear systems, a chemical reactor, and a homopolymerization reactor. The results evaluated demonstrate the better performance of the method of EKF for optimal state estimation in nonlinear chemical processes that exhibit complex dynamic behavior.

References

[1]   A. Uppal, W.H. Ray, A.B. Poore, On the dynamic behavior of continuous stirred tank reactors, Chem. Eng. Sci. 29 (4) (1974) 967−985.

[2]   R.A. Schmitz, R.R. Bautz, W.H. Ray, A. Uppal, The dynamic behavior of a CSTR: some comparisons of theory and experiment, AIChE J. 25 (2) (1979) 289−297.

[3]   L.F. Razón, R.A. Schmitz, Multiplicities and instabilities in chemically reacting systems—a review, Chem. Eng. Sci. 42 (5) (1987) 1005−1047.

[4]   L.M. Pismen, Kinetic instabilities in man-made and natural reactors, Chem. Eng. Sci. 35 (9) (1980) 1950−1978.

[5]   D.T. Lynch, S.E. Wanke, T.D. Rogers, Chaos in a continuous stirred tank reactor, Math. Model. 3 (2) (1982) 103−116.

[6]   C. Kahlert, O.E. Rössler, A. Varma, Chaos in a continuous stirred tank reactor with two consecutive first-order reactions, one exo-, one endothermic, Springer Series in Chemical Physics, Springer Berlin Heidelberg, 1981, pp. 355−365.

[7]   D.V. Jorgensen, R. Aris, On the dynamics of a stirred tank with consecutive reactions, Chem. Eng. Sci. 38 (1) (1983) 45−53.

[8]   D.C. Halbe, A.B. Poore, Dynamics of the continuous stirred tank reactor with reactions A → B → C, Chem. Eng. J. 21 (3) (1981) 241−253.

[9]   A.E. Gamboa-Torres, A. Flores-Tlacuahuac, Effect of process modeling on the nonlinear behaviour of a CSTR reactions A→ B→ C, Chem. Eng. J. 77 (3) (2000) 153−164.

[10] M.F. Doherty, J.M. Ottino, Chaos in deterministic systems: strange attractors, turbulence, and applications in chemical engineering, Chem. Eng. Sci. 43 (2) (1988) 139−183.

[11] E.J. Doedel, R.F. Heinemann, Numerical computation of periodic solution branches and oscillatory dynamics of the stirred tank reactor with A → B → C reactions, Chem. Eng. Sci. 38 (9) (1983) 1493−1499.

[12] F. Teymour, W.H. Ray, Chaos, intermittency and hysteresis in the dynamic model of a polymerization reactor, Chaos, Solit. Fractals 1 (4) (1991) 295−315.

[13] A. Flores-Tlacuahuac, V. Zavala-Tejeda, E. Saldívar-Guerra, Complex nonlinear behavior in the full-scale high-impact polystyrene process, Ind. Eng. Chem. Res. 44 (8) (2005) 2802−2814.

[14] D.R. Lewin, D. Bogle, Controllability analysis of an industrial polymerization reactor, Comput. Chem. Eng. 20, 2 (1996) S871−S876.

[15] J.C. Pinto, W.H. Ray, The dynamic behavior of continuous solution polymerization reactors—VII. Experimental study of a copolymerization reactor, Chem. Eng. Sci. 50 (4) (1995) 715−736.

[16] K.J. Kim, K.Y. Choi, J.C. Alexander, Dynamics of a cascade of two continuous stirred tank polymerization reactors with a binary initiator mixture, Polym. Eng. Sci. 31 (5) (1991) 333−352.

[17] R. Jaisinghani, W.H. Ray, On the dynamic behaviour of a class of homogeneous continuous stirred tank polymerization reactors, Chem. Eng. Sci. 32 (8) (1977) 811−825.

[18] S. Borman, Researchers find order, beauty in chaotic chemical systems, Chem. Eng. News 69 (3) (1991) 18−29.

[19] R.D. Gudi, S.L. Shah, M.R. Gray, Adaptive multirate state and parameter estimation strategies with application to a bioreactor, AIChE J. 41 (11) (1995) 2451−2464.

[20] P. de Vallière, D. Bonvin, Application of estimation techniques to batch reactors—III. Modelling refinements which improve the quality of state and parameter estimation, Comput. Chem. Eng. 14 (7) (1990) 799−808.

[21] C. Venkateswarlu, K. Gangiah, Dynamic modeling and optimal state estimation using extended Kalman filter for a kraft pulping digester, Ind. Eng. Chem. Res. 31 (3) (1992) 848−855.

[22] M.-J. Park, S.-M. Hur, H.-K. Rhee, Online estimation and control of polymer quality in a copolymerization reactor, AIChE J. 48 (5) (2002) 1013−1021.

[23] R. Li, A.B. Corripio, M.A. Henson, M.J. Kurtz, On-line state and parameter estimation of EPDM polymerization reactors using a hierarchical extended Kalman filter, J. Process Control 14 (8) (2004) 837−852.

[24] C. Sumana, C. Venkateswarlu, Optimal selection of sensors for state estimation in a reactive distillation process, J. Process Control 19 (6) (2009) 1024−1035.

[25] K.-Y. Choi, A.A. Khan, Optimal state estimation in the transesterification stage of a continuous polyethylene terephthalate condensation polymerization process, Chem. Eng. Sci. 43 (4) (1988) 749−762.

[26] C. Venkateswarlu, S. Avantika, Optimal state estimation of multicomponent batch distillation, Chem. Eng. Sci. 56 (20) (2001) 5771−5786.

[27] H. Schuler, C.U. Schmidt, Model-based measurement techniques in chemical reactor applications, Int. Chem. Eng. 33 (4) (1993).

[28] N. Pillai, S.L. Schwartz, T. Ho, A. Dokoumetzidis, R. Bies, I. Freedman, Estimating parameters of nonlinear dynamic systems in pharmacology using chaos synchronization and grid search, Journal of Pharmacokinetics and Pharmacodynamics 46 (2) (2019) 193−210.

[29] J. Hong, S. Laflamme, J. Dodson, B. Joyce, Introduction to state estimation of high-rate system dynamics, Sensors 18 (1) (2018) 217.

[30] G. Hu, Z. Zhang, A. Armaou, Z. Yan, Robust extended Kalman filter based state estimation for nonlinear dynamic processes with measurements corrupted by gross errors, Journal of the Taiwan Institute of Chemical Engineers 106 (2020) 20−33.

[31] R.R. Karri, Evaluating and estimating the complex dynamic phenomena in nonlinear chemical systems, Int. J. Chem. React. Eng. 9 (1) (2011) 94.

[32] J.C. Pinto, W.H. Ray, The dynamic behavior of continuous solution polymerization reactors—VIII. A full bifurcation analysis of a lab-scale copolymerization reactor, Chem. Eng. Sci. 50 (6) (1995) 1041−1056.

Chapter 10

Application of mechanistic model-based nonlinear filtering and observation techniques for optimal state estimation of a kraft pulping digester

10.1 Introduction

Optimal state estimation is important for monitoring and improving the performance of the pulping process. Among different pulping processes, the kraft process is the most important chemical pulping process in which the cellulose fibers are extracted from wood through the use of chemicals. In this process, the wood chips are sent to a digester, where they are mixed with cooking liquor, also known as white liquor, and the contents are heated. The white liquor is an aqueous solution mostly composed of sodium hydroxide (NaOH) and sodium sulfide (Na_2S), which dissolves most of the lignin and thus separates the cellulosic fibers from each other. The white liquor also reacts with the cellulosic fibers and affects the physical properties of the produced pulp and pulp yield. After cooking is completed, the cellulose fibers are separated from the liquid, washed, and dried to make the pulp. In the kraft pulping process, the quality of the produced pulp is very much dominated by the raw material properties and the process conditions. Therefore the pulping digester cooking conditions must be controlled so that fibers are separated without much damage while maximizing the pulp yield.

This chapter focusses on experimental aspects, dynamic modeling, and state estimation of a batch kraft pulping process [1]. Batch digesters are used for cooking wood chips by using alkali-based liquor (called white liquor) consisting of NaOH and Na_2S, which chemically dissolves the organic lignin binder and frees the individual cellulose fibers. In kraft pulping, the desired pulp quality is achieved by monitoring the degree of delignification. The degree of delignification of pulp is generally measured by an indirect laboratory analytical test known as the Kappa number. Direct determination of lignin and carbohydrates through analysis during pulping is tedious and time consuming. Thus it is difficult to control the product quality by using only infrequent and delayed laboratory data. In contrast, liquor samples during pulping can be drawn easily and can be analyzed for alkali concentration. The quality of pulp produced by the batch digester is affected by disturbances such as cooking liquor changes, production schedule changes, poor heat distribution, changes in the initial starting temperature, and changes in the length of cooking time.

Model-based estimation methods can be used to predict the operating behavior of the pulping digester by estimating certain key process variables based on the available process measurements in conjunction with the mathematical model of the process. The mathematical model that supports the estimator must incorporate the process knowledge and it should be reliable. In the past, many studies have been made to develop theoretical and experimental models of varying complexity for the purpose of prediction and control of kraft pulping processes [2–6]. Among these models, a state-space model which accounts the dynamic behaviour of the pulping process is quite useful for process design and monitoring. This kind of model can cope with the stochastic features of the process, such as modeling errors, sampling uncertainties, and process noise. Certain key variables like the lignin content and carbohydrate content are not easily measurable for monitoring and controlling the kraft pulping process. Model-based estimation techniques can be employed to estimate such unmeasured process variables using the known measurements like alkali concentration and temperature available with time. In this study, the objective is to present a dynamic model for the kraft pulping digester and apply model-based estimation techniques like extended Kalman filter (EKF) and nonlinear observer (NLO) to estimate lignin and carbohydrate contents by using effective alkali concentration as available process measurement. The performances of the state estimators are evaluated through simulation studies.

Optimal State Estimation for Process Monitoring, Fault Diagnosis and Control. DOI: https://doi.org/10.1016/B978-0-323-85878-6.00022-1

10.2 Experimental system and dynamic modeling

The details of the experimental batch digester, data generation, and dynamic model development for state estimation are explained as follows.

10.2.1 Batch pulping process

Experiments are carried out to generate the data, which can be used to build a dynamic model for the batch digester. The schematic laboratory digester shown in Fig. 10.1 is used to conduct the pulping experiments. The digester consists of a pressure vessel, liquor circulating pump, heat exchanger, pressure gauge, and temperature indicator. The raw

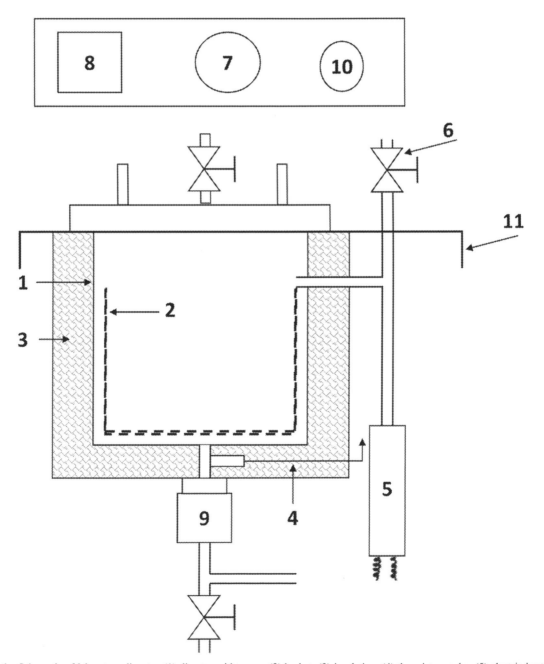

FIGURE 10.1 Schematic of laboratory digester: (1) digester with cover; (2) basket; (3) insulation; (4) thermistor probe; (5) electric heater; (6) flow control valve; (7) pressure gauge; (8) temperature indicator; (9) pump; (10) variable transformer; and (11) top.

material used is *Pinus casia* logs which are made into smaller size chips of about 25 mm length, and the screened sample of 3−5 mm width is used for pulping. The wood is analyzed for lignin, carbohydrate, and alcohol−benzene solubility, which are estimated on the dry basis of wood as 27.90%, 67.90%, and 4.37%, respectively. White liquor is prepared using 110 g L^{-1} as 96% NaOH and 65 g L^{-1} as 50% Na$_2$S in a 3:1 ratio. The liquor is analyzed by using a pH meter titration procedure. The white liquor has a composition of Na$_2$S as Na$_2$O = 27.59 g L^{-1}, NaOH as Na$_2$O = 82.305 g L^{-1}, effective alkali as Na$_2$O = 96.1 g L^{-1}, and sulfidity = 25.1%.

White liquor of specified concentration on an oven-dry wood basis is charged to the digester in each experiment. A wood sample of 250 g (moisture-free) is taken for each experiment and is added to the circulating liquor of specified concentration with a wood to liquor ratio of 1:4, and the continuous heater is immediately switched on. Liquor circulation is carried continuously through a circulating pump. Thirty-six batch experiments are carried out at three temperature levels (160°C, 170°C, and 180°C), four effective alkali concentration levels (35, 40, 45, and 50 g L^{-1}) at each temperature, and three cooking time levels in the range of 15−270 min at each alkali concentration. In each experiment, the cooking temperature is specified in the temperature range of 160°C−180°C for a specified period. The specified temperature is controlled in terms of a set point by using on−off control. The temperature rises from an initial value of around 30°C to the set-point temperature and pulping continues at that point until the specified period.

The temperature is observed by the indicator, and the effective alkali concentration is determined by drawing cooking liquor samples. Liquor samples drawn during pulping are analyzed for effective alkali concentration by using a pH meter titration procedure. At the end of the experiment, the pulp is removed from the digester, washed with water, and analyzed for lignin and carbohydrates. Thus the data of carbohydrates and lignin of the wood sample at the beginning of the experiment, and carbohydrates and lignin of the pulp sample at the end of the experiment are collected. The cooking conditions and experimental results are given in Table 10.1. In addition, for each experiment, cooking liquor composition and temperature is recorded for every 5 min during the rising period, and for every 15 min during the remaining cooking period. A sample cooking schedule of one experiment (Exp. no. 4) is given in Table 10.2. Such cooking schedules are tabulated for all the remaining 35 experiments but not shown here due to space conservation.

TABLE 10.1 Experimental conditions and results.

Exp. no.	Max temp. (°C)	Eff. alkali (%)	Time to max temp (min)	Time at max temp (min)	Lignin (%)	Kappa no.	Residual eff. alkali (g L^{-1})	Total pulp yield (%)
1	160	14	47	210	6.66	40.77	13.64	49.19
2	160	14	45	240	6.36	40.55	12.09	49.07
3	160	14	45	270	5.31	35.30	12.40	48.91
4	160	16	45	180	6.33	54.96	19.38	51.08
5	160	16	47	210	5.30	34.80	16.27	49.82
6	160	16	40	240	4.12	29.20	15.50	48.65
7	160	18	45	120	8.33	44.26	22.47	51.34
8	160	18	45	150	6.99	45.45	21.70	50.53
9	160	18	45	180	6.82	44.71	20.92	50.90
10	160	20	50	100	8.35	33.75	30.22	48.14
11	160	20	45	120	7.89	43.41	26.35	50.85
12	160	20	50	150	5.83	24.63	27.90	46.60
13	170	14	55	150	5.39	32.60	11.93	48.08
14	170	14	55	180	5.00	29.95	12.40	47.36
15	170	14	63	210	4.38	29.35	11.16	47.15
16	170	16	55	120	3.01	25.26	17.82	48.24

(Continued)

TABLE 10.1 (Continued)

Exp. no.	Max temp. (°C)	Eff. alkali (%)	Time to max temp (min)	Time at max temp (min)	Lignin (%)	Kappa no.	Residual eff. alkali (g L^{-1})	Total pulp yield (%)
17	170	16	55	150	2.19	22.00	18.60	46.26
18	170	16	55	180	2.10	21.77	16.27	46.31
19	170	18	45	90	6.90	34.99	20.15	48.66
20	170	18	50	120	4.57	28.33	22.47	48.39
21	170	18	55	150	3.75	20.94	19.37	46.57
22	170	20	60	60	4.45	25.60	31.00	46.89
23	170	20	60	90	3.50	20.80	27.90	45.87
24	170	20	55	120	2.28	15.90	25.57	43.95
25	180	14	55	90	4.70	30.30	10.85	47.61
26	180	14	67	120	4.22	27.25	10.07	46.09
27	180	14	55	150	2.76	24.60	13.17	50.22
28	180	16	60	60	4.05	29.47	17.82	49.74
29	180	16	60	90	3.83	31.78	15.52	47.02
30	180	16	60	120	2.91	23.83	13.17	45.47
31	180	18	60	45	6.53	29.54	20.92	47.55
32	180	18	63	60	4.47	27.01	19.37	47.58
33	180	18	60	90	4.18	18.02	18.60	44.86
34	180	20	62	15	6.74	41.15	32.55	48.04
35	180	20	60	30	3.63	23.20	31.00	44.90
36	180	20	60	60	2.61	16.95	26.04	42.76

TABLE 10.2 Cooking schedule of Exp. no. 4.

Time (min)	0	5	10	15	30	45	60	225
Temp (°C)	0	55	81	98	134	160	160	160
Pressure (kg cm^{-2})	–	–	–	0.5	2.5	5.25	5.50	5.50

10.2.2 Development of a dynamic model for kraft pulping digester

The statistical analysis of experimental data has made it possible to easily fit a regression equation to the cooking data [7,8]. However, linearized versions of higher-order regression models did not find much use as these models are not based on the kinetics of the pulping process. Experimental models are those which assume that the pulping reaction rates are kinetically controlled. The H-factor model [9] is one of the earliest experimental models in which the extent of delignification in kraft pulping is expressed with time and temperature. The H-factor model is more suitable, if all the charging conditions are constant, and delignification is assumed to follow zero-order kinetics. Normally charging conditions are not constant and can vary substantially. To satisfy these conditions, other experimental models [10−12] have attempted to correlate initial charge conditions with the H-factor for predicting end-point conditions. These nonrigorous models can be quite useful over a narrow range of operating conditions and are suitable for specific applications. Further, different theoretical and experimental models have been reviewed by Jutila [13]. A theoretical mathematical model consisting of a series of partial differential equations describing the combined diffusion and kinetics within a wood chip during kraft pulping has been derived and solved using the orthogonal collocation method [2]. Model predictions are compared with experimental data reported in the literature. Such a generalized, detailed theoretical model may not be very useful in practice. However, if diffusion effects within a wood chip are not accounted for and only kinetics are considered, then the distributed parameter model proposed by Gustafson et al. [2] reduces to lumped parameter model.

In this study, the development of a lumped parameter or state space model is taken up for the kraft pulping digester, since this kind of model is well suited for monitoring and controlling the process. As discussed in the earlier section, the experimentally generated data is used to develop the dynamic model of delignification, which is similar to the state space model given by Jutila [13]. The following equations describe the model:

$$dC/dt = -q_1\exp(b-a/T)CL, \quad C(0) = C_0 \tag{10.1}$$

$$dL/dt = -q_2\exp(b-a/T)CL, \quad L(0) = L_0 \tag{10.2}$$

$$dCO/dt = q_3 dL/dt, \quad CO(0) = CO_0 \tag{10.3}$$

where C is effective alkali concentration, L is lignin concentration, CO is carbohydrate concentration, and T is absolute temperature, a and b are specific constants of wood that resemble Arrhenius equation type constants such as frequency factor and energy of activation. Optimum coefficients of the pulping model are q_1, q_2, q_3, a, and b, which are computed by using the Nelder and Mead simplex method [14,15] with the following objective function:

$$J = \frac{1}{N}\sum_{i=1}^{N}\left(C_i - \hat{C}_i\right)^2 + \left(L_f - \hat{L}_f\right)^2 + \left(CO_f - \hat{CO}_f\right)^2 \tag{10.4}$$

where C is the observed value of the effective alkali concentration, $\hat{C}_i$ is the computed value of the effective alkali concentration, L_f is the observed end value of the lignin content, $\hat{L}_f$ is the computed end value of the lignin content, CO_f is the observed end value of the carbohydrate content, and $\hat{CO}_f$ is the computed end value of the carbohydrate content. The observed values of effective alkali concentrations for each experiment are determined for every 5 min during the rising period, and for every 15 min during the remaining cooking period. These alkali concentrations available during the experiment, and the lignin and carbohydrate concentrations available at the end of the experiment are used to develop the dynamic model for the kraft pulping digester.

The Nelder and Mead simplex method performs the search by tentatively expanding or contracting vertices of the simplex. For the present model, the simplex has six vertices. The computed values used in the objective function are obtained by solving the pulping model equations numerically with the help of the Runge–Kutta fourth-order technique. The initial values of model coefficients are the same for all the experiments and are given by $q_1 = 1.0 \times 10^{-5}$, $q_2 = 1.0 \times 10^{-5}$, and $q_3 = 1.0$. Since the a and b values of P. casia are not known, reported values of spruce wood [11] as $a = 16113.0$; $b = 43.2$ are used as initial values for the optimization routine. The estimated model coefficients for each of the 36 experiments are given in Table 10.3.

TABLE 10.3 Optimum model coefficients.

Expt. no	q_1	q_2	q_3	b	a
1	2.002×10^{-5}	1.705×10^{-5}	1.237	44.632	16113.32
2	1.839×10^{-5}	1.520×10^{-5}	1.180	44.816	16113.19
3	1.882×10^{-5}	1.725×10^{-5}	1.104	44.721	16113.02
4	2.001×10^{-5}	1.644×10^{-5}	1.243	44.667	16113.30
5	1.630×10^{-5}	1.559×10^{-5}	1.041	44.748	16113.23
6	1.967×10^{-5}	1.810×10^{-5}	0.982	44.671	16113.11
7	2.271×10^{-5}	1.644×10^{-5}	1.269	44.595	16112.86
8	1.660×10^{-5}	1.559×10^{-5}	1.173	44.587	16113.58

(Continued)

TABLE 10.3 (Continued)

Expt. no	q_1	q_2	q_3	b	a
9	2.176×10^{-5}	1.588×10^{-5}	1.148	44.395	16113.36
10	2.442×10^{-5}	1.881×10^{-5}	1.275	44.602	16113.75
11	2.370×10^{-5}	1.674×10^{-5}	1.346	44.436	16113.22
12	1.791×10^{-5}	1.535×10^{-5}	1.218	44.507	16113.21
13	1.845×10^{-5}	1.516×10^{-5}	1.142	44.642	16113.33
14	1.661×10^{-5}	1.420×10^{-5}	1.182	44.478	16113.21
15	1.479×10^{-5}	1.365×10^{-5}	1.055	44.550	16113.22
16	1.405×10^{-5}	1.714×10^{-5}	1.141	44.486	16113.73
17	1.997×10^{-5}	1.820×10^{-5}	1.001	44.782	16113.30
16	2.018×10^{-5}	1.812×10^{-5}	1.102	44.481	16113.47
19	2.242×10^{-5}	2.018×10^{-5}	1.112	44.644	16113.68
20	1.524×10^{-5}	1.331×10^{-5}	1.101	44.372	16113.28
21	1.794×10^{-5}	1.740×10^{-5}	0.992	44.853	16113.23
22	1.577×10^{-5}	1.519×10^{-5}	1.099	44.644	16113.50
23	1.706×10^{-5}	1.622×10^{-5}	1.024	44.512	16113.08
24	1.663×10^{-5}	1.721×10^{-5}	1.102	44.482	16113.10
25	1.833×10^{-5}	1.505×10^{-5}	1.107	44.450	16113.12
26	1.672×10^{-5}	1.697×10^{-5}	1.110	44.400	16112.48
27	1.774×10^{-5}	1.730×10^{-5}	1.100	44.753	16111.31
26	1.624×10^{-5}	1.512×10^{-5}	1.100	44.601	16112.80
29	1.599×10^{-5}	1.562×10^{-5}	1.101	44.544	16113.03
30	2.010×10^{-5}	1.820×10^{-5}	1.210	44.422	16113.48
31	2.218×10^{-5}	1.640×10^{-5}	1.277	44.688	16113.79
32	1.650×10^{-5}	1.604×10^{-5}	1.110	44.391	16113.58
33	1.570×10^{-5}	1.347×10^{-5}	1.102	44.146	16113.24
34	1.670×10^{-5}	1.770×10^{-5}	1.110	44.389	16113.24
35	1.443×10^{-5}	1.504×10^{-5}	1.102	44.479	16113.44
36	1.576×10^{-5}	1.607×10^{-5}	1.106	44.373	16113.01

The mean and standard deviation of the model coefficients of 36 experiments are shown in Table 10.4. The optimum values of model coefficients of individual experiments of Table 10.2 are closely scattered with their corresponding mean values. The a and b values obtained in Table 10.4 for the *P. casia* wood are not much different from a and b values of spruce wood [10]. Therefore the mean values of model coefficients can be reasonably employed to represent the dynamic model of the kraft pulping digester, which satisfies all the batch experiments.

To assess and compare the model developed in this work, another well-known model of delignification based on the H-factor approach [11] is also presented for this experimental system. This model is based on the Arrhenius-type equation, according to which the influence of pulping temperature on the rate of delignification is expressed by

$$-dL/dt = \exp(b - a/T) \qquad (10.5)$$

TABLE 10.4 Statistics of model coefficients.

Coefficient	Mean	SD
q_1	1.824×10^{-5}	$2.698 \times 10E-6$
q_2	1.632×10^{-5}	$1.530 \times 10E-6$
q_3	1.131	$8.380 \times 10E-2$
b	44.553	0.151
a	16113.187	0.354

The optimum coefficients a and b of Eq. (10.5) are computed for each experiment using the Nelder and Mead simplex method with $a = 16113.0$ and $b = 43.2$ as initial values, which are also used in the earlier modeling. The objective function employed for this model is given by

$$J = (L_f - \hat{L}_f)^2 \tag{10.6}$$

The mean values of 36 experiments, $a = 16113.437$ with standard deviation 0.3164 and $b = 39.143$ with standard deviation 0.1182, are employed as model coefficients. The models thus developed are used for simulation to reflect the actual operating conditions. Initial effective alkali concentration and temperature are chosen as the main operating variables. The effect of operating variables on pulp yield is studied by introducing variations in the ranges of $35-50$ g L^{-1} for effective alkali concentration, and $160°C-180°C$ for maximum temperature. The digester temperature rises from the initial value to a maximum specified value (T_{max}) for a particular period of a specified time. In actual practice, temperature versus time history will be known. However, in the simulation environment, the temperature versus time relationship is established as: [13]

$$T = \left\{ 32024.58t + 1038.82 \right\}^{1/2} \quad \text{for} \quad 0 < t < t_{rise} T_{max} \quad \text{for} \quad t > t_{rise} \tag{10.7}$$

T is the temperature in degree Celsius, and t is time in hours. The t_{rise} is the rise time corresponding to the rise temperature. The dynamic model equations, Eqs. (10.1)−(10.3) and Eq. (10.5), are solved with the fourth-order Runge−Kutta method to generate true process state trajectories. The numerical simulation results of the model of this study and the other model described by Eq. (10.5) are compared with the actual independent experimental values of lignin. Since there is no provision of drawing pulp samples from the digester during a cook, independent experiments are carried out by setting discontinuities in a cook to determine the lignin values. For this purpose, a cook is conducted with a temperature of $160°C$, 45 g L^{-1} initial effective alkali concentration, and 4 h cooking time discontinuing it at different cooking times. For each discontinuty, the starting conditions and operating conditions are kept the same. The pulp samples collected at each discontinity are analyzed for lignin and observed that the experimental values of lignin are closer to the results of the proposed model, as shown in Fig. 10.2. However, a mismatch is observed between the experimental results and the model predictions in Eq. (10.5). The reason for not matching the results of model Eq. (10.5) with the experimental lignin values may be due to the employment of only accessible end lignin values of each experiment to determine the coefficients a and b. Though the model of Eq. (10.5) is simple, this model is not suitable for state estimation as it is not a function of a dependent variable to compute filter or observer coefficient matrices. Moreover, lignin measurement is difficult and time-consuming during pulping, whereas measurements of lignin can be inferred with time through estimation.

10.3   Optimal state estimation of kraft pulping digester

The dynamic model developed for the batch kraft pulping digester is employed to estimate the states using an EKF and a NLO.

10.3.1   Design of EKF

An EKF is designed for state and parameter estimation in batch kraft pulping digester. The EKF algorithm along with the general process representation and noise relations is given in Section 4.4 of Chapter 4. The algorithm given in

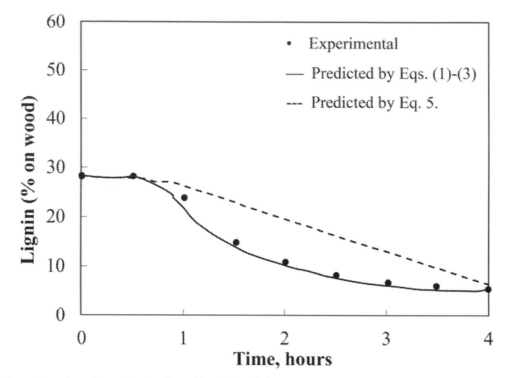

FIGURE 10.2 Comparison of experimental lignin values with model predictions.

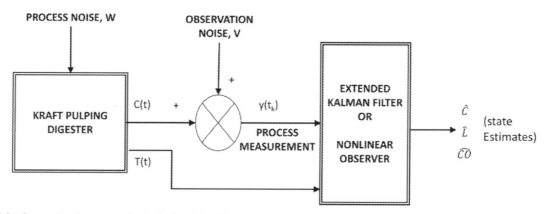

FIGURE 10.3 State estimation scheme for the kraft pulping digester.

Section 4.4.3 is implemented for optimal state estimation in the pulping process. The schematic of state estimation for kraft pulping digester is shown in Fig. 10.3.

The filter design matrices P_0, $Q(t)$, and $R(t_k)$ reflect errors in the initial state, process model, and process measurements. The $Q(t)$ and $R(t_k)$ are set as fixed design matrices and referred to as Q and R. The effective alkali concentration, C, with noise is used as known process measurement y:

$$y(t_k) = C(t_k) + v(t_k) \tag{10.8}$$

Since carbohydrate is a function of lignin (Eq. (10.3)), the EKF is implemented by considering the state vector as

$$(x)^T = (C \quad L)^T \tag{10.9}$$

The dynamic model of the pulping digester is used to represent $f(x)$ in Eq. (4.20) of Chapter 4. The measured variable, y, of Eq. (10.8), is used to estimate the states, C and L. Since Eq. (10.3) is a function of Eq. (10.2), the state estimate CO can be directly obtained from the estimated state L.

10.3.2   Design of nonlinear observer

The NLO described in Section 4.14 is adopted for state estimation of the pulping digester. Instead of using for local estimation, the observer is applied to estimate the states simultaneously. The available state x_a and the unavailable state x_u of Eq. (4.105) is represented as

$$(x_a)^T = (C)^T \tag{10.10}$$

$$(x_u)^T = (L)^T \tag{10.11}$$

Observer computations are carried out by considering the state vector given by Eqs. (10.10) and (10.11). The state carbohydrate content is computed by using the estimated lignin content. The matrix K_u of Eq. (4.112) of Chapter 4 is computed as

$$K_u = \begin{bmatrix} q_2/q_1 \\ 1 \end{bmatrix} \tag{10.12}$$

10.4   State estimation results

The model developed in this study is more suitable for estimating known and unknown variables during pulping by using easily available alkali measurement. Hence, the model represented by Eqs. (10.1)–(10.3) is employed to estimate the known and unknown states for different values of operating conditions by applying both the EKF and NLO. Random Gaussian numbers are used in the form of random Gaussian noise. A zero-mean random Gaussian noise with a standard deviation of 0.75 is used to corrupt the true values of effective alkali to obtain alkali measurement and to corrupt the true temperature.

The values of noise covariance matrices selected for the filter are given by

$$
\begin{aligned}
P_0 &= \begin{bmatrix} 0.5 & 0 \\ 0 & 0.5 \end{bmatrix} \\
Q &= \begin{bmatrix} 0.5 & 0 \\ 0 & 0.5 \end{bmatrix} \\
R &= 0.5
\end{aligned}
\tag{10.13}
$$

The design parameter of the observer K_a involved in Eq. (4.110) is selected as

$$K_a = \begin{bmatrix} 0.8 \\ 0.8 \end{bmatrix} \tag{10.14}$$

The efficiency of the EKF and NLO is analyzed by using a performance criterion, which is a sum of the squared errors of true and estimated states as given by

$$I = \sum_{k=1}^{N} \left(C(t_k) - \hat{C}(t_k) \right)^2 + \left(L(t_k) - \hat{L}(t_k) \right)^2 + \left(CO(t_k) - \widehat{CO}(t_k) \right)^2 \tag{10.15}$$

where C and $\hat{C}$ are true and estimated values of effective alkali concentration, L and $\hat{L}$ are true and estimated values of lignin content, and CO and $\hat{CO}$ are true and estimated values of carbohydrate content.

The EKF and NLO are implemented using the base-case conditions as initial effective alkali concentration is 35 g L^{-1}, cooking temperature is 160°C, and cooking time is 4 h. Initial values of lignin and carbohydrate contents used in the simulation are experimentally determined values for the wood chips. The estimated states by the EKF for various initial values of effective alkali concentration at160°C are shown in Fig. 10.4. The estimated states by the NLO for various initial values of effective alkali concentration at 180°C are shown in Fig. 10.5.

The results show that the estimated states by both the EKF and NLO are in good agreement with their corresponding true values. The results in Figs. 10.4 and 10.5 show the comparison of the measured effective alkali concentrations with the estimated concentrations. The estimated states are filtered values of known effective alkali and unknown lignin, and carbohydrate concentrations with time during pulping. Further, independent experimental results (not the ones used to determine model coefficients) are used to compare the measured and unmeasured estimation variables. A random Gaussian noise with a standard deviation of 0.75 is also employed to reflect errors in the process model by corrupting temperature and model equations. A zero mean random Gaussian noise with a standard deviation of 1.0 is also used to corrupt the true values of effective alkali to obtain alkali measurement. The estimated states by the EKF and NLO for a case of 160°C, 45 g L^{-1} initial effective alkali, and 4 h cooking time are shown in Figs. 10.6 and 10.7.

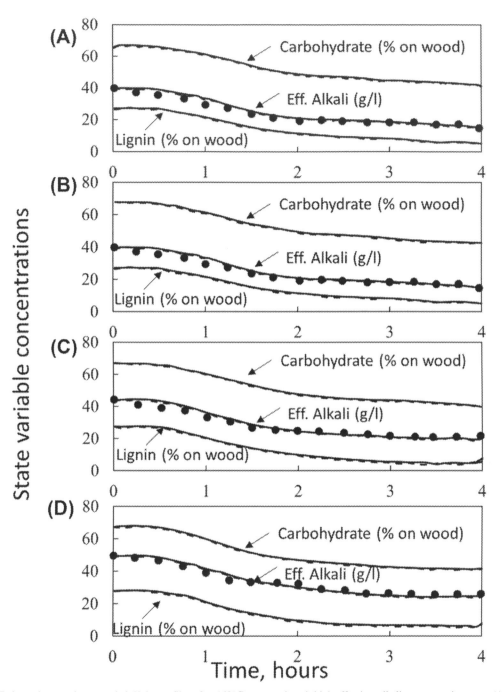

FIGURE 10.4 Estimated states by extended Kalman filter for 160°C range when initial effective alkali concentrations are (A) 35 g L^{-1}, (B) 40 g L^{-1}, (C) 45 g L^{-1}, and (D) 50 g L^{-1}; (●) experimental, (—) true, and (--) estimated.

The results show that the estimated lignin and carbohydrate are in reasonable agreement with their corresponding actual experimental values. The comparison of the initial results shows that the EKF is slightly better than the observer.

The performances of the filter and observer are evaluated in terms of the quantitative performance criteria defined by Eq. (10.15). The performances of the estimators for different operating situations are evaluated by using the same initial conditions for the observer and the filter. The performance of the estimators when initial alkali concentration varies from 35 to 50 g L^{-1} for 160°C (4 h cooking time), 170°C (3 h cooking time), and 180°C (2 h cooking time) are shown in Fig. 10.8.

The performance of the EKF is found better than the NLO for the cooking temperature at 160°C. Although the performances of both the estimators are nearly same, the EKF gave a slightly higher performance at 170°C and 180°C,

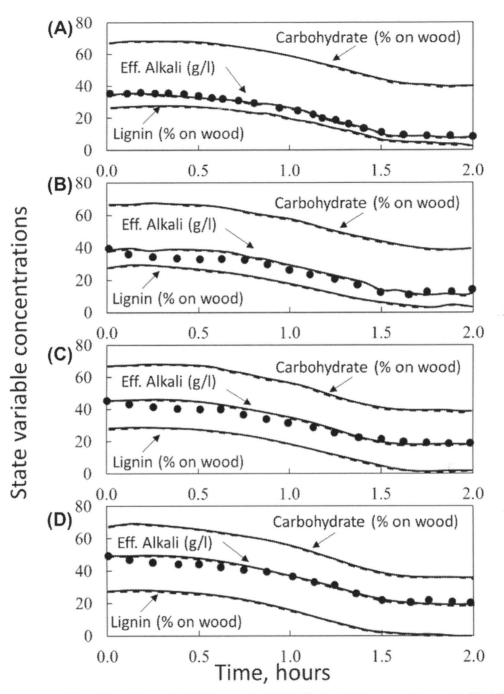

FIGURE 10.5 Estimated states by nonlinear observer for 180°C range when initial effective alkali concentrations are (A) 35 g L^{-1}, (B) 40 g L^{-1}, (C) 45 g L^{-1}, and (D) 50 g L^{-1}; (●) experimental, (—) true, and (--) estimated.

when the same initial alkali concentration is used for these cases. The trend in performance indices of the EKF for 170°C and 180°C is different from the base case of 160°C, whereas the observer has shown a similar trend for all the temperatures. The filter design matrices of P_0, Q, and R, are maintained the same for all the cases. When the temperature rises higher from the base case of 160°C, the prediction part of the filter covariance matrix increases, thus causing a rise in error and performance index. The increase of initial effective alkali concentration with the increase of temperature provides slightly faster convergence with lower values performance index. As there is no involvement of noise covariance matrices in the observer computation, this type of behavior is not found in the performance indices of the observer. The differences in performance indices of the EKF between the starting and final values of effective alkali

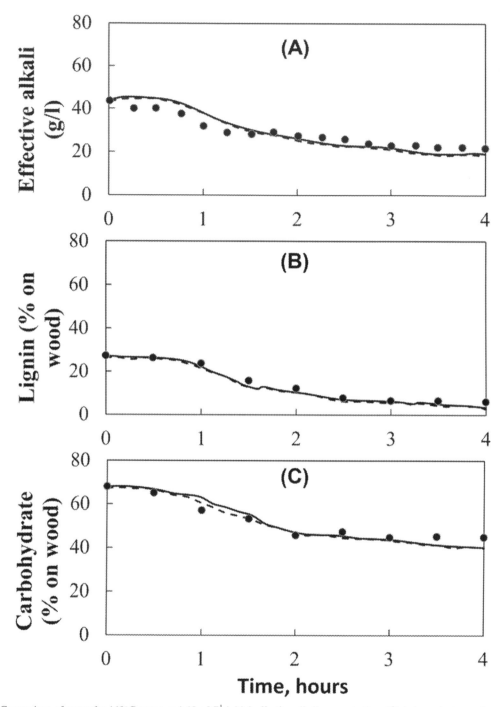

FIGURE 10.6 Comparison of states for 160°C range and 45 g L⁻¹ initial effective alkali concentration: (●) independent experiments, (—) true, and (– –) estimated by extended Kalman filter; (A) Effective alkali; (B) Lignin; (C) Cabohydrate.

are lower for the three temperature ranges when compared to that of the observer. The EKF is therefore preferred for state estimation in the kraft pulping digester.

10.5 Summary

Optimal state estimation is important for monitoring and improving the performance of a pulping process. Kraft process is the most important chemical pulping process in which the wood chips are cooked in a digester using chemical liquor to extract cellulose fibers. Certain key variables like the lignin content and carbohydrate content during cooking are not

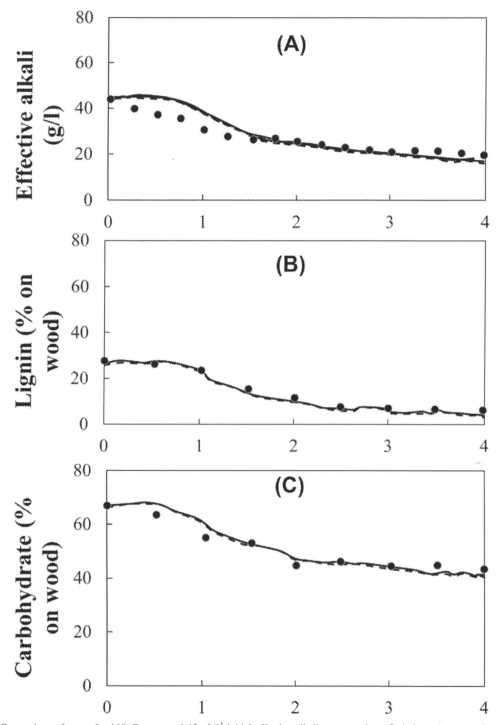

FIGURE 10.7   Comparison of states for 160°C range and 45 g L^{-1} initial effective alkali concentration: (●) independent experiments, (—) true, and (—-) estimated by the nonlinear observer; (A) Effective alkali; (B) Lignin; (C) Cabohydrate.

easily measurable for monitoring and controlling the kraft pulping process. Model-based estimation techniques can provide the estimates of such unmeasured process variables using the timely available measurements like alkali concentration and temperature during a cook. In this study, a dynamic model has been developed for the kraft pulping digester, and model-based estimation techniques, namely, EKF and NLO, are applied to estimate the lignin and carbohydrate contents by using effective alkali concentration as available process measurement. The EKF is found better for state estimation in the kraft pulping digester.

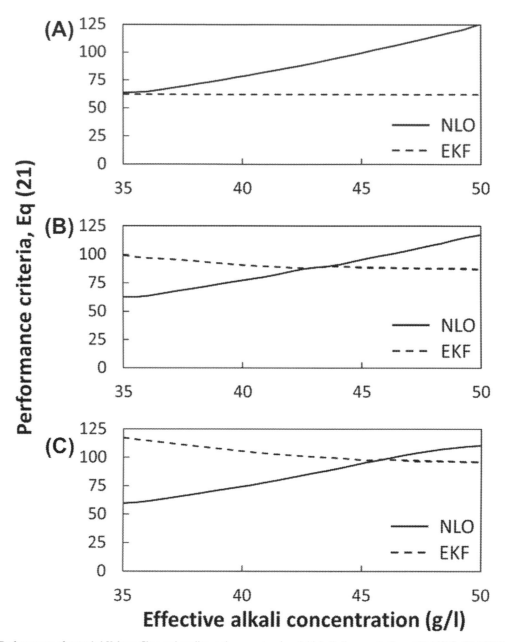

FIGURE 10.8 Performances of extended Kalman filter and nonlinear observer at various initial alkali concentrations: (A) 160°C, (B) 170°C, and (C) 180°C.

References

[1] C. Venkateswarlu, K. Gangiah, Dynamic modeling and optimal state estimation using extended kalman filter for a kraft pulping digester, Ind. Eng. Chem. Res. 31 (3) (1992) 848—855.

[2] R.R. Gustafson, C.A. Slelcher, W.T. Mckean, B.A. Finlayson, Theoritical model of the kraft pulping process, Ind. Eng. Chem. Process Des. Dev. 22 (1) (1983) 87—96.

[3] G. Jiménez, R.R. Gustafson, W.T. McKean, Modeling incomplete penetration of kraft pulping liquor, J. Pulp Paper Sci. 15 (3) (1989) 110.

[4] S. Mirams, K.L. Nguyen, Application of predictive kinetic models in eucalypt kraft pulping, in: International Pan Pacific Conference, Tappi Proceedings, 1994, 73—82.

[5] J.F. Saltin, A predictive dynamic model for continuous digesters, in: Pulping Conference, Tappi Proceedings, 1992, 261—268.

[6] B.R. Rubini, C.I. Yamamoto, Development of prediction oxygen delignification models using kinetic expressions and neural networks, Tappi J. 5 (5) (2006) 3—6.

[7] J.V. Hatton, J.L. Keays, Relationships between pulp yield and permanganate number for kraft pulps: I. Western Hemlock, White Spruce and Lodgepole Pine, Pulp Paper Mag. Can. 71 (11) (1970) 123—132.

[8] D.D. Hinricks, The effect of kraft pulping variables on delignification, Tappi 50 (4) (1967) 173−175.

[9] K.E. Vroom, The 'H' Factor: A means of expressing cooking times and temperature as a single variable, Pulp Paper Mag. Can. 58 (C) (1957) 228.

[10] J.V. Hatton, Application of empirical equations to kraft process control, Tappi 56 (8) (1973) 108−111.

[11] N.C.S. Chari, Integrated control system approach for batch digester control, Tappi 56 (7) (1973) 65−68.

[12] C. Venkateswarlu, K. Gangiah, D.D. Jatkar, S.S. Sridharan, Modeling and simulation of kraft pulping digester, Ind. J. Technol. 24 (1986) 667−670.

[13] E.A.A. Jutila, A survey of kraft cooking control models, IFAC Proc. 13 (12) (1980) 561−573.

[14] J.L. Kuester, J.H. Mize, Optimization Techniques with Fortran, McGraw-Hill, New York, 1973. Chapter 9.

[15] C. Venkateswarlu, J.S. Eswari, Chapter 3, Stochastic Global Optimization: Methods and Applications to Chemical, Biochemical, Pharmaceutical and Environmental Processes, Elsevier Inc, 2020.

Chapter 11

Application of mechanistic model-based nonlinear filtering and observation techniques for optimal state estimation of a continuous reactive distillation column with optimal sensor configuration

11.1 Introduction

The growing demand for better product quality and optimal utilization of available resources make the plant operations more challenging in terms of monitoring and control. The information available from various states of the plant has become increasingly important in devising efficient strategies for plant operation. The nonavailability of certain key process variables due to economic and practical constraints necessitates inferential estimation of unmeasured process variables highly significant in process/plant operations. Further, the optimal configuration of measurement sensors for inferential estimation of states is an important prerequisite for the successful estimation of process states. The idea of sensor configuration is to select the minimum number of sensors to obtain the maximum amount of information for reliable estimation of states in a process.

Reactive distillation is an attractive process intensification scheme involving combined operation of reaction and separation in a single unit. In reactive distillation, the integration of in situ separation function within the reactor holds the promise of increased conversion, high selectivity, and reduced capital investment. This process becomes especially useful when the system under consideration involves reversible reactions, azeotropes, and undesired product formations. The complexities caused due to dynamic interaction of reaction kinetics, mass transfer, and thermodynamics make the operation more challenging. Specifically, chemical reaction rates have a dominant influence on the performance of this process intensification scheme. The states in the process are represented by the concentrations of the components and the temperature on each stage of the column. The objective of sensor configuration for optimal state estimation is to determine the best subset of sensors for state estimation from all the possible measurements of the column.

Various techniques have been reported for inferential estimation of compositions in continuous distillation columns. These include a static estimator [1], nonlinear observers [2], partial least squares estimator [3], an extended Kalman filter (EKF) [4], modified EKF [5], nonlinear EKF [6], and dynamic partial least squares regression [7]. Also, different techniques have been reported for measurement configuration for the purpose of state estimation in continuous distillation. These include the iterative selection of temperature measurements suggested by Joseph and Brosilow [1] and the observability criterion-based selection of temperature measurements by Long and Gilles [2].

This chapter presents an inferential composition estimator with optimal sensor selection for a continuous nonlinear reactive distillation. A strategy based on empirical observability grammian is employed for sensor selection, and an EKF is applied for state estimation. The performance of this strategy is evaluated by applying it to a continuous metathesis reactive distillation column.

Optimal State Estimation for Process Monitoring, Fault Diagnosis and Control. DOI: https://doi.org/10.1016/B978-0-323-85878-6.00003-8

11.2 The process and its mathematical model

Olefin metathesis is the conversion of an olefin into lower and higher molecular weight olefins, which is a class of reactions that ideally suited for reactive distillation. This metathesis process is widely used in the petrochemical industry. In this reaction, two moles of 2-pentene react to form one mole each of 2-butene and 3-hexene,

$$2C_5H_{10} \Leftrightarrow C_4H_8 + C_6H_{12} \tag{11.1}$$
$$\underset{Pentene}{\quad} \underset{Butene}{\quad} \underset{Hexene}{\quad}$$

The normal boiling points of components in this reaction allow an easy separation between the reactant 2-pentene ($310°K$), top product 2-butene ($277°K$), and bottom product 3-hexene ($340°K$). The steady-state design aspects of the metathesis reactive distillation have been studied by Okasinski and Doherty [8]. The reaction kinetics for metathesis reaction are given by the following equations:

$$k_f = 3553.6e^{[-6.6(\text{kcal gmmol}^{-1})/\text{RT}]} \text{min}^{-1}$$
$$K_{eq} = 0.25 \tag{11.2}$$
$$R_3 = 0.5k_f \left[x_1^2 - (x_2 x_3 / K_{eq}) \right]$$

where k_f is the forward reaction rate constant, K_{eq} is the equilibrium reaction rate constant, R_3 is the rate of formation of hexene, x_1, x_2, and x_3 are the mole fractions of pentene, butene, and hexene, respectively.

In order to represent the realistic operation of the actual reactive distillation column, a rigorous nonlinear model that considers the simultaneous effect of chemical reaction, heat and mass transfer operations, and fluid flow on the plates is derived. This first principle model involves dynamic material and component balances with reaction terms and algebraic energy equations supported by vapor—liquid equilibrium and physical properties. This model assumes that the vapor phase holdup is negligible compared to the liquid-phase holdup, negligible chemical reactions in the vapor phase, negligible heat of reaction, constant column pressure during operation, negligible down comer dynamics, ideal trays and total condensation with no subcooling in the condenser. This model is described by a set of ordinary differential and algebraic equations in which differential equations represent the mass, component and energy balances around each plate of the column, and algebraic equations describe the particular variables, physical properties, and plate hydraulics. The mathematical model of the process is described by the following equations [9].

11.2.1 Total mass balance

Total condenser

$$\frac{dM_D}{dt} = V_{Nt} - (D + R) + \Delta R_D \tag{11.3}$$

Feed plate nf

$$\frac{dM_{nf}}{dt} = F_L + V_{nf-1} + L_{nf+1} - (V_{nf} + L_{nf}) + \Delta R_{nf} \tag{11.4}$$

Other plate m

$$\frac{dM_m}{dt} = V_{m-1} + L_{m+1} - (V_m + L_m) + \Delta R_m \tag{11.5}$$

Reboiler

$$\frac{dM_B}{dt} = L_1 - (V_B + B) + \Delta R_B \tag{11.6}$$

11.2.2 Component balance

Total condenser

$$\frac{d(M_D x_{D,j})}{dt} = V_{Nt} x_{Nt,j} - (D x_{D,j} + R x_{D,j}) + \Delta R_{D,j} \tag{11.7}$$

Feed plate *nf*

$$\frac{d(M_{nf}x_{nf,j})}{dt} = F_L x_{f,j} + V_{nf-1}y_{nf-1,j} + L_{nf+1}x_{nf+1,j} - (V_{nf}y_{nf,j} + L_{nf}x_{nf,j}) + \Delta R_{nf,j} \tag{11.8}$$

Other plate *m*

$$\frac{d(M_m x_{m,j})}{dt} = V_{m-1}y_{m-1,j} + L_{m+1}x_{m+1,j} - (V_m y_{m,j} + L_m x_{m,j}) + \Delta R_{m,j} \tag{11.9}$$

Reboiler

$$\frac{d(M_B x_B)}{dt} = L_1 x_{1,j} - (V_B y_{B,j} + B x_{B,j}) + \Delta R_{B,j} \tag{11.10}$$

11.2.3   Energy balance

Condenser

$$\frac{dE_D}{dt} = V_{Nt}Hv_{Nt} - (DHl_D + RHl_D) + \Delta Hr_D \tag{11.11}$$

Feed plate *nf*

$$\frac{dE_{nf}}{dt} = FLHl + V_{nf-1}Hv_{nf-1} + L_{nf+1}Hl_{nf+1} - (V_{nf}Hv_{nf} + L_{nf}Hl_{nf}) + \Delta Hr_{nf} \tag{11.12}$$

Other plate *m*

$$\frac{dE_m}{dt} = V_{m-1}Hv_{m-1} + L_{m+1}Hl_{m+1} - (V_m Hv_m + L_m Hl_m) + \Delta Hr_m \tag{11.13}$$

Reboiler

$$\frac{dE_B}{dt} = L_1 Hl_1 - (V_B Hv_B + BHl_B) + \Delta Hr_B \tag{11.14}$$

11.2.4   Tray hydraulics

Francis weir formula

$$h_l = (h_w + h_{ow}) = (Mv_m/A_m) \tag{11.15}$$

Flow of liquid over weir

$$Mv_m = \left((h_l - h_w)/1.33\right)^{3/2} l_w \tag{11.16}$$

Mole fraction normalization

$$\sum_1^{Nc} x_i = \sum_1^{Nc} y_i = 1 \tag{11.17}$$

Antoine equations

$$\log P_{m,j}^{sat} = a_j - \frac{b_j}{c_j + T_m} \tag{11.18}$$

$$y_{m,j}P_m = x_{m,j}P_{m,j}^{sat} \tag{11.19}$$

Physical properties

$$Hl_m = f(P_m, T_m, x_{m,j}) \tag{11.20}$$

$$Hv_m = f(P_m, T_m, y_{m,j}) \tag{11.21}$$

$$\rho_{liq} = f(P, T, y_j) \tag{11.22}$$

The temperature measurement model used for state estimation is given by the equation:

$$T_m = \frac{b_j}{a_j - \log(\frac{P^{sat}_{m,j} P_m}{\sum_1^{Nc} P^{sat}_{m,j} x_{m,j}})} - c_j \tag{11.23}$$

In the above equations, h_l = height of liquid on the tray, h_w = weir height, h_{ow} = height of liquid over weir, $K_{i,j}$ = equilibrium constant of jth component on tray i, D = distillate flow rate (mol h^{-1}), R = reflux flow rate (mol h^{-1}), l_w = length of weir, M_B = holdup in the reboiler (mol), M_{Bo} = initial charge to the reboiler (mol), M_D = reflux drum holdup (mol), M_i = tray liquid holdup (mol), M_m = molar liquid hold up on tray m (kg mol h^{-1}), Mv_m = volumetric flow rate of liquid over weir from tray m, N_c = number of components, N_s = number of samples, Nt = top tray, P_m = total pressure on tray m, $P_{m,j}{}^{sat}$ = saturated vapor pressure of the component j on tray m, P_i = column pressure or plate pressure, Q_i = rate of heat generation by chemical reaction activity on ith tray, $R_{i,j}$ = chemical reaction rates for jth component on ith tray (1 h^{-1}), T_m = temperature on tray i (°C), V = vapor boil up rate (mol h^{-1}), x = state vector, $x_{m,j}$ = mole fraction of liquid component j on tray m, and $y_{m,j}$ = mole fraction of vapor component j on tray m.

Antoine coefficients for all the components of the system are referred from Boublik et al. [10]. Tray hydraulic computations involve the Francis weir formula. Enthalpy calculations involve empirical relations for heat capacities, which are the functions of temperature and pressure. The coefficients of heat capacity equations for gasses are referred from Perry [11], and for liquids are taken from Yaws [12]. The dynamic simulator of multicomponent reactive distillation of this study has major computation functions for vapor flow, liquid flow, tray holdup, enthalpy, average molecular weight and density, vapor–liquid equilibrium, and reaction rate. The column specifications and steady-state details are given in Table 11.1.

11.3 Optimal sensor configuration using empirical observability grammians

Accurate selection of measurement sensors plays a significant role in the instantaneous estimation of unmeasured process variables in high-dimensional processes like a continuous multicomponent reactive distillation process.

11.3.1 Significance of optimal sensor configuration in a reactive distillation column

The success of state estimation in reactive distillation depends on the rigorous evaluation of the observability and appropriate selection of measurements that adequately characterize the process behavior. Inappropriate use of measurements for state estimation may lead to numerical problems such as singularity or over parameterization, which may drastically reduce the accuracy of the estimation scheme. Thus, for optimal state estimation, it has to be verified that the considered measurements are sufficient and the process observability criteria are satisfied. Fulfilment of observability criterion alone may not be sufficient for state estimation, but the system should exhibit a high degree of observability for the selected measurements. Thus, the development of an inferential estimator with the optimal configuration of sensors attains significance for a nonlinear reactive distillation column.

This study investigates the use of grammian covariance matrices for sensor placement for optimal state estimation in reactive distillation columns [13]. Singh and Hahn [14] have presented a method based on empirical observability grammians for the selection of a single sensor for state and parameter estimation in a binary distillation column and a fixed bed catalytic reactor. The same authors [14] have further extended this approach for the case of multiple sensors using an optimization procedure that involves a trade-off between the process information, measurement cost, and redundancy. In multicomponent reactive distillation, stage temperatures do not correspond exactly to the product compositions, and rigorous criteria are needed for the exact placement of sensors. In this study, an empirical observability grammian based approach is employed for optimal sensor configuration in multicomponent reactive distillation.

11.3.2 Measurement specific empirical observability grammians

Empirical observability grammian and grammian based sensor selection procedures for nonlinear systems are presented in Sections 6.5.1 and 6.5.2 of Chapter 6: Optimal sensor configuration methods for state estimation. The observability covariance matrix of a nonlinear system is given by

$$W_o^{nonlinear} = \sum_{l=1}^{r} \sum_{m=1}^{s} \frac{1}{rsc_m^2} \int_0^{\infty} T_l \psi_\xi^{lm}(t) T_l^T dt \tag{11.24}$$

TABLE 11.1 Column details and steady-state operating conditions.

Total trays	12
Feed tray	7
Condenser	Total condenser
Reboiler	Partial reboiler
Operating pressure	1 atm
Operating temperatures (°K)	
Reflux drum	281.1
Reboiler	333.5
Feed	308.2
Distillate composition (mole fraction)	
Pentene	0.064854
Butene	0.934597
Hexene	0.000583
Bottom product composition (mole fraction)	
Pentene	0.063969
Butene	0.000484
Hexene	0.935512
Feed composition	Liquid pentene with unit mole fraction
Flow rates (kmole h^{-1})	
Feed	100
Top product	50
Bottom product	50
Reflux	200
Vapor boil up	225
Hold ups (kmole)	
Reboiler	24.8
Reflux drum	3.1
Reboiler heat duty (kcal h^{-1})	6,450,000

where ψ_ξ^{lm} is defined as

$$\psi_\xi^{lm}(t) = (y_\xi^{ilm}(t) - y_{\xi ss}^{ilm})^T (y_\xi^{jlm}(t) - y_{\xi ss}^{jlm})$$

In the above equation, the notation r indicates the number of matrices for perturbation directions, s is the number of different perturbation sizes for each direction, c is the magnitude of perturbations, and n is the number of states of the full order system. The measurement vector $Y^{ilm}(t)$ is the output of the system corresponding to the initial perturbation condition $x(0) = c_m T_l e_l + x_{ss}$, and Y_{ss}^{ilm} is the steady-state output of the system after this perturbation. The additional notation used for the observability covariance matrix can be referred to in Section 6.5.1 of Chapter 6: Optimal sensor configuration methods for state estimation. Computation of the observability covariance matrix requires the perturbation of each state of the system at least once. The discrete form of the overall empirical observability grammian of Eq. (11.24) is expressed by

$$W_o^{nonlinear} = \sum_{l=1}^{r} \sum_{m=1}^{s} \frac{1}{rsc_m^2} \sum_{1}^{q} T_l \psi^{lm}(t) T_l^T \Delta t_k \qquad (11.25)$$

In this equation Δt is the discrete sampling time. This grammian considers the influence of all state perturbations on all measurements. The nonsingularity condition of this grammian fulfils the observability of a nonlinear system.

The empirical observability grammian defined in Eq. (11.25) can provide the observability criterion of the reactive distillation column and it is unable to provide state observability of the column with respect to each individual measurement. However, the state sensitivity of individual measurements placed on each tray of the reactive distillation column can be captured by deriving measurement-specific empirical observability grammians from the nonlinear data of the system. Observability analysis of these grammians aids to optimally configure the sensors for state estimation of the process. These grammians represent the energy of individual outputs and are denoted by,

$$W_o^{nonlinear} = \sum_{l=1}^{r} \sum_{m=1}^{s} \frac{1}{rsc_m^2} \int_0^{\infty} T_l \psi_\xi^{lm}(t) T_l^T dt \tag{11.26}$$

where ψ_ξ^{lm} is defined as

$$\psi_\xi^{lm}(t) = (y_\xi^{ilm}(t) - y_{\xi ss}^{ilm})^T (y_\xi^{jlm}(t) - y_{\xi ss}^{jlm})$$

Here y_ξ^{ilm} represents a specific process measurement and $y_{\xi ss}^{ilm}$ is the corresponding steady-state value resulting due to state perturbation. The grammian in Eq. (11.26) considers the influence of all possible state perturbations on each measurement over the entire operating region. This grammian is different from the overall empirical observability grammian defined by Eq. (11.25), where the influence of all state perturbations on all measurements are considered. The discrete form of the measurement specific empirical observability grammians of Eq. (11.26) is expressed as:

$$W_{o\xi}^{nonlinear} = \sum_{l=1}^{r} \sum_{m=1}^{s} \frac{1}{rsc_m^2} \sum_{k=0}^{q} T_l \psi_{\xi ij}^{lm}(k) T_l^T \Delta t_k \tag{11.27}$$

where the term $\psi_{\xi ij}^{lm}(k)$ is expressed by

$$\psi_{\xi ij}^{lm}(k) = (y_\xi^{ilm}(k) - y_{\xi ss}^{ilm})^T (y_\xi^{jlm}(k) - y_{\xi ss}^{jlm})$$

Here $y_\xi^{ilm}(k)$ represents the ξth output of the nonlinear process at a kth instant corresponding to the initial state perturbation condition of $x = x_{ss} + c_m T_l e_i$ and $y_{\xi ss}^{jlm}$ is the corresponding steady state value of $y_\xi^{ilm}(k)$. The term c here refers to the magnitude of initial state perturbation. The states in metathesis reactive distillation are the compositions of pentene and butene in reboiler, all stages, and in reflux drum. The temperatures on each stage represent the measurements. The states are perturbed at initial conditions and the resulting individual nonlinear measurement data from the plant is collected and used to construct the grammian in Eq. (11.27). The number of time steps q is chosen such that the system after disturbance attains a steady-state within the specified time limit for data generation. Thus, the measurement specific empirical observability grammian for each stage of the column can be computed by using individual measurement data collected for all the state perturbations. This grammian signifies the state sensitivity of each measurement in the column.

11.3.3 Optimal sensor configuration using empirical observability grammians

The metathesis reactive distillation column contains 14 trays, including a reflux drum and reboiler. There are 28 states representing the compositions of pentene and butene on each stage in the column. The measurements available from the system are 14 temperatures on each stage. The observability covariance matrix of a nonlinear system and the overall empirical observability grammian for a discrete-time nonlinear system are defined by Eqs. (6.16) and (6.20) in Chapter 6: Optimal sensor configuration methods for state estimation, along with their notation and implementation procedure. For the computation of the empirical observability grammian, the values chosen for the terms in Eq. (11.27) are: $q = 3000$, $\Delta t_L = 18$ s, $T_1 = [I]_{28 \times 28}$, $T_2 = -[I]_{28 \times 28}$, $r = 2$, $s = 1$. An initial state perturbation with a step size of $c = 0.03$ is introduced in both positive and negative directions in each of the 28 states to generate the measurement responses. The overall empirical observability grammian in Eq. (11.25) is found to be nonsingular thus fulfilling the nonlinear observability condition of the reactive distillation process. The measurement specific empirical grammian $W_{o\xi}^{nonlinear}$ of Eq. (11.27) computed for each measurement is of dimension (28 $\times$ 28). Each of the 14 empirical observability grammians is subjected to eigenvalue analysis and then treated with the quantification measures defined in Section 6.5.2 [13]. The empirical observability metrics used to find the degree of observability of nonlinear systems are evaluated for sensor configuration. The results of the quantification measures evaluated for $\max\{Trace(W_{o\xi}^{nonlinear})\}$, $\max\{\sigma_{min}(W_{o\xi}^{nonlinear})\}$, and $\min\{Condition\ Number\ (W_{o\xi}^{nonlinear})\}$ are shown in Fig. 11.1. These results enable us to rank the sensors according to the magnitudes of these measures. The aim is to select a set of two or three measurements as inputs to the composition estimator. Accordingly, the two measurement combinations (11, 12), (1, 4), and (3, 12)

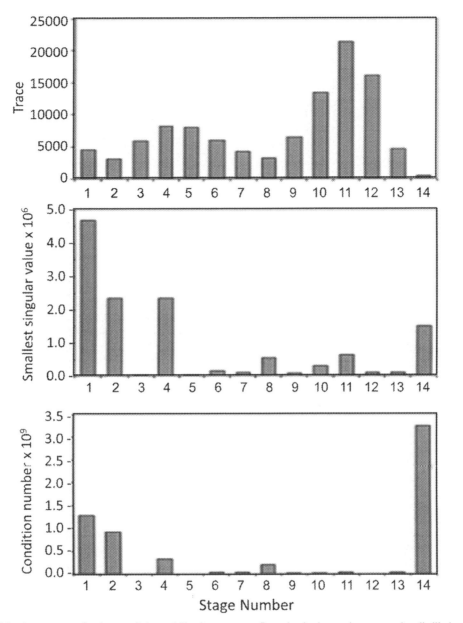

FIGURE 11.1   Quantification measures for degree of observability for sensor configuration in the continuous reactive distillation column.

suggested by trace, singular value, and condition number respectively, and the three measurement combinations (11, 12, 10), (1, 4, 2), and (3, 12, 5) suggested by these three quantification measures are found to be appropriate for sensor configuration.

11.4   State estimator design

An EKF is designed to obtain the instantaneous composition information from the temperature sensors data of the reactive distillation column. The EKF algorithm and general process representation are given in Section 4.4 of Chapter 4: Mechanistic model-based nonlinear filtering and observation techniques for optimal state/parameter estimation. The algorithm given in Section 4.4.3 is implemented for nonlinear state estimation in a continuous reactive distillation column. The schematic of the inferential estimation for metathesis reactive distillation is shown in Fig. 11.2.

The state vector for the metathesis reactive distillation column is given by

$$x = \begin{bmatrix} x_{b,1} & x_{b,2} & x_{1,1} & x_{1,2} & \dots & x_{Nt,1} & x_{Nt,2} & x_{d,1} & x_{d,2} \end{bmatrix}^T \tag{11.28}$$

Composition estimator design based on a rigorous mathematical model of multicomponent reactive distillation is not suitable for the realistic situation, since it is difficult to obtain the measured values of liquid and vapor flow rates and tray hold-ups with time. Therefore, a simplified model that assumes constant tray and reflux drum holdups and constant vapor and liquid flow rates are considered for designing the composition estimator. The elements $f_{i,j}$ of the F matrix in Eq. (4.27) in Chapter 4: Mechanistic model-based nonlinear filtering and observation techniques for optimal state/parameter estimation, are obtained by taking the partial derivatives of the component balance equations with respect to the state vector. These elements are given by

$$
\begin{aligned}
f_{1,1} &= (-V_b E_{b,11} - B + R_{b,11})/M_b \\
f_{1,2} &= (-V_b E_{b,12} + R_{b,12})/M_b \\
f_{1,3} &= (L_1)/M_b \\
f_{2,1} &= (-V_b E_{b,21} + R_{b,21})/M_b \\
f_{2,2} &= (-V_b E_{b,22} - B + R_{b,22})/M_b \\
f_{2,4} &= (L_1)/M_b \\
f_{3,1} &= (V_b E_{b,11})/M_1 \\
f_{3,2} &= (V_b E_{b,12})/M_1 \\
f_{3,3} &= (V_1 E_{1,11} - L_1 + R_{1,11})/M_1 \\
f_{3,4} &= (V_1 E_{1,12} + R_{1,12})/M_1 \\
f_{3,5} &= (L_2)/M_1 \\
f_{4,1} &= (V_b E_{b,21})/M_1 \\
f_{4,2} &= (V_b E_{b,22})/M_1 \\
f_{4,3} &= (-V_1 E_{1,21} + R_{1,21})/M_1 \\
f_{4,4} &= (-V_1 E_{1,22} - L_1 + R_{1,22})/M_1 \\
f_{4,6} &= (L_2)/M_1
\end{aligned}
\tag{11.29}
$$

Similarly, for $i = 5, 7, 9, \ldots, 23$ and $j = i - 2$, the following set of equations holds good.

$$
\begin{aligned}
f_{i,j} &= (V_{(i-4)/2} E_{(i-4)/2,11})/M_{(i-2)/2} \\
f_{i,j+1} &= (V_{(i-4)/2} E_{(i-4)/2,12})/M_{(i-2)/2} \\
f_{i,j+2} &= (-V_{(i-2)/2} E_{(i-2)/2,11} - L_{(i-2)/2} + R_{(i-2)/2,11})/M_{(i-2)/2} \\
f_{i,j+3} &= (-V_{(i-2)/2} E_{(i-2)/2,12} + R_{(i-2)/2,12})/M_{(i-2)/2} \\
f_{i,j+4} &= (L_{i/2})/M_{(i-2)/2}
\end{aligned}
\tag{11.30}
$$

And for $i = 6, 8, 10, \ldots, 24$ and $j = i - 2$, the following set of equations holds good.

$$
\begin{aligned}
f_{i+1,j} &= (V_{(i-4)/2} E_{(i-4)/2,21})/M_{(i-2)/2} \\
f_{i+1,j+1} &= (V_{(i-4)/2} E_{(i-4)/2,22})/M_{(i-2)/2} \\
f_{i+1,j+2} &= (-V_{(i-2)/2} E_{n,21} + R_{(i-2)/2,21})/M_{(i-2)/2} \\
f_{i+1,j+3} &= (-V_{(i-2)/2} E_{(i-2)/2,22} - L_{(i-2)/2} + R_{(i-2)/2,22})/M_{(i-2)/2} \\
f_{i+1,j+5} &= (L_{i/2})/M_{(i-2)/2}
\end{aligned}
$$

$$
\begin{aligned}
f_{25,23} &= (V_{Nt-1} E_{Nt-1,11})/M_{Nt} \\
f_{25,24} &= (V_{Nt-1} E_{Nt-1,12})/M_{Nt} \\
f_{25,25} &= (-V_{Nt} E_{Nt,11} - L_{Nt} + R_{Nt,11})/M_{Nt} \\
f_{25,26} &= (-V_{Nt} E_{Nt,12} + R_{Nt,12})/M_{nt} \\
f_{25,27} &= (R)/M_{Nt}
\end{aligned}
\tag{11.31}
$$

$$
\begin{aligned}
f_{26,23} &= (V_{Nt-1} E_{Nt-1,21})/M_{Nt} \\
f_{26,24} &= (V_{Nt-1} E_{Nt-1,22})/M_{Nt} \\
f_{26,25} &= (-V_{Nt} E_{Nt,21} + R_{Nt,21})/M_{Nt} \\
f_{26,26} &= (-V_{Nt} E_{Nt,22} - L_{Nt} + R_{Nt,22})/M_{Nt} \\
f_{26,28} &= (R)/M_{Nt}
\end{aligned}
\tag{11.32}
$$

$$
\begin{aligned}
f_{27,25} &= (V_{Nt} E_{Nt,11})/M_d \\
f_{27,26} &= (V_{Nt} E_{Nt,12})/M_d \\
f_{27,27} &= (-D - R + R_{d,11})/M_d \\
f_{27,28} &= (R_{d,12})/M_d
\end{aligned}
\tag{11.33}
$$

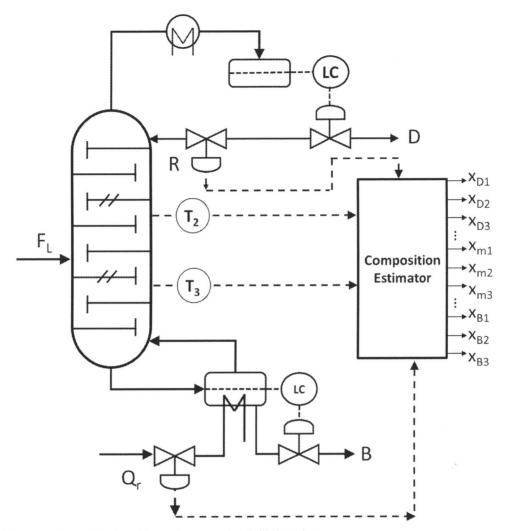

FIGURE 11.2 Schematic of state estimation of the continuous reactive distillation column.

$$
\begin{aligned}
f_{28,25} &= (V_{Nt}E_{Nt,21})/M_d \\
f_{28,26} &= (V_{Nt}E_{Nt,22})/M_d \\
f_{28,27} &= (R_{d,21})/M_d \\
f_{28,28} &= (-R - D + R_{d,22})/M_d
\end{aligned}
\tag{11.34}
$$

The values of $E_{n,11}$, $E_{n,12}$, $E_{n,21}$, and $E_{n,22}$ in the evaluation of the elements of Jacobian matrix F are given by the following equations.

$$
E_{n,11} = \frac{\partial y_{n,1}}{\partial x_{n,1}} = \frac{\left(\sum\limits_{j=1}^{Nc} P_{n,j}^{sat} x_{n,j}\right) P_{n,1}^{sat} - P_{n,1}^{sat} x_{n,1}\left(P_{n,1}^{sat} - P_{n,3}^{sat}\right)}{\left(\sum\limits_{j=1}^{Nc} P_{n,j}^{sat} x_{n,j}\right)^2}
\tag{11.35}
$$

$$
E_{n,12} = \frac{\partial y_{n,1}}{\partial x_{n,2}} = \frac{-P_{n,1}^{sat} x_{n,1}\left(P_{n,2}^{sat} - P_{n,3}^{sat}\right)}{\left(\sum\limits_{n=1}^{Nc} P_{n,j}^{sat} x_{n,j}\right)^2} \quad E_{n,21} = \frac{\partial y_{n,2}}{\partial x_{n,1}} = \frac{-P_{n,2}^{sat} x_{n,2}\left(P_{n,1}^{sat} - P_{n,3}^{sat}\right)}{\left(\sum\limits_{j=1}^{Nc} P_{n,j}^{sat} x_{n,j}\right)^2}
$$

$$E_{n,22} = \frac{\partial y_{n,2}}{\partial x_{n,2}} = \frac{\left(\sum\limits_{j=1}^{Nc} P_{n,j}^{sat} x_{n,j}\right) P_{n,2}^{sat} - P_{n,2}^{sat} x_{n,2}\left(P_{n,2}^{sat} - P_{n,3}^{sat}\right)}{\left(\sum\limits_{j=1}^{Nc} P_{n,j}^{sat} x_{n,j}\right)^2} \qquad (11.36)$$

$$\text{where } y_{n,j} = \frac{P_{n,j}^{sat} x_{n,j}}{\sum\limits_{j=1}^{Nc} P_{n,j}^{sat} x_{n,j}} \text{ and } \log P_{n,j}^{sat} = a_j - \left(\frac{b_j}{c_j + T_n}\right) \qquad (11.37)$$

The reaction terms $R_{n,11}$, $R_{n,12}$, $R_{n,2\ 1}$, and $R_{n,22}$ are given by the following equations.

$$R_{n,11} = -4\left(k_{fn} M_n x_{n,1}\right); \; R_{n,12} = 8\left(k_{fn} M_n x_{n,3}\right); \; R_{n,21} = -\left(\frac{R_{n,11}}{2}\right); \; R_{n,22} = -\left(\frac{R_{n,12}}{2}\right) \qquad (11.38)$$

where k_{fn} is given in Eq. (11.2).

The temperature model of a measurement tray is given by the equation

$$T_m = \frac{b_j}{a_j - \log\left(\dfrac{P_{m,j}^{sat} P_m}{\sum\limits_{1}^{Nc} P_{m,j}^{sat} x_{m,j}}\right)} - c_j \qquad (11.39)$$

where a, b, and c are coefficients in vapor–liquid equilibrium relation. The elements of the H matrix in Eq. (4.31) of Chapter 4 are obtained by taking the partial derivatives of Eq. (11.39) with respect to the state vector of Eq. (11.28). These elements are given by

$$h_{i,1} = \frac{-b_j(Ps_{i,1} - Ps_{i,3})}{\left(\sum\limits_{j=1}^{Nc} Ps_{i,j} x_{i,j}\right)\left\{a_j - \log\left(\dfrac{Ps_{i,j} P_i}{\sum\limits_{j=1}^{Nc} Ps_{i,j} x_{i,j}}\right)\right\}^2} \; ; \; h_{i,2} = \frac{-b_j(Ps_{i,2} - Ps_{i,3})}{\left(\sum\limits_{j=1}^{Nc} Ps_{i,j} x_{i,j}\right)\left\{a_j - \log\left(\dfrac{Ps_{i,j} P_i}{\sum\limits_{j=1}^{Nc} Ps_{i,j} x_{i,j}}\right)\right\}^2} \qquad (11.40)$$

The initial state covariance matrix Po, process noise covariance matrix Q, and measurement noise covariance matrix R are the design parameters in the estimation algorithm (Section 4.4.3 of Chapter 4, Mechanistic model-based nonlinear filtering and observation techniques for optimal state/parameter estimation), which are selected appropriately.

11.5 Estimator performance measure for optimality of sensor configuration

The veracity of the sensors configured for reactive distillation by the measurement specific empirical observability grammians is further ascertained by means of a performance index which is computed by comparing the state estimation results for different measurement combinations under different disturbance conditions with their corresponding actual resilts. The performance index is defined by

$$SumIAE_{estm} = \sum_{i=1}^{n_d} IAE_i$$

$$IAE_i = \sum_{t=0}^{tlimit} \left(\left|x_D^i(t) - \hat{x}_D^i(t)\right| + \left|x_B^i(t) - \hat{x}_B^i(t)\right|\right)\Delta t \qquad (11.41)$$

Here i refers to the disturbance condition, x_D and x_B are the actual compositions of top and bottom products in reflux drum and reboiler with $\hat{x}_D$ and $\hat{x}_B$ as their corresponding estimated compositions, IAE is the integral absolute error, $SumIAE_{estm}$ is the summated value of estimator performance. n_d represents the number of disturbance cases studied including normal operation as given in Table 11.2. Each IAE that contributes $SumIAE_{estm}$ corresponds to the specific

TABLE 11.2 Disturbance cases.

Case	Description
1	Normal operation
2	3% Decrease in Q_r
3	3% Increase in Q_r
4	3% Decrease in R
5	3% Increase in R
6	3% Decrease in F_L
7	3% Increase in F_L

TABLE 11.3 Estimator performance based on the sensors selected by the observability quantification measures.

Criterion	Sensor comb.	IAE							
		Case 1	Case 2	Case 3	Case 4	Case 5	Case 6	Case 7	$SumIAE_{estm}$
Trace	11, 12	0.0010	0.0280	0.0878	0.0635	0.0567	0.0445	0.0611	0.34294
	11, 12, 10	0.0013	0.0395	0.1048	0.0676	0.0568	0.0396	0.0570	0.36667
Min. singular value	1, 4	0.0055	0.0998	0.2700	0.2305	0.0969	0.0582	0.0385	0.79978
	1, 4, 2	0.0055	0.0933	0.2633	0.2269	0.093	0.0568	0.0364	0.77541
Condition number	3, 12	0.0037	0.0519	0.0857	0.0645	0.040	0.0066	0.0078	0.26037
	3, 12, 5	0.0043	0.0513	0.0818	0.0651	0.0399	0.0118	0.0110	0.26548

disturbance condition. The actual compositions used in the evaluation of *IAE* are those computed from numerical simulation of the rigorous mathematical model at each sample time.

11.6   Analysis of results

The design and implementation of the EKF inferential estimator with optimal sensor configuration are carried out using olefin metathesis reactive distillation column as a testbed. The design parameters of the EKF algorithm (Section 4.4.3 of Chapter 4), namely, the initial state covariance matrix P_o, the process noise covariance matrix Q, and the measurement noise covariance matrix R, are selected using the process and measurement noise information and further tuned heuristically. The matrices P_o, Q, and R with their diagonal elements of 0.0005, 0.0005, and 5.0, respectively, are found to provide effective estimator performance. The temperature data is generated by solving the rigorous mathematical model of reactive distillation (Section 11.2) using Euler's integration technique with a step size of 3.6 s. In order to reflect the real plant operation, the measurements are corrupted with zero mean random Gaussian noise by considering a 0.3% deviation in the actual data. The measurements suggested by the sensor configuration measures are further ascertained by performing state estimation with these measurements and evaluating the performance index in Eq. (11.41) for different disturbance conditions of the plant given in Table 11.2. The $SumIAE_{estm}$ results in Table 11.3 show that the set of two measurements representing stage temperatures (3, 12) and the set of three measurements representing stage temperatures (3, 12, 5) provide better estimator performance. These two and three measurement combinations correspond to those suggested by the measure $\min\left\{ConditionNumber(W_{o\xi}^{nonlinear})\right\}$. These results thus indicate better suitability of the empirical observabilty grammian based strategy for measurement selection in reactive distillation.

To find the better suitability of the measurements selected by sensor quantification measures, the estimator performance is further studied with various randomly selected two and three measurement combinations under different disturbance conditions of the plant. The IAE results of these nonoptimal measurement combinations are evaluated along with the optimally selected measurements by the condition number. The measurement set (3, 12) of the two measurement combinations and the measurement set (3, 12, 5) of the three measurement combinations are found to be optimal sets for sensor configuration. It is observed that the set with two measurements suggested by $\min\left\{ConditionNumber(W_{o\xi}^{nonlinear})\right\}$ has lower $SumIAE_{estm}$ over the set with three measurements. Since the system is weakly observable with one measurement, the selection of a single measurement for state estimation is ignored in this study. The sets with more than three measurements are also not considered since adding more measurements may introduce measurement redundancy apart from increasing computational burden.

The sensors configured by the measurement-specific empirical observability grammian are further analyzed. The $SumIAE_{estm}$ has confirmed the better suitability of the set of two sensors (3, 12) and the set of three sensors (3, 12, 5) as optimum configurations for state estimation in continuous reactive distillation. However, the condition number results of the set of three sensors (3, 12, 5) have shown the better results of individual sensors in the order, Sensor 3 > Sensor 12 > Sensor 5. Since Sensors 3, 12, and 5 are found to be optimum individually, the efficacy of the estimator is evaluated with two measurement combinations (3, 5), (3, 12), and (5, 12). The set of sensors (5, 12) has shown similar performance to that of (3, 12). The set of sensors (3, 5) has not yielded improved results, probably due to measurement redundancy because these two sensors are located closely. In this study, among the two and three optimal measurement configurations suggested by the condition number, the two measurement configuration (3, 12) is found effective over the three measurement configuration (3, 12, 5) for composition estimation in reactive distillation by the model-based state estimator.

The sensitivity and robustness of the designed composition estimator with the optimally configured measurements (3, 12) are further studied towards the effect of the estimator design parameters, Po, Q, and R, and measurement noise by considering the process operation under different disturbance conditions. The IAE results in Table 11.4

TABLE 11.4 Effect of estimator design parameters and measurement noise.

Parameter	IAE							
	Case 1	Case 2	Case 3	Case 4	Case 5	Case 6	Case 7	$SumIAE_{estm}$
Initial state covariance matrix								
P_0	0.00371	0.05188	0.08574	0.06455	0.04002	0.00662	0.00786	0.26037
$2P_0$	0.00401	0.051241	0.085423	0.0646	0.040058	0.006777	0.008139	0.26025
$P_0/2$	0.003772	0.051475	0.085503	0.064555	0.039952	0.00664	0.007925	0.25982
Process noise covariance matrix								
Q	0.00371	0.05188	0.08574	0.06455	0.04002	0.00662	0.00786	0.26037
$2Q$	0.004446	0.040445	0.072699	0.055541	0.033515	0.007397	0.006581	0.22062
$Q/2.0$	0.003381	0.067008	0.10412	0.077536	0.04949	0.007268	0.012048	0.32085
Measurement noise covariance matrix								
R	0.00371	0.05188	0.08574	0.06455	0.04002	0.00662	0.00786	0.26037
$2R$	0.003203	0.067194	0.10438	0.077609	0.049447	0.007216	0.011905	0.32095
$R/1.5$	0.004398	0.044556	0.077455	0.059008	0.036067	0.007196	0.007071	0.23575
Noise levels								
ν	0.00371	0.05188	0.08574	0.06455	0.04002	0.00662	0.00786	0.26037
2ν	0.007527	0.052732	0.088164	0.067356	0.041576	0.009576	0.011056	0.27799
4ν	0.014092	0.056084	0.094876	0.073726	0.046472	0.015976	0.017399	0.31862

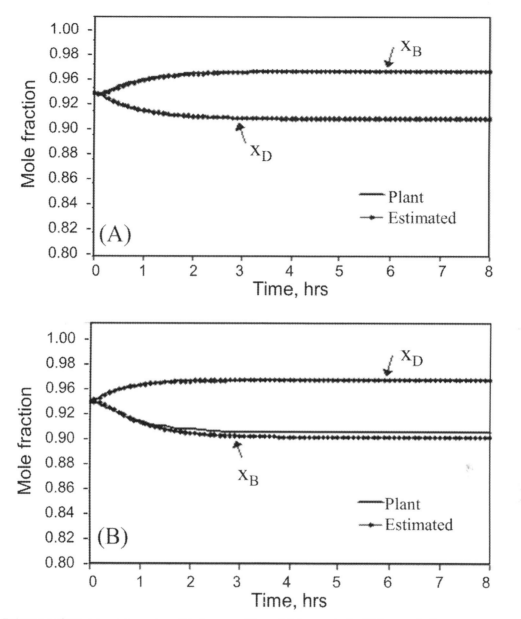

FIGURE 11.3   Estimator performance under open-loop disturbance conditions: (A) decrease in F_L; (B) increase in F_L.

show the robustness of the composition estimator towards the changes in estimator design parameters in the presence of disturbances in the process. Figs. 11.3 and 11.4 show the comparison of the estimated composition profiles with those of actual ones for different disturbance conditions when the estimator supported by the simplified model is used in conjunction with the two measurement configuration of (3, 12). A detailed, rigorous model can be used to support the estimator if resources are available in real-time. However, the state estimator supported by a detailed, rigorous model requires the measured information concerning the time-varying holdups, enthalpy, vapor and liquid flow rates, which are not available in real practice. Moreover, the detailed, rigorous model-based state estimation also requires additional computational effort. However, in this study, the estimator is also evaluated with the support of the detailed, rigorous model of the process in conjunction with the optimal set of two measurements (3, 12). The comparison of the results of the simplified model-based state estimator with that of the rigorous model-based estimator shown in Table 11.5 indicates that there is no marked difference in the performance of the estimator for these two models.

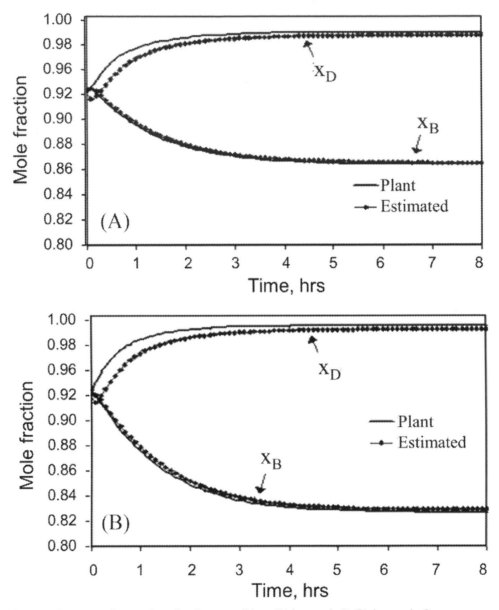

FIGURE 11.4 Estimator performance under open-loop disturbance conditions: (A) increase in R; (B) decrease in Q_r.

TABLE 11.5 Comparison of simplified and rigorous model-based estimators.

Model	IAE							
	Case1	Case2	Case3	Case4	Case5	Case6	Case7	$SumIAE_{estm}$
Rigorous	0.008099	0.04801	0.085515	0.056364	0.032946	0.016486	0.01484	0.26231
Simple	0.00371	0.05188	0.08574	0.06455	0.04002	0.00662	0.00786	0.26037

11.7 Summary

The measurement-specific empirical observability grammians provide a means of extracting the dynamic state-sensitive measurement information of the nonlinear process to evaluate its degree of observability for sensor configuration. In this chapter, the dynamic state-sensitive measurement information extracted from a nonlinear reactive distillation system is employed to configure the measurement specific empirical observability grammians, which are then subjected to various scalar quantification measures to find the degree of observability of the process in order to configure the measurement sensors for state estimation. The optimally configured measurements are incorporated in the model-based state estimation scheme for reliable estimation of compositions in the reactive distillation process. The performance of the inferential estimation scheme with optimized sensor locations is studied with respect to the effect of estimator design parameters, noisy measurements, and process disturbances. The results demonstrate the effectiveness of the inferential estimation scheme with optimized sensor locations for reliable estimation of compositions in the continuous reactive distillation process.

References

[1]   B. Joseph, C.B. Brosilow, Inferential control of processes: Part I. Steady state analysis and design, AIChE J. 24 (3) (1978) 485−492.

[2]   L. Lang, E.D. Gilles, Nonlinear observers for distillation columns, Comput. Chem. Eng. 14 (11) (1990) 1297−1301.

[3]   T. Mejdell, S. Skogestad, Composition estimator in a pilot-plant distillation column using multiple temperatures, Ind. Eng. Chem. Res. 30 (12) (1991) 2555−2564.

[4]   M.J. Olanrewaju, M.A. Al-Arfaj, Estimator-based control of reactive distillation system: application of an extended Kalman filtering, Chem. Eng. Sci. 61 (10) (2006) 3386−3399.

[5]   D.R. Yang, K.S. Lee, Monitoring of a distillation column using modified extended Kalman filter and a reduced order model, Comput. Chem. Eng. 21 (1997) S565−S570.

[6]   R. Baratti, et al., Development of a composition estimator for binary distillation columns. Application to a pilot plant, Chem. Eng. Sci. 50 (10) (1995) 1541−1550.

[7]   M. Kano, et al., Inferential control system of distillation compositions using dynamic partial least squares regression, J. Process Control 10 (2−3) (2000) 157−166.

[8]   M.J. Okasinski, M.F. Doherty, Design method for kinetically controlled, staged reactive distillation columns, Ind. Eng. Chem. Res. 37 (7) (1998) 2821−2834.

[9]   K. Alejski, F. Duprat, Dynamic simulation of the multicomponent reactive distillation, Chem. Eng. Sci. 51 (18) (1996) 4237−4252.

[10] T. Boublík, V. Fried, E. Hála, The vapor pressures of pure substances, AIChE J. 19 (5) (1973) 1085−1086.

[11] D.W. Green, M.Z. Southard, Perry's Chemical Engineers' Handbook, McGraw-Hill Education, New York, NY, 2019.

[12] C.L. Yaws, Calculate liquid heat capacity, Hydrocarb. Process. (Int. Ed.) 70 (12) (1991) 73−77.

[13] C. Sumana, C. Venkateswarlu, Optimal selection of sensors for state estimation in a reactive distillation process, J. Process Control 19 (6) (2009) 1024−1035.

[14] A.K. Singh, J. Hahn, Sensor location for stable nonlinear dynamic systems: multiple sensorc, Ind. Eng. Chem. Res. 45 (10) (2006) 3615−3623.

Chapter 12

Application of mechanistic model-based nonlinear filtering and observation techniques for optimal state estimation of a catalytic tubular reactor with optimal sensor configuration

12.1 Introduction

Optimal state estimation has vital role in monitoring and control of distributed dynamic systems. State estimator provides the estimates of unmeasured process variables based on known measurements and the process knowledge. These instantaneously estimated process variables serve as inferential measurements for process operations and control strategies. However, the estimated states critically depend on the number of available measurements and their appropriate locations in the process. For optimal state estimation, it has to be verified that the considered measurements are sufficient and the process observability criteria are satisfied. Thus besides designing efficient state estimators based on the few available data, locating sensors at appropriate positions to obtain the best information about the process dynamics has become crucial.

To place the sensors at appropriate locations, the system has to satisfy the observability criteria. For distributed dynamic systems, observability is not only affected by the choice of sensors but also by their locations. The known observations/measurements form the input to the state estimator. The measurements used for state estimation must be minimal in number to reduce the cost, and they must possess the highest sensitivity and lowest mutual interactions. The optimal sensor locations can be studied by different methods like principal component analysis, singular value decomposition (SVD) and observability Grammians, and others [1−3]. The observability Grammian-based approach is useful for optimal sensor configuration in multistage dynamic systems [4]. In this approach, the observability covariance matrix is computed from the dynamic behavior of the system outputs for different perturbations in the initial conditions of the system while the system inputs are kept at their steady values. These empirical observability Grammians are further subjected to observability metrics to quantify the degree of observability to identify the correct sensor locations [5,6].

The optimally configured sensor measurements can be used in a model-based state estimation scheme to estimate the states of the system. The extended Kalman filter (EKF) is a widely used model-based estimator for state estimation in many applications. Several applications of EKF concerning to state estimation have been reported in the literature [3,7,8]. However, in EKF implementation, the process and measurement functions are linearized around the current predicted state to compute the Jacobians. When the linearized functions represent a highly nonlinear process, the EKF may not provide effective performance. The unscented Kalman filter (UKF) is a recursive estimator that addresses some of the approximation issues of EKF [9]. Unlike the EKF, the UKF does not approximate the nonlinear process and observation models but uses the true nonlinear models. Instead of approximating a nonlinear function, the UKF approximates probability distribution by use of the unscented transform. The UKF uses the so-called sigma points and propagates them through the system model to numerically estimate the state vector. The UKF has been reported for state estimation in various nonlinear systems [10−13].

This chapter presents a sensor selection and state estimation strategy for a distributed dynamic system, in which empirical observability Grammians are used to configure the sensor locations and an UKF is employed for state

Optimal State Estimation for Process Monitoring, Fault Diagnosis and Control. DOI: https://doi.org/10.1016/B978-0-323-85878-6.00019-1

estimation. The performance of the measurement selection methodology and state estimation scheme is evaluated by applying it to a catalytic tubular reactor, in which partial oxidation of benzene takes place to produce maleic anhydride. The Grammian configured temperature measurements are employed in UKF for instantaneous estimation of reactant and product concentration profiles of the tubular reactor. The results demonstrate the usefulness of the sensor selection methodology and state estimation scheme for inferential estimation of states in the catalytic tubular reactor.

12.2 The process and its mathematical model

The process taking place in the reactor and the model describing the reactor are given as follows.

12.2.1 Process

A fixed bed tubular reactor for the partial catalytic oxidation of benzene to produce maleic anhydride has been considered [6]. The reaction is an irreversible exothermic reaction. When benzene is oxidized to maleic anhydride, the following three exothermic, irreversible gas phase reactions occur on a solid $V_2O_5-MoO_3-P_2O_5$ catalyst particle:

$$C_6H_6 + 4O_2 \xrightarrow{k_1} C_4H_2O_3 + CO + CO_2 + 2H_2O \tag{12.1}$$

$$C_6H_6 + 6O_2 \xrightarrow{k_2} 3\,CO + 3CO_2 + 3H_2O \tag{12.2}$$

$$C_4H_2O_3 + 2O_2 \xrightarrow{k_3} 2CO + 2CO_2 + H_2O \tag{12.3}$$

Reaction (12.1) is the desired path for forming maleic anhydride, which is the product from benzene. Reaction (12.2) shows the oxidation of benzene to produce carbon monoxide, carbon dioxide, and water. Reaction (12.3) shows the oxidation of the product, maleic anhydride to carbon monoxide, carbon dioxide, and water. Reactions (12.2) and (12.3) are the side reactions representing undesired oxidation of the reactant and product. Feed stream to the reactor is considered as air mixed with approximately 0.6 mol s^{-1} of benzene. Since oxygen in the air is present in excess in the feed, all reactions are assumed to be pseudo-first order for the limiting reactant. The rate constants in these reactions are considered as a function of the catalyst bed temperature. Arrhenius equation is used to calculate the reaction rate constants as a function of both the temperature and length of the reactor. The Arrhenius equation is given by

$$k_i(t, z) = A_i e^{-E_i/RT_s(t,z)} \tag{12.4}$$

k_i, A_i, and E_i are the rate constant, frequency factor, and activation energy of the ith reaction, T_s is the catalyst bed temperature, and R is the universal gas constant.

12.2.2 Mathematical model

A distributed dynamic model that considers the variation of concentration and temperature throughout the reactor is used to describe the catalytic conversion process [6]. Two mass balances are used for the reactant benzene, and the product maleic anhydride, which are expressed by partial differential equations. Dispersion is assumed to be negligible.

The molar flow of benzene, F_B (mol s^{-1}), is given by

$$\frac{\partial F_B(t, z)}{\partial t} = -v\frac{\partial F_B(t, z)}{\partial z} + D_{eff}\frac{\partial^2 F_B(t, z)}{\partial z^2} - k_1(t, z)F_B(t, z) - k_2(t, z)F_B(t, z) \tag{12.5}$$

The molar flow of maleic anhydride, F_{MA} (mol s^{-1}), is given by

$$\frac{\partial F_{MA}(t, z)}{\partial t} = -v\frac{\partial F_{MA}(t, z)}{\partial z} + D_{eff}\frac{\partial^2 F_{MA}(t, z)}{\partial z^2} + k_1(t, z)F_B(t, z) - k_3(t, z)F_{MA}(t, z) \tag{12.6}$$

where t is time, and z is axial direction. The radial direction is neglected on assuming a plug-flow reactor.

In the above equations, F_B is the benzene molar concentration, F_{MA} is the maleic anhydride concentration, D_{eff} is the effective mass diffusion coefficient, v is linear gas velocity, and k_1, k_2, and k_3 are the rate constants of the respective reactions. The partial differential equations that describe the dynamics of tubular reactor consist of diffusion–convection–reaction terms. The first term of the mass balance equations represents the dynamic term. The second term stands for the convection term. The third term signifies the diffusion term, and the other terms represent the reaction terms.

TABLE 12.1 Values of the parameter used in the model equations.

Parameter	Value (unit)
Linear gas velocity, v	2.48 (m s^{-1})
Effective mass diffusion coefficient, D_{eff}	3.17×10^{-3} (m^2 s^{-1})
Effective heat diffusion coefficient, K_{eff}	3.17×10^{-2} (m^2 s^{-1})
Effective heat transfer coefficient for fluid phase-wall temperature, U_{f-w}	10.6 (s^{-1})
Effective heat transfer coefficient for solid–fluid phase, U_{s-f}	84 (s^{-1})
Wall temperature, T_w	733 (K)
Solid-phase heat balance constant, C_s	0 0.729 (s K J^{-1})

TABLE 12.2 Values of activation energy, frequency factors, and heats of reaction.

Reaction	A_i (s^{-1})	E_i (J mol^{-1})	ΔH_i (J mol^{-1})
(12.1)	86760	71711.7	-1490×10^{-3}
(12.2)	37260	71711.7	-2322×10^{-3}
(12.3)	149.4	36026.3	-832×10^{-3}

Two heat balance equations representing the gas flow temperature T_f (K) and the temperature of the solid phase catalyst, T_s (K), are used for the reactor. The temperature of the fluid phase (gas flow) is a partial differential equation given by

$$\frac{\partial T_f(t,z)}{\partial t} = -v\frac{\partial T_f(t,z)}{\partial z} + k_{eff}\frac{\partial^2 T_f(t,z)}{\partial z^2} - U_{f-w}(T_f(t,z) - T_w) - U_{s-f}(T_s(t,z) - T_f(t,z)) \tag{12.7}$$

The catalyst bed temperature is a differential equation, which is given by

$$\frac{dT_s(t,z)}{dt} = -U_{s-f}(T_s(t,z) - T_f(t,z)) + c_s\Delta H_1 k_1(t,z)F_B(t,z)$$
$$+ c_s\Delta H_2 k_2(t,z)F_B(t,z) + c_s\Delta H_3 k_3(t,z)F_{MA}(t,z) \tag{12.8}$$

In the above equations, k_{eff} is the effective heat diffusion coefficient, U_{s-f} is the effective heat transfer coefficient of solid–fluid phase, U_{f-w} is the effective heat transfer coefficient of fluid phase-wall temperature, c_s is the solid phase heat transfer coefficient, ΔH_1, ΔH_2, and ΔH_3 are the heat of reaction of the respective reaction, and T_w is the wall temperature. Pressure drop and radial diffusion are neglected in the reactor model. The parameter values are given in Table 12.1, and the values of the heat of reaction, frequency factors and activation energies are given in Table 12.2.

12.3 Method of solution

The mathematical model of the catalytic tubular reactor is described by second order partial differential equations, which are parabolic type. To solve the parabolic system of equations, two boundary conditions (BCs) are required. The two BCs for the model equations are given as follows:

$$\text{BC 1:} F_B(t,0) = [\text{feed}]\text{mol s}^{-1}, \quad F_{MA}(t,0) = 0 \left(\text{mol s}^{-1}\right), \quad T_f(t,0) = 733 \text{ K}, \quad T_s(t,0) = 733 \text{ K} \tag{12.9}$$

$$\text{BC2:} \frac{dF_B}{dz} = 0, \quad \frac{dF_{MA}}{dz} = 0, \quad \frac{dT_f}{dz} = 0, \quad \frac{dT_s}{dz} = 0 \tag{12.10}$$

Diffusion effects at the entrance and exit are neglected. For convenience, this model is transformed into a linear finite dimensional state space form. First, the reactor length is divided into m equidistant segments indicated by z_i. The distance between two successive grid points is set as Δz. Every grid point z_i is defined by four differential (partial) equations, Eqs. (12.5)–(12.8). Then the first and second order differential terms in partial differential equations on every gird point z_i are approximated by finite difference terms according to the method of least squares (MOL). The MOL uses the finite difference technique to replace all the space derivatives to obtain a system of ordinary differential equations. Thus the original reactor model is transformed into a set of $n = 4 \times m$ ordinary differential equations, two mass and two heat balances on all m grid points over the reactor length.

To apply MOL, the first and the second order differential terms are approximated by the finite difference terms as

$$\frac{\partial f}{\partial z} = \frac{f(t, z_i) - f(t, z_{i-1})}{\Delta z} \tag{12.11}$$

$$\frac{\partial^2 f}{\partial z^2} = \frac{f(t, z_{i-1}) - 2f(t, z_i) + f(t, z_{i+1})}{\Delta z^2} \tag{12.12}$$

By applying the MOL, Eqs. (12.5)–(12.8) can be transformed to the following ordinary differential equations,

$$\frac{\partial F_B(t, z)}{\partial t} = \left(AF_{Bi-1} - BF_{Bi} + CF_{Bi+1} - F_{Bi}Af_1 \exp^{-E_1/(RT_{si})} - F_{Bi}Af_2 \exp^{-E_2/(RT_{si})} \right) \tag{12.13}$$

$$\frac{\partial F_{MA}(t, z)}{\partial t} F_{MAi,t+1} = \left(AF_{MAi-1} - BF_{MAi} + CF_{MAi+1} - F_{Bi}Af_1 \exp^{-E_1/(RT_{si})} - F_{MAi}Af_3 \exp^{-E_3/(RT_{si})} \right) \tag{12.14}$$

$$\frac{\partial T_f(t, z)}{\partial t} = \left(A_1 T_{fi-1} - B_1 T_{fi} + C_1 T_{fi+1} - U_{f-w}(T_{fi} - T_W) - U_{s-f}(T_{si} - T_{fi}) \right) \tag{12.15}$$

$$\frac{dT_s(t, z)}{dt} = \left(-U_{s-f}(T_{si} - T_{fi}) + G_1 F_{Bi} \exp^{-E_1/(RT_{si})} + G_2 F_{Bi} \exp^{-E_2/(RT_{si})} + GF_{MAi} \exp^{-E_3/(RT_{si})} \right) \tag{12.16}$$

In the above equations, Af_1, Af_2, and Af_3 are the frequency factors, and $G_j = Af_j \Delta H_j c_S$ for $j = 1-3$. The terms A, B, C, A_1, B_1, and C_1 are constants and defined as follows.

$$A = \frac{v}{\Delta z} + \frac{D_{eff}}{\Delta z^2}; \quad A_1 = \frac{v}{\Delta z} + \frac{K_{eff}}{\Delta z^2}$$

$$B = \frac{v}{\Delta z} + \frac{2D_{eff}}{\Delta z^2}; \quad B_1 = \frac{v}{\Delta z} + \frac{K_{eff}}{\Delta z^2} + U_{f-w} - U_{s-f}$$

$$C = \frac{D_{eff}}{\Delta z^2}; \quad C_1 = \frac{K_{eff}}{\Delta z^2}$$

First order differential terms in the above equations can be transformed into discrete form by Euler's method. A first order differential equation of the type

$$\frac{df}{dt} = F(x, t) \tag{12.17}$$

can be expressed according to Euler's integration formula as

$$\frac{f(x, \ t+1) - f(x, \ t)}{\Delta t} = F(x, \ t-1)$$

$$f(x, \ t+1) = f(x, \ t) + \Delta t \times F(x, \ t-1) \tag{12.18}$$

Thus using the finite difference approximations to the first and second partial derivative terms, and Euler's integration formula for the first differential term, the discrete representation of the reactor model in Eqs. (12.5)–(12.8) are expressed as follows.

$$F_{Bi,t+1} = F_{Bi,t} + \Delta t \left(AF_{Bi-1} - BF_{Bi} + CF_{Bi+1} - F_{Bi}Af_1 \exp^{-E_1/(RT_{si})} - F_{Bi}Af_2 \exp^{-E_2/(RT_{si})} \right) \tag{12.19}$$

$$F_{MAi,t+1} = F_{MAi,t} + \Delta t \left(AF_{MAi-1} - BF_{MAi} + CF_{MAi+1} - F_{Bi}Af_1 \exp^{-E_1/(RT_{si})} - F_{MAi}Af_3 \exp^{-E_3/(RT_{si})} \right) \tag{12.20}$$

$$T_{fi,t+1} = T_{fi,t} + \Delta t \left(A_1 T_{fi-1} - B_1 T_{fi} + C_1 T_{fi+1} - U_{f-w}(T_{fi} - T_W) - U_{s-f}(T_{si} - T_{fi}) \right) \tag{12.21}$$

$$T_{si,t+1} = T_{si,t} + \Delta t \left(-U_{s-f}(T_{si} - T_{fi}) + G_1 F_{Bi} \exp^{-E_1/(RT_{si})} + G_2 F_{Bi} \exp^{-E_2/(RT_{si})} + GF_{MAi} \exp^{-E_3/(RT_{si})} \right) \tag{12.22}$$

The reactor length is divided into discrete steps while accounting BCs and initial conditions to obtain the model solution.

12.4   Results of numerical solution

The reactor is of length 3.2 m, which is divided into 32 grid points with each element of length 0.1 m. The state vector at grid point z_1 is represented by

$$x = \begin{bmatrix} F_B(t, z_1), & F_{MA}(t, z_1), & T_f(t, z_1), & T_s(t, z_1) \end{bmatrix}^T \tag{12.23}$$

Thus there are 32 discrete states for each variable form the grid points. These states are initialized at all the grid points and the model equations, Eqs. (12.19)–(12.22) are solved considering a time step of 0.01 s. The solution is continued until the steady solution is reached. The solutions can be established for different initial state conditions. The steady solutions of the concentrations and temperatures are shown in Figs. 12.1–12.4 for different initial conditions of reactant concentrations. The initial conditions for the results of these figures are benzene flow rate with 0.57, 0.6, and

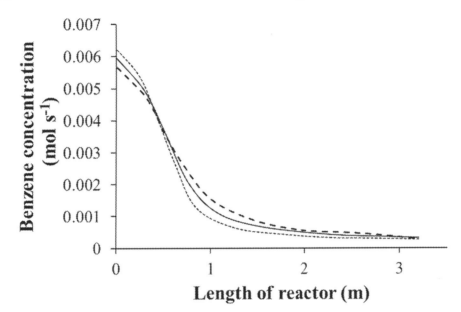

FIGURE 12.1 Benzene concentration along the length of the reactor with the initial benzene concentration of 0.0057 mol s⁻¹ (— — —), 0.006 mol s⁻¹ (———), and 0.0063 mol s⁻¹ (. . . . . .).

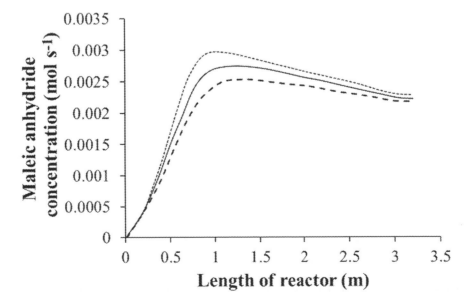

FIGURE 12.2 Maleic anhydride concentration along the length of the reactor with the initial benzene concentration of 0.0057 mol s⁻¹ (— — —), 0.006 mol s⁻¹ (———), and 0.0063 mol s⁻¹ (. . . . . .).

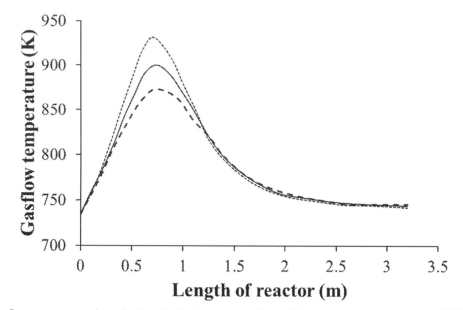

FIGURE 12.3 Gas flow temperature along the length of the reactor with the initial benzene concentration of $0.0057 \, mol \, s^{-1}$ (— — —), $0.006 \, mol \, s^{-1}$ (———), and $0.0063 \, mol \, s^{-1}$ (. . . . .).

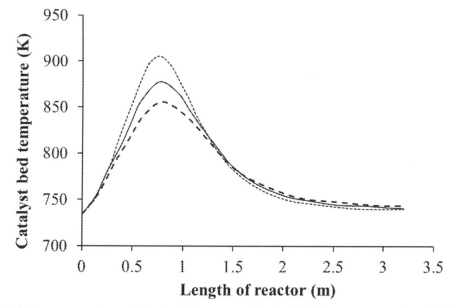

FIGURE 12.4 Catalyst bed temperature along the length of the reactor with the initial benzene concentration of $0.0057 \, mol \, s^{-1}$ (— — —), $0.006 \, mol \, s^{-1}$ (———), and $0.0063 \, mol \, s^{-1}$ (. . . . .).

0.63 mol%, maleic anhydride flow rate with 0 mol%, gas flow temperature of 733K and catalyst bed temperature of 733K. These figures indicate the steady state solutions of benzene and maleic anhydride concentrations, and the gas flow and catalyst bed temperatures with respect to the length of the reactor for the given initial conditions.

12.5   Optimal sensor configuration in a catalytic tubular reactor

Optimal state estimation of a distributed dynamic system requires rigorous selection criteria for identifying the exact sensor locations. This study employs empirical observability Grammians to select the temperature sensor locations for state estimation in a catalytic tubular reactor.

12.5.1 Significance of sensor configuration in a tubular reactor

The instruments/sensors used to measure the plant operational information should be selected to deliver the information suitable enough to monitor the status of the plant. The cost of procurement and maintenance often form an obstacle for the number of sensors that can be implemented in plant operation. For optimal state estimation in distributed dynamic systems like catalytic tubular reactor, the measurement sensors that characterize the process behavior should be placed at appropriate locations. Inappropriate placement of measurements or the use of the excess number of sensors than required may lead to numerical problems such as singularity or over parameterization, which may drastically reduce the performance of the state estimator. The success of state estimation in catalytic tubular reactor relies on the rigorous evaluation of the observability and appropriate selection of measurements that adequately characterize the process behavior. Thus for optimal state estimation, it needs to be verified that the selected measurements are adequate and the process observability criteria are satisfied. Moreover, for optimal state estimation, the system should exhibit a high degree of observability for the selected measurements. Thus the development of an inferential estimator with an optimal selection of sensors gains significance for state estimation in a catalytic tubular reactor.

12.5.2 Empirical observability Grammians

This study explores the use of Grammian covariance matrices for sensor selection for optimal state estimation in a tubular reactor. The empirical observability Grammian-based sensor selection procedure for nonlinear systems is presented in Sections 6.5.1 and 6.5.2 of Chapter 6, Optimal Sensor Configuration Methods for State Estimation. The measurement specific empirical observability Grammians approach presented in Section 11.3.2 of Chapter 11 is employed to configure the temperature sensors for the catalytic tubular reactor. Accordingly, the discrete measurement specific empirical observability Grammian of Eq. (11.27) is given as

$$W_{o\xi}^{nonlinear} = \sum_{l=1}^{r}\sum_{m=1}^{s}\frac{1}{rsc_m^2}\sum_{k=0}^{q}T_l\psi_{\xi ij}^{lm}(k)T_l^T\Delta t_k \tag{12.24}$$

where the term $\psi_{\xi ij}^{lm}(k)$ is expressed by

$$\psi_{\xi ij}^{lm}(k) = (y_\xi^{ilm}(k)-y_{\xi ss}^{ilm})^T(y_\xi^{jlm}(k)-y_{\xi ss}^{jlm})$$

Here $y_\xi^{ilm}(k)$ represents the ξth output of the nonlinear process at kth instant corresponding to the initial state perturbation condition of $x = x_{ss} + c_m T_l e_i$ and $y_{\xi ss}^{jlm}$ is the corresponding steady state value of $y_\xi^{ilm}(k)$. The notation r indicates the number of matrices for perturbation directions, s is the number of different perturbation sizes for each direction, c_m is the magnitude of state perturbations. The term e_i refers to unit vectors with respective dimensions. The states of the tubular reactor are the concentration of benzene (F_B) and maleic anhydride (F_{MA}), and the temperatures of the gas flow (T_f), and the solid phase catalyst (T_s). The temperatures are also considered as the measurements. The states are perturbed at initial conditions, and the resulting individual nonlinear measurement data from the plant is collected and used to construct the Grammians in Eq. (12.24). The number of time steps q is chosen such that the system after disturbance attains a steady state within the specified time limit. Thus the measurement specific empirical observability Grammians can be computed using individual measurement data collected for all the state perturbations. These Grammians signify the state sensitivity of each measurement location.

12.5.3 Optimal sensor configuration using empirical observability Grammians

In this study, the length of the reactor is 3.2 m, and the time is considered till the reactor reaches the steady state and the time step is taken as 0.01 s. The total reactor is divided into 32 elements, with the length of each element $\Delta z = 0.1$. Thus the solution domain of the reactor model is divided into a grid with 32 points. The states at each grid point are two concentrations and two temperatures. These two temperatures also represent the measurements of the system. The time to attain a steady state is considered 250 discrete time points, with each time step as 0.01 s. For the computation of the empirical observability Grammian in Eq. (12.24), the values chosen for the terms in the equation are: $q = 250$, $\Delta t = 0.01$ s, $T_1 = [I]_{32 \times 32}$, $T_2 = -[I]_{32 \times 32}$, $r = 2$, $s = 1$. An initial state perturbation with a step size of $c_m = 0.03$ is introduced in both positive and negative directions in each of the 32 states to generate the measurement responses. The dimensionality of each state for 250 time-steps is 250 $\times$ 32. The measurement specific empirical Grammian $W_{o\xi}^{nonlinear}$ of Eq. (12.24) computed for each measurement is of dimension (32 $\times$ 32). The $\psi_{\xi ij}^{lm}(k)$ for each measurement condition for the case of positive deviation is of dimension (32 $\times$ 32), which is denoted as ψ^{11}. Similarly, The $\psi_{\xi ij}^{lm}(k)$ for each measurement condition for the case of negative deviation is of dimension (32 $\times$ 32), which is denoted as ψ^{22}. Thus the overall empirical observability Grammian for each measurement condition is given by

$$W_{o\xi}^{\text{nonlinear}} = \frac{1}{2c^2{}_m}\left[\sum_{k=0}^{q}\underset{+ve}{\Psi}{}^{11}{}_k + \sum_{k=0}^{q}\underset{-ve}{\Psi}{}^{21}{}_k\right]\Delta t_k \tag{12.25}$$

This is also written as

$$W_{o\xi}^{\text{nonlinear}} = \frac{1}{2c^2{}_m}\left[\sum_{k=0}^{250}\underset{+ve}{\Psi}{}^{11}{}_k + \sum_{k=0}^{250}\underset{-ve}{\Psi}{}^{21}{}_k\right]\Delta t_k \tag{12.26}$$

The state sensitivity of individual measurements placed at different locations can be captured using the measurement specific empirical observability Grammians computed for the system. The overall empirical observability Grammian in Eq. (12.26) is nonsingular, thus fulfilling the nonlinear observability condition of the tubular reactor.

12.5.4 Sensor configuration results

The measurement specific empirical Grammian $W_{o\xi}^{\text{nonlinear}}$ of Eq. (12.26) computed for each measurement is of dimension (32 $\times$ 32). Each of the 32 empirical observability Grammians is subjected to eigenvalue analysis and then treated with the quantification measures [11] defined in Section 6.5.2 of Chapter 6. These quantities measures defined in Section 6.5.2 include $\max\left\{Trace(W_{o\xi}^{\text{nonlinear}})\right\}$, $\max\left\{\sigma_{\min}(W_{o\xi}^{\text{nonlinear}})\right\}$, and $\min\left\{\text{Condition number}(W_{o\xi}^{\text{nonlinear}})\right\}$. Among these measures, this study considers the measure of $Max\left\{Trace(W_{o\xi}^{\text{nonlinear}})\right\}$, defined as the sum of the diagonal elements of the singular values,

$$Trace(W_{o\xi}^{\text{nonlinear}}) = \sum_{i=1}^{n}\sigma_i \tag{12.27}$$

The larger value of this measure corresponds to a better degree of observability for sensor location. According to their magnitudes, trace values of individual measurement-specific Grammians are arranged in descending order to find the measurement ranking.

The following cases of disturbances are investigated for the placement of measurement sensors in a catalytic tubular reactor.

1. Disturbance considered in only benzene (reactant) concentration.
2. Disturbance considered in only maleic anhydride (product) concentration.
3. Disturbance considered in the effective heat transfer coefficient of solid−fluid phase and the frequency factor.
4. Disturbance considered in all the states and parameters, comprising benzene concentration, maleic anhydride concentration, effective heat transfer coefficient of solid−fluid phase, and the frequency factor.

In Case 1, a perturbation of 0.03 mol% of benzene concentration is considered in both positive and negative directions. This implies r is 2 and c_m is 0.03. For this case, the observability covariance matrix is calculated considering the disturbance in all the benzene states one by one in both directions. A total of 32 Grammians are generated. These Grammians are subjected to SVD analysis, and the optimal sensor locations are calculated using the trace method. The trace values are plotted against the discrete distance points, as shown in Fig. 12.5. The results show that the temperature sensors in the tubular reactor can be placed at discrete locations from 0.4 to 0.8 m at the grid points of 0.7, 0.6, 0.5, 0.8, and 0.4 m.

In Case 2, a perturbation of 0.03 mol% of maleic anhydride concentration is considered in both positive and negative directions. This makes $r = 2$ and $c_m = 0.03$. For this case, the observability covariance matrix is calculated considering the disturbance in all the states of maleic anhydride one by one in both directions. A total of 32 Grammians are generated. These Grammians are subjected to SVD analysis, and the optimal locations are calculated using the trace method. The trace values are plotted against the discrete distance points, as shown in Fig. 12.6. This figure shows that the temperature sensors in the tubular reactor can be placed at discrete locations from 0.6 to 1.0 m at the grid points of 0.7, 0.8, 0.6, 0.9, and 1.0 m.

In Case 3, disturbances are considered in the effective heat transfer coefficient of the solid−fluid phase and the frequency factor. In this case, the heat transfer coefficient and frequency factor values are first normalized by dividing them with 84 and 33171, respectively. After normalization, a perturbation of 0.03 is considered in both positive and negative directions of the two parameters at a time. This implies $r = 2$ and $c_m = 0.03$. For this joint case, the observability covariance matrix is calculated. A total of 32 Grammians are generated. These Grammians are subjected to SVD

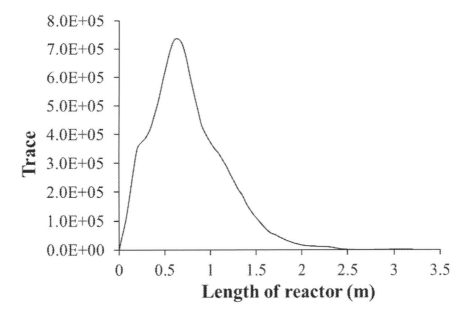

FIGURE 12.5   Trace values of the Grammians for disturbances in benzene concentration.

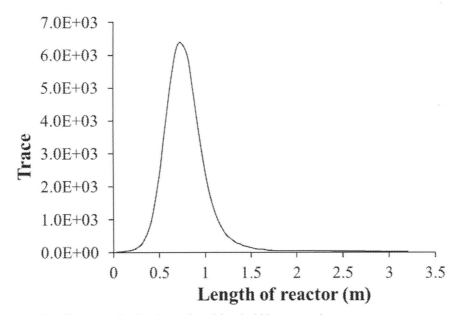

FIGURE 12.6   Trace values of the Grammians for disturbances in maleic anhydride concentration.

analysis, and the optimal locations are calculated using the trace method. The trace values are plotted against the discrete distance points, as shown in Fig. 12.7. The figure shows that the temperature sensors in the tubular reactor can be placed at discrete locations from 0.4 to 1.2 m at the grid points of 0.5, 0.4, 0.6, 1.1, and 1.2 m.

In Case 4, disturbances are considered in all the states and parameters of benzene concentration, maleic anhydride concentration, effective heat transfer coefficient of solid−fluid phase, and frequency factor. The normalized data of heat transfer coefficient and frequency factor are considered as in Case 3. Perturbation of 0.03% in states and 0.03 in parameters are considered in both directions at a time. Here again, r is considered to be 2 and c_m as 0.03. A total of 32 Grammians are generated. These Grammians are subjected to SVD analysis, and the optimal locations are calculated using the trace method. The trace values are plotted against the discrete distance points, as shown in Fig. 12.8. This figure shows that the temperature sensors in the tubular reactor can be placed at discrete locations from 0.4 to 0.8 m at the grid points of 0.5, 0.7, 0.6, 0.4, and 0.8 m.

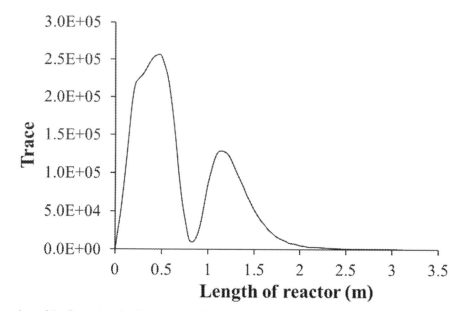

FIGURE 12.7 Trace values of the Grammians for disturbances in both heat transfer coefficient and catalyst activity.

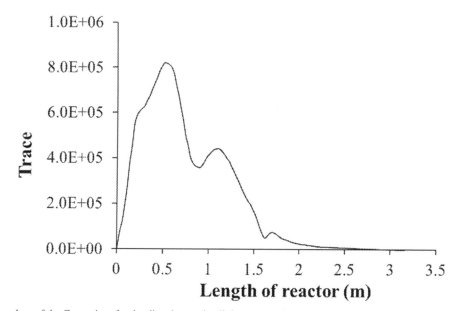

FIGURE 12.8 Trace values of the Grammians for the disturbances in all the states and parameters.

The empirical observability Grammian results in Figs. 12.5–12.8 have shown the discrete grid points at which the temperature sensors can be located along the tubular reactor. In each disturbance condition, five points are chosen to be proper points based on trace values of the Grammians. By noting the peak trace values in these figures, it can be concluded that the minimal and optimal temperature sensors for state estimation can be two sensors located at 0.5 and 0.7 m along the length of the catalytic tubular reactor.

12.6 Optimal state estimation using unscented Kalman filter

In UKF, the probability distribution of a function is approximated using an unscented transform. The unscented transformation is a method for calculating the statistics of a random variable that undergoes a nonlinear transformation and builds on the principle that it is easier to approximate a probability distribution than an arbitrary nonlinear function

[9,10]. The probability distribution is approximated by a set of deterministic points that capture the distribution's mean and covariance. These points, called sigma points, are then processed through the nonlinear model of the system, producing a set of propagated sigma points. By choosing appropriate weights, the weighted average and the weighted outer product of the transformed points, an estimate of the mean and covariance of the transformed distribution are computed.

Consider a discrete time nonlinear process, in which the process and measurement models are represented by Eq. (4.33) of Chapter 4. The variances Q_k and R_k, reflecting the process and observation noises, are given by Eq. (4.34) of Chapter 4. Consider the propagation of a random variable x (dimension $n = L$) distributed according to a Gaussian with mean x and covariance P. The aim is to obtain a Gaussian approximation of the distribution over a nonlinear function, $y = f(x)$. The unscented transform performs this approximation by selecting the so-called sigma points χ from the Gaussian estimate and passing these points through f. Assume x has mean $\hat{x}$ and covariance P_x. Specifically, $2L + 1$ sigma points are chosen to implement the algorithm described in Section 4.8 of Chapter 4.

12.6.1   State estimator design

The optimally configured temperature measurements are employed in UKF for instantaneous estimation of reactant and product concentrations of the catalytic tubular reactor. There exists a total of 64 states comprising 32 states of benzene and 32 states of maleic anhydride at 32 grid points along the length of the reactor in its discrete form is designed for state estimation in the catalytic tubular reactor. The Eqs. (12.19)−(12.22) describe the discrete mathematical model of the catalytic tubular reactor. The relations defined for the model are given as follows.

The state estimator design for the catalytic tubular reactor is carried out acording to the UKF algorithm presented in Section 4.8 of Chapter 4 (Mechanistic Model-Based Nonlinear Filtering and Observation Techniques for Optimal State/Parameter Estimation). The vector of state variables for 64 states of benzene and maleic anhydride are given as:

$$x = [F_B(t, \ z_1), \ F_{MA}(t, \ z_1), \ F_B(t, \ z_2), \ F_{MA}(t, \ z_2), \ ..., \ F_B(t, \ z_{32}), \ F_{MA}(t, \ z_{32})]^T \tag{12.28}$$

The vector of measurements are defined by:

$$y = T_f(t, \ z_5), \ T_s(t, \ z_7)]^T \tag{12.29}$$

The initial state vector is of dimension (64 × 1), and the initial measurement vector is of dimension (2 × 1). The dimensions of the initial covariance matrix P_0 is (64 × 64). The process noise covariance matrix Q is of dimension (64 × 64), and the measurement noise covariance matrix R is of dimension (2 × 2). The diagonal elements of these matrices are suitably selected such that they provide better performance of the state estimator.

The design procedure of UKF for state estimation in the tubular reactor is presented as follows:

1. The initial state vector, x_{k-1} and the mean vector $\hat{x}_{k-1}$ are of dimensions 64 × 1. The covariance matrices P_o, Q, and R, are appropriately selected.
2. According to Eq. (4.50), the dimension of sigma points generated is 64 × 129.
3. According to Eq. (4.51), the weights computed for states and covariance matrices are of dimension 64 × 129.
4. According to Eq. (4.52), the dimension of the transformed sigma points computed through projection over the nonlinear functions is 64 × 129.
5. According to Eq. (4.53), the dimension of the transformed measurement samples computed by propagating the sigma points through a nonlinear observation function is 2 × 129.
6. According to Eq. (4.54), the predicted state has a dimension of 64 × 1, and the prediction state covariance has a dimension of 64 × 64.
7. According to Eq. (4.55), the prediction measurements have a dimension of 2 × 1, the prediction measurement covariance, and the state observation cross-covariance have dimensions 2 × 2 and 64 × 2.
8. According to Eq. (4.56), in an updated step, the Kalman gain matrix, the state vector and the state covariance matrix with the dimensions of 64 × 2, 64 × 1, and 64 × 64, are computed.
9. With these computations, the UKF algorithm in Section 4.8 of Chapter 4 is implemented for state estimation in the catalytic tubular reactor.

12.6.2   State estimation results

The distributed dynamic model, Eqs. (12.5)−(12.8), representing the catalytic conversion of benzene to maleic anhydride in a fixed bed tubular reactor, is transformed into a discrete model in the form of Eqs. (12.19)−(12.22) by using the method of lines. The numerical solution of the model for the concentrations of the reactant and product and the temperatures of the gas flow rate and catalyst bed at different grid points along the length of the reactor is established for different initial conditions of reactant concentration. The optimal locations of temperature sensors for state estimation are determined by using empirical observability Grammians. The UKF in its discrete version is designed and applied for optimal state estimation in a tubular reactor using the optimally configured temperature measurements and the discrete model of the process.

The initial state covariance matrix P_0, the process noise covariance matrix Q, the observation noise covariance matrix R, and the linear measurement matrix H involved in the computation of UKF are chosen as follows:

$$P_0 = \begin{bmatrix} 10^{-6} & 0 & 0 & 0 & \ldots & \ldots & \ldots & 0 \\ 0 & 10^{-6} & 0 & \ldots & \ldots & \ldots & \ldots & 0 \\ \vdots & & & & & & & \\ 0 & 0 & 0 & \ldots & \ldots & \ldots & \ldots & 10^{-6} \end{bmatrix}_{64 \times 64}$$

$$Q_k = \begin{bmatrix} 10^{-6} & 0 & 0 & 0 & \ldots & \ldots & \ldots & 0 \\ 0 & 10^{-6} & 0 & \ldots & \ldots & \ldots & \ldots & 0 \\ \vdots & & & & & & & \\ 0 & 0 & 0 & \ldots & \ldots & \ldots & \ldots & 10^{-6} \end{bmatrix}_{64 \times 64}$$

$$R_k = \begin{bmatrix} 4 & 0 \\ 0 & 4 \end{bmatrix}$$

$$H = \begin{bmatrix} 0 & 0 & 0 & 0 & 1 & 0 & 0 & 0 & 0 & 0 & 0 & 0 & 0 & \ldots & 0 \\ 0 & 0 & 0 & 0 & 0 & 0 & 1 & 0 & 0 & 0 & 0 & 0 & 0 & 0 & 0 & \ldots & 0 \end{bmatrix}_{2 \times 64}$$

The temperature measurements are corrupted with a nominal measurement noise of 1% of their actual values. The temperature measurements derived from the model solution are considered actual temperature measurements. The temperature measurements obtained through model predictions based on the estimated states are considered as prediction measurements. The UKF algorithm described in Section 4.8 is designed for state estimation in catalytic reactor according to the procedure given in Section 12.6.1. The state estimation by UKF is carried out for different initial conditions. The results in Fig. 12.9 show the estimated and actual profiles of benzene and maleic anhydride for the initial

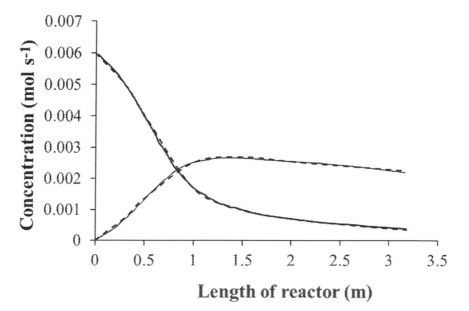

FIGURE 12.9 Benzene and maleic anhydride concentration profiles along the length of the reactor: estimated (━ ━ ━); actual (━━━━).

conditions of benzene flow rate of 0.6 mol%, maleic anhydride flow rate of 0 mol%, gas flow temperature of 733K and catalyst bed temperature of 733K. The estimation results by UKF are found in close agreement with the actual results.

12.7 Summary

Optimal state estimation has a significant role in the monitoring and control of distributed dynamic systems. This chapterpresents a sensor selection and state estimation strategy, in which an empirical observability Grammian-based methodology is used to configure the sensor locations and an UKF is employed for state estimation in distributed dynamic systems. The performance of the measurement selection methodology and state estimation scheme is evaluated by applying it to a catalytic tubular reactor, in which partial oxidation of benzene takes place to produce maleic anhydride. The Grammian configured temperature sensors are employed for instantaneous estimation of the tubular reactor's reactant and product concentration profiles. The results demonstrate the usefulness of the sensor selection methodology and state estimation scheme for inferential estimation of states in the catalytic tubular reactor.

References

[1]   J. Hahn, T.F. Edgar, A Gramian based approach to nonlinearity quantification and model classification, Ind. Eng. Chem. Res. 40 (24) (2001) 5724–5731.

[2]   A.K. Singh, J. Hahn, Determining optimal sensor locations for state and parameter estimation for stable nonlinear systems, Ind. Eng. Chem. Res. 44 (5) (2005) 5645–5659.

[3]   C. Venkateswarlu, B. Jeevan Kumar, Composition estimation of multicomponent reactive batch distillation with optimal sensor configuration, Chem. Eng. Sci. 61 (17) (2006) 5560–5574.

[4]   C. Sumana, C. Venkateswarlu, Optimal selection of sensors for state estimation in a reactive distillation process, J. Process Control 19 (6) (2009) 1024–1035.

[5]   W. Waldraff, D. Dochain, S. Bourrel, A. Magnus, On the use of observability measures for sensor location in tubular reactor, J. Process Control 8 (5–6) (1998) 497–505.

[6]   F.W.J. van den Berg, H.C.J. Hoefsloot, H.F.M. Boelens, A.K. Smilde, Selection of optimal sensor position in a tubular reactor using robust degree of observability criteria, Chem. Eng. Sci. 55 (4) (2000) 827–837.

[7]   E.L. Haseltine, J.B. Rawlings, Critical evaluation of extended kalman filtering and moving-horizon estimation, Ind. Eng. Chem. Res. 44 (8) (2005) 2451–2460.

[8]   D. Dochain, State and parameter estimation in chemical and biochemical processes: a tutorial, J. Process Control 13 (8) (2003) 801–818.

[9]   S.J. Julier, J.K. Uhlmann, New extension of the Kalman filter to nonlinear systems, in: Signal Processing, Sensor Fusion, and Target Recognition VI. SPIE, 1997.

[10] S. Julier, J. Uhlmann, H.F. Durrant-Whyte, A new method for the nonlinear transformation of means and covariances in filters and estimators, IEEE Trans. Autom. Control 45 (3) (2000) 477–482.

[11] A.K. Singh, J. Hahn, Sensor location for stable nonlinear dynamic systems: multiple sensor case, Ind. Eng. Chem. Res. 45 (10) (2006) 3615–3623.

[12] A.K. Singh, J. Hahn, State estimation for high-dimensional chemical processes, Comput. Chem. Eng. 29 (11–12) (2005) 2326–2334.

[13] A. Romanenko, J.A.A.M. Castro, The unscented filter as an alternative to the EKF for nonlinear state estimation: a simulation case study, Comput. Chem. Eng. 28 (3) (2004) 347–355.

Chapter 13

Applications of data-driven model-based methods for process state estimation

13.1 Introduction

State estimation methods can be classified into two broad approaches, namely, the first principle model-based approach and the empirical model-based approach. In the first principle model-based approach, a rigorous/semirigorous mathematical model of the process is used in conjunction with an estimation algorithm and the available process measurements to provide reliable real time estimates for the state variables. In the empirical/data-driven model-based approach, a state estimator can be built by utilizing heuristic or operational data of the process. Empirical models can be developed very quickly without requiring detailed insight into the process.

In a modern industrial environment, process data contains a great deal of information that provides the basis for process monitoring, estimation, and control. The recent advances in sensors and data measurement techniques enable collecting large amounts of multivariable data from chemical processes. An important step in understanding the behavior of the process is the extraction of the salient features encapsulated within the process data. Data-driven modeling is an approach by which empirical models are derived from the data that allows estimating unmeasured process states from measurements. Concerning this issue, different data-driven modeling tools such as projection to latent structures (PLS), artificial neural networks (ANN), and radial basis function networks (RBFN) are used to extract the relevant information from the correlated data [1−3]. The PLS is a technique that simultaneously extracts information (latent variables) from both input and output data. The PLS can summarize the underlying structures of the process as a linear combination of the latent variables and relate the predictor and response variables by means of regression models. ANN's are computer systems that mimic the human brain's operations by mathematically modeling its neurophysiological structure. They consist of a large number of computational units connected in a massively parallel structure. The interconnections between the layers of the network and the transfer functions of the neuron processing functions represent the distributed relationships between input and output data. ANN's have wide applications in data modeling of several disciplines. RBFN is an alternative modeling tool to ANN. The advantage of RBFN is that it can be configured automatically, while optimally updating its parameters. RBFN has been proved as an effective tool in several applications.

Batch processes are widely used in many chemical, biochemical, and pharmaceutical industries, of which batch distillation is the most important separation process. The separation of chemicals by batch distillation involves the collection over time of a series of product cuts of different compositions. If the instantaneous compositions during the operation are known, they can serve to improve the process performance. However, direct and instantaneous composition measurement during operation is not economical and involves significant time lags and technical limitations. This difficulty can overcome through online inference of compositions by means of known measurements by using a state estimation scheme. The data-driven state estimators derived based on PLS, ANN, and RBFN are useful alternatives to mechanistic model based state estimators for online estimation of compositions in batch processes like batch distillation.

This chapter presents the development of various data-driven state estimators and their application for composition estimation in multicomponent batch separation system.

13.2 Projection to latent structures model-based compositions estimator for multicomponent batch distillation

The PLS is a technique that simultaneously extracts information from both input and output data [4]. The PLS can summarize the underlying structures of the process as linear combination of the latent variables and relate the predictor and

Optimal State Estimation for Process Monitoring, Fault Diagnosis and Control. DOI: https://doi.org/10.1016/B978-0-323-85878-6.00008-7

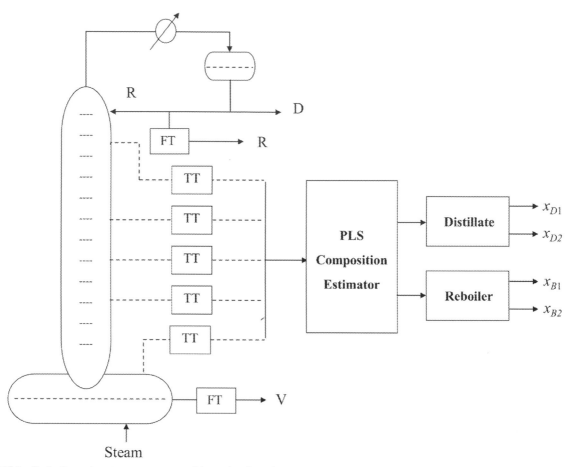

FIGURE 13.1 Projection to latent structures composition estimation scheme.

response variables by means of regression models. This study presents a PLS model-based scheme for inferential estimation of product compositions in batch distillation. The detailed description of batch distillation along with its mathematical model is reported in Chapter 7, Application of Mechanistic Model-Based Nonlinear Filtering and Observation Techniques for Optimal State Estimation in Multicomponent Batch Distillation. The simulation data of temperatures and compositions to develop the PLS estimator can be obtained by solving the mathematical model of batch distillation given in Chapter 7 for different operating conditions. The temperature measurements that serve as inputs to the PLS estimator are suitably selected to provide improved estimates of product compositions. The design and implementation of the PLS state estimator is studied by applying it to a batch distillation column involving the separation of a hydrocarbon mixture, cyclohexane/heptane/toluene. The schematic of the PLS estimator is shown in Fig. 13.1. In figure, x_{D1} and x_{D2} refer to cyclohexane and heptane in the distillate, and x_{B2} and x_{B3} denote heptane and toluene in reboiler.

13.2.1 State estimator development

A brief description of the batch distillation, data processing for state estimation, and the state estimator development is given as follows:

13.2.1.1 Distillation process

A 20 tray batch distillation column involving the separation of cyclohexane/heptane/toluene mixture is considered a simulation testbed to develop and implement the proposed soft sensor. The realistic operation of the actual batch distillation column is represented by a rigorous nonlinear model derived from first principles involving dynamic material and component and algebraic energy equations supported by vapor−liquid equilibrium and physical properties. The detailed description of the operation of the batch distillation column and the equations describing the process can be referred to elsewhere [5].

13.2.1.2   Data generation

The details of the batch separation process and its mathematical model are discussed in Section 7.2 of Chapter 7. The data needed to develop the PLS composition estimator is generated by using the mathematical model of batch distillation column under various conditions of initial mixture compositions, specified product compositions, reflux flow rate, and reboiler heat duty. The mathematical model representing the separation of a hydrocarbon mixture, cyclohexane/heptane/toluene in a 20 tray column [5,6], is considered for data generation. A vapor boil up rate of 2690 moles, condenser holdup of 50 moles and tray holdup of 10 moles are employed. In this system, cyclohexane and heptane are separated as distillate products, and toluene is separated as a product in the still pot. The Euler method with a step size of 0.001 h is used to integrate the differential equations of the system. The temperature measurements are obtained from the Antoine equation through the bubble point calculation procedure. The temperature data at every 0.001 h are corrupted with random Gaussian noise, and these measurements are used for state estimation.

13.2.1.3   Development of projection to latent structures composition estimator

A description of PLS is provided in Section 5.3 of Chapter 5. The inferential model development in PLS is based on two data matrices, X and Y, associated with the process measurements and the quality variables. A detailed description of the PLS algorithm and its mathematical formulation can be found elsewhere [1]. PLS extracts the latent variables from the respective data matrices. Each pair of latent variables accounts for a certain amount of variability in the input and output data sets. Usually, most of the variance of the two data blocks can be accounted for by the first few latent variables, whilst the higher order latent variables are typically associated with the random noise in the data. The appropriate number of latent variables required to describe the latent structures is generally identified through cross-validation procedure.

13.2.1.4   Data preprocessing

This is an important step in PLS model development. A proper data normalization can favor the determination of the most representative PLS model structure. For this, a suitable scaling procedure for the available data sets needs to be adopted. If X denotes the input data matrix of temperatures and Y denotes the output data matrix of compositions, the normalization of data of these matrices is required. Each column is centered in each of these matrices, by subtracting its column mean and scaled by its standard deviation. The data matrix X is first scaled in such a way that each row is normalized to zero mean and unit variance:

$$\hat{x}_{ij} = \frac{x_{ij} - \overline{x}_i}{\sigma x_i} \tag{13.1}$$

with

$$\overline{x}_i = \frac{1}{n} \sum_{j=1}^{n} x_{ij} \tag{13.2}$$

$$\sigma x_i^2 = \frac{\sum_{j=1}^{n} \left(x_{ij} - \overline{x}_i\right)^2}{n - 1} \tag{13.3}$$

where x_{ij} are the actual elements and $\hat{x}_{ij}$ are the normalized elements of the X matrix, and n is the number of elements in each row of X matrix. $\overline{x}_i$ and σx_i are the mean and standard deviation of the ith row of X, respectively. Similarly, normalization of data in the Y matrix can be performed. The normalized data matrices are used for PLS processing.

The PLS algorithm decomposes the input and output data matrices X and Y of dimensions $m \times n_x$ and $m \times n_y$, respectively, into two lower dimensional score matrices T and U of dimensions $n_x \times k$ and $n_y \times k$, respectively. The score matrices T and U represent the projection of the original matrices X and Y onto the latent variable space along with two residual matrices E and F of dimensions $m_x \times n_x$ and $m_y \times n_y$ that contain part of X and Y that left of the regression:

$$Y = TP^T + E \tag{13.4}$$

$$X = UQ^T + F \tag{13.5}$$

The latent variables are aligned along with the k columns of the two score matrices, T ($n_x \times k$) and U ($n_y \times k$), and are ordered in such a way that the amount of variance of the original data described by each variable decrease as the number of latent variables increases. In this study, the scores of input and output matrices are computed using the NIPALS algorithm described in Section 5.6.1 of Chapter 5.

13.2.2 State estimation results

The input and output data matrices X and Y of batch distillation are collected and arranged in three dimensional arrays ($I \times J \times K$), where I represents the number of batches, J refers to the number of measured variables, K denotes the number of fixed time intervals and time represents the third dimension. Each horizontal slice in the three-dimensional array contains the trajectories of all the variables from a single batch. Each vertical slice collects the values of all the variables for all batches at the same time instant. These matrices need to be unfolded and arranged in two dimensional arrays to be used with the PLS algorithm. The unfolding is done variable wise by taking the vertical slices along the time axis of the matrix and ordering them side by side to the right to generate two dimensional matrices of dimensions ($KI \times J$). Because the simulated batches have different durations, the number of samples in each batch varies depending on the duration of the operation.

Thus 455 samples of different batches generated under different operating conditions are considered for PLS modeling. After studying different measurement combinations, the temperature data corresponding to Tray 1, Tray 19, and Tray 21 are found appropriate for PLS model development. The temperature data represent the X matrix of dimension (455×3). The respective composition data of cyclohexane (x_{D1}) and heptane in distillate (x_{D2}) and heptane (x_{B2}) and toluene (x_{B3}) in reboiler forms the Y matrix of dimension (455×4). The mean and standard deviation of the actual input temperature data of T1, T19, and T21 are 373.5689, 363.6615, 362.9791 and 4.300239, 6.435311, 6.257244, respectively. The mean and standard deviation of the actual output composition data of x_{D1}, x_{D2}, x_{B2}, x_{B3} are 0.132564, 0.464783, 0.435772, 0.55031 and 0.130551, 0.098088, 0.386745, 0.399134, respectively. The actual input and output data are normalized according to Eq. (13.1), and the covariance matrices of normalized data are used to compute the eigenvalues and eigenvectors. The covariance matrix of temperature data is of dimension (3×3), and the composition data is of dimension (4×4). From these results, two eigenvectors are retained as principal components (PCs) for temperature data, and two eigenvectors are retained as PCs for composition data. These PCs are the eigenvectors corresponding to the largest eigenvalue of the covariance matrix arranged in decreasing order. The normalized data of inputs and outputs with their respective PCs provide the input and output scores. The dimension of each of these score matrices is (455×2). The input and output scores are related by means of a linear regression model. The regression model coefficients are determined based on the least squares algorithm. The regression models that relates the input and output scores are of the form

$$Y_1 = a_0 + a_1 X_1 + a_2 X_2 \tag{13.6}$$

$$Y_2 = b_0 + b_1 X_1 + b_2 X_2 \tag{13.7}$$

Here, Y_1 and Y_2 represent the composition scores, and X_1 and X_2 represent the temperature scores. The coefficients of Eq. (13.6) are computed as $a_0 = 0.00298514$, $a_1 = -0.0766086$, and $a_2 = -0.1763127$. The coefficients of Eq. (13.7) are obtained as $b_0 = 0.00573887$, $b_1 = -0.3049061$, and $b_2 = -0.2022484$. These models are used to predict the composition scores from the temperature scores. At every sample time, the actual temperature data in conjunction with the PCs derived from the test data is used to compute the temperature scores. These temperature scores are used in Eqs. (13.6) and (13.7) to compute the composition scores. These compositions scores, along with their corresponding PCs, are used in Eq. (13.8) to predict the normalized compositions using the following equation:

$$Y_{\text{pred}} = Y_{\text{sc}} \left(Y_{\text{pc}}^T Y_{\text{pc}} \right)^{-1} \left(Y_{\text{pc}}^T \right) \tag{13.8}$$

where Y_{pc} represents the PCs for compositions, Y_{sc} represents composition scores, and Y_{pred} denoted the predicted compositions. These normalized compositions with their respective means and standard deviations provide the actual compositions.

The PLS estimator thus developed is used to estimate the compositions based on the temperature measurements. At each sampling instant, the temperature measurements enter as inputs to the estimator, and their respective scores are computed. The regression relation provides composition scores from temperature scores, which are then inverted to provide actual compositions. Fig. 13.2 shows the actual and PLS estimator compositions of cyclohexane in distillate for samples of different batches. Fig. 13.3 shows the actual and PLS estimator compositions of cyclohexane in reboiler for samples of different batches. The actual and estimated composition profiles of cyclohexane and heptane for a specific single batch (x_{int} = 0.407/0.394/0.199; $\hat{x}_{int}$ = 0.42/0.37/0.21; x_{spec} = 0.85/0.91/0.92) are shown in Fig. 13.4. These results indicate that the PLS estimator predicted results are in near agreement with the actual results. The PLS estimator thus serves as a useful tool for inferential estimation of compositions in batch distillation.

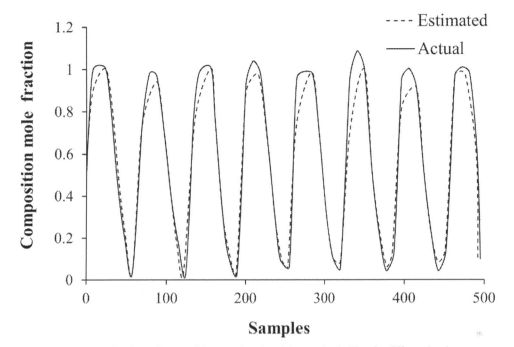

FIGURE 13.2   Comparison of actual and estimated composition samples of cyclohexane in distillate for different batches.

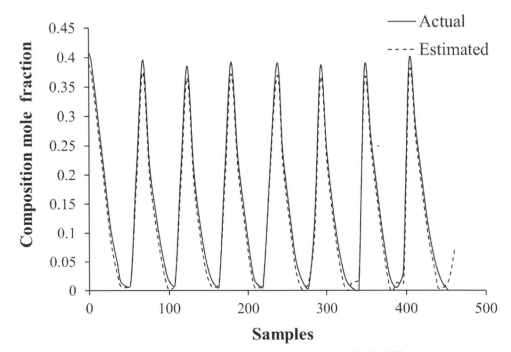

FIGURE 13.3   Comparison of actual and estimated composition samples of cyclohexane in reboiler for different batches.

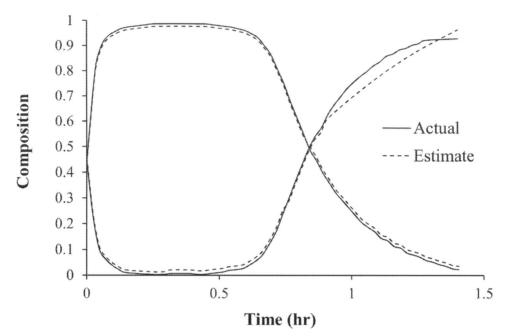

FIGURE 13.4 Actual estimated composition profiles of cyclohexane and heptane in distillate for a single batch.

13.3 Artificial neural network model-based compositions estimator for multicomponent batch distillation

This study aims at developing an ANN soft sensor/state estimator for instantaneous estimation of compositions in multicomponent batch distillation. The scheme involves two multilayered feed forward neural networks to estimate the distillate stream and reboiler product compositions. The development and evaluation of the state estimator are studied for compositions estimation of a multicomponent batch distillation involving the separation of cyclohexane/heptane/toluene mixture.

13.3.1 State estimator development

The description of the batch distillation considered for design and implementaton ANN based state estimator is the same as given in Section 13.2.1.1, and can be referred elsewhere [5]. The details of measurement configuration for state estimation and the development of state estimator are given as follows. Section 13.2.1.

13.3.1.1 Measurement sensors selection

A multivariate statistical technique, called principal component analysis (PCA), is used to select the temperature measurements and to use them as inputs to the composition estimator. The configuration of temperature measurements for state estimation in multicomponent batch distillation using PCA is described in Section 7.5 of Chapter 7: Application of Mechanistic Model-Based Nonlinear Filtering and Observation Techniques for Optimal State Estimation in Multicomponent Batch Distillation.

13.3.1.2 Development of artificial neural network composition estimator

The widely used ANN paradigm is a multilayered feed-forward network (MFFN) with multilayered perceptron, mostly comprising three sequentially arranged layers of processing units. An MFFN provides a mapping between an input (x) and an output (y) through a nonlinear function f, that is, $y = f(x)$. The three layered MFFN has input, hidden, and output layers, each layer comprising of nodes. All the nodes in the input layer are connected using weighted links to the hidden layer nodes; similar links exist between the hidden and output layer nodes. Usually, the input and hidden layers also contain a bias node possessing constant output of one. The nodes in the input layer do not perform any numerical processing, whereas the hidden and output layer nodes perform all numerical processing, and they are termed as active nodes. The details concerning the structure, processing functions, training, and information processing of feed-forward neural networks are elaborated in Section 5.4 of Chapter 5. In this study, two individual neural networks are configured for composition estimation in distillate stream and reboiler. The schematic of the artificial neural composition soft sensor for batch distillation is shown in Fig. 13.5. The notation for the symbols in batch distillation can be referred to Chapter 7.2.,

The problem of neural network modeling is to obtain a set of weights such that the summation of the squared prediction error defined by the difference between the network predicted outputs and the desired outputs is minimum. The iterative training makes the network recognize patterns in the data and creates an internal model, which provides predictions for the new input condition. The inputs for both the distillate stream and reboiler networks are the temperature measurements of different operating conditions and a unit bias. The distillate section network has the outputs of product compositions for cyclohexane and heptane, and the reboiler network has the outputs of product compositions for toluene and heptane. The data sets representing the inputs and the corresponding outputs for training the neural networks can be obtained through simulation of the batch distillation model. The details of a MFFN with back-propagation training by generalized delta rule are reported in the literature [7].

Network training is an iterative procedure that begins with initializing the weight matrix randomly. The network learning process involves two types of passes: a forward pass and a reverse pass. In the forward pass, an input pattern from the example data set is applied to the input nodes, the weighted sum of the inputs to the active node is calculated, which is then transformed into output using a nonlinear activation function such as a sigmoid function. The outputs of the hidden nodes computed in this manner form the inputs to the output layer nodes whose outputs are evaluated similarly. In the reverse pass, the pattern-specific squared error is computed and used to update the network weights according to the gradient strategy. The repetition of the weight updating procedure for all the patterns in the training set, completes one training iteration. The distillate and bottom ANN estimators are trained iteratively until convergence in the objective is achieved. The trained and learned networks are used to directly infer the product compositions based on temperature measurements of the batch distillation column.

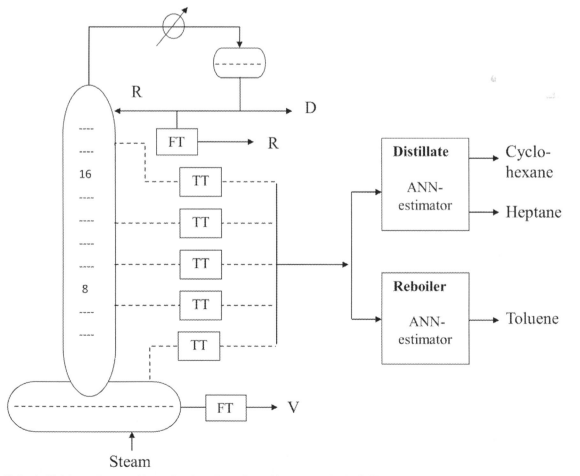

FIGURE 13.5 Artificial neural composition estimation scheme for multicomponent batch distillation.

13.3.2 State estimation results

The development of ANN soft sensor for composition estimation in multicomponent batch distillation is carried out using the data of a hydrocarbon system involving the separation of a cyclohexane—heptane—toluene mixture into products and slop cuts in a 20-tray column [5,6]. The data generation procedure is the same as briefed in Section 13.2.1.2. The input and output data used to train the networks is generated with respect to the changes in initial mixture compositions, specified product purities, distillate flow rate and vapor boil up. Thus a total of 1840 data sets are generated in which 920 data sets are used for training purpose, and the remaining data sets are used to test the developed network model. A multivariate statistical tool called PCA is used to select the temperature measurements. The procedure for sensor configuration using PCA can be referred to in Section 6.3.3 of Chapter 6, Optimal Sensor Configuration Methods for State estimation. Accordingly, the temperatures of reboiler, 2nd, 11th, 18th, and 20th trays are selected as measurements. The number of hidden nodes selected for both the networks of distillate stream and reboiler is 12. The learning rate α and the momentum factor η are selected to be 0.5 and 0.1 for the distillate stream network, whereas the values of α and η selected for the reboiler network are 0.7 and 0.01, respectively. The number of iterations to obtain the convergence in distillate stream and reboiler network models are 25,000 and 40,000, respectively.

The composition soft sensor thus developed is studied towards the effect of various initial mixture compositions, noise levels in the measurements, reflux, and distillate flow rates. The amounts and compositions estimated by the estimator are compared with those obtained from the base case studies. Fig. 13.6 shows the actual and the estimated compositions in the distillate and reboiler for x_{int} = 0.340/0.335/0.325, $\overline{x_{int}}$ = 0.333/0.333/0.333 and x_{spec} = 0.9/0.8/0.8. Fig. 13.7 shows the actual and the estimated compositions in the distillate and reboiler for x_{int} = 0.407/0.394/0.199, $\overline{x_{int}}$ = 0.42/0.37/0.21 and x_{spec} = 0.85/0.91/0.92. These results correspond to the temperature measurements with the random Gaussian noise in the range of -0.5 to 0.5. These results show the effectiveness of the NN soft sensor for composition estimation in multicomponent batch distillation. The results demonstrate the usefulness of the neural network soft sensor for inferential estimation of instantaneous compositions in multicomponent batch distillation.

13.4 Radial basis function network model-based compositions estimator for multicomponent batch distillation

This study presents a RBFN soft sensor/state estimator for instantaneous estimation of compositions in multicomponent batch distillation. The development and implementation of the RBFN estimator are studied by using a multicomponent batch distillation as a simulation testbed.

13.4.1 State estimator development

The description of the batch distillation process and its data generation for developing the RBFN estimator is the same as briefed in Sections 13.2.1.1, 13.2.1.2, Sections 13.3.1.1, and 13.3.1.2. The details concerning the structure of RBFN, its automatic configuration and implementation procedure are given in Section 5.5 of Chapter 5. Further details on automatic configuration RBFN can be referred to elsewhere [3]. As discussed in Section 13.3.1, Sections 13.2.1.1 and 13.2.1.2, a 20-tray batch distillation column involving the separation of cyclohexane/heptane/toluene mixture is considered as a simulation testbed to develop and implement the RBFN soft sensor. In this study, the temperature measurements used as inputs to the RBFN estimator are heuristically selected. Cyclohexane and heptane in the distillate, and the same cyclohexane and heptane in the reboiler are considered as the estimator outputs. The data matrix corresponding to temperatures is denoted as X, which has a dimension $m \times n_x$, where n_x refers to the input variables. The data matrix corresponding to compositions is denoted as Y, which has a dimension $m \times n_y$, where n_y refers to the output variables. The data of the X matrix is normalized using the normalization procedure given by Eqs. (13.1)—(13.3). The same procedure is followed to normalize the data of the Y matrix. The normalized data of the X matrix is used to compute the covariance matrices of the input data according to Eq. (5.5). In the same way, normalized data of the Y matrix is used to compute the covariance matrix of the output data. The covariance matrices of the X and Y data matrix are obtained as $n_x \times n_x$ and $n_y \times n_y$, respectively. The PC matrices that are determined from X and Y covariance matrices are of dimensions $n_x \times k_1$ and $n_y \times k_2$, respectively. The projection of X and Y data matrices onto the PC matrices gives the respective score matrices of X and Y with the dimensions $m \times k_1$ and $m \times k_2$, respectively. The RBFN algorithm described in Section 5.5 of Chapter 5, is trained using the scores of temperature data as its inputs and the scores of corresponding composition data as its outputs. The iteratively converged RBFN represents the RBFN estimator for multicomponent o31.1. and 13.3.1.2

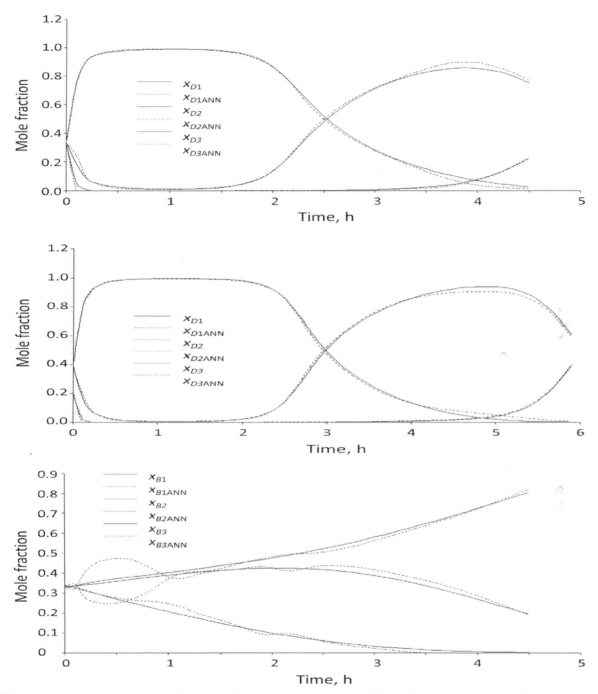

FIGURE 13.6   Actual and estimated compositions in distillate and reboiler for x_{int} = 0.333/0.333/0.334; $\overline{x_{int}}$ = 0.340/0.335/0.325 and x_{spec} = 0.9/0.8/0.8.

13.4.2   State estimation results

The development of an RBFN soft sensor for composition estimation in multicomponent batch distillation is carried out using the data of a hydrocarbon separation system cyclohexane—heptane—toluene mixture into products and slop cuts in a 20-tray batch distillation column. The data generation procedure is the same as briefed in Section 13.2.1.2. The temperature measurements of the 5th, 9th, and 12th trays are heuristically selected and used as inputs to the RBFN estimator. The composition data of cyclohexane and heptane in the distillate, and cyclohexane and heptane in reboiler are considered as the estimator outputs. The input and output data used to train the networks is generated with respect to the changes in initial mixture compositions, specified product purities, distillate flow rate, and vapor boil up. Thus 455

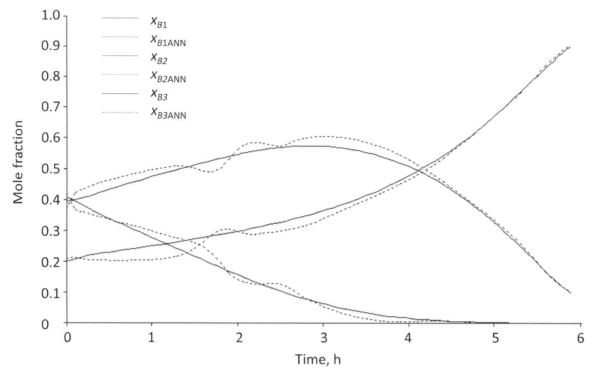

FIGURE 13.7 Actual and estimated compositions in distillate and reboiler for $x_{\text{int}} = 0.407/0.394/0.199$; $\overline{x_{\text{int}}} = 0.42/0.37/0.21$ and $x_{\text{spec}} = 0.85/0.91/0.92$.

data sets of inputs along with the corresponding outputs are used to develop an RBFN estimator. The temperature data of different operating conditions are arranged in X matrix of dimension (455 × 3). The composition data corresponding to the temperature data is arranged in the Y matrix of dimension (455 × 4). Data normalization of input and output data is performed using the mean centered data. For this, the mean values of the temperature data are obtained as 370.866518, 368.909099, and 367.488202, respectively, and the corresponding standard deviations are computed as 5.254509, 5.859859, and 6.287717, respectively. Similarly, the mean values of the composition data are obtained as 0.132564, 0.464783, 0.435772, and 0.55031, respectively, and the corresponding standard deviations are computed as 0.130551, 0.098088, 0.386745, and 0.399134, respectively. From the temperature data, the covariance matrix of dimension (3 × 3) is computed. The eigenvectors of the input and output data covariance matrices are computed, and the PCs are selected. The PCs corresponding to the input and output data are of dimensions (3 × 1) and (4 × 2), respectively. The eigenvectors and the selected PCs of input and output matrices are given in Table 13.1.

The score matrices of input and output data of dimensions (455 × 1) and (455 × 2) are obtained by multiplying the data matrices with the PCs.

The RBFN described in Section 5.5 of Chapter 5 is trained by using the scores of temperature data as its inputs and the corresponding scores of composition data as its outputs. The parameters used to configure the RBFN model are given in Table 13.2. The notation can be referred to Section 5.5. The network has one input node for input scores and two output nodes for output scores. The number of RBF's recruited is five. The number of input and output score samples used to train the network is 455. The network training procedure can be referred to in Section 5.5. The network is trained for 12,000 iterations, by which its convergence is achieved. The resulting weights, means, and standard deviations due to network convergence are given in Table 13.3.

The trained RBFN model is used to predict the composition scores for the corresponding temperature scores. At every sampling time, the normalized temperature data of dimension (1 × 3) is multiplied by the PCs of dimension (3 × 1) to obtain the input score of dimension (1 × 1), which becomes the input to trained RBFN. The trained RBFN model takes the input scores and provides the output scores of dimension (1 × 2). These output scores are transformed into actual compositions of dimension (1 × 4) by using the following relation:

$$Y_{\text{pred}} = Y_{\text{sc}} \left(Y_{\text{pc}}^{T} Y_{\text{pc}} \right)^{-1} \left(Y_{\text{pc}}^{T} \right)$$

TABLE 13.1 Eigenvectors and selected principal components of input and output data.

Input data

Eigen vectors			Selected principal component	
0.742914	−0.341416	0.575774	0.575774	
−0.076473	0.811229	0.579703	0.579703	
−0.665005	−0.474703	0.576556	0.576566	

Output data

Eigen vectors				Selected principal component	
0.743494	−0.436342	−0.505752	0.032204	−0.4363	−0.5058
0.331932	0.897868	−0.288219	−0.024189	0.8979	−0.2882
0.394257	0.058070	0.574986	0.714549	0.0581	0.5750
−0.434340	−0.008195	−0.574922	0.698425	−0.0082	−0.5749

TABLE 13.2 Radial basis function network parameters.

Parameter	Value
Learning rate, α	1.0×10^{-5}
Error margin, ε_m	2.0×10^{-5}
Initial nominal variance, σ_o	0.1
Initial effective radius, r_o	1.0
Lower bound on radius, r_L	0.001
Radius decrement rate, r_d	0.00055
Decay factor for error gradient, α	5.5×10^{-4}
Increment rate for saturation criterion, β	1.0×10^{-9}

TABLE 13.3 Optimum radial basis function networks parameters.

Weights		Mean	Standard deviation
0.35191	0.41625	−0.1475	0.49288
0.74886	0.53141	−0.6772	0.65513
0.31519	0.03231	0.33456	0.28538
−0.1239	0.40257	0.91332	0.21831
−0.0211	0.83078	1.15757	0.20646

where Y_{pc} represents the PCs for compositions, Y_{sc} represents composition scores, and Y_{pred} denoted the predicted compositions. Since in distillate and reboiler, two product compositions are obtained by RBFN estimator, the third component concentration can be obtained by subtracting the sum of these two from unity.

The predictive performance of the RNFN estimator is evaluated by using the scores data corresponding to the temperature data of different batches with different operating conditions. The compositions predicted by the RBFN estimator in distillate and reboiler based on the temperature data of different batches with different operating conditions are shown in Figs. 13.8 and 13.9. Fig. 13.8 represents the composition estimates of cyclohexane composition in the distillate, and Fig. 13.9 represents the composition estimates of cyclohexane compositions in reboiler. In the figures, the abscissa represents the samples in numbers used for prediction, and the ordinate represents the composition mole fractions. The results in these figures indicate that though the predicted compositions are in close agreement with their corresponding actual ones, there is a slight overshoot and under shoot in the estimation results. This can be avoided from

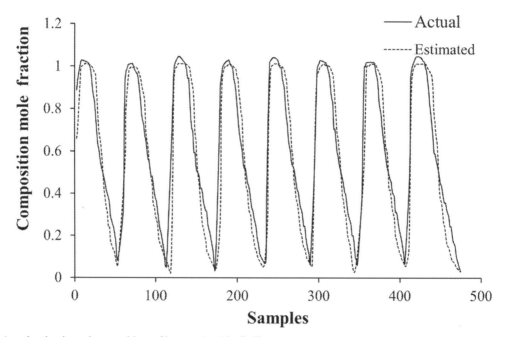

FIGURE 13.8 Actual and estimated compositions of heptane (x_{D2}) in distillate.

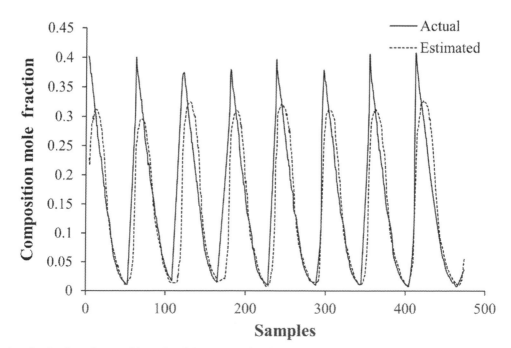

FIGURE 13.9 Actual and estimated compositions of cyclohexane (x_{B1}) in reboiler.

the physical reality on considering the estimated composition value above unity as one and below zero as zero, respectively. From the results, it is observed that the estimated compositions deviate slightly from the actual compositions. The results in aggregate indicate the better suitability of the RBFN estimator for compositions prediction in multicomponent batch distillation.

13.5 NIPALS—RBFN model-based compositions estimator for multicomponent batch distillation

The NIPALS algorithm presented in Section 5.6.3 of Chapter 5 is combined with the RBFN algorithm described in Section 5.5 to formulate the NIPALS—RBFN algorithm and is employed for composition estimation in batch distillation.

13.5.1 State estimator development

The PLS summarizes the underlying structures of the process as linear combinations of the original variables and relate the predictor and response variables by means of linear regression models between the latent variables. However, linear PLS is inappropriate for modeling nonlinear systems. A number of methods have been proposed in the literature to integrate nonlinear features within the linear PLS framework and thus provide a nonlinear algorithm [8—12]. In this study, a nonlinear iterative partial least squares (NIPALS) is integrated with the RBFN to form a NIPALS—RBFN state estimator. The automatic configuration of RBFN is reported elsewhere [3] and in Section 5.5 of this book. A detailed description of the NIPALS algorithm and its mathematical formulation can be found in the literature [13] and also in Section 5.6.3 of Chapter 5. As discussed in Section 13.3.1, 13.2.1.1 a 20-tray batch distillation column involving the separation of cyclohexane/heptane/toluene mixture is considered as a simulation testbed to develop and implement the NIPALS—RBFN soft sensor. The data generation procedure of batch distillation process is the same as discussed in Section 13.2.1.2.

13.5.2 State estimation results

455 samples of input and output data of different batches are generated under different operating conditions are treated with the NIPALS algorithm in Section 5.6.3 to compute the loadings and scores of the input and output data. The input data correspond to the temperatures of Tray 16, Tray 17, and Tray 20. These temperature data represent the X matrix of dimension (455 $\times$ 3). The corresponding composition data of cyclohexane (x_{D1}) and heptane in distillate (x_{D2}) and cyclohexane (x_{B1}) and heptane (x_{B2}) in reboiler is taken as the output data. This data forms a Y matrix of dimension (455 $\times$ 4). Data preprocessing is more important for the NIPALS algorithm. A suitable scaling procedure for the available data sets needs to be adopted, as proper data normalization can favor the determination of the most representative scores by the NIPALS algorithm. The arrangement of input and output data matrices for scores calculation is similar to that of the PLS procedure described in Section 13.2.1.4.

The normalization of data on the elements in X and Y matrices is performed as discussed in Section 13.2.1, 13.2.1.4 and according to Eq. (13.1). The mean and standard deviation values corresponding to the actual data of input temperatures of T16, T17, and T20 are 373.5689, 363.6615, 362.9791, and 4.300239, 6.435311, 6.257244, respectively. The values of the mean and standard deviation corresponding to the actual data of compositions of x_{D1}, x_{D2}, x_{B1}, x_{B2} are 0.132564, 0.464783, 0.435772, 0.55031, and 0.130551, 0.098088, 0.386745, 0.399134, respectively. The normalized data of inputs and outputs are used in the NIPALS—RBFN algorithm in Section 5.6.3 to compute the loadings and scores of the input and output data matrices. The RBFN parameters involved in NIPALS—RBFN are given in Table 13.4. The network has two input nodes and two output nodes. The number of RBF's recruited is nine. The number of input and output data samples used to train the network is 455. The network is trained for 10,000 iterations, by which its convergence is achieved.

The loading matrices corresponding to the input and output data are of the size (3 $\times$ 3) and (4 $\times$ 4), respectively. Two of the loading columns corresponding to the input data and two of the loading columns corresponding to the output data are considered PCs. These PCs of input and output data are given in Table 13.5. The resulting scores of input data from NIPALS—RBFN are of the dimension (455 $\times$ 2). In the same way, the resulting scores of output data are of the dimension (455 $\times$ 2). These scores are computed by the NIPALS—RBFN algorithm so as to satisfy iterative convergence of the result with a smaller tolerance.

The trained NIPALS—RBFN is used for instantaneous estimation of compositions from temperature measurements. The normalized input temperature data and the PCs obtained from the NIPALS algorithm are used to compute the input

TABLE 13.4 Radial basis function networks parameters.

Parameter	Value
Learning rate, α	1.0×10^{-5}
Error margin, ε_m	2.0×10^{-5}
Initial nominal variance, σ_o	0.1
Initial effective radius, r_o	1.0
Lower bound on radius, r_L	0.001
Radius decrement rate, r_d	0.995
Decay factor for error gradient, α	5.5×10^{-4}
Increment rate for saturation criterion, β	1.0×10^{-9}

TABLE 13.5 Selected principal components for radial basis function networks—projection to latent structures.

Input PCs		Output PCs	
−0.577012	−0.646485	0.545690	0.398848
−0.578720	−0.097753	0.253056	0.916684
−0.576363	0.757559	−0.564421	−0.020437
		0.565345	−0.013931

PC, Principal components.

data scores. These input scores are fed to the trained RBFN model to obtain the output scores. The output scores thus obtained provide the output compositions according to the formula

$$Y_{\text{pred}} = Y_{\text{sc}} \left(Y_{\text{pc}}^T Y_{\text{pc}} \right)^{-1} \left(Y_{\text{pc}}^T \right) \tag{13.9}$$

where Y_{pc} represents the PCs for compositions, Y_{sc} represents composition scores, and Y_{pred} denotes the estimated compositions. These normalized estimated compositions are denormalized to obtain the actual compositions, y_{ij} from the following formula

$$\hat{y}_{\text{ij}} = \frac{y_{\text{ij}} - \overline{y}_i}{\sigma y_i} \tag{13.10}$$

Here y_{ij} are the actual compositions and $\hat{y}_{\text{ij}}$ are the normalized compositions, and i refers the distillate D and bottoms B. The mean $\overline{y}_i$ and standard deviation σ in the above equation are the values that represent the trained composition data. The distillate and reboiler compositions estimated by the NIPALS—RBFN estimator for the test data of input temperatures of different batches are shown in Figs. 13.10 and 13.11. From these results, it is observed that the estimated compositions are in close agreement with the actual compositions. These results thus indicate the better performance of the NIPALS—RBFN estimator for inferential estimation of compositions in batch distillation.

Different data-driven model-based state estimators that are developed in this study for compositions estimation in multicomponent batch distillation are the multivariate statistical model-based PLS estimator, artificial intelligence-based ANN estimator, RBFN estimator, and NIPALS—RBFN estimator. The data to develop the state estimators are generated by solving the mathematical model of batch distillation under various conditions of initial mixture compositions, specified product compositions, reflux flow rate and reboiler heat duty. The effectiveness of these estimators is evaluated by performing state estimation using the temperature measurements of the batch distillation column under various operating conditions and comparing the estimated compositions with the actual compositions. The performance of these state

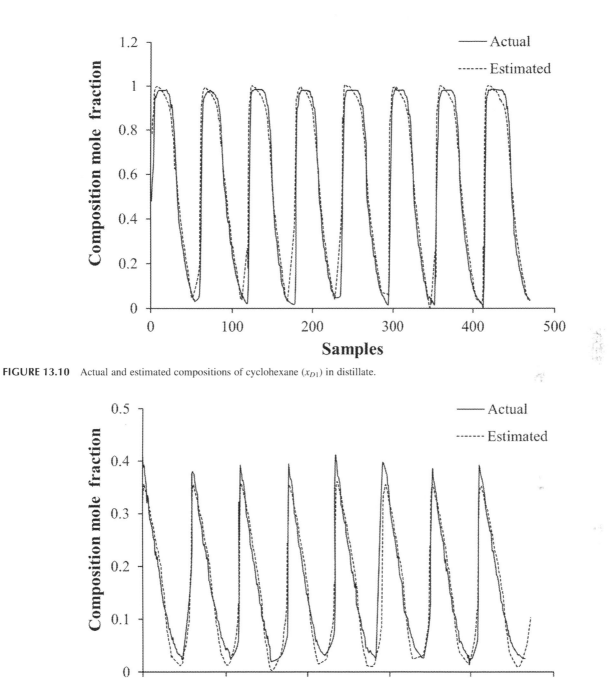

FIGURE 13.10   Actual and estimated compositions of cyclohexane (x_{D1}) in distillate.

FIGURE 13.11   Actual and estimated compositions of cyclohexane (x_{B1}) in reboiler.

estimators is further evaluated in terms of a quantitative performance measure, mean integral square error (MISE), which is defined based on the estimated and actual compositions. The MISE is defined as

$$\text{MISE} = \frac{1}{n} \sum_{i=1}^{n} (x_i - \hat{x}_i)^2 \tag{13.10}$$

Here x_i refers to the actual compositions and $\hat{x}_i$ refers to the estimated compositions of cyclohexane (x_{B1}) and heptane (x_{B2}) in reboiler, and cyclohexane (x_{D1}) and heptane (x_{D2}) in the distillate. The MISE values computed for different estimators based on the data of 455 samples are shown in Table 13.6. These results show the better performance of the

TABLE 13.6 Comparison of estimators.

Estimation method	MISE
PLS estimator	0.01495
ANN estimator	0.03610
RBFN estimator	0.01896
NIPALS-RBFN estimator	0.01290

ANN, Artificial neural networks; *MISE*, mean integral square error; *NIPALS−RBFN*, nonlinear iterative partial least squares−radial basis function networks; *PLS*, projection to latent structures; *RBFN*, radial basis function networks.

NIPALS−RBFN estimator followed by the PLS estimator. The improved performance of the NIPALS−RBFN estimator can be attributed to the automatic recruitment of RBFs as well as using the NIPALS algorithm for the optimal determination of loadings and scores.

13.6 Summary

Different data data-driven model-based state estimators are developed for compositions estimation in multicomponent batch distillation. These are the multivariate statistical model-based PLS estimator, artificial intelligence-based ANN estimator, RBFN estimator, and NIPALS−RBFN estimator. These estimators are designed to provide the distillate and reboiler compositions based on the temperature measurements of the column. A rigorous nonlinear mathematical model of a batch distillation column involving the separation of a hydrocarbon mixture, cyclohexane/heptane/toluene, is considered as the simulation testbed to generate the data and to develop and implement the data-driven model-based composition estimators. The temperature measurement sensors used for each of these estimators are selected using the measurement configuration methods as well as heuristic selection criteria. The effectiveness of these estimators is evaluated by performing state estimation using the temperature measurements of the batch distillation column under various operating conditions and comparing the estimated compositions with the actual compositions, and further in terms of a quantitative performance measure, mean integral square error (MISE), defined based on estimated and actual compositions. These results show the better performance of the NIPALS−RBFN estimator followed by the PLS estimator.

References

[1] P. Geladi, B.R. Kowalski, Partial least squares regression: a tutorial, Anal. Chim. Acta 185 (1) (1986) 1−17.

[2] N.V. Bhatt, T.J. McAvoy, Use of neural networks for dynamic modeling and control of chemical process systems, Comput. Chem. Eng. 14 (4/5) (1990) 573−583.

[3] S. Lee, R.M. Kil, A Gaussian potential function network with hierarchically self organizing learning, Neural Netw. 4 (2) (1991) 207−224.

[4] A. Hoskuldsson, PLS regression methods, J. Chemom. 2 (3) (1988) 211−228.

[5] C. Venkateswarlu, S. Avantika, Optimal state estimation of multicomponent batch distillation, Chem. Eng. Sci. 56 (20) (2001) 5771−5786.

[6] S. Farhat, M. Czernicki, L. Pibouleau, S. Domench, Optimization of multiple-fraction batch distillation by nonlinear programming, AIChE J. 36 (9) (1990) 1349−1360.

[7] J. Thibault, V. Berugesen, A. Cheruy, On-line prediction of fermentation variables using neural networks, Biotechnol. Bioeng. 36 (10) (1990) 1041.

[8] S. Wold, N. Kettaneh-Wold, B. Skagerberg, Nonlinear PLS modeling with fuzzy inference system, Chemom. Intell. Lab. Syst. 7 (1−2) (1989) 53−65.

[9] G. Baffi, E.B. Mortin, A.J. Morris, Nonlinear projection to latent structures revisited: the quadratic PLS algorithm, Comput. Chem. Eng. 23 (3) (1999) 395−411.

[10] S.J. Qin, T.J. McAvoy, Nonlinear PLS modeling using neural networks, Comput. Chem. Eng. 16 (4) (1992) 379−391.

[11] Z. Dong, N. Ma, A novel nonlinear partial least square integrated with error-based extreme learning machine, IEEE Access 7 (2019) 59903−59912.

[12] H. Liu, C. Yang, B. Carlsson, S.J. Qin, C. Yoo, Dynamic nonlinear partial least squares modeling using Gaussian process regression, Industrial & Engineering Chemistry Research 58 (36) (2019) 16676−16686.

[13] H. Wold, Estimation of principal components and related models by iterative least squares, in P. R. Krishnajah (Ed.), Multivariate Analysis, (pp. 391−420), Academic Press, New York, 1966.

Part III

Application of data driven model-based methods for process state estimation

Chapter 14

Optimal state and parameter estimation for fault detection and diagnosis in continuous stirred tank reactor

14.1 Introduction

The state estimation problem for dynamic systems is a major fundamental problem for process monitoring, fault detection, and diagnosis. Early and accurate detection and diagnosis of process faults are essential in the operation of modern chemical plants to reduce downtime and costs, increase safety and product quality, and minimize the impact on the environment. In general, the problem of fault detection in dynamic systems can be formulated as the identification of changes that occur in the process, sensors, controllers, and other components with unknown magnitudes at unknown times. Faults in the initial stage only decrease the performance of the plant. Therefore, early detection of faults is highly beneficial, otherwise, undetected and uncorrected faults can lead to failure of some part of the process or process components. Failure increases the operating cost, results in loss of production, and may lead to a catastrophic event. Diagnosis is determining the magnitude, location and time of occurrence of the faults in process, equipment, and other process components.

Fault detection and diagnosis have received a lot of attention in the literature, and many methods have been proposed by various researchers for stochastic and deterministic dynamical systems. Several surveys and books have been published on this theme [1−8]. The approaches of process fault detection and diagnosis can be broadly classified into the analytical redundancy approach, knowledge-based approach, and data driven modeling approach [3,9,10]. The analytical redundancy approach can also be referred as the quantitative model-based approach. This approach emphasizes the estimation of process variables and parameters based on a model of the process. It can detect, locate, and identify the fault as soon as it occurs in the plant. This approach can tolerate modeling uncertainties and incompleteness in measurements to a certain extent. Methods based on Kalman filters and Luenberger observers and their extended versions can be included under the category of analytical redundancy approach. The knowledge-based approach utilizes qualitative models of the plant combined with experienced based heuristics for fault inference. This approach requires the construction and maintenance of a comprehensive knowledge base. Techniques like expert systems, neural networks, Petri nets, and fuzzy logic can be considered under this category. In data driven modeling approach, the operational data of the plant is treated with different statistical methods to extract the state of the system that enables to detect the abnormal behavior and identify an assignable cause for the out-of-control status of the process. Different feature extraction methods such as principal component analysis, partial least squares, Fisher discriminant analysis as well as their kernel versions belong to the data driven modeling approach.

This study presents a state and parameter estimation based quantitative model-based approach for fault detection and diagnosis in nonlinear systems. Quantitative methods of process fault detection and diagnosis based on filtering and parameter estimation utilize dynamic process models and process measurements to compute the measured and unmeasured process variables and process model parameters. Quantitative fault diagnosis is mostly carried out using the extended Kalman filter (EKF) [9,11−14]. Despite its importance, the EKF is not free from disadvantages, as discussed by other researchers [15,16]. This study presents the method of EKF and various two-level methods for fault detection and diagnosis in nonlinear, time-varying, and stochastic chemical processes. Various two-level methods are formulated using different versions of EKF, recursive least squares, and a reduced-order extended Luenberger observer. The two-level methods are specified for state estimation in the first level and fault diagnosis via parameter identification in the second level. The performances of these methods are evaluated by applying them to a nonlinear continuous stirred tank reactor (CSTR) with a heat exchanger.

Optimal State Estimation for Process Monitoring, Fault Diagnosis and Control. DOI: https://doi.org/10.1016/B978-0-323-85878-6.00001-4

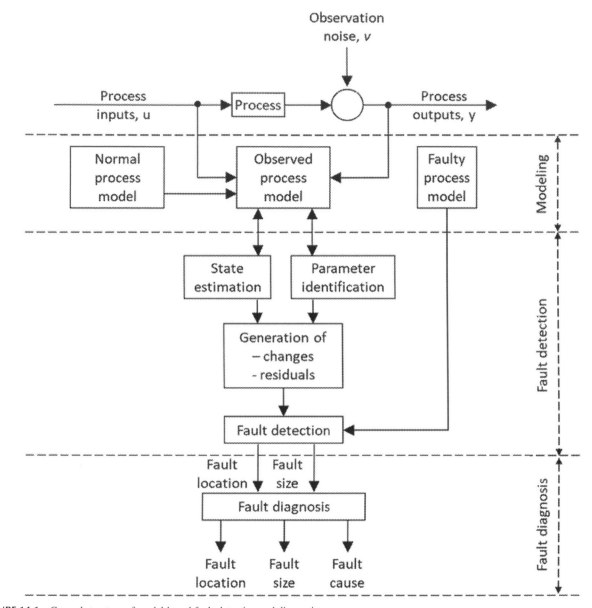

FIGURE 14.1 General structure of model-based fault detection and diagnosis.

14.2 General structure of model-based fault detection and diagnosis

In quantitative model-based fault diagnosis, both the states and parameters in the process model are estimated using accessible process measurements. The application of quantitative model-based fault detection and diagnosis techniques to time dependent processes requires dynamic models that describe their quantitative performance. The status of an operational process depends not only on the steady state conditions, but also on the dynamics of the disturbances encountered during operation. Process faults, which occur during the operations, appear as undesirable state and parameter variations. Therefore, fault detection by state and parameter estimation is desirable as they are relatively more sensitive in detecting changes in the operating behavior of the process. Although fault detection can be done by examining the estimated states and parameters, fault diagnosis is only concerned with the estimated parameters. This is because possible faults in the process can be associated with the specific parameters of the process model and the parameters can be related to the physical features of the process. The general structure of quantitative model-based fault diagnosis is shown in Fig. 14.1.

14.3   General process description for fault detection and diagnosis

A nonlinear, time variant, and stochastic processes with process faults is considered as a general problem for process fault detection and diagnosis. The following dynamic model describes the operating behavior of the process:

$$\dot{x}(t) = f(x(t), \ t) + w(t), \ x(0) = x_0 \tag{14.1}$$

where $x(t)$ is the n-dimensional state vector, f is the nonlinear function of state $x(t)$, and $w(t)$ is the n-dimensional additive Gaussian noise input into the process.

The discrete time linear measurement model is given by:

$$y(t_k) = Hx(t_k) + v(t_k) \tag{14.2}$$

where $y(t_k)$ is a m-dimensional observation vector at time, t_k, $v(t_k)$ is a m-dimensional vector of random Gaussian noise with zero mean, and H is a $m \times n$ dimensional observation matrix.

The observation vector $y(t_k)$ is related to the state vector by the nonlinear relationship:

$$y(t_k) = h[x(t_k)] + v(t_k) \tag{14.3}$$

where h is a nonlinear function of state $x(t_k)$.

Process model Eq. (14.1) considers the parameters as known constant values. Therefore, unknown time varying parameters can be treated as additional state variables, and the form of the model Eq. (14.1) becomes:

$$\dot{x}(t) = f(x(t), \theta(t), t) + w(t), \ x(0) = x_0 \tag{14.4}$$

where $\theta(t)$ is the p-dimensional parameter vector. Both $x(t)$ and $\theta(t)$ of the model in Eq. (14.4) are to be estimated from the noise-corrupted measurements $y(t_k)$ of Eq. (14.2).

The exact nature of the time variation of parameters is usually unknown in process fault detection and diagnosis problems. Therefore, it is assumed that the time deviation of parameters is driven by $\theta = w_2(t)$, where $w_2(t)$ is random noise with covariance Q_2. The states $x(t)$ and parameters $\theta(t)$ are combined into an augmented state vector denoted by $x_a(t)$ as given by

$$x_a(t) = [x(t)^T \theta(t)^T]^T \tag{14.5}$$

The augmented process model is:

$$\dot{x}_a(t) = f_a(x_a(t), t) + w(t) \tag{14.6}$$

The augmented process model can also be expressed as:

$$\dot{x}_a(t) = \begin{bmatrix} \dot{x}(t) \\ \dot{\theta}(t) \end{bmatrix} = \begin{bmatrix} f(x(t), \ \theta(t), \ t) \\ 0 \end{bmatrix} + \begin{bmatrix} w_1(t) \\ w_2(t) \end{bmatrix} \tag{14.7}$$

where $w_1(t)$ is the noise associated with states.

The augmented measurement relation is:

$$y(t_k) = [H(t_k) \vdots 0] \begin{bmatrix} x(t_k) \\ \theta(t_k) \end{bmatrix} + v(t_k) \tag{14.8}$$

where $H(t_k)$ is measurement coefficient matrix and $v(t_k)$ is observation noise.

The nonlinear measurement model for the augmented vector can also be expressed as

$$y(t_k) = h_a(x_a(t_k)) + v(t_k) \tag{14.9}$$

The initial state x_0 is a Gaussian random vector with mean $\bar{x}_0$. The expected initial state covariance matrix P_0 is represented by

$$E[(x_0 - \bar{x})(x_0 - \bar{x})^T] = P_0 \tag{14.10}$$

The process noise w and the measurement noise v are uncorrelated zero-mean white-noise processes with known covariance matrices, Q and R respectively, which are given by the following relations:

$$E[w(t)w^T(t)] = Q(t)$$

$$E[v(t_k)v^T(t_k)] = R(t_k) \tag{14.11}$$

where $Q(t)$ refers process noise covariance matrix and $R(t_k)$ refers observation noise covariance matrix. The initial state covariance matrix P_0 reflects the uncertainty in the initial states, and the process and observation noise covariance matrices Q and R reflect the uncertainty in the model and measurements.

14.4 Nonlinear CSTR, its mathematical model and fault cases considered

A second-order exothermic reaction in which a product B_m is produced from a reactant A_m is considered in the reactor. A schematic of the stirred tank chemical reactor with heat exchanger is shown in Fig. 14.2.

The reaction is given by

$$2A_m \overset{K_r}{\to} B_m \tag{14.12}$$

The reaction rate coefficient is given by

$$K_r(T_0) = K_{ro}\exp(-E_a/R_g T_o) \tag{14.13}$$

where K_{ro} is a frequency factor (8.42×10^8 m^3 kgmol^{-1} s^{-1}), E_a is an activation energy (1.04×10^8 J kgmol^{-1}), R_g is the gas constant (8.319×10^3 J kgmol^{-1} K^{-1}), T_o is a reaction temperature (K).

The dynamic behavior of the reactor is described by the following ordinary differential equations.

The mass balance on reactant A_m is given by

$$V_r \frac{dC_o}{dt} = qC_i - qC_o - 2V_r C_o^2 K_r(T_0), \quad C_o(0) = 0.3412 \tag{14.14}$$

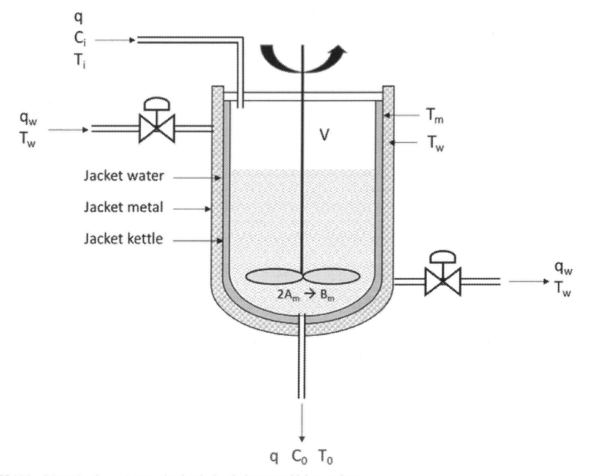

FIGURE 14.2 Schematic of a continuous stirred tank chemical reactor with heat exchanger.

where V_r is a reactor volume (8.50 m³), C_i is an inlet concentration of reactant (9.62 kg mol m⁻³), C_o is a outlet concentration of reactant (kg mol m⁻³), q is an inlet and outlet volumetric flowrate (8.5 × 10⁻³ m³ s⁻¹).

The energy balance for the process is represented by

$$\rho C_p V_r \frac{dT_o}{dt} = \rho C_p q(T_i - T_o) - \Delta H V_r C_o^2 K_r(T_0) - ah_{mo}(T_0 - T_m), \quad T_o(0) = 525.7 \tag{14.15}$$

where ρC_p is a volumetric heat capacity of reactor contents (4.35 × 106 J m⁻³), ΔH is a heat of reaction (−9.30 × 107 J kg mol⁻¹), a is an effective heat exchanger area (46.5 m²), T_i is an inlet temperature (456 K), T_m is a jacket wall temperature (K), and h_{mo} is an overall heat transfer coefficient (799 J m⁻² s⁻¹ K⁻¹).

The heat balance of jacket water is represented by

$$\rho_w C_w V_w \frac{dT_w}{dt} = \rho_w C_w q_w(T_{wi} - T_m) - ah_{wm}(T_w - T_m), \quad T_w(0) = 472.2 \tag{14.16}$$

where T_w is a jacket water temperature (K), $\rho_w C_w$ is a volumetric heat capacity of water jacket (2.24 × 106 J m⁻³), V_w is a water jacket volume (2.83 m³), q_w is a coolant water volumetric flowrate (0.0283 m³ s⁻¹), T_{wi} is a coolant inlet temperature (456 K), h_{wm} is a heat transfer coefficient from jacket water to jacket metal (980 J m⁻² s⁻¹ K⁻¹), and a is a heat exchanger area from jacket water to jacket metal (46.5 m²).

The heat balance of jacket metal is represented by

$$\rho_m C_m V_m \frac{dT_m}{dt} = ah_{wm}(T_w - T_m) - ah_{mo}(T_m - T_o), \quad T_m(0) = 496.2 \tag{14.17}$$

where $\rho_m C_m$ is a volumetric heat capacity of jacket wall (7.45 × 106 J m⁻³) and V_m is a metal jacket volume (0.283 m³).

Fault cases considered: A change in the feed flowrate q caused by pipe plugging and deviation of heat transfer coefficient h_{mo} caused by scaling of the heat exchanger surface are considered as fault modes in the reactor. Measured outputs of the reactor are concentration of the chemical product C_o, temperature of the product T_0 and temperature of the jacket water of heat exchanger T_w. All the measurements include noise with zero mean. Various methods are employed to filter the measurements, to estimate unknown wall temperature and identify fault parameters, feed flowrate, and heat transfer coefficient.

14.5 Method of extended Kalman filter

A strategy based on an EKF is presented for fault detection and diagnosis in nonlinear CSTR with a heat exchanger. Continuous reactor dynamics and discrete observations are used to estimate state variables and identify process parameters. The identified process parameters can be used for detecting faults in the reactor and diagnosing their causes.

14.5.1 Design strategy

The EKF algorithm described in Section 4.4.3 (Chapter 4, Mechanistic model-based nonlinear filtering and observation techniques for optimal state/parameter estimation) is designed and applied for the combined estimation of states and fault parameters in the nonlinear CSTR system. State estimation and fault parameter identification using the method of EKF is shown in Fig. 14.3.

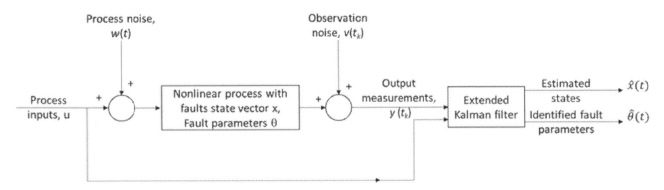

FIGURE 14.3   State estimation and fault parameter identification using extended Kalman filter.

The relationships defined in the reactor model for the design of EKF aregiven as follows:
The augmented state vector as in Eq. (14.5):

$$x_a(t) = [x_1(t)\, x_2(t)\, x_3(t)\, x_4(t)\, \theta_1(t)\, \theta_2(t)]^T = [C_0(t)\, T_0(t)\, T_w(t)\, T_m(t)\, q(t)\, h_{m0}(t)]^T \tag{14.18}$$

The observation vector,

$$y(t_k) = [y_1(t_k)\, y_2(t_k)\, y_3(t_k)]^T = [C_0(t_k)\, T_0(t_k)\, T_w(t_k)]^T \tag{14.19}$$

where $C_0(t_k)$, $T_0(t_k)$, and $T_w(t_k)$ are observed states with zero mean random Gaussian noise.

The function f_a in Eq. (14.6) is formed by augmenting the reactor model Eqs. (14.14)−(14.17) with process parameters:

$$f_a = \begin{bmatrix} C_1\theta_1(C_i - x_i) - C_2x_1^2\exp(-C_3/x_2) \\ C_1\theta_1(T_i - x_2) - C_4x_1^2\exp(-C_3/x_2) - C_5\theta_2(x_2 - x_4) \\ C_6(T_{wi} - x_3) - C_7(x_3 - x_4) \\ C_8(x_3 - x_4) - C_9\theta_2(x_4 - x_2) \\ 0 \\ 0 \end{bmatrix} \tag{14.20}$$

where $C_1 = 1/V_r$; $C_2 = 2K_{r0}$; $C_3 = E_a/R_g$; $C_4 = DHK_{r0}/rc_p$; $C_5 = a/rc_pV_r$; $C_6 = q_w/V_w$; $C_7 = ah_{wm}/r_wc_wV_w$; $C_8 = ah_{wm}/r_mc_mV_m$; $C_9 = a/r_mc_mV_m$

The nonlinear measurement model h_a in Eq. (14.9) is expressed as

$$h_a = \begin{bmatrix} x_1 + \left[C_1\theta_1(C_i - x_i) - C_2x_1^2\exp(-C_3/x_2)\right]\Delta t \\ x_2 + \left[C_1\theta_1(T_i - x_2) - C_4x_1^2\exp(-C_3/x_2) - C_5\theta_2(x_2 - x_4)\right]\Delta t \\ x_3 + [C_6(T_{wi} - x_3) - C_7(x_3 - x_4)]\Delta t \end{bmatrix} \tag{14.21}$$

where Δt is the sampling time.

The state transition matrix F in Eq. (4.27) is computed for the model in Eq. (14.20) as

$$F = \begin{bmatrix} f_{11} & f_{12} & 0 & 0 & f_{15} & 0 \\ f_{21} & f_{22} & f_{24} & 0 & f_{25} & f_{26} \\ 0 & 0 & f_{33} & f_{34} & 0 & 0 \\ 0 & f_{42} & f_{43} & f_{44} & 0 & f_{46} \\ 0 & 0 & 0 & 0 & 0 & 0 \\ 0 & 0 & 0 & 0 & 0 & 0 \end{bmatrix} \tag{14.22}$$

where $f_{11} = -C_1q_1 - 2C_2x_1\exp(-C_3/x_2)$; $f_{12} = -C_2x_1^2(C_3/x_2^2)\exp(-C_3/x_2)$; $f_{15} = C_1(C_i - x_1)$; $f_{21} = -2C_4x_1\exp(-C_3/x_2)$; $f_{22} = -C_1q_1 - C_4x_1^2(C_3/x_2^2)\exp(-C_3/x_2) - C_5q_2$; $f_{24} = C_5q_2$; $f_{25} = C_1(T_i - x_2)$; $f_{26} = -C_5(x_2 - x_4)$; $f_{33} = -C_6 - C_7$; $f_{34} = C_7$; $f_{42} = C_9q_2$; $f_{43} = C_8$; and $f_{44} = -C_8 - C_9q_2$; $f_{46} = -C_9(x_4 - x_2)$.

The measurement Jacobian matrix H in Eq. (4.31) is computed for the model in Eq. (14.21) as

$$H = \begin{bmatrix} h_{11} & h_{12} & 0 & 0 & h_{15} & 0 \\ h_{21} & h_{22} & 0 & h_{24} & h_{25} & h_{26} \\ 0 & 0 & h_{33} & h_{34} & 0 & 0 \end{bmatrix} \tag{14.23}$$

where $h_{11} = 1.0 + \left[-C_1q_1 - 2C_2x_1\exp(-C_3/x_2)\right]\Delta t$; $h_{12} = \left[-C_2x_1^2(C_3/x_2^2)\exp(-C_3/x_2)\right]\Delta t$; $h_{15} = [C_1(C_i - x_1)]\Delta t$; $h_{21} = \left[-2C_4x_1\exp(-C_3/x_2)\right]\Delta t$; $h_{22} = 1.0 + \left[-C_1q_1 - C_4x_1^2(C_3/x_2^2)\exp(-C_3/x_2) - C_5q_2\right]\Delta t$; $h_{24} = [C_5q_2]\Delta t$; $h_{25} = [C_1(T_i - x_2)]\Delta t$; $h_{26} = [-C_5(x_2 - x_4)]\Delta t$; $h_{33} = 1.0 + [-C_6 - C_7]\Delta t$; $h_{34} = C_7\Delta t$.

The EKF described in Section 4.4.3 (Chapter 4, Mechanistic model-based nonlinear filtering and observation techniques for optimal state/parameter estimation) is used to estimate states and identify fault parameters. The state and fault parameter estimation scheme by EKF in CSTR system is shown in Fig. 14.4.

14.5.2 Analysis of results

The EKF designed above is applied for state estimation and fault parameter identification in nonlinear CSTR. The cases of faults considered in the reactor are given in Table 14.1.

The measurements of the reactor are the concentration of the chemical product C_o, the temperature of the product T_0 and the temperature of the jacket water of heat exchanger T_w. Simulation measurements are generated for normal operation and all the cases of faults by solving the reactor model equations given in Section 14.4 numerically and by corrupting the states with zero-mean random Gaussian noise. The standard deviations of the measurement noises are: $v_1(t_k) = 0.0009$ (kg mol m^{-3}), $v_2(t_k) = 0.30$ (K) and $v_3(t_k) = 0.30$ (K). The simulated process measurements for the case of faults in q and h_{mo} are shown in Fig. 14.5.

The initial augmented state vector, the initial state covariance matrix, P_o, the process noise covariance matrix, Q, the observation noise covariance matrix, R and the linear measurement matrix, H chosen for the method of EKF are given as follows:

$$\begin{bmatrix} C_o(0) \\ T_o(0) \\ T_w(0) \\ T_m(0) \\ q_o(0) \\ h_{mo}(0) \end{bmatrix} = \begin{bmatrix} 0.3412 \\ 525.7 \\ 472.2 \\ 496.2 \\ 0.0085 \\ 799.0 \end{bmatrix} \tag{14.24}$$

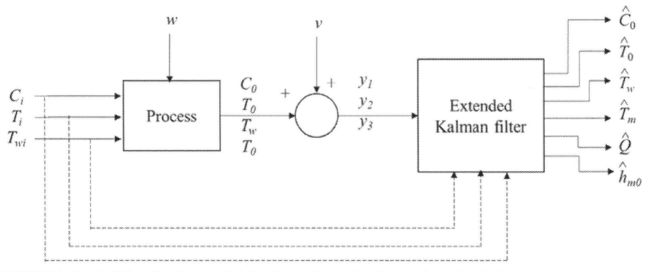

FIGURE 14.4 Extended Kalman filter for process fault detection and diagnosis in nonlinear continuous stirred tank reactor.

TABLE 14.1 Cases of faults in nonlinear continuous stirred tank reactor.

S. no.	Case studied
i	A step change of feed flowrate q from 0.0085 m³ s^{-1} to 0.0068 s^{-1} at 300 s.
ii	A ramp change of h_{mo} from 799 J m^{-2} s^{-1} K^{-1} to 559 J m^{-2} s^{-1} K^{-1} during 200–500 s.
iii	A step change of q from 0.0085 m³ s^{-1} to 0.0068 m³ s^{-1} at 300 s and a ramp change of h_{mo} from 700 J m^{-2} s^{-1} K^{-1} to 559 J m^{-2} s^{-1} K^{-1} during 200–500 s.

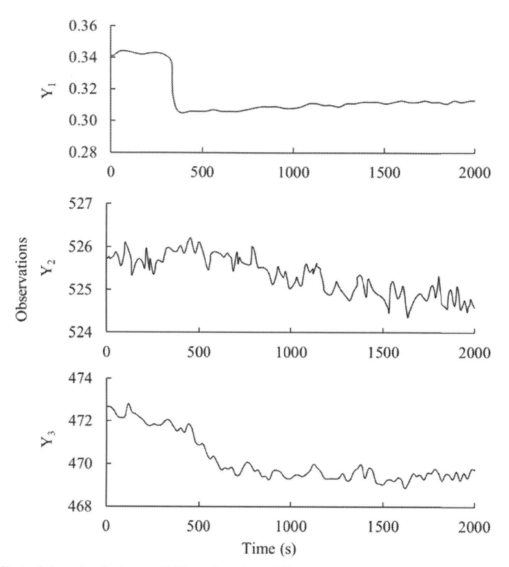

FIGURE 14.5   Simulated observations for the case of 20% step change in q and 30% ramp change in h_{mo}.

$$P_0 = \begin{bmatrix} 0.001 & 0 & 0 & 0 & 0 & 0 \\ 0 & 1 & 0 & 0 & 0 & 0 \\ 0 & 0 & 1 & 0 & 0 & 0 \\ 0 & 0 & 0 & 1 & 0 & 0 \\ 0 & 0 & 0 & 0 & 1 \times 10^{-6} & 0 \\ 0 & 0 & 0 & 0 & 0 & 1 \end{bmatrix} \tag{14.25}$$

$$Q = \begin{bmatrix} 0.0001 & 0 & 0 & 0 & 0 & 0 \\ 0 & 0.01 & 0 & 0 & 0 & 0 \\ 0 & 0 & 0.5 & 0 & 0 & 0 \\ 0 & 0 & 0 & 1 & 0 & 0 \\ 0 & 0 & 0 & 0 & 1 \times 10^{-6} & 0 \\ 0 & 0 & 0 & 0 & 0 & 1 \times 10^4 \end{bmatrix} \tag{14.26}$$

$$H = \begin{bmatrix} 1 & 0 & 0 & 0 \\ 0 & 1 & 0 & 0 \\ 0 & 0 & 1 & 0 \end{bmatrix} \tag{14.27}$$

The performance index J for EKF is expressed by

$$J = \sum_{k=1}^{N} \left\{ [x(0) - \hat{x}(0)]^T P_o^{-1} [x(0) - \hat{x}(0)] \right\} + \left\{ y(t_k) - h_a[\hat{x}(t_k)] \right\}^T R^{-1} \left\{ y(t_k) - h_a[\hat{x}(t_k)] \right\}$$
$$+ \left\{ [x(t_k) - \hat{x}(t_k)]^T Q^{-1} [x(t_k) - \hat{x}(t_k)] \right\} \tag{14.28}$$

The EKF is applied for state estimation and fault diagnosis in the CSTR system using the above conditions. The known states $\hat{C}_o$, $\hat{T}_o$, and $\hat{T}_w$, the unknown state $\hat{T}_m$, and fault parameters $\hat{q}$ and $\hat{h}_{mo}$ are estimated recursively for each set of measurements for normal operation and all the fault cases. The performance indices evaluated using Eq. (14.28) for fault situations indicate that they are markedly different from normal operation. The sensitivity of the estimator is also studied towards the effect of noise covariance matrices and observation noise. The increase in process noise covariance matrix, Q increases the fluctuations in estimated states and identified fault parameters. The decrease in observation noise covariance matrix, R is found to increase the fluctuations in estimated responses, but increasing R is found to reduce the fluctuations. The estimated states and the identified fault parameters by EKF for 20% step change of fault parameter q and 30% ramp change of fault parameter h_{mo} are shown in Figs. 14.6 and 14.7.

14.6 Method of reduced order extended Luenberger observer and extended Kalman filter

This is a two-level method, in which the first level provides state estimates and the second level identifies fault parameters [9]. In the first level, a reduced-order extended Luenberger observer is used for estimating state variables in the presence of faults using accessible process measurements. In the second level, the state estimates resulting from the observer are used in the EKF to identify process parameters for fault diagnosis. State estimation and fault parameter identification using this method are shown in Fig. 14.8.

14.6.1 Design strategy

The details of the CSTR system and its mathematical model are illustrated in Section 14.4. The fault cases considered in CSTR are given in Table 14.1. The reduced order extended Luenberger observer described in Section 4.13 (Chapter 4, Mechanistic model-based nonlinear filtering and observation techniques for optimal state/parameter estimation) is designed and applied for state estimation in nonlinear CSTR.

The following relationships are defined for the observer in Section 4.13 and the process model:

$$\delta x(t) = [\delta x_1(t), \delta x_2(t), \delta x_3(t), \delta x_4(t)]^T = [C_o(t) - C_o^*, T_o(t) - T_o^*, T_w(t) - T_w^*, T_m(t) - T_m^*]^T \tag{14.29}$$

$$\delta u(t) = [\delta u_1(t), \delta u_2(t), \delta u_3(t)]^T = [C_i(t) - C_i^*, T_i(t) - T_i^*, T_{wi}(t) - T_{wi}^*]^T \tag{14.30}$$

$$\delta y(t) = [\delta y_1(t), \ \delta y_2(t), \ \delta y_3(t)]^T = [\delta x_1(t), \ \delta x_2(t), \ \delta x_3(t)]^T \tag{14.31}$$

The notation C_o^*, T_o^*, T_w^*, and T_m^* are considered state variables at the initial operating point corresponding to the process inputs C_i^*, T_i^*, and T_{wi}^*.

The linearized reactor state space model used for state estimation in the presence of faults is represented by

$$\delta \dot{x}(t) = \begin{bmatrix} a_{11} & a_{12} & 0 & 0 \\ a_{21} & a_{22} & 0 & 0 \\ 0 & 0 & a_{33} & a_{34} \\ 0 & 0 & a_{43} & a_{44} \end{bmatrix} \delta x(t) + \begin{bmatrix} b_{11} & 0 & 0 \\ 0 & b_{22} & 0 \\ 0 & 0 & b_{33} \\ 0 & 0 & 0 \end{bmatrix} \delta u(t) + \begin{bmatrix} 1 & 0 & 0 \\ 0 & 1 & -1 \\ 0 & 0 & 0 \\ 0 & 0 & \dfrac{\rho C_p V_r}{\rho_m C_m V_m} \end{bmatrix} f_e(t, \theta) \tag{14.32}$$

$$\delta y(t) = \begin{bmatrix} 1 & 0 & 0 & 0 \\ 0 & 1 & 0 & 0 \\ 0 & 0 & 1 & 0 \end{bmatrix} \delta x(t) + v(t) \tag{14.33}$$

where a_{ij}'s are linearized reactor model coefficients obtained by linearizing the reactor model Eqs. (14.14)–(14.17) with the respective state vector, and b_{ij}'s are computed by linearization of reactor model with the respective input vector. The process fault vector of Eq. (4.103) in Section 4.13 (Chapter 4) is obtained as

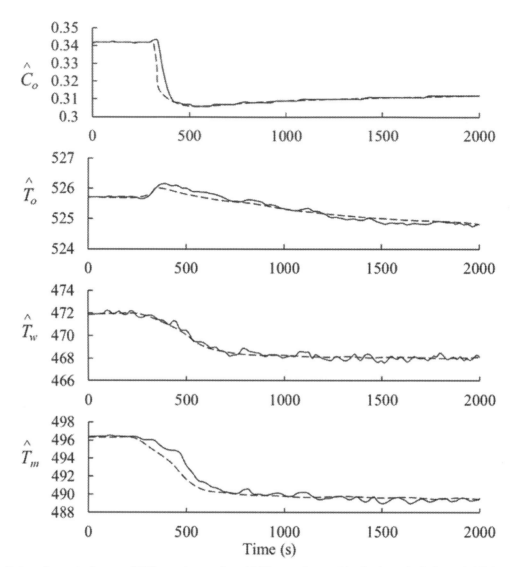

FIGURE 14.6   Estimated states in the case of 20% step change of q and 30% ramp change of h_{mo} by the method of extended Kalman filter.

$$[f(t, \theta)]^T = \left[-\frac{-q}{v_r} \delta x_1 \quad \frac{a h_{mo}}{\rho C_p v_r} (\delta x_2 - \delta x_4) \right]^T \tag{14.34}$$

The linearized reactor model is in the general form of Eq. (4.101) in Section 4.13 (Chapter 4). To obtain the observer coefficient matrices in Section 4.12.1, first, an arbitrary matrix W is assigned, and then matrices V and G are obtained by using the relation:

$$\begin{bmatrix} C \\ \cdots \\ W \end{bmatrix}^{-1} \begin{bmatrix} 1 & 0 & 0 & 0 \\ 0 & 1 & 0 & 0 \\ 0 & 0 & 1 & 0 \\ \cdots & \cdots & \cdots & \cdots \\ 0 & 0 & 0 & 1 \end{bmatrix}^{-1} = [V \vdots G] = \begin{bmatrix} 1 & 0 & 0 & \vdots & 0 \\ 0 & 1 & 0 & \vdots & 0 \\ 0 & 0 & 1 & \vdots & 0 \\ 0 & 0 & 0 & \vdots & 1 \end{bmatrix} \tag{14.35}$$

The notation can be referred to in Section 4.12. The rank of the CD matrix is obtained as

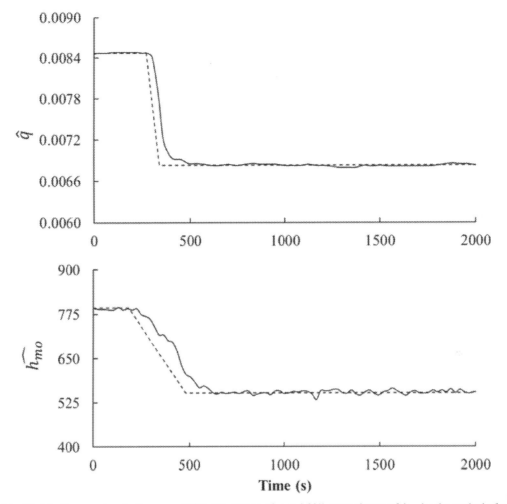

FIGURE 14.7   Identified fault parameters in the case of 20% step change of q and 30% ramp change of h_{mo} by the method of extended Kalman filter.

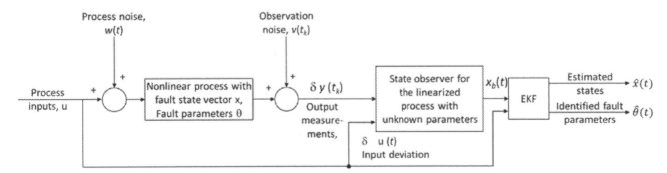

FIGURE 14.8   State estimation and fault parameter identification using a reduced-order Luenberger observer and extended Kalman filter.

$$\text{rank}[CD] = \begin{bmatrix} 1 & 0 & 0 \\ 0 & 1 & -1 \\ 0 & 0 & 0 \end{bmatrix} = 2 \tag{14.36}$$

where C and D are the coefficient matrices of linearized reactor model as in Eq. (4.101) in Section 4.13. Since the second and third columns of the CD matrix are linearly dependent, the first and third elements of $f(t, \theta)$ and the corresponding columns of the D-matrix are considered for observer computation. With these design calculations of CSTR system, the reduced order extended Luenberger observer, Eq. (4.104) in Section 4.13 is employed to compute the state vector. The output vector resulting from the observer is expressed as

$$\hat{x}_b(t) = \begin{bmatrix} \hat{C}_{ob}(t) & \hat{T}_{ob}(t) & \hat{T}_{wb}(t) & \hat{T}_{mb}(t) \end{bmatrix}^T \tag{14.37}$$

The observer output state vector, Eq. (14.37) is used as the observation vector to the EKF for fault parameter identification.

14.6.2 Analysis of results

The initial state vector is given by

$$\begin{bmatrix} C_o{}^* \\ T_o{}^* \\ T_w{}^* \\ T_m{}^* \end{bmatrix} = \begin{bmatrix} 0.3412 \\ 525.7 \\ 472.2 \\ 496.2 \end{bmatrix} \tag{14.38}$$

The linearized reactor model equations, Eq. (14.32), are integrated numerically for normal operation and faulty operation for the situations given in Table 14.1. The state deviations obtained from Eq. (14.32) are corrupted with the zero mean random Gaussian noise to obtain the measurement deviations according to Eq. (14.33). These measurement deviations are given by:

$$\begin{bmatrix} \delta y_1 \\ \delta y_2 \\ \delta y_3 \end{bmatrix} = \begin{bmatrix} 1 & 0 & 0 & 0 \\ 0 & 1 & 0 & 0 \\ 0 & 0 & 1 & 0 \end{bmatrix} \begin{bmatrix} \delta x_1 \\ \delta x_2 \\ \delta x_3 \\ \delta x_4 \end{bmatrix} + \begin{bmatrix} 0.003 \\ 1.0 \\ 1.0 \end{bmatrix} [v] \tag{14.39}$$

where v is zero mean random Gaussian noise. Reduced order extended Luenberger observer defined by Eq. (4.104) in Section 4.13 is used to estimate the unknown state, the heat exchanger jacket metal wall temperature along with the other states by using the reactor concentration, reactor temperature, and coolant water temperature as deviation measurements. The states resulting from the observer, Eq. (14.37), are treated as inferential measurements for fault parameter identification by EKF.

Thus, the inferential measurements to the EKF are expressed as

$$y(t_k) = [y_1(t_k), \ y_2(t_k), \ y_3(t_k), \ y_4(t_k)]^T = [\hat{C}_{ob}(t_k), \ \hat{T}_{ob}(t_k), \ \hat{T}_{wb}(t_k), \ \hat{T}_{mb}(t_k)]^{\hat{T}} \tag{14.40}$$

The estimated states by the observer along with the fault parameters represent the augmented state vector x_a for the EKF as given by

$$\hat{x}_a = [\hat{C}_{ob}(t_k), \ \hat{T}_{ob}(t_k), \ \hat{T}_{wb}(t_k), \ \hat{T}_{mb}(t_k), \ \hat{q}, \ \hat{h}_{mo}]^T \tag{14.41}$$

The state transition matrix F in Eq. (4.27) is denoted as F_a and is computed by taking the partial derivatives of the model, Eqs. (14.14−14.17) concerning the augmented state vector.

The nonlinear measurement model h_a in Eq. (14.9) for the reactor system is expressed as

$$h_a = \begin{bmatrix} x_1 + \left[C_1\theta_1(C_i - x_i) - C_2 x_1^2 \exp\left(-C_3/x_2\right)\right]\Delta t \\ x_2 + \left[C_1\theta_1(T_i - x_2) - C_4 x_1^2 \exp\left(-C_3/x_2\right) - C_5\theta_2(x_2 - x_4)\right]\Delta t \\ x_3 + \left[C_6(T_{wi} - x_3) - C_7(x_3 - x_4)\right]\Delta t \\ x_4 + \left[C_8(x_3 - x_4) - C_9\theta_2(x_4 - x_2)\right]\Delta t \end{bmatrix} \tag{14.42}$$

where Δt is sampling time. The measurement Jacobian matrix, H_a of Eq. (4.31), is obtained by taking the partial derivatives of Eq. (14.42) with the augmented state vector given by Eq. (14.41). With these relations, the EKF is applied for state estimation and fault parameter identification in nonlinear CSTR for normal operation and different fault cases given in Table 14.1.

The same initial state covariance matrix, P_o and the process noise covariance matrix, Q, as in the method of EKF in Section 14.5.2, along with the following observation noise covariance matrix, R are employed.

$$R = \begin{bmatrix} 0.001 & 0 & 0 & 0 \\ 0 & 5 & 0 & 0 \\ 0 & 0 & 5 & 0 \\ 0 & 0 & 0 & 5 \end{bmatrix} \tag{14.43}$$

For problems of fault detection and diagnosis, states are to be estimated correctly, otherwise, parameter estimation based on estimated states may lead to incorrect results. In the first level of this method, a reduced order Luenberger observer is employed to estimate the unknown state, that is, the heat exchanger jacket metal wall temperature, T_m in the presence of faults in conjunction with the known measurements. The observer outputs are used as inferential measurements for fault diagnosis. The estimated states by the observer can also be helpful for fault detection. The identified fault parameters by the reduced order extended Luenberger observer and EKF for 20% step change of fault q and 30% ramp change in fault h_{mo} are shown in Fig. 14.9.

14.7   Method of two-level extended Kalman filter

This method involves optimal state estimation, and uncertain process parameter identification separately by a state EKF and a parameter EKF. States and parameters are exchanged between the two estimators for each new value of measurement.

14.7.1   Design strategy

The general process representation for the method of two-level EKF is illustrated in Section 4.4.2. The two-level EKF algorithm for separate estimation of states and process parameters is described in Section 4.6 (Chapter 4, Mechanistic model-based nonlinear filtering and observation techniques for optimal state/parameter estimation). The schematic of state estimation and fault parameter identification using the two-level EKF is shown in Fig. 14.10.

14.7.1.1   Design of state estimation filter

The following relations are defined in the reactor model.
The state vector:

$$x(t) = [x_1(t)\ x_2(t)\ x_3(t)\ x_4(t)]^T = [C_0(t)\ T_0(t)\ T_w(t)\ T_m(t)]^T \tag{14.44}$$

The input vector:

$$u(t) = [u_1(t)\ u_2(t)\ u_3(t)]^T = [C_i(t)\ T_i(t)\ T_{wi}(t)]^T \tag{14.45}$$

The fault vector:

$$\theta(t) = [\theta_1(t)\quad \theta_2(t)]^T = [q(t)\quad h_{m0}(t)]^T \tag{14.46}$$

The observation vector:

$$y(t_k) = [y_1(t_k)\ y_2(t_k)\ y_3(t_k)]^T = [C_0(t_k)\ T_0(t_k)\ T_w(t_k)]^T \tag{14.47}$$

The f_x involved in Eq. (4.40) of Section 4.6.1 is defined by the reactor model equations as:

$$f_x = \begin{bmatrix} C_1\theta_1(C_i - x_i) - C_2x_1^2\exp(-C_3/x_2) \\ C_1\theta_1(T_i - x_2) - C_4x_1^2\exp(-C_3/x_2) - C_5\theta_2(x_2 - x_4) \\ C_6(T_{wi} - x_3) - C_7(x_3 - x_4) \\ C_8(x_3 - x_4) - C_9\theta_2(x_4 - x_2) \end{bmatrix} \tag{14.48}$$

The quantities of C_i are defined under Eq. (14.20). The state transition matrix F_x in Eq. (4.41) of Section 4.6.1 is computed by taking the partial derivatives of the reactor model with respect to state vector. This matrix is expressed as

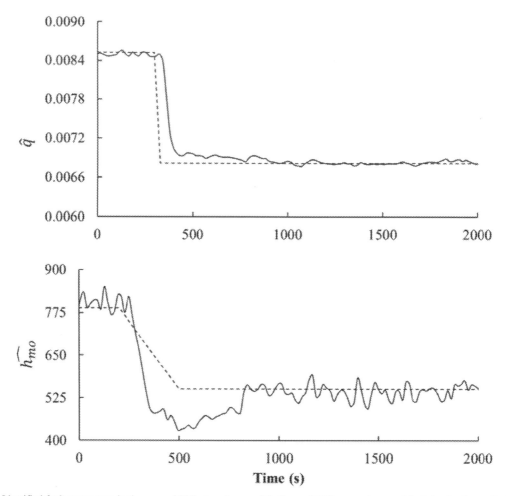

FIGURE 14.9 Identified fault parameters in the case of 20% step change of fault q and 30% ramp change of fault h_{mo} by the method of the reduced-order Luenberger observer and extended Kalman filter.

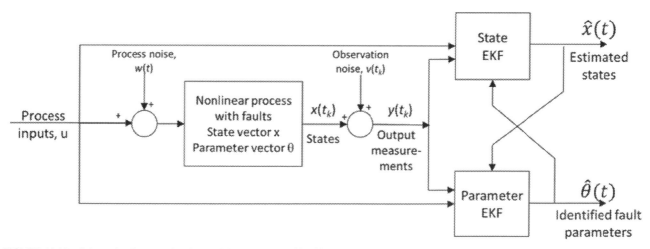

FIGURE 14.10 Schematic of state estimation and fault parameter identification using a two-level extended Kalman filter.

$$F_x = \begin{bmatrix} f_{11} & f_{12} & 0 & 0 \\ f_{21} & f_{22} & 0 & f_{24} \\ 0 & 0 & f_{33} & f_{34} \\ 0 & f_{42} & f_{43} & f_{44} \end{bmatrix}$$

(14.49)

The elements of this matrix are defined under Eq. (14.22).

The nonlinear measurement model h_k involved in Eq. (4.24) represents the reactor measurement model in Eq. (14.21). The measurement Jacobian matrix, H_x in Eq. (4.41) is computed by taking the partial derivatives of the reactor measurement model in Eq. (14.42) with respect to the state vector.

14.7.1.2 Design of parameter estimation filter

The following relations are defined in the reactor model.

The parameter vector:

$$\theta(t) = [\theta_1(t)\; \theta_2(t)]^T = [q(t)\quad h_{m0}(t)]^T \tag{14.50}$$

The $g(\theta)$ in Eq. (4.23) of Section 4.4.2 for the reactor model is expressed as:

$$g_\theta = \begin{bmatrix} 0 \\ 0 \end{bmatrix} \tag{14.51}$$

The parameter Jacobian matrix G_θ involved in Eq. (4.43) of Section 4.6.2 assumes zero value. The parameter gradient matrix H_θ involved in Eq. (4.42) for the reactor model is computed using Eq. (4.44) of Section 4.6.2. These relations are used in the parameter identification filter algorithm in Section 4.6.2 to estimate the process fault parameters.

The performance index J for two-level EKF is expressed by:

$$\begin{aligned}
J = \sum_{k=1}^{N} &\Big\{ [x(0) - \hat{x}(0)]^T P_{xo}^{-1} [x(0) - \hat{x}(0)] + [\theta(0) - \hat{\theta}(0)]^T P_{\theta o}^{-1} [\theta(0) - \hat{\theta}(0)] \Big\} + \\
&\Big\{ y(t_k) - h_a[\hat{x}(t_k), \hat{\theta}(t_k)] \Big\}^T R^{-1} \Big\{ y(t_k) - h_a[\hat{x}(t_k), \hat{\theta}(t_k)] \Big\} + \\
&\Big\{ [x(t_k) - \hat{x}(t_k)]^T Q_x^{-1} [x(t_k) - \hat{x}(t_k)] + [\theta(t_k) - \hat{\theta}(t_k)]^T Q_\theta^{-1} [\theta(t_k) - \hat{\theta}(t_k)] \Big\}
\end{aligned} \tag{14.52}$$

14.7.2 Analysis of results

The two-level EKF designed above is used for state estimation and fault diagnosis in the CSTR system for normal operation and faulty situations given in Table 14.1. The measurements of the reactor are the reactant concentration, reactor temperature, and jacket water temperature. These measurements are represented by

$$\begin{bmatrix} y_1(t_k) \\ y_2(t_k) \\ y_3(t_k) \end{bmatrix} = \begin{bmatrix} 1 & 0 & 0 & 0 \\ 0 & 1 & 0 & 0 \\ 0 & 0 & 1 & 0 \end{bmatrix} \begin{bmatrix} x_1(t_k) \\ x_2(t_k) \\ x_3(t_k) \\ x_4(t_k) \end{bmatrix} + \begin{bmatrix} 0.003 \\ 1.0 \\ 1.0 \end{bmatrix} [v(t_k)] \tag{14.53}$$

where $v(t_k)$ is a random Gaussian noise generated using random numbers within -1.0 to 1.0.

The initial conditions, noise covariance matrices, and measurement coefficient matrix selected for implementation of the two level EKF algorithm in Section 4.6 (Chapter 4, Mechanistic model-based nonlinear filtering and observation techniques for optimal state/parameter estimation) are given as follows.

$$\begin{bmatrix} C_o(0) \\ T_o(0) \\ T_w(0) \\ T_m(0) \end{bmatrix} = \begin{bmatrix} 0.3412 \\ 525.7 \\ 472.2 \\ 496.2 \end{bmatrix} \tag{14.54}$$

$$P_{x0} = \begin{bmatrix} 0.001 & 0 & 0 & 0 \\ 0 & 1 & 0 & 0 \\ 0 & 0 & 1 & 0 \\ 0 & 0 & 0 & 1 \end{bmatrix} \tag{14.55}$$

$$Q_X = \begin{bmatrix} 0.0001 & 0 & 0 & 0 \\ 0 & 0.01 & 0 & 0 \\ 0 & 0 & 0.5 & 0 \\ 0 & 0 & 0 & 1 \end{bmatrix} \tag{14.56}$$

$$R = \begin{bmatrix} 0.001 & 0 & 0 \\ 0 & 5 & 0 \\ 0 & 0 & 5 \end{bmatrix} \tag{14.57}$$

$$H = \begin{bmatrix} 1 & 0 & 0 & 0 \\ 0 & 1 & 0 & 0 \\ 0 & 0 & 1 & 0 \end{bmatrix} \tag{14.58}$$

$$\begin{bmatrix} q(0) \\ h_{mo}(0) \end{bmatrix} = \begin{bmatrix} 0.0085 \\ 799 \end{bmatrix} \tag{14.59}$$

$$P_{\theta 0} = \begin{bmatrix} 1.0 \times 10^{-6} & 0 \\ 0 & 1 \end{bmatrix} \tag{14.60}$$

$$Q_\theta = \begin{bmatrix} 1.0 \times 10^{-6} & 0 \\ 0 & 1 \times 10^4 \end{bmatrix} \tag{14.61}$$

In this study, a method of two-level EKF is presented for process fault detection and diagnosis to circumvent the problems of convergence, instability, and computational burden associated with the EKF. The method is applied for state estimation and fault parameter identification for normal operation and different fault situations in Table 14.1. The separate estimation of states and identification of fault parameters by this method reduces the computational effort due to the involvement of reduced size matrices in computations. The quantitative performance of two-level EKF is expressed by Eq. (14.52). The method's sensitivity is studied for the effect of observation noise, v, measurement noise covariance matrix, R, the initial state covariance matrices, P_{xo} and $P_{\theta o}$, and process noise covariance matrices Q_x and Q_θ. This method is found more stable towards changes in noise covariance matrices and observation noise than the method of EKF. The quantitative performances of this method for normal situation and various faulty situations are lower than the method of EKF. The estimated states and identified fault parameters of this method are found to agree well with their corresponding true values. The identified fault parameters by the method of two-level EKF for 20% step change of fault q and 30% ramp change in fault h_{mo} are shown in Fig. 14.11.

14.8 Method of a discrete version of extended Kalman filter and sequential least squares

By this method, states are estimated using a discrete version of the EKF and parameters are identified using sequential least squares.

14.8.1 Design strategy

In this study, a discrete version of the EKF described in Section 4.4.4 is employed for state estimation, and a sequential least squares algorithm is employed for parameter identification [17].

14.8.1.1 Design of state estimation filter

The same relations that are defined for the design of state estimating filter of two-level EKF (Eqs. (14.44)–(14.48)) are also used for state estimation filter of this method. The general discrete model expression in Eq. (4.35) of Section 4.4.4 involved in the discrete version of Kalman filter is expressed by the following reactor model:

$$f_x = \begin{bmatrix} x_1 + \left[C_1\theta_1(C_i - x_i) - C_2 x_1^2 \exp(-C_3/x_2) \right] \Delta t \\ x_2 + \left[C_1\theta_1(T_i - x_2) - C_4 x_1^2 \exp(-C_3/x_2) - C_5\theta_2(x_2 - x_4) \right] \Delta t \\ x_3 + \left[C_6(T_{wi} - x_3) - C_7(x_3 - x_4) \right] \Delta t \\ x_4 + \left[C_8(x_3 - x_4) - C_9\theta_2(x_4 - x_2) \right] \Delta t \end{bmatrix} \tag{14.62}$$

The quantities of C_i are defined under Eq. (14.20). The elements of the state transition matrix, F_k, involved in Eq. (4.35) of Section 4.4.4 are obtained by taking the partial derives of the discrete reactor model, Eq. (14.62), concerning the state vector.

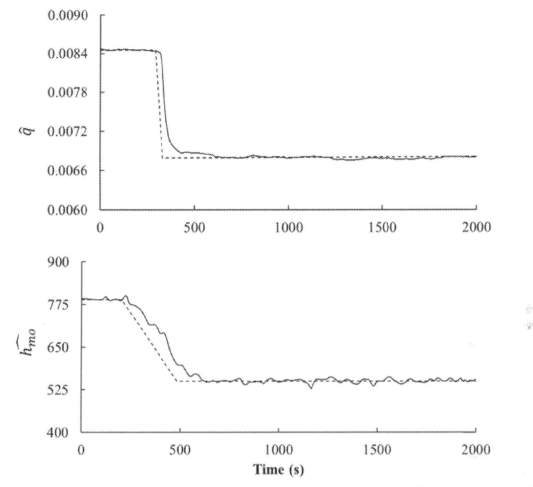

FIGURE 14.11 Identified fault parameters in the case of 20% step change of fault q and 30% ramp change of fault h_{mo} by the two-level extended Kalman filter method.

14.8.1.2   Design of parameter identifier

Sequential least squares algorithm for parameter estimation and its design to the reactor system is explained. The linear relationships assumed for state model and observations are given by

$$\hat{x}(t+1,\theta) = F(\theta)\hat{x}(t,\theta) + G(\theta)y(t) \tag{14.63}$$

$$\hat{y}(t,\theta) = H\hat{x}(t,\theta) + v(t) \tag{14.64}$$

where $F(\theta)$ is a linear discrete model matrix, $G(\theta)$ is a linear discrete measurement matrix, and t is discrete time. The sequential least squares algorithm for parameter identification is described as follows.

$$\varepsilon(t) = y(t) - H\hat{x}(t,\theta(t-1))$$

$$\Lambda(t) = \Lambda(t-1) + \gamma(t)[\varepsilon(t)\varepsilon^T(t) - \Lambda(t-1)]$$

$$S(t) = \psi^T(t)P_\theta(t-1)\psi(t) + \lambda(t)\Lambda(t)$$

$$L(t) = P_\theta(t-1)\psi(t)S^{-1}(t)$$

$$\hat{\theta}(t) = \hat{\theta}(t-1) + L(t)\varepsilon(t)$$

$$P_\theta(t) = P_\theta(t-1)\psi(t) - L(t)S(t)^L T(t)/\lambda(t) \tag{14.65}$$

In the above equations, $\varepsilon(t)$ is the error between the process measurements and model measurements, $\Lambda(t)$ is the covariance of the error between the process measurements and model measurements, $S(t)$ is the error covariance of the predicted parameter, $L(t)$ is the parameter gain matrix, $\theta(t)$ is estimated parameter, $P_\theta(t)$ is estimated parameter covariance matrix, $\psi(t)$ is gradient matrix, $\lambda(t)$ is forgetting factor and $\Upsilon(t)$ is scalar gain sequence. The matrix $\psi(t)$ is computed by the relation,

$$\psi^T(t) = Hw(t, \theta) \tag{14.66}$$

where $w(t, \theta)$ is recursive gradient matrix, whose updating relation is given by

$$w(t + 1, \theta) = F(\theta)w(t, \theta) + M(\theta) \tag{14.67}$$

where $M(\theta)$ is a parameter gradient matrix, which is computed by the relation,

$$M(\theta) = \frac{\partial}{\partial \theta}[F(\theta)x] \tag{14.68}$$

The forgetting factor $\lambda(t)$ is obtained by the relation,

$$\lambda(t) = \frac{\gamma(t-1)}{\gamma(t)}[1 - \gamma(t)] \tag{14.69}$$

The initial conditions for the implementation of sequential least squares algorithm are set as follows:
The initial gradient matrix:

$$w(0, \theta) = 0 \tag{14.70}$$

The initial parameter estimate:

$$\theta(0) = \theta_0 \tag{14.71}$$

The covariance for model uncertainty:

$$P_\theta(0) = P_0 \tag{14.72}$$

The covariance of the initial local error between the data and the measurement model:

$$\Lambda(0) = E(\varepsilon_0 \varepsilon_0^T) \tag{14.73}$$

The state estimation and parameter identification by this method is explained as follows. First, consider the current values of states and parameters to be $\hat{x}(t-1)$ and $\hat{\theta}(t-1)$. Then, by using the incoming measurements $y(t)$ and the current state estimate, the parameter value is updated from $\hat{\theta}(t-1)$ to $\theta(t)$. This new value is then fed to the Kalman filter to obtain the new state estimate $\hat{x}(t)$. In this way, the states and parameters are updated recursively. The schematic of this method for state estimation and process parameter identification is shown in Fig. 14.12.

14.8.2 Analysis of results

The design parameters such as the initial conditions, noise covariance matrices, and measurement coefficient matrix used in the state estimating filter of this method are the same as the state estimating filter of the method of two-level EKF. This method is applied for state estimation and fault parameter identification for normal operation and for all the fault situation considered in Table 14.1. The sequential least squares method for parameter identification involves the following additional design parameters.
The initial covariance of error between the data and measurement model:

$$\Lambda(0) = \begin{bmatrix} 0.01 & 0 & 0 \\ 0 & 10 & 0 \\ 0 & 0 & 10 \end{bmatrix} \tag{14.74}$$

The constant forgetting factor:

$$\lambda = 0.99 \tag{14.75}$$

The constant scalar gain sequence:

$$\gamma = 0.02 \tag{14.76}$$

The initial gradient matrix:

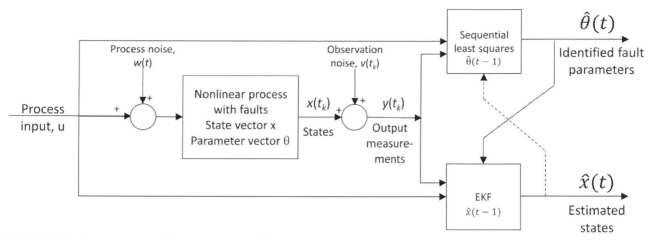

FIGURE 14.12   State estimation and fault parameter identification using a discrete version of extended Kalman filter and sequential least squares.

$$w(0, \theta) = [0] \tag{14.77}$$

In this method, states are estimated using a discrete version of the EKF and parameters are identified by sequential least squares. For each measurement data set, state estimation and parameter identification are carried out recursively for normal operation and different faulty situations. The performance of this is evaluated towards the effect of observation noise, initial state noise covariance matrix, process and observation noise covariance matrices. The performance of this method is found slightly lower when compared with the method of two-level EKF. In this method, computations are simplified due to the involvement of lower dimensional matrices. The identified fault parameters in the case of 20% step change of fault q and 30% ramp change of fault h_{mo} by the method of discrete version of EKF and sequential least squares are shown in Fig. 14.13.

14.9   Method of discrete version of extended Kalman filter and simultaneous least squares

By this method, states are estimated using a discrete version of the EKF and parameters are identified using simultaneous least squares. The schematic for state estimation and fault parameter identification using the method of discrete version of EKF and simultaneous least squares is shown in Fig. 14.14.

14.9.1   Design strategy

14.9.1.1   Design of state estimator

The state estimator design is the same as the method of EKF and sequential least squares given in Section 14.8.1.1.

14.9.1.2   Design of parameter identifier

In simultaneous least squares, the gradient matrix, $\psi(t)$ in Eq. (14.66) is computed using the following procedure for simultaneous identification of model parameters [17]. First $F(\theta)$ involved in Eq. (14.67) is computed by the relation,

$$F(\theta) = \phi(t_k) - K_k H \tag{14.78}$$

where $\phi(t_k)$ is the discrete model state transition matrix, which is F_k in Eq. (4.35), K_k is the Kalman gain matrix involved in Eq. (4.37) and H is the measurement matrix. The discrete model state transition matrix F_k in Eq. (4.35) is computed by the equation:

$$F_k = (I + \frac{\partial f}{\partial x} \Delta t)x = \hat{x}\big|_{(t_k/t_{k-1})} \tag{14.79}$$

The $F(\theta)$ computed by Eq. (14.78) acts as a coupling matrix between the discrete EKF and recursive least squares. The elements of the matrix, $M(\theta)$ involved in Eq. (14.67) are computed by taking the partial derivatives of the discrete dynamic model of the reactor, Eq. (14.62), with respect to the parameters. The matrices $F(\theta)$ and $M(\theta)$ thus computed

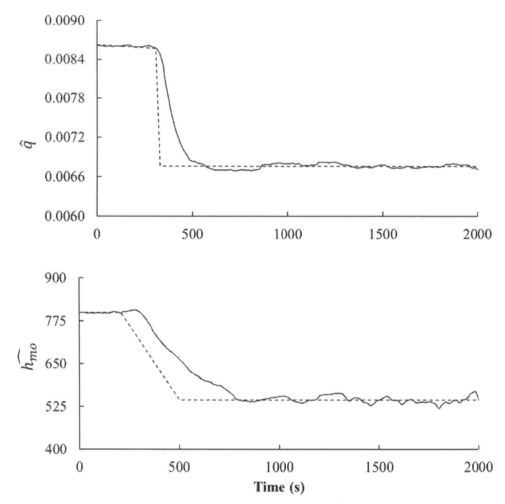

FIGURE 14.13 Identified fault parameters in the case of 20% step change of fault q and 30% ramp change of fault h_{mo} by the method of discrete version of extended Kalman filter and sequential least squares.

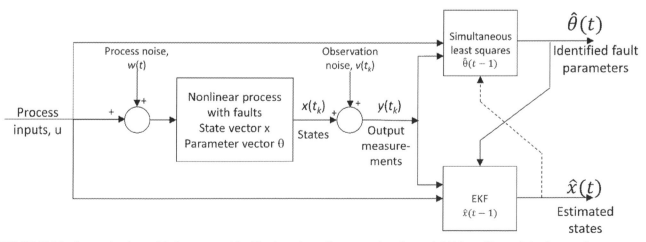

FIGURE 14.14 State estimation and fault parameter identification using a discrete version of extended Kalman filter and simultaneous least squares.

are used in Eq. (14.67) to update $w(t,\theta)$. The $w(t,\theta)$ is used in Eq. (14.66) to compute $\psi(t)$ involved in recursive least squares algorithm in Section 14.8.1.2 for simultaneous identification of process parameters.

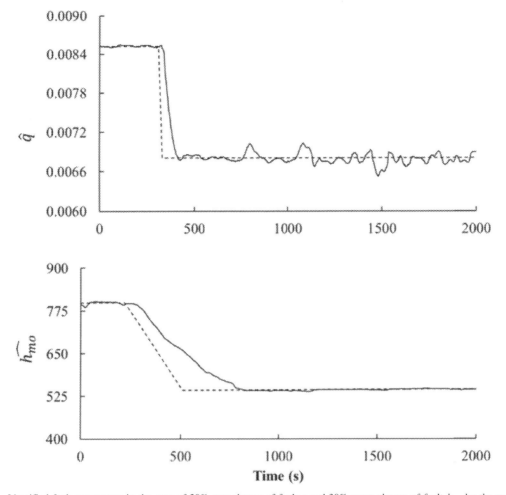

FIGURE 14.15 Identified fault parameters in the case of 20% step change of fault q and 30% ramp change of fault h_{mo} by the method of discrete version of extended Kalman filter and simultaneous least squares.

14.9.2   Analysis of results

The design parameters such as the initial conditions, noise covariance matrices, and measurement coefficient matrix used for state estimating filter of this method are the same as the state estimating filter of the method of two-level EKF. The initial conditions used for parameter estimation are the same as in the case of sequential least squares. This method is applied for state estimation and fault parameter identification in the CSTR system for normal operation and for all the fault situations considered in Table 14.1. In this method, certain delay is observed in initially identified parameters to converge to their true values, but the identified parameters converged to their true values after the delay. The performance of this method is found lesser to the method of two-level EKF. However, the results indicate slightly better performance of this method over the discrete version of EKF and sequential least squares. The identified fault parameters in the case of 20% step change of fault q, and 30% ramp change of fault h_{mo} by the method of the discrete version of EKF and simultaneous least squares are shown in Fig. 14.15.

14.10   Summary

Optimal state estimation is the most important problem for model-based fault detection and diagnosis of nonlinear dynamic systems. In quantitative model-based fault diagnosis, both the states and parameters in the process model are estimated by using accessible measurements of the process. This study presents the method of EKF and various two-level methods for fault detection and diagnosis in nonlinear, time-varying, and stochastic chemical processes. The method of EKF is formulated for the combined estimation of states and fault parameters. The two-level methods for fault detection and diagnosis are formulated using different versions of EKF, recursive least squares, and a reduced-order extended Luenberger observer. These methods are specified for state estimation in the first level and fault diagnosis via parameter identification in the second level. The performances of these methods are evaluated by applying them to a nonlinear CSTR with a heat exchanger. The reduced order extended Luenberger observer-based approach has the advantage of estimating the unmeasured process states in the presence of faults. Among various methods, the method of two-level EKF is found to exhibit better performance with lower computational effort.

References

[1] A.S. Willsky, A survey of design methods for failure detection in dynamic systems, Automatica 12, 6 (1976) 601−611.

[2] P.M. Frank, Fault diagnosis in dynamic systems using analytical and knowledge-based redundancy: a survey and some new results, Automatica 26, 3 (1990) 459−474.

[3] R. Isermann, Process fault detection based on modeling and estimation methods—a survey, Automatica 20, 4 (1984) 387−404.

[4] D.M. Himmelblau, Fault Detection and Diagnosis in Chemical and Petrochemical Processes, Elsevier, New York, 1986.

[5] M. Basseville, Detecting changes in signals and systems—a survey, Automatica 24, 3 (1988) 309−326.

[6] Detection of abrupt changes in signals and dynamical systems, in: M. Basseville, A. Benveniste (Eds.), Lecture Notes on Control and Information Sciences, 77, Springer-Verlag, Berlin, 1986, p. 1986.

[7] R. Patton, P. Frank, R. clark, Fault Diagnosis in Dynamic Systems, Theory and Applications, Prentice Hall, New York, 1989.

[8] V. Venkatasubramanian, R. Rengaswamy, S.N. Kavuri, K. Yin, A review of process fault detection and diagnosis. Part III: Process history based methods, Comput. Chem. Eng. 27, 3 (2003) 327−346.

[9] C. Venkateswarlu, K. Gangiah, M. Bhagavatha Rao, Two-level methods for incipient fault diagnosis in nonlinear chemical processes, Comput. Chem. Eng. 16, 5 (1992) 463−476.

[10] J.J. Macgregor, T. Kourti, Statistical process control of multivariate processes, Contr. Eng. Pract. 3 (1995) 403−414.

[11] S. Park, D.M. Himmelblau, Fault detection and diagnosis via parameter estimation in lumped dynamic systems, Ind. Eng. Chem. Process Des. Dev. 22, 3 (1983) 482−487.

[12] K. Watanabe, D.M. Himmelblau, Fault diagnosis in nonlinear chemical processes, AIChE J. 29, 2 (1983) 243−260.

[13] K. Watanabe, D.M. Himmelblau, Incipient fault diagnosis of nonlinear processes with multiple causes of faults, Chem. Eng. Sci. 39, 3 (1984) 491−508.

[14] D.T. DalleMolle, D.M. Himmelblau, Fault detection in a single-stage evaporator via parameter estimation using the Kalman filter, Ind. Eng. Chem. Res. 26, 12 (1987) 2482−2489.

[15] R.J. Litchfield, K.S. Campbell, A. Locke, The application of several Kalman filters to the control of a real chemical reactor, Trans. IChemE 57 (1979) 113−120.

[16] B. Bellingham, F.P. Lees, The detection of malfunction using a process control computer: a Kalman filtering technique for general control loops, Trans. IChemE 55 (1977) 253−265.

[17] W. Famirez, Optimal state and parameter identification: an application to batch beer fermentation, Chem. Eng. Sci. 42, 11 (1987) 2749−2756.

Chapter 15

Optimal state and parameter estimation for fault detection and diagnosis of a nonlinear batch beer fermentation process

15.1 Introduction

Optimal state estimation is an integral part of bioprocess systems engineering with applications related to process monitoring, fault diagnosis, and control. There has been an increasing interest in applying state and parameter identification techniques to increase the efficiency and performance of bioprocess systems. Process monitoring and fault diagnosis of bioprocesses typically require reliable real-time process variable information. However, many of the important bioprocess variables cannot be measured online and require the need for online state and parameter estimation techniques. In addition, most biochemical systems of practical interest are inherently nonlinear. The use of a process model that reflects this essential nonlinear structure would prove to be more beneficial for bioprocess state estimation. Consequently, the estimation algorithms developed should either inherently be based on the nonlinear model structure or alternatively approximate the nonlinearities using a linear model with adaptation.

Extended Kalman filters (EKFs) and nonlinear observers employ process models which reflect the essential nonlinear structure in a nonlinear or linearized fashion. Since most biotechnical processes exhibit nonlinear dynamic behavior, EKF and nonlinear observers are most useful in bioprocess state estimation. EKF is a standard nonlinear estimation technique that has been used for the combined estimation of states and parameters of nonlinear systems. Several applications of EKF have been reported for state and parameter estimation of biotechnical processes [1−3]. A variety of linear, adaptive, nonlinear, exponential observers are also used for state estimation in bioprocess systems [4−11]. Adaptive/nonlinear observers and EKF have also been reported for parameter estimation and fault diagnosis problems in bioprocesses [10−13].

This chapter presents a state and parameter estimation-based quantitative model-based approach for fault detection and diagnosis in nonlinear systems. Quantitative methods of process fault detection and diagnosis based on filtering and parameter estimation utilize dynamic process models and process measurements to compute measured and unmeasured process variables and process model parameters. This study presents the method of EKF and various two-level methods for fault detection and diagnosis in nonlinear and time varying systems. Various two-level methods are formulated using different versions of EKF, recursive least squares and a reduced-order extended Luenberger observer. These methods are specified for state estimation in the first level and fault diagnosis via parameter identification in the second level. The performances of these methods are evaluated by applying them to a nonlinear batch beer fermentation process.

15.2 General structure and general process description for model-based fault detection and diagnosis

Quantitative model-based fault diagnosis of transient dynamic systems requires dynamic models to describe their quantitative performance. Process faults, which occur during the operations, appear as undesirable state and parameter variations. In a quantitative model-based approach, both the states and parameters in the process model are estimated using accessible process measurements. Fault detection can be done by examining the estimated states and parameters.

Optimal State Estimation for Process Monitoring, Fault Diagnosis and Control. DOI: https://doi.org/10.1016/B978-0-323-85878-6.00016-6

However, fault diagnosis is performed based on the estimated parameters only, because possible faults in the process are associated with the specific parameters of the process model, and the parameters can be related to physical features of the process. The general structure for model-based fault detection and diagnosis is described in Section 14.2. The general process description for fault detection and diagnosis is explained in Section 14.3.

15.3   Batch beer fermentation process, its mathematical model and fault cases

Batch beer fermentation is a nonlinear and unsteady state process with unpredictable parameter uncertainties. In this process, the fermenting medium consists of three sugars, glucose, maltose, and maltotriose, as limiting nutrients. The products are biomass and ethanol. A transient model of batch beer fermentation is used for fault detection and diagnosis [14]. The model describes the rate of change of glucose, maltose and maltotriose concentrations, and the rate of change of fermenting medium temperature.

15.3.1   Mathematical model

The model is described by the following equations:
   The material balance on glucose,

$$\frac{dG_c}{dt} = \frac{-V_G G_c}{K_G + G_c} X_c \tag{15.1}$$

where G_c = glucose concentration (mol m^{-3}), V_G = velocity rate constant for glucose (h^{-1}), K_G = Michaelis constant for glucose (mol m^{-3}), and X_c = biomass concentration (mol m^{-3}).
   The material balance on maltose,

$$\frac{dM_c}{dt} = \frac{-V_m M_c}{K_m + M_c} \frac{K'_g}{K'_g + G_c} X_c \tag{15.2}$$

where M_c = maltose concentration (mol m^{-3}), V_M = velocity rate constant for maltose (h^{-1}), K_M = Michaelis constant for maltose (mol m^{-3}), and K'_G = inhibition rate constant for glucose (mol m^{-3}).
   The material balance on maltotriose,

$$\frac{dN_c}{dt} = \frac{-V_N N_c}{K_N + N_c} \frac{K'_g}{K'_g + G_c} \frac{K'_m}{K'_m + M_c} X_c \tag{15.3}$$

where N_c = maltotriose concentration (mol m^{-3}), V_N = velocity rate constant for maltotriose (h^{-1}), K_N = Michaelis constant for maltotriose (mol m^{-3}), and K'_M = inhibition rate constant for maltotriose (mol m^{-3}).
   The energy balance of the process,

$$\frac{dT}{dt} = \frac{1}{\rho C_p} \left[\Delta H_{FG} \frac{dG_c}{dt} + \Delta H_{FM} \frac{dM_c}{dt} + \Delta H_{FN} \frac{dN_c}{dt} - u(T - T_c) \right] \tag{15.4}$$

where T = fermentation temperature ($^\circ$C), u = fermenter cooling rate constant (kJ h^{-1} m^{-3} $^\circ$C^{-1}), C_p = mixture heat capacity (kJ kg^{-1} $^\circ$C^{-1}), r = mixture density (kJ m^{-3}), T_c = coolant temperature ($^\circ$C), and DH_{Fi} = heat of fermentation (kJ m^{-1}).
   Fermentation kinetic rate expressions are given by

$$V_i = V_{io} \exp\left[-E_{vi}/R_G(T + 273.15)\right] \tag{15.5}$$

$$K_i = K_{io} \exp\left[-E_{ki}/R_G(T + 273.15)\right] \tag{15.6}$$

$$K'_i = K'_{io} \exp\left[-E'_{ki}/R_G(T + 273.15)\right] \tag{15.7}$$

where V_{io} = velocity rate frequency factor (h^{-1}), K_{io} = Michaelis constant frequency factor (mol m^{-3}), K'_{io} = inhibitionconstant frequency factor (mol m^{-3}), E_{vi} = velocity rate activation energy (cal mol^{-1}), E_{ki} = Michaelis constant activation energy (cal mol^{-1}), and E'_{ki} = inhibition constant activation energy (cal mol^{-1}).
   The production rates of biomass and ethanol are given by

$$X_c = X_o - R_{XG}(G_c - G_o) - R_{XM}(M_c - M_o), \tag{15.8}$$

$$E_c = -R_{EG}(G_c - G_o) - R_{EM}(M_c - M_o) - R_{EN}(N_c - N_o) \tag{15.9}$$

where R_{Xi} = the stoichiometricyield of biomass per mole of sugar reacted (mol biomass/mol sugar), and R_{Ei} = the stoichiometricyield of ethanol per mole of sugar reacted (mol ethanol/mol sugar).

The proportional feedback control law u is given by

$$u = K_p(T - T_{set}) \tag{15.10}$$

where K_p = controller gain constant (kJ h^{-1} °C^{-1} m^{-3}), and T_{set} = setpoint temperature (°C).

15.3.2   Fault cases in batch beer fermentation

Fault detection and diagnosis is usually carried out by identifying the model parameters. The sources of faults are related to the process parameters. In batch beer fermentation, the velocity rate frequency factor V_{io} can change from batch to batch during fermentation due to genetic changes in different generations of yeast organisms. Actual glucose consumption under normal operation is about 36 h. A slight rise in glucose frequency factor V_{Go} can cause to consume all the glucose within a few hours of fermentation. Sometimes, the parameters used in the model are imprecise or incorrect. Correct values of model parameters can be determined through their identification using the known process data. The possible sources of faults in batch beer fermentation are improperly determined initial incorrect values of glucose and maltose reaction rate frequency factors (V_{io}) and deterioration in the performance of yeast organisms because of genetic and environmental changes. To diagnose the causes of faults, estimation of states and identification of unknown parameters are needed. The measurements of the fermentation are glucose (G_c), maltose (M_c) and maltotriose (N_c) concentrations, and fermentation temperature (T). Glucose is considered a nonmeasurable state for the observer.

15.4   Method of extended Kalman filter

A strategy based on an EKF is presented for fault detection and diagnosis in transient nonlinear batch beer fermentation. Continuous reactor dynamics and discrete observations are used to estimate state variables and identify process parameters. The identified process parameters can be used for detecting faults in the process and diagnosing their causes.

15.4.1   Design strategy

The EKF algorithm described in Section 4.4.3 is designed and applied for the combined estimation of states and fault parameters in the nonlinear batch beer fermentation process. State estimation and fault parameter identification using the method of EKF are shown in Fig. 14.3 of Section 14.5.1.

The relationships defined in the nonlinear model of batch beer fermentation for the design of EKF are given as follows:

The augmented state vector:

$$x_a(t) = [x_1(t) \, x_2(t) \, x_3(t) \, x_4(t) \, \theta_1(t) \, \theta_2(t)]^T = [C_c(t) \, M_c(t) \, N_c(t) \, T(t) \, ln \, V_{Go}(t) \, ln \, V_{Mo}(t)]^T \tag{15.11}$$

The observation vector:

$$y(t_k) = [y_1(t_k) \, y_2(t_k) \, y_3(t_k) \, y_4(t_k)]^T$$
$$= [G_c(t_k) \, M_c(t_k) \, N_c(t_k) \, T(t_k)]^T \tag{15.12}$$

The $G_c(t_k)$, $M_c(t_k)$, $N_c(t_k)$, and $T(t_k)$ are observations with zero-mean random Gaussian noise.

The augmented model f_a in Eq. (14.6) is represented by

$$f_a = \begin{bmatrix} f_{G_c}(t) \\ f_{M_c}(t) \\ f_{N_c}(t) \\ f_T(t) \\ f_{\theta_1}(t) \\ f_{\theta_2}(t) \end{bmatrix} = \begin{bmatrix} \dot{G}_c \\ \dot{M}_c \\ \dot{N}_c \\ \dot{T} \\ 0 \\ 0 \end{bmatrix} \tag{15.13}$$

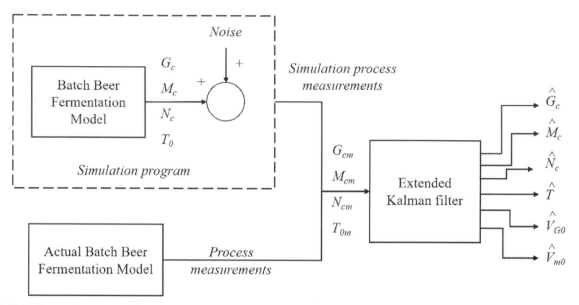

FIGURE 15.1   State and parameter estimation scheme in batch beer fermentation by extended Kalman filter.

The nonlinear measurement model h_a in Eq. (14.9) is expressed as

$$h_a = \begin{bmatrix} x_1 + f_{G_c}(t_k)\Delta t \\ x_2 + f_{M_c}(t_k)\Delta t \\ x_3 + f_{N_c}(t_k)\Delta t \\ x_4 + f_T(t_k)\Delta t \end{bmatrix} \tag{15.14}$$

where f_{Gc}, f_{Mc}, f_{Nc}, and f_T represent the dynamic model of batch beer fermentation described by Eqs. (15.1–15.4), and Δt is sampling time. The state transition matrix F in Eq. (4.27) is computed as F_a by taking the partial derivatives of the model in Eq. (15.13) with respect to the augmented state vector:

$$F_a = \begin{bmatrix} f_{11} & 0 & 0 & f_{14} & f_{15} & 0 \\ f_{21} & f_{22} & 0 & f_{24} & 0 & f_{26} \\ f_{31} & f_{32} & f_{33} & f_{34} & 0 & 0 \\ f_{41} & f_{42} & f_{43} & f_{44} & f_{45} & f_{46} \\ 0 & 0 & 0 & 0 & 0 & 0 \\ 0 & 0 & 0 & 0 & 0 & 0 \end{bmatrix} \tag{15.15}$$

The measurement Jacobian matrix H in Eq. (4.31) is computed as H_a by taking the partial derivatives of the model in Eq. (15.14) with respect to the augmented state vector:

$$H_a = \begin{bmatrix} h_{11} & 0 & 0 & h_{14} & h_{15} & 0 \\ h_{21} & h_{22} & 0 & h_{24} & 0 & h_{26} \\ h_{31} & h_{32} & h_{33} & h_{34} & 0 & 0 \\ h_{41} & h_{42} & h_{43} & h_{44} & h_{45} & h_{46} \end{bmatrix} \tag{15.16}$$

The EKF described in Section 4.4.3 is used to estimate states and to identify fault parameters. The state and fault parameter estimation scheme by EKF in the batch beer fermentation process is shown in Fig. 15.1.

15.4.2   Analysis of results

The EKF is applied for state estimation and fault parameter identification according to the design strategy given in the above section. The cases of faults considered in the batch beer fermentation process are given in Table 15.1.

Simulation measurements are generated for normal operation and all the cases of faults by solving the fermenter model equations numerically using the Runge–Kutta fourth-order method with a step size of 0.2 h. The states are corrupted with zero-mean random Gaussian noise in the range of -1.0 to 1.0 to obtain the measurements. The standard deviations of the measurement noises are: $v_1(t_k) = 0.6$ gmol m^{-3}, $v_2(t_k) = 0.6$ gmol m^{-3}, $v_3(t_k) = 0.6$ gmol m^{-3}, and $v_4(t_k) = 0.075°$C.

TABLE 15.1 Cases of faults in nonlinear beer fermentation process.

S. no.	Fault cases studied
i	Incorrect initial value of $\ln V_{Go}$.
ii	Incorrect initial value of $\ln V_{Mo}$.
iii	Incorrect initial values of both $\ln V_{Go}$ and $\ln V_{Mo}$.
iv	Ramp increase of $\ln V_{Go}$ during 10–20 h period.
v	Ramp increase of $\ln V_{Mo}$ during 10–20 h period.
vi	Ramp increase of $\ln V_{Go}$ and $\ln V_{Mo}$ during 10–20 h period, respectively.

The measurement data is obtained as follows.

$$y(t_k) = Hx(t_k) + v(t_k) \tag{15.17}$$

$$\begin{bmatrix} y_1(t_k) \\ y_2(t_k) \\ y_3(t_k) \\ y_4(t_k) \end{bmatrix} = \begin{bmatrix} 1 & 0 & 0 & 0 \\ 0 & 1 & 0 & 0 \\ 0 & 0 & 1 & 0 \\ 0 & 0 & 0 & 1 \end{bmatrix} \begin{bmatrix} x_1(t_k) \\ x_2(t_k) \\ x_3(t_k) \\ x_4(t_k) \end{bmatrix} + \begin{bmatrix} 2.0 \\ 2.0 \\ 2.0 \\ 0.25 \end{bmatrix} [v(t_k)] \tag{15.18}$$

The y_1 (t_k), y_2 (t_k), y_3 (t_k), and y_4 (t_k) are the measurements of glucose, maltose, and maltotriose concentrations, and temperature, respectively.

The initial augmented state vector, initial product vector, initial state noise covariance matrix, P_o, process noise covariance matrix, Q, and observation noise covariance matrix, R are chosen for the method of EKF as follows.

$$\begin{bmatrix} G_c(0) \\ M_c(0) \\ N_c(0) \\ T(0) \\ \ln V_{Go}(0) \\ \ln V_{Mo}(0) \end{bmatrix} = \begin{bmatrix} 70.0 \\ 220.0 \\ 40.0 \\ 8.0 \\ 35.77 \\ 16.4 \end{bmatrix} \tag{15.19}$$

$$\begin{bmatrix} X_c(0) \\ E_c(0) \end{bmatrix} = \begin{bmatrix} 175.0 \\ 0.0 \end{bmatrix} \tag{15.20}$$

$$P_0 = \begin{bmatrix} 4 & 0 & 0 & 0 & 0 & 0 \\ 0 & 4 & 0 & 0 & 0 & 0 \\ 0 & 0 & 0 & 0.0625 & 0 & 0 \\ 0 & 0 & 0 & 0 & 1 & 0 \\ 0 & 1 & 0 & 0 & 0 & 1 \end{bmatrix} \tag{15.21}$$

$$Q = \begin{bmatrix} 4 & 0 & 0 & 0 & 0 & 0 \\ 0 & 4 & 0 & 0 & 0 & 0 \\ 0 & 0 & 0 & 0.0625 & 0 & 0 \\ 0 & 0 & 0 & 0 & 1 & 0 \\ 0 & 1 & 0 & 0 & 0 & 1 \end{bmatrix} \tag{15.22}$$

$$R = \begin{bmatrix} 4 & 0 & 0 & 0 \\ 0 & 4 & 0 & 0 \\ 0 & 0 & 4 & 0 \\ 0 & 0 & 0 & 0.0625 \end{bmatrix} \tag{15.23}$$

The EKF is applied for state estimation and fault diagnosis in batch beer fermentation using the above conditions for normal operation, and faulty situations given in Table 15.1. The estimated states and product responses along with their corresponding true values for normal operation are shown in Figs. 15.2 and 15.3.

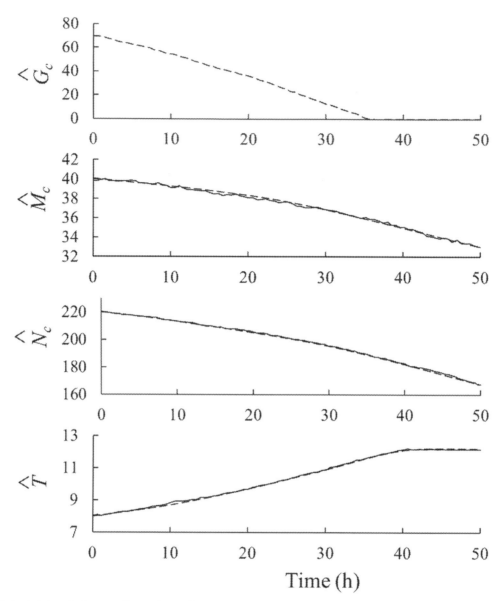

FIGURE 15.2 Estimated states in the case of normal operation.

15.5 Method of reduced-order extended Luenberger observer and extended Kalman filter

This is a two-level method in which the states are estimated in the first level, and fault parameters are identified in the second level [13]. In the first level, a reduced-order extended Luenberger observer is used for estimating state variables in the presence of faults using accessible process measurements. In the second level, the state estimates resulting from the observer are used in the EKF to identify the process parameters for fault diagnosis. The schematic state estimation and fault parameter identification using this method is shown in Fig. 14.8.

15.5.1 Design strategy

The batch beer fermentation and its mathematical model are illustrated in Section 15.3. The fault cases considered in the nonlinear fermentor are given in Table 15.1. The reduced-order extended Luenberger observer described in Section 4.13 is designed and applied for state estimation in nonlinear transient beer fermentation.

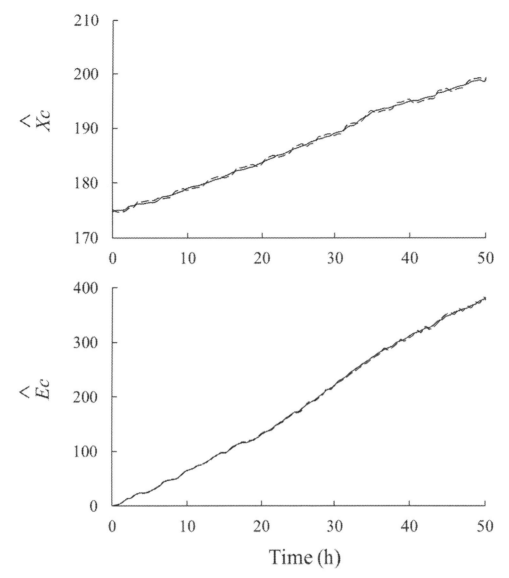

FIGURE 15.3   Estimated product responses in the case of normal operation.

The following relationships are defined for the observer in Section 4.13 and the process model:

$$\delta x(t) = [\delta x_1(t),\ \delta x_2(t),\ \delta x_3(t),\ \delta x_4(t)]^T \\ = \left[G_c(t)-G_i^*,\ M_c(t)-M_i^*,\ N_c(t)-N_i^*,\ T(t)-T_i^*\right]^T \tag{15.24}$$

$$\delta u(t) = u(t) - u_i^* \tag{15.25}$$

$$\delta y(t) = [\delta y_2(t),\ \delta y_3(t),\ \delta y_4(t)]^T \\ = [\delta x_2(t),\ \delta x_3(t),\ \delta x_4(t)]^T \tag{15.26}$$

The notation G_i^*, M_i^*, N_i^*, and T_i^* denotes the state variables at the initial operating point. The corresponding input variable is u_i^*. The linearized batch beer fermentation state-space model used for state estimation in the presence of faults is represented by

$$\delta \dot{x}(t) = \begin{bmatrix} 0 & 0 & 0 & 0 \\ 0 & 0 & 0 & 0 \\ a_{31} & a_{32} & a_{33} & a_{34} \\ a_{41} & a_{42} & a_{33} & a_{34} \end{bmatrix} \delta x(t) + \begin{bmatrix} 0 \\ 0 \\ 0 \\ b_{41} \end{bmatrix} \delta u(t) + \begin{bmatrix} 1 & 0 \\ 0 & 1 \\ 0 & 0 \\ d_{41} & d_{42} \end{bmatrix} f_e(t,\theta) \tag{15.27}$$

$$\delta y(t) = \begin{bmatrix} 0 & 1 & 0 & 0 \\ 0 & 0 & 1 & 0 \\ 0 & 0 & 0 & 1 \end{bmatrix} \delta x(t) + v(t) \tag{15.28}$$

where a_{ij}'s are linearized reactor model coefficients obtained by linearizing the beer fermentation model equations in Section 15.3 with respect to state vector, and b_{ij}'s are computed by linearization of fermentation model with the respective input vector. The elements d_{ij}'s represent the linearized model coefficients corresponding to the fault parameters. The process fault vector of Eq. (4.103) in Section 4.13 is obtained as

$$[f_e(t, \theta)]^T = [f_{11}\delta x_1 + f_{14}\delta x_4 \quad f_{21}\delta x_1 + f_{22}\delta x_2 + f_{24}\delta x_4]^T \tag{15.29}$$

The elements of the state transition matrix computed in Eq. (15.15) represent the f_{ij}'s of the above equation. The linearized model is in the general form of Eq. (4.101) in Section 4.13. To obtain the observer coefficient matrices in Section 4.12.1, first, an arbitrary matrix W is assigned, and the matrices V and G are obtained by using the relation:

$$\begin{bmatrix} C \\ \cdots \\ W \end{bmatrix}^{-1} = \begin{bmatrix} 0 & 1 & 0 & 0 \\ 0 & 0 & 1 & 0 \\ 0 & 0 & 0 & 1 \\ \cdots & \cdots & \cdots & \cdots \\ 1 & 0 & 0 & 0 \end{bmatrix}^{-1} = [V \vdots G] = \begin{bmatrix} 0 & 0 & 0 & \vdots & 1 \\ 1 & 0 & 0 & \vdots & 0 \\ 0 & 1 & 0 & \vdots & 0 \\ 0 & 0 & 1 & \vdots & 0 \end{bmatrix} \tag{15.30}$$

The notation can be referred in Section 4.12.
The rank of the CD matrix is obtained as

$$\text{rank}[CD] = \begin{bmatrix} 0 & 1 \\ 0 & 0 \\ d_{41} & d_{42} \end{bmatrix} = 2 \tag{15.31}$$

where C and D are coefficient matrices in Eq. (4.101) in Section 4.13.

The model is found to satisfy the conditions specified in Section 4.12.1, thus the coefficients of the reduced-order Luenberger observer can be determined. The relations that define the observer coefficient matrices in Eq. (4.97) of Section 4.12.1 are computed as follows:

$$L = \begin{bmatrix} -d_{42}/d_{41} & K_{02} & 1/d_{41} \end{bmatrix} \tag{15.32}$$

$$M = \begin{bmatrix} 1 & d_{42}/d_{41} & -K_{02} & -1/d_{41} \end{bmatrix} \tag{15.33}$$

$$H_b = \begin{bmatrix} -d_{42}/d_{41} & K_{02} & 1/d_{41} \\ 1 & 0 & 0 \\ 0 & 1 & 0 \\ 0 & 0 & 1 \end{bmatrix} \tag{15.34}$$

$$F = \begin{bmatrix} -a_{41}/d_{41} - K_{02} \cdot a_{31} \end{bmatrix} \tag{15.35}$$

$$K = [(d_{42}/d_{41})a_{31}K_{02} - a_{32}K_{02}(d_{42}/d_{41})^2 a_{41} - a_{42}/d_{41} \quad K_{02}{}^2 a_{31} - a_{33}K_{02} - a_{41}/d_{41}K_{02}a_{43}/d_{41} \\ - a_{31}/d_{41}K_{02}a_{34}K_{02} - (a_{41}/d_{41})^2 - a_{44}/d_{41}] \tag{15.36}$$

$$MB = \begin{bmatrix} -b_{41}/d_{41} \end{bmatrix} \tag{15.37}$$

The above computed coefficient matrices are used in the reduced-order extended Luenberger observer defined by Eq. (4.104) of Section 4.13 to estimate the unknown state variable. The resulting outputs from the observer equations are,

$$\delta\hat{x}_1 = Z(t) - d_{42}/d_{41}\delta y_2 + K_{02}\delta y_3 + 1/d_{41}\delta y_4$$
$$\delta\hat{x}_2 = \delta y_2$$
$$\delta\hat{x}_3 = \delta y_3 \tag{15.38}$$

$$\delta\hat{x}_4 = \delta y_4$$

The estimated observer states and the resulting output responses are given as follows:

$$\hat{G}_b = \delta\hat{x}_1 + G_i^*$$
$$\hat{M}_b = \delta\hat{x}_2 + M_i^*$$
$$\hat{N}_b = \delta\hat{x}_3 + N_i^*$$
$$\hat{T}_b = \delta\hat{x}_4 + T_i^*$$

(15.39)

$$X_c = X_o - R_{XG}(\hat{G}_b - G_o) - R_{XM}(\hat{M}_b - M_o),$$

(15.40)

$$E_c = - R_{EG}(\hat{G}_b - G_o) - R_{EM}(\hat{M}_b - M_o) - R_{EN}(\hat{N}_n - N_o)$$

(15.41)

The observer output state vector is used as the observation vector to the EKF for fault parameter identification. The design strategy given in Section 15.4.1 is employed for combined estimation states and fault parameters by EKF.

15.5.2   Analysis of results

The inferential measurement vector to the method of EKF is given by

$$y(t_k) = [y_1(t_k)\ y_2(t_k)\ y_3(t_k)\ y_4(t_k)]^T$$
$$= \left[\hat{G}_b(t_k)\ \hat{M}_b(t_k)\ \hat{N}_b(t_k)\ \hat{T}_b(t_k)\right]^T$$

(15.42)

The augmented state vector to the EKF is represented by

$$\hat{x}_a = \left[\hat{G}_b\ \hat{M}_b\ \hat{N}_b\ \hat{T}_b\ \ln\hat{V}_{Go}\ \ln\hat{V}_{Mo}\right]^T$$

(15.43)

The numerical values of initial augmented states are same as given in Eq. (15.19). The state transition matrix F in Eq. (4.27) is denoted as F_a, and it is represented in Eq. (15.15). The nonlinear measurement model in Eq. (4.31) is denoted as h_a, and it is represented in Eq. (15.14). The measurement Jacobian matrix, H_a of Eq. (4.31), is obtained by taking the partial derivatives of Eq. (15.14) with respect to the augmented state vector in Eq. (15.43). The same initial state covariance matrix, P_o, process noise covariance matrix, Q, and observation noise covariance matrix, R, given in Section 15.4.2 are employed in EKF computation. With these relations and data, the EKF is applied for state estimation and fault parameter identification in the transient beer fermentation process for normal operation and different fault cases given in Table 15.1. The estimated glucose concentration by the observer along with the other observer outputs is shown in Fig. 15.4. The fault parameters identified by the method of reduced-order extended Luenberger observer and EKF in the case of ramp changes in $\ln V_{GO}$ and $\ln V_{MO}$ are shown in Fig. 15.5. The observer of this study based on a linearized state-space model of the process is useful in situations where all the state variables are not directly measurable, or their determination through analytical methods is not suitable for process monitoring. The method of reduced-order extended Luenberger observer and EKF can be applied for process fault detection and diagnosis even if all the states in the process are not available as measurements.

15.6   Method of two-level extended Kalman filter

This method involves in estimating the states and identifying uncertain process parameters separately in two levels by a state EKF and a parameter EKF. States and parameters are exchanged between the two estimators for each new value of measurement. The state and parameter estimators are designed by using continuous process dynamics and discrete process measurements.

15.6.1   Design strategy

The general process representation for the method of two-level EKF is illustrated in Section 4.4.2. The two-level EKF algorithm for separate estimation of states and process parameters is described in Section 4.6. The schematic of state estimation and fault parameter identification using the two-level EKF is shown in Fig. 14.10.

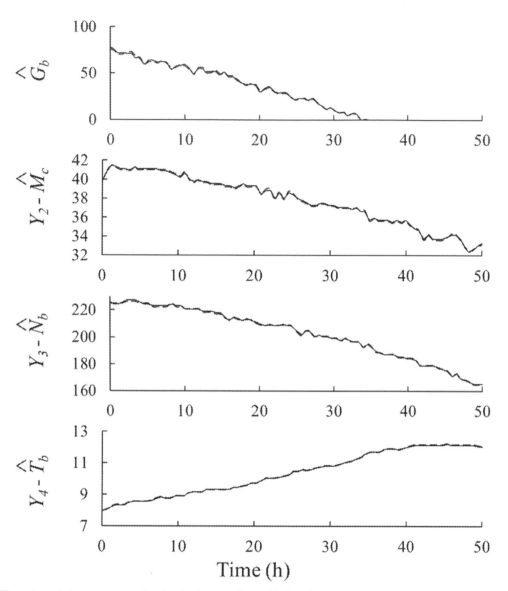

FIGURE 15.4 The estimated glucose concentration by the observer along with the other observer outputs in the case of initial incorrect values of lnV_{Go} and lnV_{Mo}.

15.6.1.1 Design of state estimation filter

The following relations are defined for the transient fermentation model.

The state vector:

$$x(t) = [x_1(t) \ x_2(t) \ x_3(t) \ x_4(t)]^T$$
$$= [G_c(t) \ M_c(t) \ N_c(t) \ T(t)]^T \tag{15.44}$$

The product vector:

$$x_p(t) = [X_c(t) \ E_c(t)]^T \tag{15.45}$$

The parameter vector:

$$\theta(t) = [\theta_1(t) \ \theta_2(t)]^T = [lnV_{Go}(t) \quad lnV_{mo}(t)]^T \tag{15.46}$$

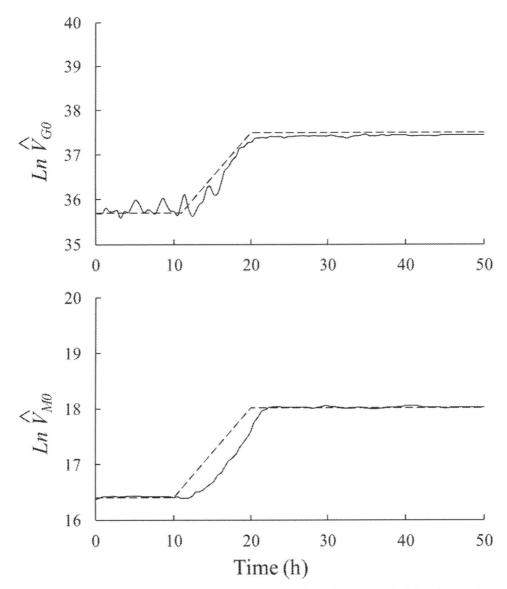

FIGURE 15.5 Identified fault parameters in the case of ramp changes in $\ln V_{GO}$ and $\ln V_{MO}$ by the method of the reduced-order Luenberger observer and EKF.

The observation vector:

$$y(t_k) = [y_1(t_k)\ y_2(t_k)\ y_3(t_k)\ y_4(t_k)]^T$$
$$= [G_c(t_k)\ M_c(t_k)\ N_c(t_k)\ T(t_k)]^T \tag{15.47}$$

The control input:

$$u(t) = K_p(T - T_{set}) \tag{15.48}$$

The state-space model of the fermentor is given by

$$f_x = \begin{bmatrix} f_{G_c}(t) \\ f_{M_c}(t) \\ f_{N_c}(t) \\ f_T(t) \end{bmatrix} = \begin{bmatrix} \dot{G}_c \\ \dot{M}_c \\ \dot{N}_c \\ \dot{T} \end{bmatrix} \tag{15.49}$$

The state transition matrix f_x involved in Eq. (4.40) of Section 4.6 is computed by taking the partial derivatives of the fermentation model, Eq. (15.49) with respective state vector. This matrix is represented by

$$F_x = \begin{bmatrix} f_{11} & 0 & 0 & f_{14} \\ f_{21} & f_{22} & 0 & f_{24} \\ f_{31} & f_{32} & f_{33} & f_{34} \\ f_{41} & f_{42} & f_{43} & f_{44} \end{bmatrix} \tag{15.50}$$

The nonlinear measurement model, h_k involved in Eq. (4.24) represents the fermentation measurement model in Eq. (15.14). The measurement transition matrix H_x in Eq. (4.41) is obtained by taking the partial derivatives of the measurement model in Eq. (15.14) with respect to the state vector, and this matrix is represented by

$$H_x = \begin{bmatrix} h_{11} & 0 & 0 & h_{14} \\ h_{21} & h_{22} & 0 & h_{24} \\ h_{31} & h_{32} & h_{33} & h_{34} \\ h_{41} & h_{42} & h_{43} & h_{44} \end{bmatrix} \tag{15.51}$$

These relations are used in the state estimation filter algorithm in Section 4.6.1 to estimate the process states.

15.6.1.2 Design of parameter estimation filter

The following relations are defined for the fermentation model.

The parameter vector is given by Eq. (15.46). The $g(\theta)$ in Eq. (4.23) of Section 4.4.2 is expressed as

$$g_\theta(t) = \begin{bmatrix} 0 \\ 0 \end{bmatrix} \tag{15.52}$$

The parameter Jacobian matrix G_θ involved in Eq. (4.43) of Section 4.6.2 is given as $G_\theta = [0]$. The parameter gradient matrix H_θ for the batch beer fermentation model is computed using Eq. (4.44) of Section 4.6.2 These relations are used in the parameter identification filter algorithm in Section 4.6.2, to estimate the process parameters.

15.6.2 Analysis of results

As designed above, the two-level EKF is used for state estimation and fault parameter identification in the batch fermentation process for normal operation and faulty situations. The process measurements are the glucose, maltose and maltotriose concentrations, and fermentation temperature. These measurements of batch beer fermentation are expressed by Eq. (15.18). The initial conditions chosen for states and parameters, state covariance matrices, and process noise covariance matrices are given as follows:

$$\begin{bmatrix} G_c(0) \\ M_c(0) \\ N_c(0) \\ T(0) \end{bmatrix} = \begin{bmatrix} 70.0 \\ 220.0 \\ 40.0 \\ 8.0 \end{bmatrix} \tag{15.53}$$

$$P_{x0} = \begin{bmatrix} 4 & 0 & 0 & 0 \\ 0 & 4 & 0 & 0 \\ 0 & 0 & 4 & 0 \\ 0 & 0 & 0 & 0.0625 \end{bmatrix} \tag{15.54}$$

$$Q_x = \begin{bmatrix} 4 & 0 & 0 & 0 \\ 0 & 4 & 0 & 0 \\ 0 & 0 & 4 & 0 \\ 0 & 0 & 0 & 0.0625 \end{bmatrix} \tag{15.55}$$

$$\begin{bmatrix} \theta_1(0) \\ \theta_2(0) \end{bmatrix} = \begin{bmatrix} 35.77 \\ 16.4 \end{bmatrix} \tag{15.56}$$

$$P_{\theta0} = \begin{bmatrix} 1 & 0 \\ 0 & 1 \end{bmatrix} \tag{15.57}$$

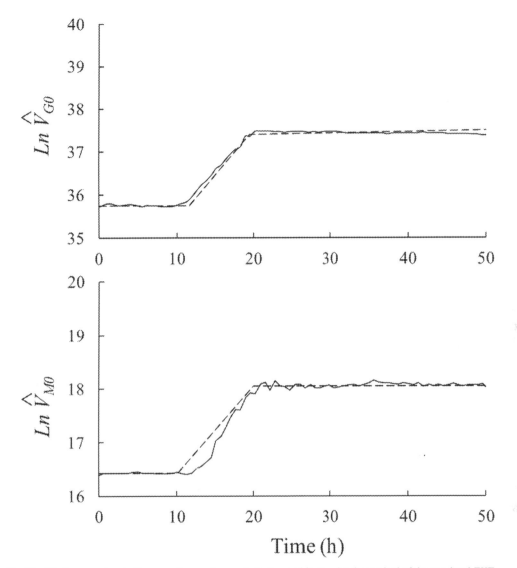

FIGURE 15.6   Identified fault parameters in the case of ramp changes in ln V_{GO} and ln V_{MO} by the method of the two-level EKF.

$$Q_\theta = \begin{bmatrix} 1 & 0 \\ 0 & 1 \end{bmatrix} \tag{15.58}$$

The measurement noise covariance matrix R is same as Eq. (15.23). The linear measurement matrix is given by

$$H = \begin{bmatrix} 1 & 0 & 0 & 0 \\ 0 & 1 & 0 & 0 \\ 0 & 0 & 1 & 0 \\ 0 & 0 & 0 & 1 \end{bmatrix} \tag{15.59}$$

The two-level EKF algorithm in Section 4.6 is applied for state estimation and fault parameter identification in batch beer fermentation using the above specified conditions. The estimated states, product responses, and identified fault parameters are compared with their corresponding true values for normal operation and all the fault situations given in Table 15.1. The identified fault parameters for the case of ramp changes in $\ln V_{Go}$ and $\ln V_{Mo}$ are shown in Fig. 15.6. The fault parameters identified by the method of two-level EKF are closely agreed with their corresponding true values. The estimated responses by the two-level EKF are found more stable toward the effects of noise and noise covariance matrices.

15.7 Method of discrete version of extended Kalman filter and sequential least squares

By this method, states are estimated using a discrete version of the EKF and parameters are identified using sequential least squares.

15.7.1 Design strategy

In this study, a discrete version of the EKF described in Section 4.4.4 is employed for state estimation, and a sequential least-squares algorithm is employed for parameter identification [13].

15.7.1.1 Design of state estimation filter

The same relations defined for the design of state estimating filter of two-level EKF (Eqs. 15.44−15.51) are also used for state estimation by the discrete version of the EKF of this method. The general discrete model expression in Eq. (4.35) of Section 4.4.4 of the discrete version of the Kalman filter is represented by the fermenter model as follows:

$$f_k = \begin{bmatrix} x_1 + f_{G_c}(t)\,\Delta t \\ x_2 + f_{M_c}(t)\,\Delta t \\ x_3 + f_{N_c}(t)\,\Delta t \\ x_4 + f_T(t)\,\Delta t \end{bmatrix} \tag{15.60}$$

where f_{Gc}, f_{Mc}, f_{Nc}, and f_T represent the dynamic model equations of batch beer fermentation described by Eqs. (15.1−15.4), and Δt is sampling time. The elements of the state transition matrix, F_k, in Eq. (4.36) of Section 4.4.4 are obtained by taking the partial derives of the discrete model of fermentation, Eq. (15.60) with respect to the state vector.

15.7.1.2 Design of parameter identifier

The linear relationships assumed for state and observation models are the same as Eqs. (14.63) and (14.64). The sequential least-squares algorithm described by Eqs. (14.65−14.69) in Section 14.8.1.2 is used for parameter identification. The schematic of state estimation and fault parameter identification using the discrete version of EKF and sequential least squares is depicted in Fig. 14.12.

15.7.2 Analysis of results

The design parameters such as the initial conditions, noise covariance matrices, and measurement coefficient matrix for the state estimating filter of this method are the same as the state estimating filter of the method of two-level EKF. The sequential least squares for parameter identification involve the following design parameters and initial conditions.

The initial covariance of error between the data and measurement model:

$$\Lambda(0) = \begin{bmatrix} 4 & 0 & 0 & 0 \\ 0 & 4 & 0 & 0 \\ 0 & 0 & 4 & 0 \\ 0 & 0 & 0 & 0.0625 \end{bmatrix} \tag{15.61}$$

The initial parameter covariance matrix:

$$P_{\theta 0} = \begin{bmatrix} 1 & 0 \\ 0 & 1 \end{bmatrix} \tag{15.62}$$

The constant forgetting factor:

$$\lambda = 0.99 \tag{15.63}$$

The constant scalar gain sequence:

$$\gamma = 0.01 \tag{15.64}$$

The initial gradient matrix:

$$w(0, \theta) = [0] \tag{15.65}$$

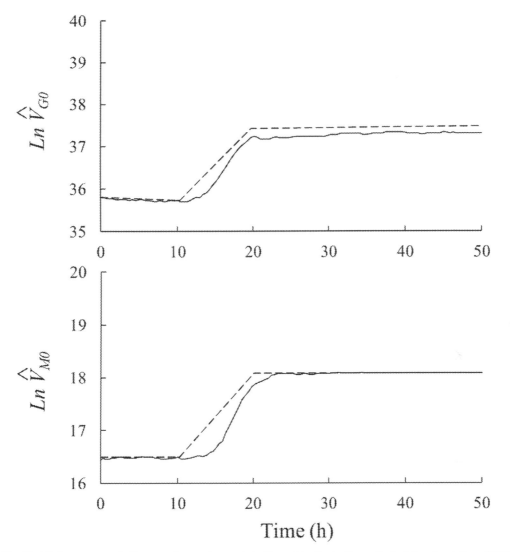

FIGURE 15.7   Identified fault parameters in the case of ramp changes in ln V_{Go} and ln V_{Mo} by the method of discrete version of EKF and sequential least squares.

This method is applied for state estimation and fault parameter identification in the transient beer fermentation process with the above conditions. By using the initial conditions and incoming measurements, parameter estimates are updated by sequential least squares. These parameter estimates and previous state estimates are used in EKF to obtain updated state estimates. Updated temperature is used to compute the control law. By augmenting the control law with the process model, a new set of measurements are generated for the next instant of estimation. For each measurement data set, state estimation by the discrete version of EKF and parameter identification by sequential least-squares are carried out recursively for normal operation and the cases of faults given in Table 15.1. The performance of this is evaluated towards the changes in observation noise, initial state noise covariance matrix, process and observation noise covariance matrices. The performance of this method is found slightly lower when compared with the method of two-level EKF. In this method, computations are simplified due to the involvement of lower-dimensional matrices. The identified fault parameters in the case of ramp changes in $\ln V_{Go}$ and $\ln V_{Mo}$ by the method of the discrete version of EKF and sequential least-squares are shown in Fig. 15.7.

15.8   Method of discrete version of extended Kalman filter and simultaneous least squares

By this method, states are estimated using a discrete version of the EKF and parameters are identified by using simultaneous least squares. The schematic for state estimation and fault parameter identification using the discrete version of EKF and simultaneous least squares is shown in Fig. 14.14.

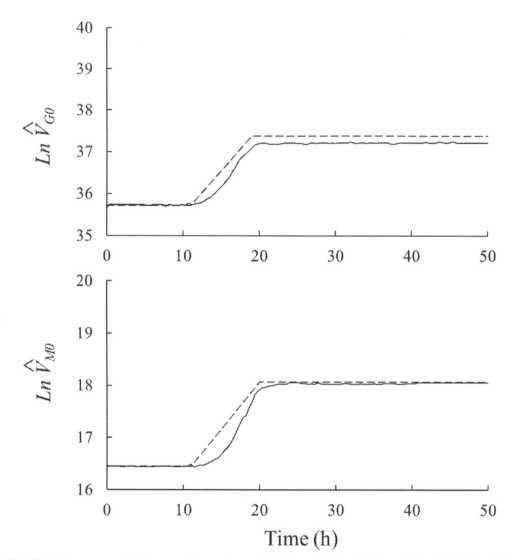

FIGURE 15.8 Identified fault parameters in the case of ramp changes in $\ln V_{GO}$ and $\ln V_{MO}$ by the method of the EKF and simultaneous least squares.

15.8.1 Design strategy

The design strategy is briefed as follows.

Design of state estimator: The state estimator design is the same as the method of EKF and sequential least-squares in Section 15.7.1.1.

Design of parameter identifier: In simultaneous least squares, the parameter identification procedure is similar to the design of parameter identifier of Section 14.9.1.2. The updating of the recursive gradient matrix $w(t,\theta)$ in Eq. (14.67) involves $F(\theta)$, which is computed by

$$F(\theta) = \phi(t_k) - K_k H \tag{15.66}$$

Here $\phi(t_k)$ is the same as the discrete model state transition matrix F_k in Eq. (4.35), K_k is the Kalman gain matrix involved in Eq. (4.37), and H is the measurement matrix. The equation that computes F_k in Eq. (4.35) is

$$F_k = (I + \frac{\partial f}{\partial x}\Delta t)\big|_{x=\hat{x}(t_k/t_{k-1})} \tag{15.67}$$

The $F(\theta)$ in Eq. (15.66) acts as a coupling matrix between the discrete EKF and recursive least squares. The elements of the matrix, $M(\theta)$ in Eq. (14.67), are computed by taking the partial derivatives of the discrete dynamic model,

Eq. (15.60), with respect to the parameters. The matrices $F(\theta)$ and $M(\theta)$ thus computed are used in Eq. (14.67) to update $w\ (t,\theta)$. The $w\ (t,\theta)$ is used in Eq. (14.66) to compute $\psi(t)$, which is used in the least squares algorithm in Section 14.8.1.2 for simultaneous identification of process parameters.

15.8.2   Analysis of results

The initial conditions, noise covariance matrices, and measurement coefficient matrix used for the state estimating filter of this method are the same as the state estimating filter of the method of two-level EKF. The initial conditions used for parameter estimation are the same as in the case of sequential least squares. This method is applied for state estimation and fault parameter identification in the transient beer fermentation process with these conditions. By using the initial conditions and incoming measurements, parameter estimates are updated by simultaneous least squares. These parameter estimates and previous state estimates are used in EKF to obtain updated state estimates. Updated temperature is used to compute the control law. By augmenting the control law with the process model, a new set of measurements are generated for the next instant of estimation. State estimation and parameter identification are carried out by this method for normal operation and the cases of faults given in Table 15.1. From the identified parameters, it is observed that there is certain delay in identified parameters initially but after the delay, the identified parameters converged to their true values. The performance of this method is found to be lower than that of two-level EKF. The results indicate slightly better performance of this method over the discrete version of EKF and sequential least squares. The identified fault parameters in the case of ramp changes in $\ln V_{Go}$ and $\ln V_{Mo}$ by the method of the discrete version of EKF and simultaneous least-squares are shown in Fig. 15.8. This method requires lower computational effort due to the involvement of lower-dimensional matrices.

In the beer fermentation model, the natural logarithm of glucose reaction rate frequency factor, $\ln V_{Go}$ and maltose reaction rate frequency factor, $\ln V_{Mo}$ are actually 35.77 and 16.4, respectively. When $\ln V_{Go}$ is 40.0, the model consumes all the glucose within one hour of fermentation. The actual consumption time of $\ln V_{Go}$ is a little over 35 h. The identified initial values of $\ln V_{Go}$ and $\ln V_{Mo}$ can determine, whether they are correct or not. The identified values of $\ln V_{Go}$ and $\ln V_{Mo}$ during fermentation can determine the deterioration in the performance of yeast organisms.

15.9   Summary

Different two-level methods and the method of EKF are presented for fault detection and diagnosis in nonlinear processes. The performances of these methods are evaluated for process fault detection and diagnosis by applying them to a transient beer fermentation process. The EKF is computationally intensive due to the involvement of large size matrices because of the augmentation of states and parameters. The initial state covariance matrix, process noise, and observation noise covariance matrices in the first level of all the two-level methods are kept the same. The computational effort required for the method of EKF and sequential least squares is lower due to the involvement of smaller size matrices. However, this method has exhibited lower performance when compared with the other two-level methods. The observer of this study based on a linearized state-space model of the process is useful in situations where all the state variables are not directly measurable, or their determination through analytical methods is not suitable for process monitoring. The fault parameters identified by the method of two-level EKF are more closely agreed with their corresponding true values when compared with the other methods. The results indicate that the method of two-level EKF is preferred for fault detection and diagnosis in transient beer fermentation.

References

[1] K.H. Bellgardt, W. Kuhlman, H.D. Meyer, K. Schugerl, M. Thoma, Application of extended Kalman filter for state estimation of a yeast fermentation, IEE(D) Proc. 133 (1986) 226.

[2] G.A. Montague, A.J. Morris, A.R. Wright, M. Aynsley, A. Ward, Modelling and adaptive control of fed-batch penicillin fermentation, Can. J. Chem. Eng. 64 (1986) 567.

[3] J. Staniskis, R. Simutis, A measuring systems for biotechnical processes based on discrete methods of estimation, Biotech. Bioeng. 28 (1986) 362.

[4] J. Chu, M. Ohshima, I. Hoshimoto, T. Takamatsu, J. Wanj, A design method for a class of roust nonlinear observers, J. Chem. Eng. Jpn. 22 (1989) 228.

[5] G. Bastin, D. Dichain, On-line Estimation and Adaptive Control of Bioreactors, Elsevier, Amsterdam, 1990.

[6] Y. Pomerleau, M. Peerier, Estimation of multiple specific growth rates in bioprocesses, AIChE J. 36 (1990) 207.

[7]   M. Farza, H. Hammouri, S. Othman, K. Busawon, Nonlinear observers for parameter estimation in bioprocesses, Chem. Eng. Sci. 52 (1997) 4251.

[8]   F.S. Wang, W.C. Lee, L.L. Chang, On-line estimation of biomass based on acid production in *Zymomonas mobilis* cultures, Bioproc. Eng. 18 (1998) 329.

[9]   J.E. Claes, J.F. Van Impe, On-line estimation of the specific growth rate based on variable biomass measurements: experimental validation, Bioproc. Eng. 21 (1999) 389.

[10] P. Bogaerts, A hybrid asymptotic Kalman observers for bioprocesses, Bioproc. Eng. 20 (1999) 249.

[11] V.N. Lubenova, Stable adaptive algorithm for simultaneous estimation of time-varying parameters and state variables on aerobic bioprocess, Bioproc. Eng. 21 (1999) 219.

[12] V. Holmberg, G. Olsson, Simultaneous on-line estimation of oxygen transfer rate and respiration rate, in: A. Johnson (Ed.), IFAC Proc Modelling and Control of Biotechnological Processes, 10, Pergamon Press, Oxford, 1986, p. 205.

[13] C. Venkateswarlu, K. Gangaiah, M. Bhagavantha Rao, Two level methods for incipient fault diagnosis in nonlinear chemical processes, Comput. Chem. Eng. 16 (1992) 463.

[14] W.F. Amirez, Optimal state and parameter identification: an application to batch beer fermentation, Chem. Eng. Sci. 42 (1987) 2749−2756.

Chapter 16

Optimal state and parameter estimation for fault detection and diagnosis of a high-dimensional fluid catalytic cracking unit

16.1 Introduction

Optimal state and parameter estimation-based fault detection and diagnosis via analytical redundancy/quantitative model-based techniques are gaining considerable importance in plant operations. Plants in chemical process industries are typically complex, highly nonlinear and many variables are usually not measurable, or their measurement becomes expensive. Faults can disturb the normal process operation, thus reducing the plant's performance or even leading to its breakdown or catastrophic failure. There has been an increasing interest in fault detection and diagnosis in industry and academia in recent years. This is due to the increased degree of automation and the growing demand for higher performance, efficiency, reliability, and safety in industrial systems. The need for effective management of detection, isolation, and localization of malfunctions has prompted the development of a number of online fault detection and isolation (FDI) techniques over the years.

Efficient diagnosis of process plants can lead to considerable incentives in terms of high performance, reliability, and safety. Model-based fault diagnosis via state and parameter estimation is an analytical redundancy technique that relies on the principle that possible faults in the monitored process can be associated with specific parameters and states of the mathematical model of the process. Model-based methods employ appropriate transformations and deploy stochastic estimators for state and parameter estimation. In these methods, the difference between the prediction and actual measurements provides the information required to judge the deviation of the residues from their healthy Gaussian white noise. The variations in the residues thus generated by one or a combination of parameters and state estimates are used to characterize the faults. The main benefit of the model-based approach is that the existing inputs and information can be utilized for estimator processing without requiring additional sensors.

When the model-based approach is used for fault diagnosis of complex nonlinear plants, the diagnosis system has to be made substantially robust and adaptive. Model linearization that causes unnecessary parameter variations during the process operation should be avoided by using the nonlinear model of the plant. Over the years, various model-based estimation methods for fault detection and diagnosis have been developed for dynamical systems [1−10]. In analytical redundancy-based approach, different versions of extended Kalman filter (EKF) are mostly employed for fault detection and diagnosis [8,11,12]. The EKF based methods provide the estimates of states and parameters of the mathematical model of a system by which the abnormal behavior of a system can be identified. However, EKF has certain limitations when it is applied to complex systems. One of the problems is the computation of the state transition matrix, which poses difficulties in complex systems. The Jacobian computation for state transition matrix is a computationally expensive operation, and there is no universal and robust numerical way to carry it out [13]. The first-order linearization of the model in EKF can introduce large errors in mean and covariance of the state vectors and even cause divergence of the filter [14]. Recently, new derivative free filtering algorithms called sigma point Kalman filters are introduced in state estimation field as improvements to EKF. The unscented Kalman filter (UKF) introduced by Julier et al. [15] and the SRUKF developed by Rudolph van der Merwe and Eric [16] are referred to as sigma point Kalman filters. These methods are based on unscented transformation, a mechanism for propagating mean and covariance through a nonlinear transformation. The UKF determines the mean and covariance accurately to the second order, while the EKF can only

Optimal State Estimation for Process Monitoring, Fault Diagnosis and Control. DOI: https://doi.org/10.1016/B978-0-323-85878-6.00012-9

obtain first-order accuracy [17]. Thus the UKF provides better state estimates for nonlinear systems [18]. However, UKF requires calculation of new-set of sigma points at each sample time, requiring a matrix square root (SR) of the state covariance matrix. While the SR of the covariance matrix is an integral part of the UKF, it is still the full covariance that is recursively updated. In SRUKF, a SR matrix of state covariance will be propagated directly, avoiding the need to refactorize it at each time step [16]. The SRUKF uses the QR factorization method and Cholesky factor rank 1 updating mechanism to achieve numerical stability and guarantees positive semidefiniteness of the Kalman filter covariances.

In this chapter, the methods of discrete EKF and UKF are presented for model-based fault diagnosis of nonlinear chemical plant. The performance of these methods is evaluated by applying to a high-dimensional fluid catalytic cracking (FCC) process. In particular, FCC unit (FCCU) involves complicated hydrodynamics and complex kinetics of both cracking and coke burning reactions. Fault diagnosis of such multivariable, high-dimensional, nonlinear, and strongly interacting plant with several subsystems like feed and preheat systems, cracking riser, reactor, regenerator, air blowers, and catalyst transport system is a challenging task. The mathematical representation of FCCU comprising the reactor/regenerator sections is considerably large and is reported elsewhere [19]. This process has long been recognized as a good candidate to develop and validate control schemes and fault diagnosis techniques, thus, it serves as a typical complex testbed to evaluate the proposed fault diagnosis methodologies.

16.2 Process representation

A large family of real chemical processes can be described by dynamic models, which are composed of differential and algebraic system of equations (DAE), where the differential equations describe the dynamic state behavior of the system and the algebraic equations represent the thermodynamic and the kinetic relationships of the system. The DAE model of the system is represented by

$$\dot{x} = f(x, z, \theta, u) + w \tag{16.1}$$

$$0 = g(x, z, \theta, u) \tag{16.2}$$

$$y = h(x, z, \theta) + v \tag{16.3}$$

where x is the state vector, z is the algebraic variable vector, θ is the parameter vector, u is the input vector, y is the measured output vector, and w is the modeling error/noise affecting the process.

The discrete time state space version of Eqs. (16.1)−(16.3) can be written as

$$x_{k+1} = f(x_k, z_k, \theta_k, u_k, k) + w_k \tag{16.4}$$

$$0 = g(x_k, z_k, \theta_k, u_k, k) \tag{16.5}$$

$$y_k = h(x_k, z_k, \theta_k, k) + v_k \tag{16.6}$$

where w and v denote the process noise and observation noise, respectively. The function $f(x_k, z_k, \theta_k, u_k, k)$ denotes the state of the system at time t_{k+1} obtained by integrating Eq. (16.1) and solving Eq. (16.2) from t_k to t_{k+1} with the initial conditions of x_k, z_k, and θ_k for a given u_k.

16.3 Fluid catalytic cracking unit

FCCU receives a number of hydrocarbon feed streams with different molecular weight distributions from several different refinery units and cracks the heavier components in these streams to more valuable lighter ones in the presence of a catalyst. The FCCU consists of various subsystems including feed system, preheat systems, cracking riser, reactor, regenerator, air blowers, and catalyst transport system. In FCCU operation, a preheating furnace pretreats the feed for cracking. Most of the endothermic reaction takes place in the reactor riser. As a result of cracking, carbonaceous products are deposited on the catalyst. This deposition causes reduction of the effectiveness and lifetime of the catalyst. The catalyst is continuously regenerated by blowing hot air. Thus coke is combusted to CO, CO_2, and H_2O. There is a constant flow of regenerated and spent catalyst between the reactor and regenerator. A detailed description of FCCU is reported in the literature [19]. The schematic diagram of FCCU is shown in Fig. 16.1.

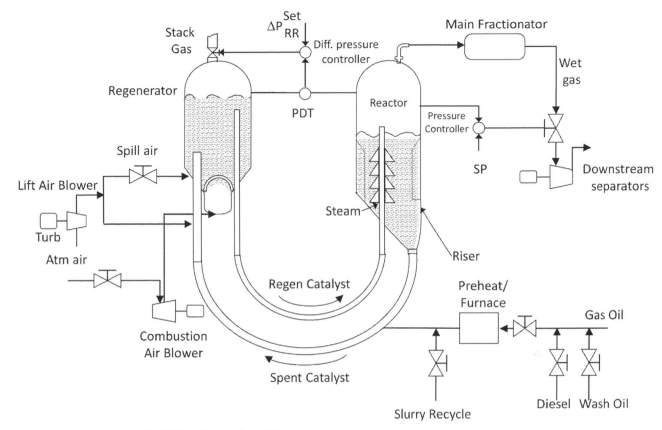

FIGURE 16.1   Schematic of Model IV fluid catalytic cracking unit.

16.4   Mathematical model of fluid catalytic cracking unit

The dynamic behavior of FCCU is represented by ordinary nonlinear differential and algebraic equations. The mathematical model of the process involves 17 differential state equations and 34 algebraic state equations. Altogether 20 measurements are available from the process. The parameters used to represent the faults in the process include combustion air blower discharge pipe flow resistance factor, furnace overall heat transfer coefficient, atmospheric temperature, pressure drop across reactor main fractionator and flow of fuel to preheat furnace. The mathematical model is reported in the literature [20]. The description of subsystems of FCCU along with their mathematical relations is briefed as follows.

16.4.1   Feed and preheat systems

These systems, along with their schematics, are described as follows:

16.4.1.1   Feed system

The feed system is shown in Fig. 16.2.

As shown in Fig. 16.2, three fresh feeds are available to the FCCU system, gas oil from tankage (F_{go}), wash oil (F_1), and diesel (F_2). The sum of these flows is the total fresh feed (F_3). In addition, a slurry recycle from the bottom of the fractionator (F_4) is added to the fresh feed stream after the preheat furnace. Flow controllers are available on F_1, F_2, F_3, and F_4. The model ignores the dynamics of the controllers and the flow streams, so that the actual flows are equal to the controller set points at all times:

$$F_1 = F_1{}^{set}; F_2 = F_2{}^{set}; F_3 = F_3{}^{set}; F_4 = F_4{}^{set} \qquad (16.7)$$

16.4.1.2   Preheat system

The fresh feed stream (F_3) is assumed to enter the preheat furnace at a specified temperature T_1 (Fig. 16.2). This system is not modeled. Instead, the temperature T_3 of the stream F_3 is measured as a disturbance variable.

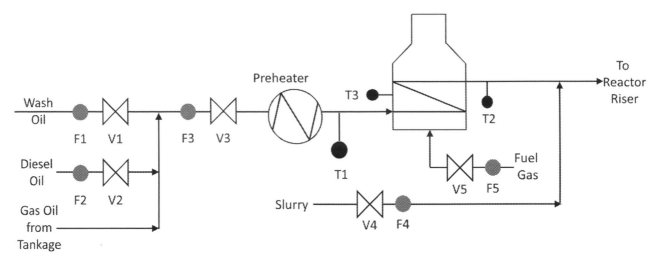

FIGURE 16.2 Feed system with furnace.

The furnace firebox temperature and outlet temperature are modeled with the following dynamic energy balances:

$$\frac{dT_3}{dt} = \frac{1}{\tau_{fb}} \left(F_5 \Delta H_{fu} - U_{Af} T_{lm} - Q_{loss} \right) \tag{16.8}$$

where,

$$T_{lm} = \frac{(T_3 - T_1) - (T_3 - T_2)}{\ln \frac{(T_3 - T_1)}{(T_3 - T_2)}}$$

$$Q_{loss} = a_1 F_5 T_3 - a_2$$

$$F_5 = F_5^{set}$$

$$\frac{dT_2}{dt} = \frac{1}{\tau_{fo}} (T_{2SS} - T_2) \tag{16.9}$$

$$T_{2ss} = T_1 + \frac{U_{Af} T_{lm}}{F_3}$$

16.4.2 Reactor

The schematic of the reactor is shown in Fig. 16.3. The reactor and main fractionator models contain the greatest degree of simplification in the FCCU model. The reactor riser (Fig. 16.3) is simulated as a dilute phase passport line. The hot catalyst from the regenerator mixes with the feed oil from the preheat system and the recycle oil from the main fractionator. The hot catalyst provides sensible heat, heat of vaporization, and heat of reaction for the endothermic cracking reactions. All of the cracking reactions are assumed to occur only in the reactor riser. The riser yield models predict only the amount of coke deposited at the catalyst and amount of wet gas (C_6 and lighter product). A constant coke composition is assumed. Despite the simplicity of the coke yield model, the simulation predicts the main effects of changes in various feed flow rate at the regenerator temperature. The yield model also provides for simulating feed composition changes through the reactor, which alters coke producing characteristics of various feeds. The composition of wet gas is not predicted. Since the compositions are not predicted, a detailed main fractionator is not provided in the simulation. However, pressures in the main fractionator are modeled, which allows the simulation to predict compressor loading, an important aspect of the overall control problem for the FCCU system.

16.4.2.1 Coke and wet gas yield

The yield of coke deposited on the catalyst in the riser is assumed to be affected by the weight hourly space velocity (WHSV) in the riser, the concentration of carbon on the regenerated catalyst (C_{rgc}), catalyst residence time in the riser

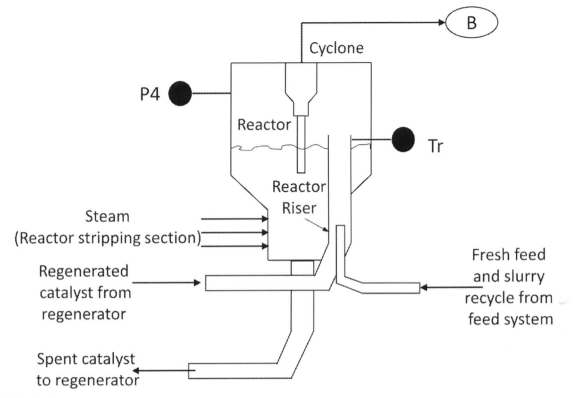

FIGURE 16.3   Reactor.

(τ_r) and the coking characteristics of the various feeds. The effect of riser temperature is not modeled. The coke production is given by

$$F_{coke} = \frac{58.52(F_3 + F_4)F_C\tau_r^{-0.69}}{100WHSV} \tag{16.10}$$

where

$$\tau_r = \frac{W_{ris}}{60F_{rgc}}$$

$$WHSV = \frac{3600(F_3 + F_4)}{W_{ris}} \quad F_C = \frac{\psi_f F_3 + 5F_4 + 3F_1 - 0.8F_2}{F_3 + F_4}$$

$\psi_f = 1$ for normal gas oil; >1 for heavier than normal gas oil; <1 for lighter than gas oil.
    The yield of wet gas is assumed to be affected only by the riser temperature and is given by:

$$F_{wg} = (F_3 + F_4)(C_1 + \dot{C}_2(T_r - T_{ref})) \tag{16.11}$$

16.4.2.2   Reactor mass balances

Coke and catalyst balances are written for the reactor. For simplicity, stirred tank dynamics are assumed. The coke balance includes the carbon entering the reactor on the regenerated catalyst and is given by

$$\frac{dC_{sc}}{dt} = \left\{ F_{rgc}C_{rgc} + F_{coke} - F_{sc}C_{sc} - C_{sc}\frac{dW_r}{dt} \right\}\frac{1}{W_r} \tag{16.12}$$

The catalyst balance is given by

$$\frac{dW_r}{dt} = F_{rgc} - F_{sc} \tag{16.13}$$

16.4.2.3 Reactor riser energy balance

The riser energy balance also assumes stirred tank dynamics, with negligible heat loss to the environment. The heat of cracking is assumed to be simply proportional to riser temperature. The energy balance is written as

$$M_{cpeff}\frac{dT_r}{dt} = Q_{in} - Q_{out} \tag{16.14}$$

where

$$Q_{in} = Q_{rgc} + F_3 C_{pfl}(T_2 - T_{base,f})$$

$$Q_{out} = Q_{cat,\ out} + Q_{slurry} + Q_{cracking} + Q_{ff}$$

$$Q_{catout} = F_{rgc}C_{pc}(T_r - T_{base})$$

$$Q_{slurry} = F_4\left[C_{psv}(T_r - T_{ref}) + Q_{sr}\right]$$

$$Q_{ff} = F_3\left[C_{pfv}(T_r - T_{ref}) + Q_{fr}\right]$$

$$Q_{cracking} = (F_3 + F_4)\Delta H_{crack}$$

$$\Delta H_{crack} = 172.7 + 3(T_r - T_{ref})$$

A constant temperature drop across the reactor stripper is assumed,

$$T_{sc} = T_r - \Delta T_{stripper} \tag{16.15}$$

16.4.2.4 Reactor riser pressure balances

The pressure at the bottom of the reactor riser requires the force balance on the regenerated catalyst U-bend. It is given by:

$$P_{rb} = P_4 + \frac{\rho_{ris}h_{ris}}{144} \tag{16.16}$$

$$\rho_{ris} = \frac{F_3 + F_4 + F_{rgc}}{v_{ris}}$$

$$v_{ris} = \frac{F_3 + F_4}{\rho_v} + \frac{F_{rgc}}{\rho_{part}}$$

The inventory of the catalyst in the riser needs the calculation of the catalyst residence time in the riser and is given by:

$$W_{ris} = \frac{F_{rgc}A_{ris}h_{ris}}{v_{ris}} \tag{16.17}$$

16.4.2.5 Reactor and main fractionator pressure balances

The main fractionator is modeled as a single large volume that includes the volume of piping and intermediate vessels between the main fractionator and the wet gas compressor. It is modeled as

$$\frac{dP_5}{dt} = 0.833(F_{wg} - F_{v11} - F_{v12} + F_{v13}) \tag{16.18}$$

where

$$F_{v12} = K_{12}V_{12}\sqrt{P_5 - P_{atm}}$$

A constant pressure drop between the main fractionator and the reactor is assumed, given the reactor pressure as:

$$P_4 = P_5 + \Delta P_{frac} \tag{16.19}$$

Fig. 16.4 shows the schematic of the main fractionator and wet gas compressor.

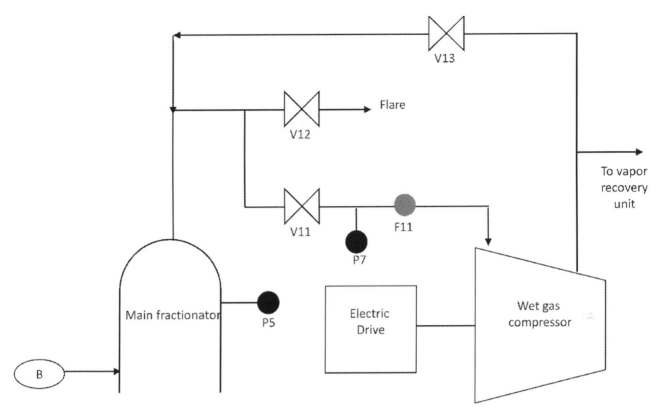

FIGURE 16.4   Main fractionator and wet gas compressor.

16.4.2.6   Wet gas compressor

The wet gas compressor is modeled as a single stage centrifugal compressor driven by a constant electric motor. It is assumed that the compressor is pumping against a constant pressure in the vapor recovery unit. The compressor performance equation relates suction flow to the polytrophic head. A single surge point is specified. Two bypass lines with valves are available to implement antisurge control. The valves are a flare valve (V_{12}) on the suction line of the compressor and a bypass line between the compressor discharge and inlet lines (V_{13}). However, the wet gas compressor is normally operated far from surge, and in simulation, antisurge controllers can be excluded.

The suction to head relation of the compressor is given by

$$F_{sucn,wg} = 11600 + \sqrt{1.366*10^8 - 0.1057*H_{wg}^2} \tag{16.20}$$

where

$$H_{wg} = 182,922.1\{C_{rw}^{0.0942} - 1\}$$

$$C_{rw} = \frac{P_{vru}}{P_7}$$

The molar flow rate through the compressor is obtained from the ideal gas law:

$$F_{11} = \frac{520*F_{sucn,wg}*P_7}{379*60*590*14.7} \tag{16.21}$$

The pressure balance around the compressor is given by assuming

$$\frac{dP_7}{dt} = 5(F_{v11} - F_{11}) \tag{16.22}$$

where

$$F_{v11} = K_{11}f_{pp}(V_{11})\sqrt{P_5 - P_7}$$

$$F_{v13} = K_{13}V_{13}P_{vru}$$

Flow characteristics are assumed nonlinear and are given by:

$$f_{pp}(x) = e^{2\ln(0.15)(1-x)} \qquad x > 0.5$$
$$= 0.3x \qquad x < = 0.5$$

(16.23)

16.4.3 Regenerator

A schematic of the regenerator is shown in Fig. 16.5. Spent catalyst from the reactor is transported through the U-bend and enters the regenerator through a lift pipe. Pressurized air is injected into the bottom of the lift pipe to assist in catalyst circulation. Coke is burnt off in the regenerator by contacting the catalyst with air in a fluidized bed. Combustion air is delivered directly to the bottom of the regenerator by the combustion air blower, and lift air both contribute to the fluidization and combustion reactions. Regenerated catalyst flows over a weir into the regenerator standpipe for transport back to the cracking riser. The fluidized bed is assumed to consist of two phases: a bubble phase of gaseous reactants and products moving up the bed in plug flow and a perfectly mixed dense phase containing gas and solid catalyst. Mass transfer occurs between the two phases as gas moves up the bed.

A detailed description of the regenerator mass and energy balances and the kinetic equations for the reactions can be found elsewhere [19]. A summary of model equations and assumptions is provided here.

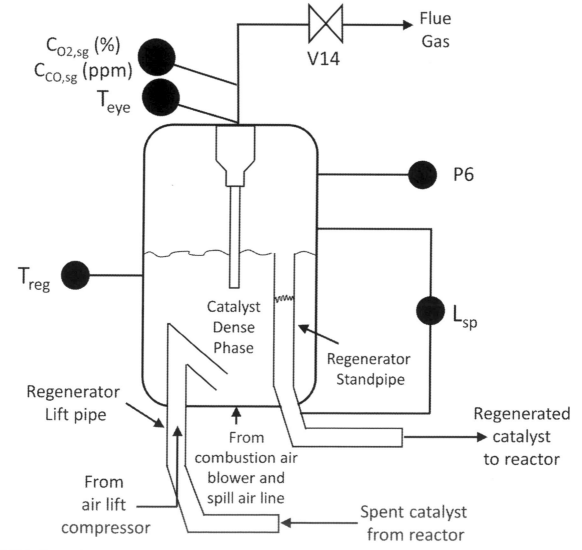

FIGURE 16.5 Regenerator.

16.4.3.1   Energy balance

The catalyst phase of the fluidized bed is assumed to be perfectly mixed. Coke deposited on the catalyst in the reactor consists of carbon and hydrogen. It is assumed that all of the hydrogen in the coke is burned off in the regenerator. The carbon remaining on the catalyst after regeneration is returned to the reactor with the regenerated catalyst, where additional coke is deposited. The concentration of coke on the spent catalyst represents the sum of carbon on the regenerated catalyst and new deposited coke. The following energy balance equations represent the regenerator.

$$((W_{reg} + W_{sp})C_{pc} + M_1)\frac{dT_{reg}}{dt} = Q_{in} - Q_{out} \tag{16.24}$$

$$Q_{in} = Q_{air} + Q_H + Q_c + Q_{sc} \tag{16.25}$$

$$Q_{out} = Q_{fg} + Q_{rgc} + Q_e \tag{16.26}$$

$$Q_{air} = F_{air}Cp_{air}(T_{air} - T_{base}) \tag{16.27}$$

$$Q_H = F_H \Delta H_H \tag{16.28}$$

$$Q_c = F_{air}(X_{CO,\ sg}\Delta H_1 + X_{CO2,\ sg}\Delta H_2) \tag{16.29}$$

$$Q_{sc} = F_{sc}Cp_c(T_{sc} - T_{base}) \tag{16.30}$$

$$Q_{fg} = (F_{air}(X_{O2,\ sg}Cp_{O2} + X_{CO,\ sg}Cp_{CO} + X_{CO2,\ sg}Cp_{CO2} + 0.79Cp_{N2}) + 0.5F_HCp_{H2O})(T_{cyc} - T_{base}) \tag{16.31}$$

$$Q_{rgc} = F_{rgc}Cp_c(T_{reg} - T_{base}) \tag{16.32}$$

$$F_H = F_{sc}(C_{sc} - C_{rgc})C_H \tag{16.33}$$

The heat generated by hydrogen combustion is Q_H. The quantity of hydrogen present due to the new deposition of coke in the reactor and the amount of hydrogen burned off in the regenerator, assuming total conversion is F_H. The following equations give the temperature profile above the bed. Catalysts present above the bed due to carry over, and therefore, temperature increases above the bed due to the heat generated. The heat balance between the top of the bed and the inlet of the cyclones is affected by catalysts in this zone. Empirical equations describing catalyst entrainment are given below:

$$\frac{dT_{reg}(Z)}{dz} = 0, \ \ 0 \le Z \le Z_{bed}$$

$$\frac{dT_{reg}(Z)}{dz} = (\Delta H_1 \frac{dX_{CO}(Z)}{dz} + \Delta H_2 \frac{dX_{CO2}(Z)}{dz})\frac{1}{C_p(Z)}, \ \ Z_{bed} < Z \le Z_{top} \tag{16.34}$$

$$C_p(Z) = 0.79C_{pN2} + X_{CO}(Z)C_{pCO} + X_{CO2}(Z)C_{pCO2} + X_{O2}(Z)C_{pO2} + \{0.5C_{pH2O}F_H + \delta_zC_{pc}M_e\}\frac{1}{F_{air}} \tag{16.35}$$

$$\delta_z = 0 \ \ Z \ge Z_{cyc} \\ = 1 \ \ Z < Z_{cyc} \tag{16.36}$$

16.4.3.2   Carbon balance

The carbon balance expressions are given by the following equations:

$$\frac{dC_{rgc}}{dt} = \left(\frac{dW_c}{dt} - C_{rgc}\frac{dW_{reg}}{dt}\right)\frac{1}{W_{reg}} \tag{16.37}$$

$$\frac{dW_{reg}}{dt} = F_{sc} - F_{sp} \tag{16.38}$$

$$\frac{dW_c}{dt} = (F_{sc}C_{sc} - F_H) - (F_{sp}C_{rgc} + 12F_{air}(X_{CO,\ sg} + X_{CO2,\ sg})) \tag{16.39}$$

The first term in the parenthesis in the regenerator carbon balance in Eq. (16.39) represents the quantity of carbon entering the regenerator with the spent catalyst. The quantity $F_{sc}C_{sc}$ is the quantity of coke, while F_H is the amount of hydrogen. The remaining terms describe the amount of carbon leaving the regenerator on catalyst flowing over the weir into the standpipe and carbon in the flue gas.

Catalyst flows over a weir and into the regenerator standpipe is represented by

$$\frac{dW_{sp}}{dt} = F_{sp} - F_{rgc} \tag{16.40}$$

$$F_{sp} = fof \sqrt{A_{sp}}(Z_{bed} - Z_{sp}) - 4925 - 20(3-d) \tag{16.41}$$

$$d = \min(3, h_{sp} - \frac{W_{sp}}{\rho_c A_{sp}}) \tag{16.42}$$

$$L_{sp} = \frac{W_{sp}}{\rho_c A_{sp}} \tag{16.43}$$

16.4.3.3 Mass balances

The concentrations of carbon monoxide and carbon dioxide in the air entering into the regenerator are assumed to be zero. The amount of oxygen available for the reactions in the regenerator is the amount of oxygen available in the air-flow minus the oxygen consumed in the combustion of hydrogen. The reaction rates are affected by the volume fraction of the catalyst in the bed and the disengaging zone above the bed. These balances are expressed by

$$\frac{dX_{O2}(Z)}{dz} = (100(-0.5K_1 - K_2)\rho_B(Z)C_{rgc} - K_3 X_{CO}(Z))\frac{X_{O2}(Z)}{v_s} \tag{16.44}$$

$$\frac{dX_{CO}(Z)}{dz} = \left(100K_1\rho_B(Z)C_{rgc} - 2K_3 X_{CO}(Z)\right)\frac{X_{O2}(Z)}{v_s} \tag{16.45}$$

$$\frac{dX_{CO2}(Z)}{dz} = \frac{-dX_{O2}(Z)}{dz} - 0.5\frac{dX_{CO}(Z)}{dz} \tag{16.46}$$

$$X_{CO2}(Z) = X_{O2}(Z=0) - X_{O2}(Z) - 0.5X_{CO}(Z) \tag{16.47}$$

$$X_{O2}(Z=0) = \frac{1}{F_{air}}\{0.21F_{air} - 0.25F_H\} \tag{16.48}$$

$$K_1 = 6.9547*\exp\left(19.88 - \frac{34000}{T_{reg}(Z) + 459.6}\right) \tag{16.49}$$

$$K_2 = 0.69148*\exp\left(15.06 - \frac{25000}{T_{reg}(Z) + 459.6}\right) \tag{16.50}$$

$$K_3 = 0.6412*\exp\left(25.55 - \frac{45000}{T_{reg}(Z) + 459.6}\right) \tag{16.51}$$

The output concentrations are converted to the appropriate units as given below:

$$C_{O2,sg} = \frac{100F_{air}X_{O2}}{F_{sg}} \tag{16.52}$$

$$C_{CO,sg} = \frac{10^6 * 28 * X_{CO}}{28X_{CO} + 44X_{CO2} + 32X_{O2} + 22.12} \tag{16.53}$$

16.4.3.4 Void fraction of the catalyst

Void fraction in the regenerator bed is a function of superficial velocity. In the disengaging section above the bed, catalyst volume fraction drops off vertically. It is assumed that any entrained catalyst entering the cyclones is removed and returned to the bed so that $\rho_B(Z)$ is zero in the cyclones.

$$\frac{d\rho_B(z)}{dz} = 0; \rho_B(Z) = (1 - \varepsilon_e) \qquad 0 \leq Z \leq Z_{bed} \tag{16.54}$$

$$\frac{d\rho_B(z)}{dz} = \frac{-1000 F_{air}\rho_B(Z)}{A_{reg}*v_s*\rho_{c,dilute}} \quad Z_{bed} < Z \leq Z_{cyc} \tag{16.55}$$

$$= 0; \rho_B(Z) = 0 \quad Z_{cyc} < Z$$

$$\varepsilon_e = \min\left[1, \max\left(\varepsilon_f, \varepsilon_f + \frac{1.904 + 0.363 v_s - 0.048 v_s^2}{Z_{bed}}\right)\right] \tag{16.56}$$

$$\varepsilon_f = 0.332 + 0.06 v_s \tag{16.57}$$

16.4.3.5   Catalyst entrainment

The mass flow rate of entrained catalyst leaving the bed is described empirically by the following equation:

$$M_e = A_{reg}*v_s*\rho_{c,dilute} \tag{16.58}$$

$$\rho_{c,dilute} = -0.8780 + 0.582*v_s \tag{16.59}$$

The superficial velocity v_s is given by

$$v_s = \frac{(F_{air} + F_{sg})}{2}\left(\frac{1}{\rho_g A_{reg}}\right) \tag{16.60}$$

where

$$\rho_g = \frac{520*P_6}{(379*14.7)(T_{reg} + 459.6)}$$

16.4.3.6   Pressure balance

The pressure balance in the regenerator assumes ideal gas behavior. The pressure at the bottom of the regenerator P_{rgb} is required to calculate airflow into the regenerator from the air blowers.

$$\frac{dP_6}{dt} = \frac{1}{V_{reg,g}}\left\{R\left(n\frac{dT_{reg}}{dt} + (T_{reg} + 459.6)\frac{dn}{dt}\right)\right\} \tag{16.61}$$

$$\frac{dn}{dt} = F_{air} - F_{sg} \tag{16.62}$$

$$V_{reg,g} = A_{reg}Z_{cyc} - A_{reg}Z_{bed}(1 - \varepsilon_e) \tag{16.63}$$

$$P_{rgb} = P_6 + \frac{W_{reg}}{144 A_{reg}} \tag{16.64}$$

$$\Delta P_{RR} = P_6 - P_4 \tag{16.65}$$

Stack gas flows from the regenerator through the stack valve V_{14} as given by:

$$F_{sg} = K_{14}V_{14}\sqrt{P_6 - P_{atm}} \tag{16.66}$$

16.4.3.7   Bed height

The regenerator bed height is given empirically by:

$$Z_{bed} = \min\left\{Z_{cyc}, \left(2.85 + 0.8 v_s + \frac{W_{reg} - \rho_{c,dilute}*A_{reg}*Z_{cyc}}{A_{reg}*\rho_{c,dilute}}\right)\left(\frac{1}{1 - \frac{\rho_{c,dilute}}{\rho_{c,dense}}}\right)\right\} \tag{16.67}$$

where

$$\rho_{c,dense} = \rho_{part}(1 - \varepsilon_f)$$

16.4.3.8 Airlift calculations

Spent catalyst from the reactors enters the regenerator through the spent catalyst U-bend lift pipe. Air from the lift air blower is injected into the bottom of the lift pipe to assist the circulation of the catalyst. An increase in lift air-flow rate lowers the density of catalyst in the lift pipe, which further lowers the head, resulting in an increase of catalyst flow through the U-bend. Changes in density in the lift pipe occur with an assumed time constant τ_{fill} and are given by:

$$\frac{d\rho_{lift}}{dt} = \left(\frac{F_{sc}}{v_{cat,lift}*A_{lp}} + \rho_{air,g} - \rho_{lift} \right) \frac{1}{\tau_{fill}} \tag{16.68}$$

$$\rho_{air,g} = \frac{29*P_6}{R(T_{sc} + 459.6)} \tag{16.69}$$

$$v_{air,lift} = \frac{F_9}{A_{lp}*\rho_{airg}} \tag{16.70}$$

$$v_{cat,lift} = \max \left(v_{air,lift} - v_{slip}, \frac{F_{sc}}{A_{lp}*\rho_{part}} \right) \tag{16.71}$$

The pressure at the bottom of the lift pipe is the sum of the regenerator pressure P_6 and the heads in the lift pipe and the regenerator bed above the top of the lift pipe.

$$P_{blp} = P_6 + \frac{\rho_{lift}h_{lift}}{144} + \frac{(Z_{bed} - Z_{lp})*\rho_{c,dense}}{144} \tag{16.72}$$

16.4.4 Air blowers

Air is supplied to the regenerator by two centrifugal compressors. The bulk of regenerator air is supplied by the combustion air blower, which supplies air directly to the bottom of the regenerator. At a given catalyst rate, excess lift air blower capacity may be available. In this situation, air may be bypassed from the lift air blower directly to the bottom of the regenerator through the spill airline.

16.4.4.1 Combustion air blower

A combustion air blower is a centrifugal compressor driven by a constant speed electric motor. Throughput is controlled by a throttling valve (V_6) on the compressor inlet line. An atmospheric vent line and valve (V_7) are available on the discharge line for use in the antisurge control system. A head capacity performance equation is provided that relates suction volume as a function of discharge pressure when the suction is at normal atmospheric pressure. The relation to obtain suction flow rates at different suction pressures is the same curve for one atmosphere at the same polytrophic head. Single surge point on the performance curve is specified. The regenerator combustion air blower is shown in Fig. 16.6.

The head capacity performance equation is given by:

$$F_{sucn,comb} = 45000 + \sqrt{1.581*10^9 - 1.249*10^6 p_{base}^2} \tag{16.73}$$

where

$$P_{base} = \frac{14.7P_2}{P_1}$$

$$\frac{dP_1}{dt} = \frac{R(T_{atm} + 459.6)}{29V_{comd,d}}(F_{v6} - F_6) = 0 \tag{16.74}$$

$$\frac{dP_2}{dt} = \frac{R(T_{comb,d} + 459.6)}{29V_{comb,d}}(F_6 - F_{v7} - F_7) \tag{16.75}$$

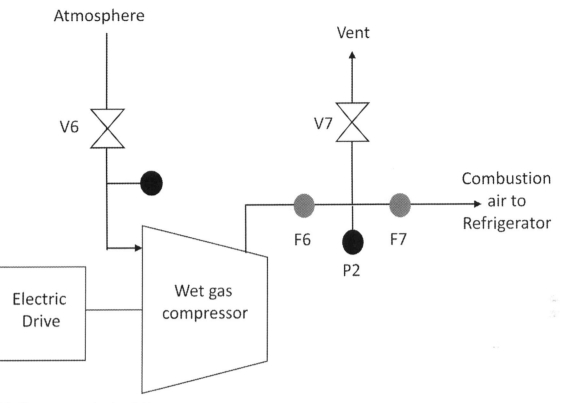

FIGURE 16.6 Regenerator combustion air blower.

$$F_6 = \frac{0.04511 P_1 F_{sucn,comb}}{T_{atm} + 459.6} \tag{16.76}$$

$$F_7 = K_{comb}\sqrt{P_2 - P_{rgb}} \tag{16.77}$$

$$F_{v6} = K_6 f_{pp}(V_6)\sqrt{P_{atm} - P_1} \tag{16.78}$$

$$F_{v7} = K_7 f_{pp}(V_7)\sqrt{P_2 - P_{atm}} \tag{16.79}$$

Suction and discharge pressures are given by Eqs. (16.74) and (16.75).

16.4.4.2 Lift air blower

The lift air blower is a single stage compressor driven by a variable speed steam turbine with a speed control governor. A local controller manipulates a steam valve for speed adjustment. An atmospheric vent line and valve are available on the discharge line for use in the antisurge control system. A single performance equation is provided that relates polytrophic head, speed, and suction volume. Suction flow for a given discharge pressure is returned at a specific base speed and then corrected for the actual compressor speed. A surge line is defined, relating surge flow rate to discharge pressure. The regenerator lift air blower is shown in Fig. 16.7.

The performance equation is given by:

$$F_{base} = 8600 + \sqrt{2.582*10^8 - 1.068*10^5 P_{base,d}^2} \tag{16.80}$$

where

$$P_{base,d} = \left[\left(P_3^M - P_{atm}^M\right)\left(\frac{S_b}{S_a}\right)^2 + P_{atm}^M \right]^{\frac{1}{M}}$$

$$M = \frac{K_{avg} - 1}{K_{avg}\eta_p}$$

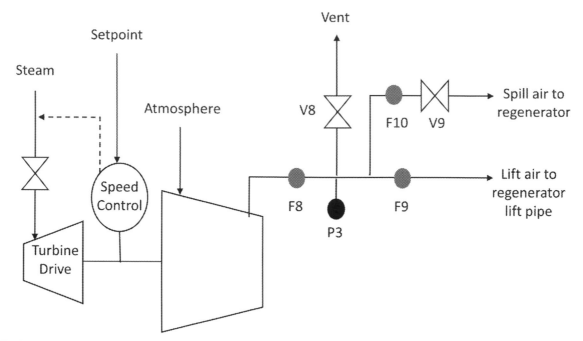

FIGURE 16.7 Regenerator lift air blower.

Lift air flow rate is obtained from:

$$F_8 = \frac{0.04511 P_{atm} F_{sucn,lift}}{T_{atm} + 459.6} \tag{16.81}$$

$$F_{sucn,lift} = F_{base} * \left(\frac{S_a}{S_b}\right) \tag{16.82}$$

$$F_{surge,lift} = 5025 + 112 * P_3 \tag{16.83}$$

$$F_9 = K_{lift} \sqrt{P_3 - P_{blp}} \tag{16.84}$$

$$F_{10} = K_9 V_9 \sqrt{P_3 - P_{rgb}} \tag{16.85}$$

$$F_{v8} = K_8 V_8 \sqrt{P_3 - P_{atm}} \tag{16.86}$$

$$F_{air} = \frac{F_7 + F_9 + F_{10}}{29} \tag{16.87}$$

$$F_t = F_7 + F_9 + F_{10} \tag{16.88}$$

Flow rates associated with the lift air blower system and spill line are given by Eqs. (16.84)−(16.88). The discharge pressure and speed of the lift air blower are given by:

$$\frac{dP_3}{dt} = \frac{R(T_{lift,d} + 459.6)}{29 V_{lift,d}} (F_8 - F_{v8} - F_9 - F_{10}) \tag{16.89}$$

$$S_a = S_a^{\min} + 1100 V_{lift} \tag{16.90}$$

16.4.5 Catalyst circulation

Circulation of spent and regenerated catalyst is modeled as single-phase flow governed by simple force balances. Factors that affect catalyst circulation in a real FCCU, such as concentration of carbon on regenerated catalyst and injection of steam at various points in the U-bend, have been ignored in this model. Because the dynamics of the catalyst circulation lines are orders of magnitude faster than other dynamic elements in the FCCU system. A pseudo steady state assumption is applied to the force balances on the spent and regenerated catalyst lines. Constant friction factors are assumed for each circulation line.

The force balance on the regenerated catalyst U-bend is given by:

$$\frac{dv_{rgc}}{dt} = \frac{f_{rgc}}{M_{rgc}} = 0 \tag{16.91}$$

$$M_{rgc} = \left[W_{sp} + A_{ubend,rgc} * L_{ubend,rgc} * \rho_c \right] \left(\frac{1}{g} \right) \tag{16.92}$$

$$f_{rgc} = \Delta P_{rgc} * A_{ubend,rgc} - v_{rgc} * L_{ubend,rgc} * f_{ubend,rgc} \tag{16.93}$$

$$\Delta P_{rgc} = 144(P_6 - P_{rb}) + \left[\frac{W_{sp}}{A_{sp}} + \left(E_{tap} - E_{oil\ inlet} \right) \rho_c \right] \tag{16.94}$$

$$F_{rgc} = v_{rgc} * A_{ubend,rgc} \rho_c \tag{16.95}$$

The force balance on the spent catalyst circulation line is written as

$$\frac{dv_{sc}}{dt} = \frac{f_{sc}}{M_{sc}} = 0 \tag{16.96}$$

$$M_{sc} = \left[W_\tau + A_{ubend,sc} * L_{ubend,sc} * \rho_c \right] \left(\frac{1}{g} \right) \tag{16.97}$$

$$f_{sc} = \Delta P_{sc} * A_{ubend,sc} - v_{sc} * L_{ubend,sc} * f_{ubend,sc} \tag{16.98}$$

$$\Delta P_{sc} = 144 \left(P_4 - P_{blp} \right) + \left[\frac{W_r}{A_{stripper}} + \left(E_{stripper,tap} - E_{lift,air} \right) \rho_c \right] \tag{16.99}$$

$$F_{sc} = v_{sc} * A_{ubend,sc} \rho_c \tag{16.100}$$

16.4.6 Nomenclature and numerical data

The nomenclature and the numerical values to some of the quantities of the above equations are given as follows [19]: A_{lp} = cross sectional area of lift pipe (8.73 ft^2), A_{reg} = cross sectional area of regenerator (590 ft^2), A_{ris} = cross sectional area of reactor riser (9.6 ft^2), A_{sp} = cross sectional area of stand pipe (7 ft^2), $A_{stripper}$ = cross sectional area of reactor stripper (60 ft^2), $A_{ubend,rgc}$ = cross sectional area of regenerator catalyst U-bend (3.7 ft^2), $A_{ubend,sc}$ = cross sectional area of spent catalyst U-bend (5.2 ft^2), a_1 = furnace heat loss parameter (0.15 B.t.u/s°F), a_2 = furnace heat loss parameter (200B.t.u/s), $C_{CO,sg}$ = concentration of carbon monoxide in stack gas (ppm), C_H = weight fraction of hydrogen in coke (0.075 lb H$_2$/lb coke), $C_{O2,\ sg}$ = concentration of oxygen in regenerator stack gas (mol%), C_p (z) = average heat capacity (B.t.u/mol°F), C_{pair} = heat capacity of air (7.08 B.t.u/mol°F), C_{pc} = heat capacity of catalyst (0.31 B.t.u/lb°F), C_{pCO} = heat capacity of carbon monoxide (7.28 B.t.u/mol°F), C_{pCO2} = heat capacity of carbon dioxide (11.0 B.t.u/mol°F), C_{pfl} = heat capacity of fresh feed liquid (0.82 B.t.u/lb°F), C_{pfv} = heat capacity of fresh feed vapor (0.81 B.t.u/lb°F), C_{pH20} = heat capacity of steam (8.62 B.t.u/mol°F), C_{pN2} = heat capacity of nitrogen (7.22 B.t.u/mol°F), C_{po2} = heat capacity of oxygen (7.62 B.t.u/mol°F), C_{psv} = heat capacity of slurry vapor (0.8 B.t.u/lb°F), C_{rgc} = weight fraction of coke on regenerated catalyst (lb coke/lb catalyst), C_{rw} = wet gas compressor compression ratio, C_{sc} = weight fraction of coke on spent catalyst (lb coke/lb catalyst), C_1 = wet gas product constant (0.0088438 mol/lb feed), C_2 = wet gas product constant (0.00004 mol/lb feed), $E_{lift\ air}$ = elevation of regenerator lift air injection (134 ft), $E_{oil\ inlet}$ = elevation of oil inlet into reactor riser (124.5 ft), $E_{stripper\ tap}$ = elevation of reactor stripper tap (130 ft), E_{tap} = pressure tap elevation on stand pipe (155 ft), F_{air} = air flow rate in to regenerator (mol/s), F_B = effect of feed type on coke production, F_{base} = air lift compressor inlet suction flow at base conditions (ICFM), F_{coke} = production of coke in reactor riser (lb/s), F_{go} = flow of gas oil to reactor riser (lb/s), F_H = burning rate of hydrogen (lb/s), fof = overflow factor (424), $f_{pp}(x)$ = nonlinear valve flow rate function, f_{rgc} = force exerted by regenerated catalyst (lb$_f$), F_{rgc} = flow rate of regenerated catalyst (lb$_f$), f_{sc} = force exerted by spent catalyst (lb$_f$), F_{sc} = flowrate of spent catalyst (lb/s), F_{sg} = stack gas flow (mol/s), F_{sp} = flow into atmosphere (lb/s), $F_{sucn,comb}$ = combustion air blower inlet suction flow (ICFM), $F_{sucn,lift}$ = lift air blower inlet suction flow (ICFM), $F_{sucn,wg}$ = wet gas compressor inlet suction flow (ICFM), $F_{surge,comb}$ = combustion air blower surge flow (45100 ICFM), $F_{surge,lift}$ = lift air blower surge flow (ICFM), $F_{surge,wg}$ = wet gas compressor surge flow (11700 ICFM), F_T = air flowrate into regenerator (lb/s), $F_T^{set=}$ setpoint to total air flow controller (lb/s), $f_{ubend,rgc}$ = regenerated catalyst friction factor (17 lb s/ ft^2), $f_{ubend,sc}$ = spent catalyst friction factor (47 lb s/ ft^2), F_{v6} = flow through combustion air blower suction valve (V_6) (lb/s), F_{v7} = flow through combustion air blower vent valve (V_7) (lb/s), F_{v8} = flow through lift air blower vent valve (V_8) (lb/s), F_{v11} = flow through wet gas compressor suction valve (V_{11}) (mol/s), F_{v12} = flow through wet gas flare valve (V_{12}) (mol/s), F_{v13} = flow through wet gas compressor antisurge valve (V_{13}) (mol/s), F_{wg} = wet gas production in reactor (mol/s), F_1 = flow of wash oil to reactor riser (lb/s), F_1^{set} = setpoint to wash oil flow controller (lb/s), F_2 = flow of diesel oil to reactor riser (lb/s), F_2^{set} = setpoint to diesel oil flow controller, F_3 = flow of fresh feed to reactor riser (lb/s), F_3^{set} = set point to fresh feed flow controller (lb/s), F_4 = flow of slurry flow rate to controller (lb/s), $F_4^{set=}$ setpoint to slurry flow controller (lb/s), F_5 = flow of fuel to furnace (lb/s), F_5^{set} = setpoint to furnace fuel oil flow controller, F_6 = combustion air blower throughput (lb/s), F_7 = combustion air flow to the regenerator (lb/s), F_8 = lift air blower throughput (lb/s), F_9 = lift air flow to the regenerator (lb/s), F_9^{set} = lift air flow setpoint (lb/s), F_{10} = spill air flow to the regenerator (lb/s), F_{11} = wet gas flow to the vapor recovery unit (mol/s), g = acceleration due to gravity (32.2 ft lb$_m$/s^{2}lb$_f$), h_{lift} = height of lift pipe (34 ft), h_{ris} = height of reactor riser (60 ft), h_{sp} = height of regenerator standpipe (20 ft), H_{wg} = wet gas compressor head (psia), K_{avg} = average ratio of specific heats (1.39), k_{comb} = combustion air blower discharge pipe flow resistance factor (40.0 lb/s$\sqrt{}$psia), K_{lift} = lift air blower discharge pipe flow resistance factor (5 lb/s$\sqrt{}$psia), K_1 = reaction rate constant for reaction I (1/s), K_2 = reaction rate constant for reaction II (1/s), K_3 = reaction rate constant for reaction III (mol air/s mol CO), K_6 = combustion air blower suction valve flow rating (250 lb/s $\sqrt{}$psia), K_7 = combustion air blower vent valve flow rating (15 lb/s $\sqrt{}$psia), K_8 = lift air blower vent valve flow rating (5 lb/s $\sqrt{}$psia), K_9 = lift air blower spill valve flow rating (10 lb/s$\sqrt{}$psia), K_{11} = wet gas compressor suction valve flow rating (1.5 mol/s$\sqrt{}$psia), K_{12} = wet gas flare valve flow rating (0.5 mol/s$\sqrt{}$psia), K_{13} = wet gas compressor antisurge valve flow rating (0.01 mol/s$\sqrt{}$psia), K_{14} = regenerator stack gas valve flow rating (1.1 mol/s$\sqrt{}$psia), L_{sp} = level of catalyst in stand pipe, $L_{ubend,rgc}$ = length of regenerated catalyst U-bend (56 ft), $L_{ubend,sc}$ = length of spent catalyst U-bend (56 ft), M = polytrophic exponent, M_{cpeff} = effective heat capacity of riser vessel and catalyst (10,000 B.t.u/°F), M_e = flow rate of entrained catalyst from dense bed into dilute phase (lb/s), M_I = effective heat capacity of regenerator mass (200000 B.t.u/°F), M_{rgc} = inertial mass of regenerated catalyst (lbfs^2/ft), M_{sc} = inertial mass of spent catalyst (lbfs^2/ft), n = amount of gas (mol), P_{atm} = atmospheric pressure (14.7 psia), P_{base} = combustion air blower base discharge pressure (psia), $P_{base,d}$ = lift air blower base discharge pressure (psia), P_{blp} = pressure at bottom of lift pipe (psia), P_{rb} = pressure at bottom of reactor riser (psia), P_{rgb} = pressure at bottom of regenerator (psia), P_{vru} = discharge pressure of wet gas compressor to vapor recovery unit (101 psia), P_1 = combustion air blower suction pressure (psia), P_2 = combustion air blower discharge pressure (psia), P_3 = lift air blower discharge pressure (psia), P_4 = reactor pressure (psia), P_4^{set} = reactor pressure controller set point (psia), P_5 = reactor fractionator pressure (psia), P_6 = regenerator pressure (psia), P_7 = wet gas compressor suction pressure (psia), Q_{air} = enthalpy of incoming air (B.t.u/s), Q_C = total heat of burning carbon (B.t.u/s), $Q_{cat\ out}$ = enthalpy of catalyst out of reactor riser (B.t.u/s), $Q_{cracking}$ = heat generated from cracking (B.t.u/s), Q_e = total heat lost from regenerator to environment (556 B.t.u/s), Q_{ff} = heat required to bring fresh feed to reactor riser temperature (B.t.u/s), Q_{fg} = enthalpy of outgoing stack gas (B.t.u/s), Q_{fr} = heat required to raise temperature of slurry from 500°F to 1000°F (vapor) (300B.t.u/lb), Q_H = enthalpy of incoming hydrogen (B.t.u/s), Q_{in} = enthalpy of regenerator, reactor (B.t.u/s), Q_{out} = enthalpy out of regenerator, reactor (B.t.u/s), Q_{loss} = heat loss from furnace (B.t.u/s), Q_{rgc} = enthalpy of regenerated catalyst (B.t.u/s), Q_{sc} = enthalpy of spent catalyst (B.t.u/s), Q_{slurry} = heat required to bring slurry to reactor riser temperature (B.t.u/s), Q_{sr} = heat required to raise temperature of fresh feed from 700°F to 1000°F (vapor) (412 B.t.u/s), R = universal gas constant (10.73 ft^3 psia/lb mol°R), S_a = actual speed of the lift air blower (rpm), S_a^{max} = maximum speed of the lift air blower (6320 rpm), S_a^{min} = minimum speed of the lift air blower (5000 rpm), S_b = base speed of the lift air blower (5950 rpm), t = time, T_{air} = temperature of air entering regenerator (270°F), T_{atm} = temperature of atmosphere (75°F), T_{base} = base temperature (1100°F), $T_{base,f}$ = base temperature of reactor fresh feed (700°F), $T_{comb,d}$ = combustion air blower discharge temperature (190°F), T_{cyc} = regenerator stack gas temperature at cyclone, T_{diff} = regenerator bed and cyclone temperature difference (°F), $T_{lift,d}$ = lift air blower discharge temperature (225°F), T_{lm} = furnace log mean temperature difference (°F), T_{ref} = base temperature for reactor riser energy balance (999°F), T_{reg} = temperature of regenerator bed (°F), T_{sc} = temperature of spent catalyst entering regenerator (°F), T_1 = temperature of fresh feed entering furnace (460.9°F), T_2 = temperature of fresh feed entering reactor riser (°F), T_{2ss} = steady state furnace outlet temperature (°F), T_3 = furnace fire box temperature (°F), U_{Af} = furnace overall heat transfer coefficient (25 B.t.u/s), $V_{air,lift}$ = velocity of air in lift pipe (ft/s), $V_{cat,lift}$ = velocity of catalyst in lift pipe (ft/s), $V_{comb,d}$ = combustion air blower discharge system volume (1000 ft^3), V_{lift} = manipulated variable for lift air blower steam valve (200 ft^3), $V_{reg,g}$ = regenerator volume occupied by gas (ft^3), V_{rgc} = velocity of regenerated catalyst (ft/s), V_{ris} = volumetric flow rate in reactor riser (ft^3/s), V_s = superficial velocity in regenerator (ft/s), V_{sc} = velocity of spent catalyst (ft/s), V_{slip} = slip velocity (2.2 ft/s), V_6 = combustion air blower suction valve position (0−1), V_7 = combustion air blower vent valve position (0−1), V_8 = lift air blower vent valve position (0−1), V_9 = spill air valve position (0−1), V_{11} = wet gas compressor suction valve position (0−1), V_{12} = wet gas flare valve position (0−1), V_{13} = wet gas compressor vent valve position (0−1), V_{14} = stack gas valve position (0−1), W_C = inventory of carbon in reactor (lb), W_r = inventory of catalyst in reactor (lb), W_{reg} = inventory of carbon in regenerator (lb), W_{ris} = inventory of carbon in reactor riser (lb), W_{sp} = inventory of carbon in regenerator stand pipe (lb), $WHSV$ = weight hourly space velocity (lb oil/(lb cat/h)), X_{CO} = molar ratio of CO to air (mol CO/mol air), $X_{CO,sg}$ = molar ratio of CO to air in stack gas (mol CO/mol air), X_{CO2} = molar ratio of CO_2 to air (mol CO_2/mol air), $X_{CO2,sg}$ = molar ratio of CO_2 to air in stack gas (mol CO_2/mol air), X_{N2} = molar ratio of N_2 to air (mol air), X_{O2} = molar ratio of O_2 to air (mol O_2/mol air), $X_{O2,sg}$ = molar ratio of O_2 to air in stack gas (mol O_2/mol air), Z_{bed} = regenerator dense bed height (ft), Z_{cyc} = height of cyclone inlet (45 ft), Z_{lp} = height of lift pipe discharge (11 ft), Z_{sp} = stand pipe exit height from bottom of regenerator (13 ft), and Z_{top} = O_2, CO measurement point (49.5 ft). δ_z = Dirac delta function, ΔH_{crack} = heat of cracking (B.t.u/lb), ΔH_{fu} = heat of combustion of furnace fuel (1000 B.t.u/scf), ΔH_H = heat of combustion of hydrogen (60960 B.t.u/lb), ΔH_1 = heat of formation of CO (46368 B.t.u/mol CO produced), ΔH_2 = heat of formation of CO_2 (B.t.u/mol CO_2 produced), ΔP_{frac} = pressure drop across reactor main fractionator (9.5 psia), ΔP_{rgc} = pressure drop from bottom of stand pipe to oil inlet elevation of reactor riser (lb/ft^2), ΔP_{RR} = differential pressure between regenerator and reactor (psi), ΔP_{RR}^{set} = differential pressure controller set point (psi), ΔP_{sc} = pressure drop from bottom of reactor to lift air injection elevation (lb/ft^2), $\Delta T_{stripper}$ = temperature drop across reactor stripper (35°F), ε_e = effective void fraction in regenerator dense phase bed, ε_f = apparent void fraction in regenerator dense phase bed, η_p = polytrophic efficiency (1), $\rho_{air,g}$ = density of air at regenerator conditions (lb/ft^3), ρ_B = volume fraction of catalyst, ρ_c = density of catalyst in U-bend and regenerator stand pipe (45 lb/ft^3), $\rho_{c,dilute}$ = density of catalyst in the dilute phase (lb/ft^3), $\rho_{c,dense}$ = density of catalyst in the dense phase (lb/ft^3), ρ_g = density of exit gas (mol/ft^3), ρ_{lift} = density of catalyst in lift pipe (lb/ft^3), ρ_{part} = selected density of catalyst (68 lb/ft^3), ρ_{ris} = average density of material in reactor riser (lb/ft^3), ρ_v = vapor density at reactor riser conditions (0.57 lb/ft^3), τ_{fb} = furnace firebox time constant (200 s), τ_{fill} = riser fill time (40 s), τ_{fo} = furnace time constant (60 s), τ_r = catalyst residence time in riser, and ψ_f = effective coke factor for gas−oil feed.

16.5   Fluid catalytic cracking unit system variables

In the FCCU system, there are 17 state variables, 34 algebraic variables, and 20 measurements.

16.5.1   State variables

The 17 state variables are combustion air blower discharge pressure P_2, lift air blower discharge pressure P_3, main fractionator pressure P_5, regenerator pressure P_6, wet gas compressor suction pressure P_7, fresh feed temperature T_2, furnace firebox temperature T_3, reactor riser temperature T_r, regenerator temperature T_{reg}, regenerator catalyst inventory W_{reg}, regenerator standpipe catalyst inventory W_{sp}, reactor catalyst inventory W_r, regenerator carbon inventory W_c, the concentration of coke on the spent catalyst C_{sc}, the concentration of carbon on the regenerated catalyst C_{rgc}, moles of gas in the regenerator n, and catalyst density in lift pipe ρ_{lift}.

16.5.2   Algebraic variables

The 34 algebraic variables of the model are regenerator gas superficial velocity v_s, pressure at the bottom of the regenerator p_{rgb}, regenerator dense bed height z_{bed}, stack gas temperature T_{cyc}, stack gas O_2 concentration $X_{O2,\ sg}$, stack gas CO concentration $X_{CO,sg}$, stack gas CO_2 concentration $X_{CO2,sg}$, stack gas flow F_{sg}, pressure at the bottom of lift pipe P_{blp}, combustion air blower suction pressure P_1, combustion air blower flow F_6, combustion air blower suction flow F_{v6}, combustion air flow to regenerator F_7, combustion air blower vent flow F_{v7}, lift air blower flow F_8, lift air blower vent flow F_{v8}, lift air flow to regenerator F_9, spill air flow to the regenerator F_{10}, regenerated catalyst flow F_{rgc}, carbon on the regenerated catalyst C_{rgc}, level in stand pipe L_{sp}, spent catalyst flow F_{sc}, pressure at riser bottom P_{rb}, inventory of catalyst in the riser W_{ris}, wet gas production F_{wg}, coke production F_{coke}, wet gas flow F_{11}, wet gas comp. suction flow F_{v11}, wet gas flare valve flow F_{v12}, wet gas antisurge flow F_{v13}, wash oil flow F_1, diesel fuel flow F_2, fresh feed flow F_3, slurry flow F_4, and furnace fuel flow F_5.

16.5.3   Measured variables

The 20 measured outputs are lift air flow rate F_9, combustion airflow rate F_7, spill air flow rate F_{10}, total airflow rate F_T, fresh feed temperature T_2, furnace firebox temperature T_3, riser temperature T_r, reactor pressure P_4, main fractionator pressure P_5, wet gas compressor suction flow rate $F_{sucn,wg}$, stack gas temperature T_{cyc}, stack gas O_2 concentration

$C_{O2,sg}$, stack gas CO concentration $C_{CO,sg}$, regenerator bed temperature T_{reg}, regenerator pressure P_6, standpipe catalyst level L_{sp}, combustion air blower suction flow rate $F_{sucn,comb}$, lift air blower suction flow $F_{sucn,lift}$, lift air blower speed S_a, and lift air blower surge point $F_{surge,lift}$.

16.6 Fault cases considered in fluid catalytic cracking unit

The fault sources and the effects associated with faults in FCCU are considered as follows.

1. Change in k_{comb} is due to leaking in valve (V_7). Therefore, variation in k_{comb} will affect P_2, which in turn affects F_7 and the total flow rate to the regenerator. This results in incomplete combustion, and the outlet concentrations of combustion gases will change.
2. Decrease in heat transfer coefficient is due to fouling of heat exchanger. A decrease in the overall heat transfer coefficient results in a decrease in the fresh feed temperature and an increase in the flue gas temperature leaving the furnace.
3. Lower feed temperature decreases the endothermic reaction and decreases the riser temperature.
4. Change in atmospheric temperature T_{atm} due to climatic conditions. This will affect the combustion air blower throughput F_6 and the combustion air blower discharge pressure P_2. The change in P_2 results in a change in airflow into regenerator F_T and cyclone temperature T_{cyc}.
5. Change in ΔP_{frac} is due to pressure upsets in downstream units and changes in valve positions of v_{12}, v_{13}, and v_{11}. Pressure upsets in downstream units propagating back to the FCC section are the most severe disturbances for this system. The catalyst circulation rate and reactor pressure are also affected. Increased reactor pressure leads to an increased catalyst circulation rate. That means, there is a positive gain between changes in catalyst circulation rate and changes in reactor pressure.
6. Change in fuel flow rate can occur due to partial plugging of pipe or control valve or leakage failure. This will affect the furnace firebox's outlet temperature (T_2) and the reactor temperature (T_r). Thus all the variables related to reactor and regenerator will be affected.

16.7 Design of discrete version of extended Kalman filter

A discrete version of the EKF described in Section 4.4.4 is designed and applied for state estimation and fault parameter identification in FCCU with the computations adjudged to the DAE system of equations.

The DAE of FCCU are of the form in Eqs. (16.1)–(16.3). The discrete time state space version of Eqs. (16.1)–(16.3) are in the form of Eqs. (16.4)–(16.6) with the notation for state variable vector $= x$, algebraic variable vector $= z$, fault parameter vector $= \theta$, input vector $= u$, measured output vector $= y$, and modeling error/noise affecting the process $= w$. In this study, a discrete Kalman filter described in Section 4.4.4 with the discrete state space model form given in Eqs. (16.4)–(16.6) is employed for state estimation and fault parameter identification in FCCU. The relations defined for the model are given as follows.

The vector of state variables:

$$x = \begin{bmatrix} P_2, & P_3, & P_5, & P_6, & P_7, & T_2, & T_3, & T_r, & T_{reg}, & W_{reg}, & W_{sp}, & W_r, & W_c, & C_{sc}, & C_{rgc}, & n, & \rho_{lift} \end{bmatrix}^T \quad (16.101)$$

The vector of algebraic variables:

$$z = \begin{bmatrix} v_s, & P_{rgb}, & Z_{bed}, & T_{cyc}, & X_{O2,\,sg}, & X_{CO,\,sg}, & X_{CO2,\,sg}, & F_{sg}, & P_{blp}, & P_1, & F_6, & F_{v6}, & F_7, & F_{v7}, & F_8, & F_{v8}, F_9, \\ F_{10}, & F_{rgc}, & C_{rgc}, & L_{sp}, & F_{sc}, & P_{rb}, & W_{ris}, & F_{wg}, & F_{coke}, & F_{11}, & F_{v11}, & F_{v12}, & F_{v13}, & F_1, & F_2, & F_3, & F_4, & F_5 \end{bmatrix}^T \quad (16.102)$$

The vector of measured variables:

$$y = \begin{bmatrix} F_9, & F_7, & F_{10}, & F_T, & T_2, & T_3, & T_r, & P_4, & P_5, & F_{sucn,wg}, & T_{cyc}, & C_{O2,sg}, C_{CO,sg}, \\ T_{reg}, & P_6, & L_{sp}, & F_{sucn,comb}, F_{sucn,lift}, & S_a, & F_{surge,lift} \end{bmatrix}^T \quad (16.103)$$

The fault parameter vector considered in FCCU is given by

$$\theta = \begin{bmatrix} k_{comb}, & UA_f, & T_{atm}, & \Delta P_{frac}, & F_5 \end{bmatrix}^T \quad (16.104)$$

The augmented vector with states and parameters is represented by

$$x_a = [x, z, \theta]^T \quad (16.105)$$

The augmented vector thus becomes

$$x_a = \begin{bmatrix} P_2, & P_3, & P_5, & P_6, & P_7, & T_2, & T_3, & T_r, & T_{reg}, & W_{reg}, & W_{sp}, & W_r, & W_c, & C_{sc}, & C_{rgc}, & n, & \rho_{lift}, & k_{comb}, & UA_f, & T_{atm}, & \Delta P_{frac}, & F_5 \end{bmatrix}^T$$

(16.106)

The state transition matrix F_k in Eq. (4.36) is computed as

$$F_k = \frac{\partial x_{k+1}}{\partial x_k} = \frac{\partial f}{\partial x_k} + \frac{\partial f}{\partial z_k}\frac{\partial z_k}{\partial x_k} = \begin{bmatrix} \frac{\partial f}{\partial x_k} & \frac{\partial f}{\partial z_k} \end{bmatrix}\begin{bmatrix} I \\ \frac{\partial z_k}{\partial x_k} \end{bmatrix}$$

(16.107)

The measurement transition matrix H_k in Eq. (4.38) is computed as

$$H_k = \frac{\partial y_k}{\partial x_k} = \frac{\partial h}{\partial x_k} + \frac{\partial h}{\partial z_k}\frac{\partial z_k}{\partial x_k} = \begin{bmatrix} \frac{\partial h}{\partial x_k} & \frac{\partial h}{\partial z_k} \end{bmatrix}\begin{bmatrix} I \\ \frac{\partial z_k}{\partial x_k} \end{bmatrix}$$

(16.108)

The matrix $\frac{\partial z_k}{\partial x_k}$ can be computed using the following formula

$$\frac{\partial z_k}{\partial x_k} = -\left(\frac{\partial g}{\partial z_k}\right)^{-1}\frac{\partial g}{\partial x_k}$$

(16.109)

The x in the above equations can be treated as x_a. The matrix F_k measures the sensitivity of the system with respect to the initial condition x_k. The matrix H_k measures the sensitivity of the outputs with respect to the states. With these computations, the discrete EKF presented in Section 4.4.4 is implemented for state estimation and fault parameter identification in FCCU.

16.8   Design of unscented Kalman filter

The UKF algorithm described in Section 4.8 of Chapter 4 is implemented for state estimation and fault parameter identification in FCCU. The vectors of variables and parameters defined in Eqs. (16.101)–(16.106) are also used in UKF computation. The initial augmented state vector is of dimension (22×1) and the dimension of initial covariance matrix P_0 is (22×22). The process noise covariance matrix Q is of dimension (22×22) and the measurement noise covariance matrix R is of dimension (20×20). The dimension of algebraic variables is (34×1). The diagonal elements of these matrices are heuristically tuned to provide better performance of the fault diagnostic methods. The nonlinear algebraic equations are solved based on the states and parameters. The design procedure of UKF for FCCU is as follows:

1. The initial state vector, x_{k-1} and the mean vector $\hat{x}_{k-1}$ are of dimensions, 22×1. The initial state covariance matrix, P_0 and the process noise covariance matrix Q are of dimensions, 22×22. The measurement covariance matrix R has a dimension of 20×20.
2. According to Eq. (4.50), the dimension of sigma points generated is 22×45.
3. According to Eq. (4.51), the weights computed for states and covariance matrices are of dimension 22×45.
4. According to Eq. (4.52), the dimension of the transformed sigma points computed through projection over the nonlinear functions is 22×45.
5. According to Eq. (4.53), the dimension of the transformed measurement samples computed by propagating the sigma points through a nonlinear observation function is 20×45.
6. According to Eq. (4.54), the predicted state has a dimension of 22×1, and the prediction state covariance has a dimension of 22×22.
7. According to Eq. (4.55), the prediction measurements have a dimension of 20×1, and the prediction measurement covariance, and the state observation cross covariance are of dimension 20×20 and 22×20.
8. According to Eq. (4.56) in the update step, the Kalman gain matrix, the state vector, and the state covariance matrix have the dimension of 22×20, 22×1, and 22×22, respectively.

The algebraic variable vector of dimension, 34×1 is computed based on the updated state vector at every time step. With these computations, the UKF algorithm in Section 4.8 of Chapter 4 is implemented iteratively for state estimation and fault parameter identification in FCCU.

16.9 Analysis of results

The dynamics of FCCU is represented by DAE system of equations. The differential equations and the algebraic equations describing the FCCU are solved for every time step. The ordinary differential equations representing the mass and energy and balances of the regenerator are integrated along the vertical distance yielding composition and temperature profiles. The compositions and temperature at the exit of the cyclones at the top of the regenerator are used to compute the overall energy balance of the regenerator. The ODEs in the vertical distance are reinitialized based on the values at the top, and the integration begins a new from the bottom of the dense bed. The data for solving FCCU model and implementation of estimation algorithms can be taken from nomenclature and numerical values given in the above section.

The initial state vector in Eq. (16.101) is chosen as

$$x = \begin{bmatrix} P_2 = 35.20187, & P_3 = 40.28494, & P_5 = 23.52, & P_6 = 29.64, & P_7 = 22.67028, \\ T_2 = 667.2611, & T_3 = 16.07.55, & T_r = 995.13, & T_{reg} = 1271.874, & W_{reg} = 273689.2, \\ W_{sp} = 3567.831, & W_r = 101411.9, & W_c = 1297.124, & C_{sc} = 0.007841, & C_{rgc} = 0.000871, \\ n = 245.9322, & \rho_{lift} = 3.299055 \end{bmatrix}^T$$

The initial fault parameter vector in Eq. (16.104) is taken as

$$\theta = \begin{bmatrix} k_{comb} = 40, & UA_f = 25, & T_{atm} = 75, & \Delta P_{frac} = 9.5, & F_5 = 34 \end{bmatrix}^T$$

Additional data can be obtained from the reported literature [20]. The diagonal elements of P_0 and Q are chosen such that they are around 1% initial state deviations, and the diagonal elements of R are chosen around 2% of measurement deviations.

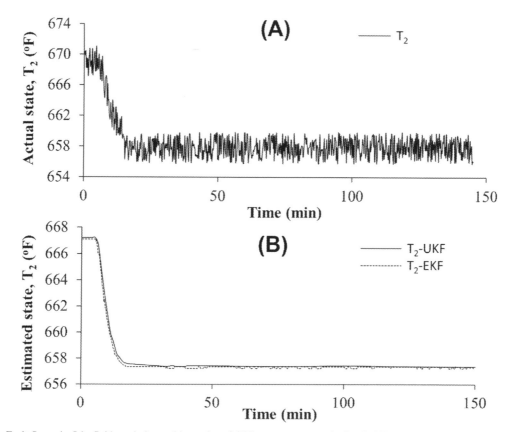

FIGURE 16.8 Fault Scenario 5 in fluid catalytic cracking unit at 0.25% measurement noise level: (A) sensor measurement, T_2 (°F); (B) estimated temperatures, T_2 (°F) by extended Kalman filter and unscented Kalman filter.

The state estimation and noise filtering efficiency of the estimators are analyzed by corrupting the FCCU measurements with a random Gaussian noise in the form random numbers in the range of -1.0 to 1.0. The measurements are corrupted using the noise at two levels as Level 1 of 0.25% and Level 2 of 1.5% of the magnitudes of the measured variables. The fault scenarios of Section 16.6 are studied by using the measurements at these noise levels and performing state estimation by applying EKF and UKF. For instance, in the case of fault Scenario 5 in Section 16.6, a change in fuel flow rate occurs due to partial plugging of pipe. T_2 denotes the measured temperature of fresh feed entering the reactor riser. When this temperature is corrupted with the noise at two levels and used as a measurement to the estimators, the filtered (estimated) temperatures, T_2-EKF and T_2-UKF are shown in Figs. 16.8 and 16.9, respectively. These results show the improved noise filtering efficiency of the estimators. These studies can be performed for other fault cases in Section 16.6.

The results of the estimators are also analyzed by choosing a random Gaussian noise within the range of -1.0 to 1.0 and corrupting the measurements with the noises of 0.25%, 0.5%, 1%, 1.25%, and 1.5% of their actual magnitudes. The initial state covariance matrix, P_0, process noise covariance matrix, Q, and observation noise covariance matrix, R, are appropriately tuned in implementing the algorithm. The EKF and UKF are applied for state estimation and fault parameter identification for normal operation and the cases of faults in Section 16.6. For instance, consider fault case 5 in Section 16.6 as a step decrease in fuel flow rate F_5 to the furnace. The decrease in F_5 can be caused by partial plugging of pipe or leakage or failure of the flow controller. This affects the temperature of the fresh feed entering to reactor riser, T_2, and thus the reactor temperature. This effect can be carried to the regenerator affecting all the related variables. When EKF and UKF are applied for state estimation and fault parameter identification for this case, by using the measurements at a noise level of 1.25%, the estimation results are shown in Figs. 16.10 and 16.11. The results in these figures indicate the comparison of actual results with the estimated results. The performance of the estimators is also

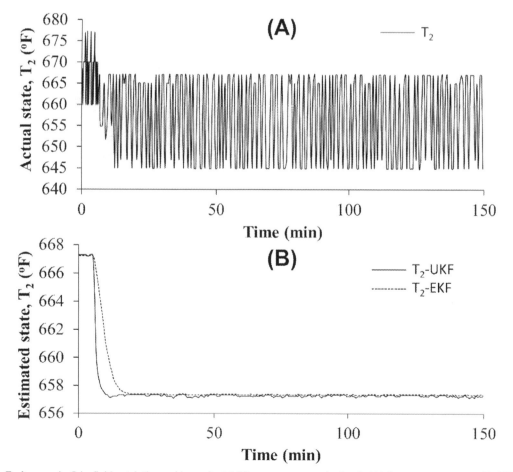

FIGURE 16.9 Fault scenario 5 in fluid catalytic cracking unit at 1.5% measurement noise level: (A) Sensor measurement, T_2 (°F); (B) estimated temperatures, T_2 (°F) by extended Kalman filter and unscented Kalman filter.

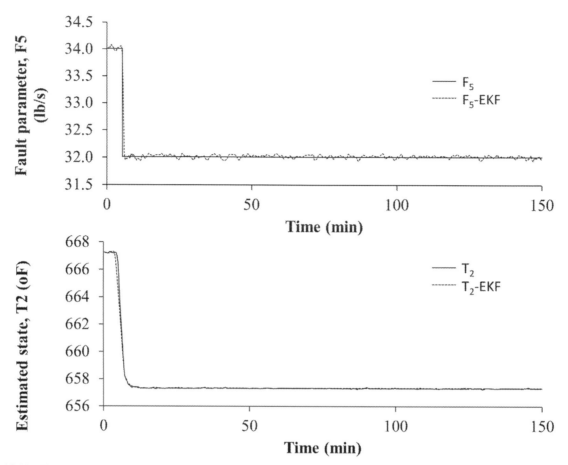

FIGURE 16.10 The estimated state, T_2 (°F) and the identified fault parameter, F_5 (lb/s) by extended Kalman filter for the fault case 5 in Section 16.6 when a measurement noise of 1.25% is considered.

studied for a combination of two fault situations in Section 16.6. The results analyzed for different fault scenarios exhibit the effectiveness of the stochastic quantitative model-based estimators, especially, the method of UKF for state estimation and fault diagnosis of FCCU.

16.10   Summary

Optimal state and parameter estimation-based fault detection and diagnosis via quantitative model-based techniques are gaining considerable importance in plant operations. Efficient diagnosis of process plants can lead to considerable incentives in terms of high performance, reliability and safety. Model-based fault diagnosis via state and parameter estimation relies on the principle that possible faults in the monitored process can be associated with specific parameters and states of the mathematical model of the process. In several processes, the full state of the plant is not directly measurable, and the process measurements are usually corrupted with noise. Therefore, reducing the noise in the measurements and reconstructing the state of the plant is a crucial issue from the process monitoring and control point of view. Designing a state estimator for fault diagnosis of a high-dimensional, nonlinear, and strongly interacting FCCU with several subsystems like feed and preheat systems, cracking riser, reactor, regenerator, air blowers, and catalyst transport system is a challenging task. FCCU involves complicated hydrodynamics and complex kinetics of both cracking and coke burning reactions. In this study, stochastic model-based estimators, namely, nonlinear discrete Kalman filter and UKF are designed and applied for state estimation and fault parameter identification of the complex FCCU plant. The results analyzed for several fault scenarios exhibit the usefulness of stochastic quantitative model-based estimators, especially, the method of UKF for state estimation and fault diagnosis of FCCU.

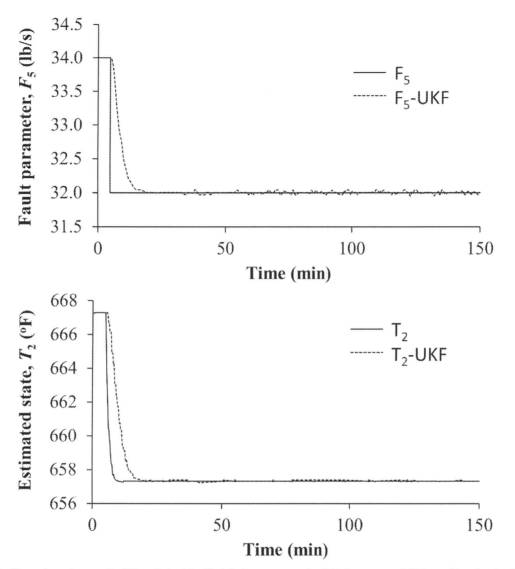

FIGURE 16.11 The estimated state, T_2 (°F) and the identified fault parameter, F_5 (lb/s) by unscented Kalman filter for the fault case 5 in Section 16.6 when a measurement noise of 1.25% is considered.

References

[1]   A.S. Willsky, A survey of design methods for failure detection in dynamic systems, Automatica 12 (1976) 601–611.

[2]   P.K. Andow, Fault detection and diagnosis in chemical and petrochemical processes, Chem. Eng. J. 20 (1980) 79.

[3]   R. Isermann, Process fault detection based on modeling and estimation methods—a survey, Automatica 20 (1984) 387–404.

[4]   M. Basseville, Detection of abrupt changes in signals and dynamical systems, Lecture Notes in Control and Information Sciences, Springer-Verlag, 1985.

[5]   M. Basseville, Detecting changes in signals and systems—a survey, Automatica 24 (1988) 309–326.

[6]   R.J. Patton, P.M. Frank, R.N. Clarke, Fault Diagnosis in Dynamic Systems: Theory and Application, Prentice-Hall, Inc., 1989.

[7]   P.M. Frank, Fault diagnosis in dynamic systems using analytical and knowledge-based redundancy, Automatica 26 (1990) 459–474.

[8]   C. Venkateswarlu, K. Gangiah, M.B. Rao, Two-level methods for incipient fault diagnosis in nonlinear chemical processes, Comput. Chem. Eng. 16 (1992) 463–476.

[9]   J.F. MacGregor, T. Kourti, Statistical process control of multivariate processes, Control Eng. Pract. 3 (1995) 403–414.

[10] V. Venkatasubramanian, R. Rengaswamy, S.N. Kavuri, K. Yin, A review of process fault detection and diagnosis, Comput. Chem. Eng. 27 (2003) 327–346.

[11] S. Park, D.M. Himmelblau, Fault detection and diagnosis via parameter estimation in lumped dynamic systems, Ind. Eng. Chem. Process Des. Dev. 22 (1983) 482–487.

[12]   D.T. Dalle Molle, D.M. Himmelblau, Fault detection in a single-stage evaporator via parameter estimation using the Kalman filter, Ind. Eng. Chem. Res. 26 (1987) 2482–2489.

[13]   C. Moler, C. Van, Loan, nineteen dubious ways to compute the exponential of a matrix, SIAM Rev. 20 (1978) 801–836.

[14]   E.A. Wan, R. Van Der Merwe, The unscented Kalman filter for nonlinear estimation, in: Proceedings of the IEEE 2000 Adaptive Systems for Signal Processing, Communications, and Control Symposium (Cat. No.00EX373), IEEE, Lake Louise, Alberta, Canada, 2000, pp. 153–158.

[15]   S.J. Julier, J.K. Uhlmann, H.F. Durrant-Whyte, A new approach for filtering nonlinear systems, in: Proceedings of 1995 American Control Conference—ACC'95, American Autom Control Council, Seattle, WA, 1995, pp. 1628–1632.

[16]   R. Van der Merwe, E.A. Wan, The square-root unscented Kalman filter for state and parameter-estimation, in: 2001 IEEE International Conference on Acoustics, Speech, and Signal Processing. Proceedings (Cat. No.01CH37221), IEEE, Salt Lake City, UT, 2001.

[17]   E.A. Wan, R. van der Merwe, The unscented Kalman filter, Kalman Filtering and Neural Networks, John Wiley & Sons, Inc., 2001, pp. 221–280.

[18]   S.J. Julier, J.K. Uhlmann, New extension of the Kalman filter to nonlinear systems, Signal Processing, Sensor Fusion, and Target Recognition VI, SPIE, 1997.

[19]   R.C. McFarlane, R.C. Reineman, J.F. Bartee, C. Georgakis, Dynamic simulator for a Model IV fluid catalytic cracking unit, Comput. Chem. Eng. 17 (1993) 275–300.

[20]   W. Ford, R. Reineman, I. Vasalos, R. Fahrig, Modeling catalytic cracking regenerators, in: NPRA Annual Meeting, 1976.

Part IV

Optimal state estimation for process control

Chapter 17

Optimal state estimator-based inferential control of continuous reactive distillation column

17.1 Introduction

Online state estimation techniques provide the estimates for the state variables defining a process based on a dynamic process model and the incoming data from process measurement sensors. Stochastic filters address the problem of estimating process states in the presence of random process disturbances and measurement errors. Stochastic filters have been applied for process state estimation in various processes, including distillation [1−7]. Reactive distillation is the process of performing chemical reaction and multistage distillation simultaneously. This combined operation is especially useful for the chemical reactions in which chemical equilibrium limits the conversion and can be advantageously used for reactions at temperatures and pressures suitable to the distillation of components. In addition, reactive distillation significantly enhances the overall conversion of certain equilibrium reactions. These advantages of reactive distillation over conventional configurations of reactors followed by separators have motivated a renewed interest in the use of reactive distillation technology to produce important chemicals. However, reactive distillation exhibits a much more complex behavior than the conventional process due to interaction of simultaneous reaction and distillation. As a result, it leads to challenging problems in design, optimization, and control.

In distillation/reactive distillation, temperature control is widely used, because temperatures can be readily measured and directly incorporated into the control scheme. Temperature controllers can satisfactorily maintain the product compositions in binary columns and certain multicomponent columns, where nonlinearities and interactions are not severe. However, this is not always satisfactory for many multicomponent distillation and reactive distillation processes, as stage temperatures do not correspond exactly to the product compositions. Furthermore, reactive distillation poses additional difficulties due to the dynamic interaction of reaction kinetics, mass transfer, and thermodynamics, and these interactions often cause counteracting influences on the process. Such an interactive and nonlinear reactive separation process can be better controlled by using composition controllers. However, composition control based on direct composition measurement is difficult because online composition analysers suffer from measurement delays and high investment and maintenance costs. Thus, inferential estimation of compositions and estimator-based control have become highly relevant for reactive distillation.

This chapter presents an extended Kalman filter (EKF)-based composition estimator for composition control of continuous reactive distillation column [8]. The estimated product compositions serve as inferential composition measurements for the classical proportional-integral (PI) controllers and genetically tuned PI controllers to control the column's desired top and bottom product compositions. The details of controller design and controller tuning, and the analysis of the results of the state estimator-based controller are presented by using a continuous metathesis reactive distillation column as a simulation test bed.

17.2 Process and the dynamic model

Here, the process considered is an olefin metathesis that finds wide applications in the petrochemical industry. In this process, the olefin is converted into lower and higher molecular weight olefin products. When simultaneous reaction and separation of top and bottom products with high purities are carried out in a reactive distillation column, the column typically exhibits very strong interactions and nonlinear behavior. Such high purity reactive separations impose constraints on the column operation, which makes the controller design crucial. Therefore, before considering the design of

Optimal State Estimation for Process Monitoring, Fault Diagnosis and Control. DOI: https://doi.org/10.1016/B978-0-323-85878-6.00009-9

an inferential control scheme, it is important to explore the characteristics of the process operation in terms of its nonlinearities, interactions, and stability. The schematic of the estimator supported composition control scheme is shown in Fig. 17.1.

17.2.1 Process description

In olefin metathesis reaction, two moles of 2-pentene react to form one mole each of 2-butene and 3-hexene,

$$2C_5H_{10} \Leftrightarrow C_4H_8 + C_6H_{12}$$
$$\text{Pentene} \quad \text{Butene} \quad \text{Hexene} \tag{17.1}$$

The normal boiling points of components in this reaction allow an easy separation between the reactant 2-pentene (310°K), top product 2-butene (277°K), and bottom product 3-hexene (340°K). The steady-state design aspects of this metathesis reactive distillation have been studied by Okasinski and Doherty [9]. The following equations give the reaction kinetics for metathesis reaction,

$$k_f = 3553.6 * e^{(-6.6/\text{RT})}$$
$$K_{eq} = 0.25 \tag{17.2}$$
$$R_3 = 0.5k_f\left[x_1^2 - (x_2x_3/K_{eq})\right]$$

where R_3 is the rate of formation of hexane, x_1, x_2, and x_3 are the mole fractions of pentene, butene, and hexene, respectively.

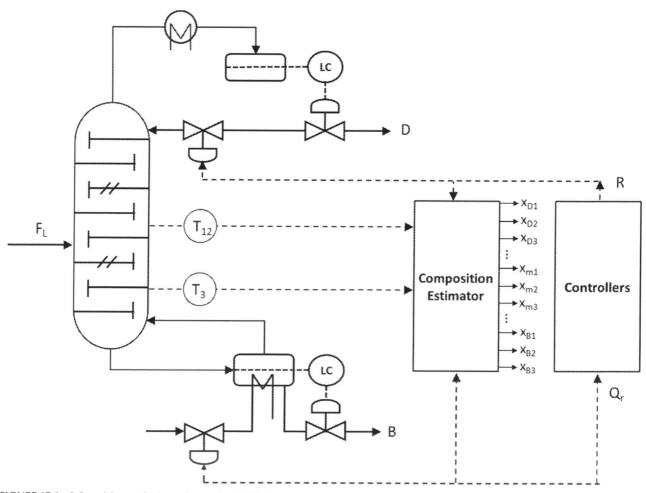

FIGURE 17.1 Inferential control scheme for reactive distillation.

17.2.2 The dynamic model of the process

The dynamic model representing the process involves mass and component balance equations with reaction terms and algebraic energy equations supported by vapor—liquid equilibrium and physical properties. The assumptions made in the formulation of the model include adiabatic column operation, negligible heat of reaction, negligible vapor holdup, liquid phase reaction, and physical equilibrium in streams leaving each stage. The assumption of negligible energy accumulation on the plates changes differential energy balance equations to the algebraic equations. More details of the model can be referred from Alejski and Duprat [10]. These model equations are also given in Section 11.2 of Chapter 11. Antoine coefficients for all the components of the system are obtained from Boublik et al. [11]. Tray hydraulic computations involve the Francis weir formula. Enthalpy calculations involve empirical relations for heat capacities, which are the functions of temperature and pressure. The coefficients of heat capacity equations for gasses are referred to from Perry's chemical engineers' handbook [12], and for liquids from Yaws [13]. The column specifications and steady-state details are given in Table 11.1. The rigorous mathematical model of the reactive distillation is solved using Euler's method with a step size of 0.001 h.

17.3 The process characteristics

To design a state estimator-based composition controller for a continuous reactive distillation column, it is important to understand the dynamic characteristics of the process. In this section, the characteristics of the process are studied in terms of its nonlinearities, interactions, and stability.

17.3.1 Nonlinearity analysis

The total degrees of freedom available with the column are reboiler heat load, reflux flow rate, and top and bottom product flow rates. The reflux drum and reboiler levels are maintained by manipulating the top and bottom product flow rates. The top and bottom product compositions can be controlled by manipulating the reflux flow rate and reboiler heat load. The multifunctional nature of reactive distillation causes strong nonlinearities, which must be assessed and analyzed concerning changes in the operating conditions. The nonlinearity of the metathesis reactive distillation can be evaluated by examining the open-loop responses of the top and bottom product compositions toward the changes in varying magnitudes of input conditions. Fig. 17.2(A) and (B) shows the response behavior of the top product (2-butene) and bottom product (3-hexene) compositions for input disturbances of different step magnitudes in both positive and negative directions. These results show that an increase of reflux flow rate increases the response of top product composition, but decreases the bottom product composition response.

Similarly, the increase of reboiler heat load increases the bottom product composition response, but decreases the response of top product composition. The wide disproportionality in the steady-state deviations of the responses in both directions for the proportional changes in input conditions indicates the violation of the principle of superposition, thus depicting the system's nonlinear behavior. From these results, it is observed that the magnitude of steady-state response deviation from 1% to 3% in R is found to be considerably different from that of 3% to 5%. A similar trend is observed for the case of changes in reboiler heat load. Thus, the violation of the principle of superposition and the highly unsymmetrical behavior of the system, as exhibited in Fig. 17.2(A) and (B) reveals the nonlinearity of the metathesis reactive distillation.

The steady-state gains of top and bottom product responses toward positive and negative step changes in inputs, reflux flow rate and reboiler heat load, are also plotted as shown in Fig. 17.2(C) and (D). These steady-state gains of top and bottom product responses for the same magnitude of positive and negative step disturbances are quite different and exhibit large variations for different step changes in input conditions. The gains have shown an increasing trend for the increase of step disturbance in a positive direction and a decreasing trend for the increase of step disturbance in negative direction. A large variation in gains is observed concerning the magnitude and direction of the disturbance, and the variation appeared to be more in the bottom loop. The higher magnitude of gains for positive step disturbances than those of negative disturbances indicate the column sensitivity toward the input conditions. These studies thus signify the presence of strong nonlinearities in the metathesis reactive distillation column.

17.3.2 Interaction and stability aspects

To design multiloop controllers for a multivariable nonlinear reactive distillation, it is important to examine the interaction effects in the column. Relative gain analysis (RGA) provides a measure for process interactions and suggests the

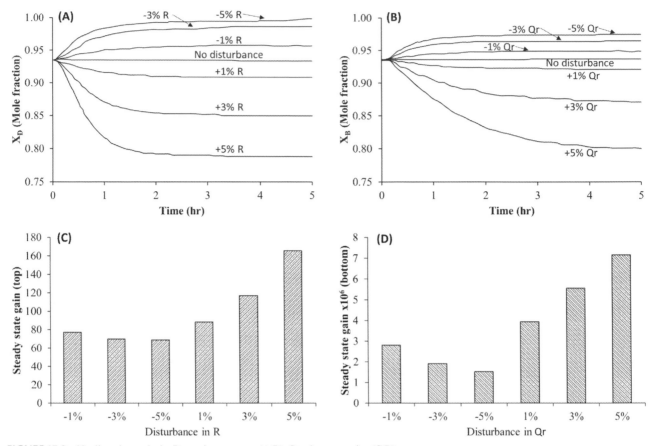

FIGURE 17.2 Nonlinearity analysis: Dynamic responses (A,B); Steady stage gains (C,D).

most effective pairing of controlled and manipulated variables [14]. The steady-state gains of the top and bottom responses are used to evaluate the RGA of the column. For a system with two loops, the steady-state gain matrix, K is given by

$$K = \begin{bmatrix} K_{11} & K_{12} \\ K_{21} & K_{22} \end{bmatrix}$$ (17.3)

where

$$K_{11} = \left(\frac{\partial x_D}{\partial R}\right)_{Q_r}; \quad K_{12} = \left(\frac{\partial x_D}{\partial Q_r}\right)_{R}; \quad K_{21} = \left(\frac{\partial x_B}{\partial R}\right)_{Q_r}; \quad K_{22} = \left(\frac{\partial x_B}{\partial Q_r}\right)_{R}$$

The RGA is defined as

$$\Lambda = \begin{bmatrix} \lambda & 1-\lambda \\ 1-\lambda & \lambda \end{bmatrix}$$ (17.4)

where,

$$\lambda = \frac{1}{1 - \frac{K_{11}K_{12}}{K_{21}K_{22}}}$$

The top product (2-butene) composition is controlled by manipulating the reflux flow rate, and the bottom product (3-hexene) composition is controlled by reboiler heat load. The RGA results of metathesis reactive distillation column evaluated for $\pm 5\%$ step changes in Q_r and R are given by

$$\Lambda = \begin{bmatrix} 7.089 & -6.089 \\ -6.089 & 7.089 \end{bmatrix}$$ (17.5)

The higher the value of λ from unity indicates the more severity of loop interactions. The RGA matrix evaluated above signifies that the reactive distillation column is highly interactive. The stability of the control structure is evaluated by using the Niederlinski Index (NI) [14], which is given by

$$\text{NI} = \frac{|K|}{\prod\limits_{j=1}^{N} K_{jj}} \tag{17.6}$$

where N is the number of control loops in the system, and $|K|$ is the determinant of steady-state gain matrix K and K_{jj} is its diagonal element. For a control structure to be stable, it is necessary that the index should be positive. For the 2×2 control of the metathesis reactive distillation column, the NI is found to be 1.8589. This fulfills the stability condition for the control structure of the metathesis reactive distillation column.

17.4 Classical proportional-integral/proportional-integral-derivative controllers for distillation column

The objective is to present a simple and effective composition control scheme for a highly interactive and nonlinear reactive distillation column. In this section, a classical PI/proportional-integral-derivative (PID) controllers used for the control of distillation column are briefly discussed. The PI and PID controllers are extensively used in many industrial control systems despite significant advancements in modern control theory. Decentralized (multiloop) PI/PID controllers are used in many multi-input multi-output (MIMO) processes, because their structure is simple, design is fast, and principle is easier to understand. These PI/PID controllers provide satisfactory performance for many MIMO systems. Various multiloop PI/PID controllers with different tuning techniques have been reported for composition and temperature control in distillation process [15–17]. Conventional PI/PID controllers have also been reported for direct composition control and estimator-based composition control of reactive distillation processes [18–21]. Recently, PI controller tuning methods based on relay feedback and step tests have been studied for reactive distillation [22] and found that relay feedback provides better results. However, these methods consider single loop tuning rules and require the process should exhibit sustained oscillations to determine the ultimate gain and period. In these methods, the detuning of controllers is performed to preserve the loop stability or meet some performance specifications. The detuning of controllers accounting nonlinearities and interactions for all disturbance conditions in complex dynamic multivariable systems are not easy to achieve.

17.5 Brief description of genetic algorithms

The basics of genetic algorithms (GA's) used in the development of the genetically tuned PI controllers for the control of metathesis reactive distillation columns are briefly presented:

The GA's are stochastic search algorithms that are based on the mechanics of natural selection and genetics. GA's can effectively handle nonconvex and nonlinear optimization problems and have a high potential for finding the global optimum. A GA aims at evolving an optimal solution by a randomized information exchange through a sequence of probabilistic transformations governed by a selection scheme biased toward high-quality solutions. Since GA's use a population of potential solutions and the probabilistic transition rules to create new solutions, they usually have a higher probability of finding the global optimum [23]. Many applications of GA's have been reported to solve different optimization problems [24–26].

The GA's encode the candidate solutions of an optimization algorithm as a string of characters which are usually binary digits. The string is called a chromosome, and the variables that are coded into a chromosome form a substring. The length of the string is usually determined according to the solution's accuracy. The genetic algorithm considers several random strings for the variables, which together form a population. GA modifies and updates the population iteratively, searching for good solutions to the optimization problem. Each iteration step is called a generation. The GA begins with a population of random strings representing the variables and then evaluates the strings fitness by using a fitness function, $F(x)$, defined by

$$F(x) = \frac{1}{1 + f(x)} \tag{17.7}$$

where $f(x)$ is an objective function. The specification of the fitness function is an important aspect of GA search because the solution of the optimization problem and the algorithm's performance depend on this function. Inspired by the "survival of the fittest" idea, the GA maximizes/minimizes the fitness value. The fitness function evaluates the fitness of every individual in the population until the end condition, such as the generation number or degree of convergence, is satisfied. If the end condition is not satisfied, the next step will be population evolution using genetic operators.

The algorithm starts with the generation of an initial population. This population contains individuals which represent initial estimates for the optimization problem. The fitness of every individual in the population is obtained through a cost function evaluation, and new individuals are generated for the next generation. Then, the population evolution for the next generation is performed using genetic operators such as selection, crossover, and mutation. Selection chooses individuals from the previous population for reproduction according to their fitness values. Individuals with better fitness values survive to form a mating pool. The essential idea of these operators is to pick up the above average strings from the current population and insert their multiple copies in the mating pool in a probabilistic manner. Crossover is applied after selection to produce new individuals by merging the chromosomes of two individuals (parents) to obtain two other individuals (children). Parent pairs are randomly chosen from the selected population, and a crossover operator is used for merging. The mutation is the last operation in which a mutation operator is applied to perform bit-wise mutation with a specified mutation probability. In mutation, better strings are created by carrying out a local search around the current solution and diversity in the population is maintained. Selection, crossover, and mutation are repeated for a fixed number of generations.

17.6 Design of genetically tuned proportional-integral controllers

Despite the abundance of PI and PID tuning techniques, the development of new methods continues unabated. The design of stable and robust PI/PID controllers for composition control of highly nonlinear and interactive processes like reactive distillation that operate in a wide range of operating conditions is a major concern. In recent years, stochastic search methods such as GA are receiving considerable interest for achieving high efficiency and locating global optimum in problem space. Due to its high potential for global optimization, GA is recognized as a powerful tool in many control-oriented applications such as parameter identification and control system design [27–30] The GA is also used to derive a feedback feed-forward temperature control system for a distillation column based on a linearized transfer function model of the column [28].

17.6.1 Controller tuning

The PI controllers for top and bottom product compositions of reactive distillation column are defined by

$$R(t) = R_0 + k_{1D}(x_D(t) - x_D{}^{set}) + k_{2D} \int_0^\tau (x_D(t) - x_D{}^{set})d\tau$$

$$Q_r(t) = Q_{r0} + k_{1B}(x_B(t) - x_B{}^{set}) + k_{2B} \int_0^\tau (x_B(t) - x_B{}^{set})d\tau \tag{17.8}$$

where $R(t)$ and $Q_r(t)$ are the manipulated reflux flow rate and reboiler heat loads with R_0 and Q_{r0} as their initial steady-state values. The tuning parameters k_{1D}, k_{2D} are the proportional and integral constants of the top loop, and k_{1B}, k_{2B} are the bottom loop's parameters. Each candidate in the GA population represents a set of these four parameters, which are referred to as genes. These genes form a chromosome that will self-evolve by reproduction, cross over, and mutation operations sequentially to generate a new population with the improved objective. The basic aim of GA is to identify a single set of optimal or near-optimal controller settings that can yield good controller performance for any disturbance condition.

17.6.2 Formulation of the objective function

The objective function for the GA search is formulated such that it quantifies the controller performance by accounting for the nonlinear dynamics of the process operating under different disturbance modes. For the PI controllers tuning problem, the x in the objective function $f(x)$ in Eq. (17.7) represents the proportional and integral parameters of both top and bottom controllers. Thus, the $f(x)$ in Eq. (17.7) is denoted as J_o, which is defined by

$$J_o = f(k_{1D}, \ k_{2D}, \ k_{1B}, \ k_{2B}) \tag{17.9}$$

For n number of process disturbance modes, J_o is evaluated as

$$J_o = \sum_{i=1}^{n} J_i \tag{17.10}$$

where J_i is the performance criterion corresponding to the ith disturbance mode, which is the weighted sum of the individual loop performances as given by,

$$J_i = w_D J_i^D + w_B J_i^B \tag{17.11}$$

Here J_i^D, J_i^B are the top and bottom loops objectives, and w_D, w_B are the weights assigned to them.

GA searches for the candidate solutions by exploring the search space stochastically at several points and evaluating J_o at each of these points. Each set of updated candidate solutions of the GA population in the search process represents the parameters of the top and bottom controllers, which are implemented on the model of the process to evaluate J_o. The function J_o quantifies the deviation of the actual response with the desired response and is minimized to establish the controller parameters optimal. Thus, the GA tuning accounts for the information concerning all kinds of process disturbances and leads to the design of a robust decentralized PI control scheme for the multivariable process.

17.6.3 Desired response specifications for genetic-algorithm tuning

The desired response for each disturbance condition is specified based on certain characteristics such as peak overshoot, undershoot, settling time, response time, and offset. The time-bound limits for the desired response trajectories are set such that any response falling within these bounds can be considered the desired response. The objective function in the GA controller tuning problem can be explicitly expressed utilizing the response deviations from the prespecified time-bound limits for each disturbance condition. The individual objectives in Eq. (17.11) are evaluated as

$$J_i^D = f(k_{1D}, k_{2D}) = \sum_{t=0}^{t_{limit}} \left(\max\left[(LL_i^D(t) - x_D(t)), \ 0 \right] + \max\left[(x_D(t) - UL_i^D(t)), \ 0 \right] \right) \Delta t$$

$$J_i^B = f(k_{1B}, k_{2B}) = \sum_{t=0}^{t_{limit}} \left(\max\left[(LL_i^B(t) - x_B(t)), \ 0 \right] + \max\left[(x_B(t) - UL_i^B(t)), \ 0 \right] \right) \Delta t \tag{17.12}$$

LL_i^D, UL_i^D, LL_i^B, and UL_i^B are the user-defined continuous functions specifying the lower and upper bounds for the top and bottom product compositions for i^{th} disturbance case, and their corresponding time limits are t_1, t_2, and t_3. These limits are specified based on studying the response characteristics of the respective disturbance condition. Due to the interactive nature of the column, any process disturbance will result in an opposite effect in both the responses. Based on the analysis of the responses resulting from the model of the process, the time-bound limits that specify the lower and upper boundaries for the responses of individual disturbance cases can be set as

$$\left. \begin{array}{l} UL(t) = x^{set} + l_1 \\ LL(t) = x^{set} - l_2 \end{array} \right\} \quad ; \quad t \leq t_1$$

$$\left. \begin{array}{l} UL(t) = x^{set} + l_3 \\ LL(t) = x^{set} - l_4 \end{array} \right\} \quad ; \quad t_1 < t \leq t_2$$

$$\left. \begin{array}{l} UL(t) = x^{set} + l_5 \\ LL(t) = x^{set} - l_6 \end{array} \right\} \quad ; \quad t_2 < t \leq t_3 \tag{17.13}$$

$$\left. \begin{array}{l} UL(t) = x^{set} + l_7 \\ LL(t) = x^{set} - l_7 \end{array} \right\} \quad ; \quad t > t_3$$

The above equation gives the general representation of the time-bound upper and lower limits for top and bottom loop responses. For i^{th} disturbance condition, these bounds for the desired responses of top and bottom loops can be denoted by UL_i^D, LL_i^D and UL_i^B, LL_i^B, respectively. The magnitudes of $l_1 - l_7$ and the time limits $t_1 - t_3$ are chosen from the shape of the desired response curve. More specifically $l_1 - l_4$ depends on the allowable peak overshoot and undershoots. The l_5, l_6 signify the decay ratio and l_7 stand for the allowable offset value. The time limits t_1, t_2, t_3 depend on the rise time, response time, and settling time of the desired response. These limits are specific to individual responses and vary for each disturbance case.

The GA procedure for computing the optimal tuning parameters of PI controllers is illustrated as a flow chart in Fig. 17.3. The tuning procedure is carried out offline using the process model. This procedure considers the dynamic state information of the process for different disturbance conditions and incorporates this information in the performance function, J_0, for evaluating the tuning parameters. The dotted block in the flow chart of Fig. 17.3 describes the objective function evaluation in the GA search procedure. For each set of updated controller parameters, J_0 evaluation requires the simulation of the closed-loop system for all disturbance conditions. For instance, in run i corresponding to the ith disturbance case, the top and bottom responses are generated, and the individual objectives J_i are evaluated.

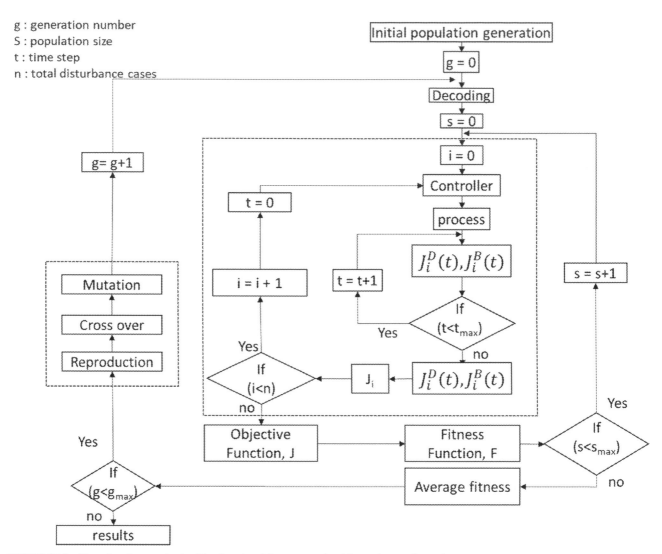

FIGURE 17.3   Flow chart for genetic algorithm-based multiloop proportional-integral controller tuning.

The J_0 is then obtained from the weighted summation of these individual objectives. For illustration, consider the response curves and time-bound limits in Figs. 17.4 and 17.5 that represent the cases of a setpoint change in top product composition and load disturbance, respectively. Fig. 17.4 illustrate the setpoint tracking of top loop response and the coupling rejection of bottom loop response, respectively. The shaded area formed by parts of the response curve that lie outside the specified bounds contributes to the objective function, which has to be minimized by the GA search procedure to determine the optimal parameters of the controllers. The shaded area shown in Fig. 17.5 represents the top and bottom loop objectives J_{Di} and J_{Bi} for the respective disturbance conditions. These objectives provide the quantification of how far the actual response from the desired response is over the specified period. The parameters of the controllers are to be determined such that the process output response should lie within the bounds specified for the desired response for any disturbance condition. Since the controller design accounts for the nonlinear dynamics of the process to establish a unique set of controller parameters under different disturbance conditions, the controller parameters are not specific to a particular disturbance and are valid for all types of disturbances.

17.7   Design of composition estimator

An EKF is designed to obtain the instantaneous composition estimates from the temperature measurements, serving as inferential measurements for composition control of continuous reactive distillation column.

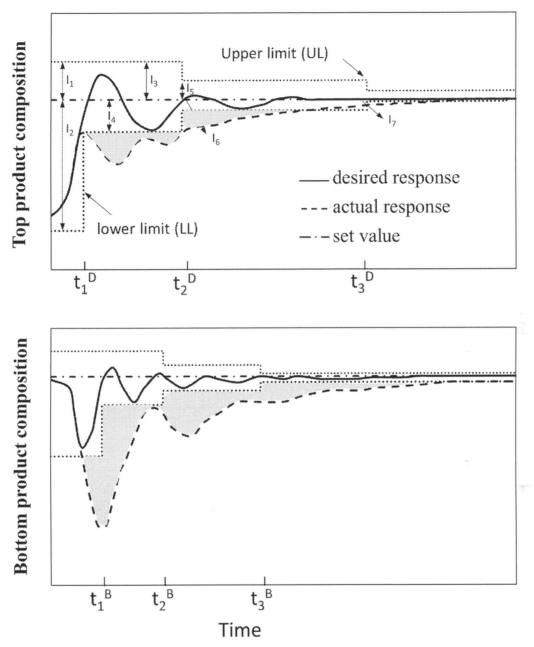

FIGURE 17.4   Performance evaluation for genetic algorithm tuning for x_D setpoint change.

The general process and measurement representations, and noise relations are given in Section 4.4.1 of Chapter 4. The EKF algorithm given in Section 4.4.3 of Chapter 4 is implemented for nonlinear state estimation for a continuous reactive distillation column.

The state vector for the metathesis reactive distillation column is given by

$$x = \begin{bmatrix} x_{b,1} & x_{b,2} & x_{1,1} & x_{1,2} \cdots & x_{Nt,1} & x_{Nt,2} & x_{d,1} & x_{d,2} \end{bmatrix}^T \tag{17.14}$$

Composition estimator design based on a rigorous mathematical model of multicomponent reactive distillation is not suitable for a realistic situation, since it is difficult to obtain the measured values of liquid and vapor flow rates and tray holdups with time. Moreover, implementing a rigorous model-based composition estimator for high dimensional reactive distillation also requires more computational effort. Therefore, a simplified model that assumes constant tray and reflux drum holdups, and constant vapor and liquid flow rates is considered for designing the composition estimator. The elements $f_{i,j}$ of the F matrix in Eq. (4.27) of the estimation algorithm are obtained by taking the partial derivatives

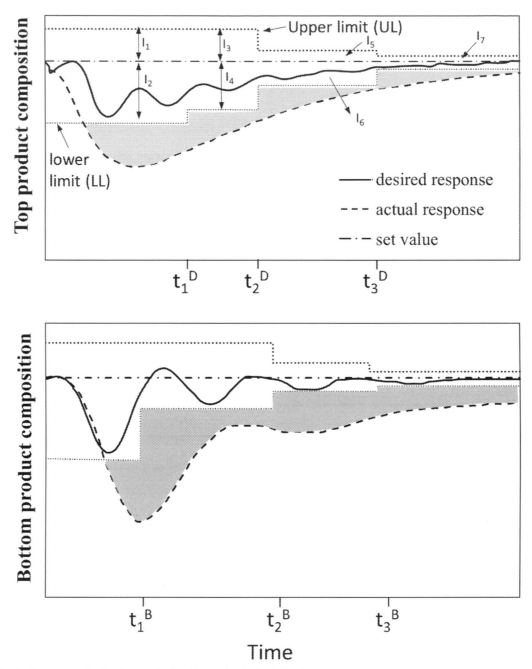

FIGURE 17.5 Performance evaluation for genetic algorithm tuning for load disturbance.

of the component balance equations concerning the state vector. The dynamic model of the continuous reactive distillation column is described in Section 11.2 of Chapter 11. The equation that gives the temperature model of a measurement tray is

$$T_m = \frac{b_j}{a_j - \log\left(\dfrac{P^{sat}_{m,j} P_m}{\sum_1^{Nc} P^{sat}_{m,j} x_{m,j}}\right)} - c_j \qquad (17.15)$$

where a, b, and c are coefficients in vapor−liquid equilibrium relation. The elements of the H matrix in Eq. (4.31), of the estimation algorithm are obtained by taking the partial derivatives of Eq. (17.15) with respect to the state vector.

The EKF design parameters, namely, the initial state covariance matrix, P_o, process noise covariance matrix, Q, and measurement noise covariance matrix, R, are selected appropriately. More details concerning the algorithm can be referred to elsewhere [31].

17.8 Analysis of results

The design and implementation results of the GA tuned PI strategy for inferential control of continuous metathesis reactive distillation column are discussed as follows.

The GA search procedure described in the earlier section is employed to optimize the tuning parameters of the top and bottom controllers k_{1D}, k_{2D}, k_{1B}, and k_{2B}. The optimization involves binary coding for the controller tuning parameter vector with a string length of 48 bits accounting for four substrings, each of length 12 bits. The initial population size is kept as 40 and the generations considered are 200. Based on the steady-state gains of the process, it has been observed that the proportional and integral parameters for the top loop are in the order of 10^2 and 10^3, respectively, and both these parameters for the bottom loop are in the order of 10^9. These parameters are appropriately scaled in the range of [0−10] to facilitate a uniform search space for GA implementation. The crossover and mutation probabilities are selected as 0.85 and 0.005, respectively. For every GA evaluation, the population string parameters are decoded and multiplied by their respective scaling factors and implemented on the process model to compute the objective function. The disturbance cases considered for objective function evaluation include steady-state operation, setpoint changes of top and bottom loops, load disturbances such as increase and decrease in feed flow rate. The time-bound limits for the desired top and bottom responses are realized by analyzing the response characteristics for each disturbance condition. On considering equal importance for both top and bottom products of metathesis reactive distillation column, the weightings w_B and w_D are assigned to be 0.5. Fig. 17.6 depicts the convergence of the fitness function, objective function, and the scaled GA controller parameters concerning the number of generations. The results of Fig. 17.6 show that the integral parameters of top and bottom loops have converged in less than 100 generations. In contrast, the convergence of the proportional parameters took more than 120 generations. The GA evaluated

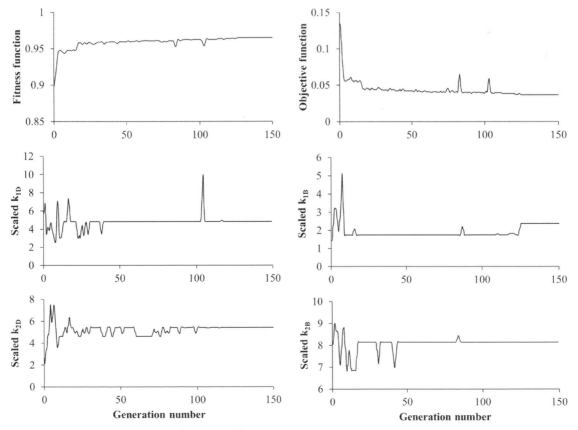

FIGURE 17.6 Genetic algorithm tuning results for the controllers.

proportional and integral parameters of the top loop are 487.67 and 5431.01, and those of the bottom loop are 23,833,940 and 81,587,300.

The composition vector to be estimated in metathesis reactive distillation is of dimension $[28 \times 1]$. The state estimator supported by the simplified dynamic model of reactive distillation considers the temperature measurements as its inputs and provides the estimates of compositions instantly. The temperature measurements of the reactive distillation column are obtained through bubble point calculation procedure by solving the rigorous model of the process. To reflect the real situation, the temperature data of every sampling instant is corrupted with random Gaussian noise of zero mean and a standard deviation of $0.2°C$. For the metathesis reactive distillation column considered in this study, the optimal temperature measurements that serve as inputs to the estimator are established using an empirical observability Grammian based methodology [32]. Based on this methodology, the temperatures of Trays 3 and 12 are found to be the best measurements, and these are used for composition estimation in the metathesis reactive distillation column. The elements of the matrices P_o, Q, and R involved in the estimator are heuristically selected, and these matrices with their respective diagonal elements as 0.0005, 0.0005, and 5.0 are found to provide effective estimator performance. The estimator is studied towards the effect of different disturbances, and the results in Fig. 17.7 illustrate the state tracking efficiency of the estimator for multiple-step changes in R and Q_r under open-loop conditions.

The estimated compositions are used as inferential measurements to the GA tuned PI controllers. The composition of the top product, butene, is controlled by manipulating the reflux flow rate, and the bottom product, hexene, is controlled by manipulating the reboiler heat load. The liquid levels in the reboiler and reflux drum are maintained by adjusting the top and bottom product flow rates using conventional PI controllers. The proportional and integral terms for both the level controllers are selected as -1250 and -10, respectively. Fig. 17.8 compares the inferential measurement tracking efficiency of the GA tuned PI controllers with the actual compositions for simultaneous setpoint changes in both top and bottom product compositions.

The composition control results of the GA tuned PI controllers are compared with the results of conventional controllers. The conventional tuning methods considered are based on step test and relay feedback test [22]. In the step test method, the ultimate gain (K_u) and the ultimate period (P_u) for each loop are obtained from the respective open-loop

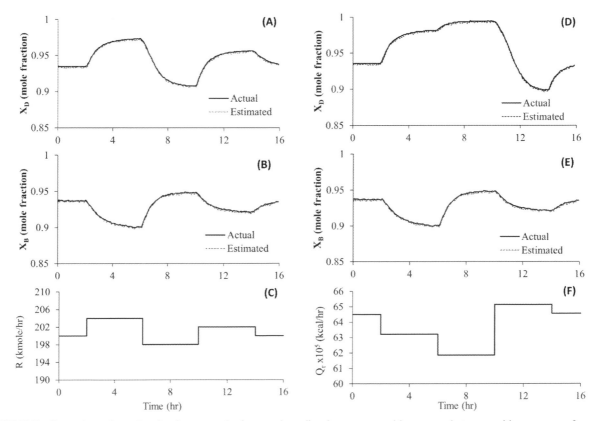

FIGURE 17.7 Comparison of actual and estimator results for open-loop disturbances: top and bottom product composition responses for multiple step changes in reflux flow rate (A, B, C); top and bottom product composition responses for multiple step changes in reboiler heat load (D, E, F).

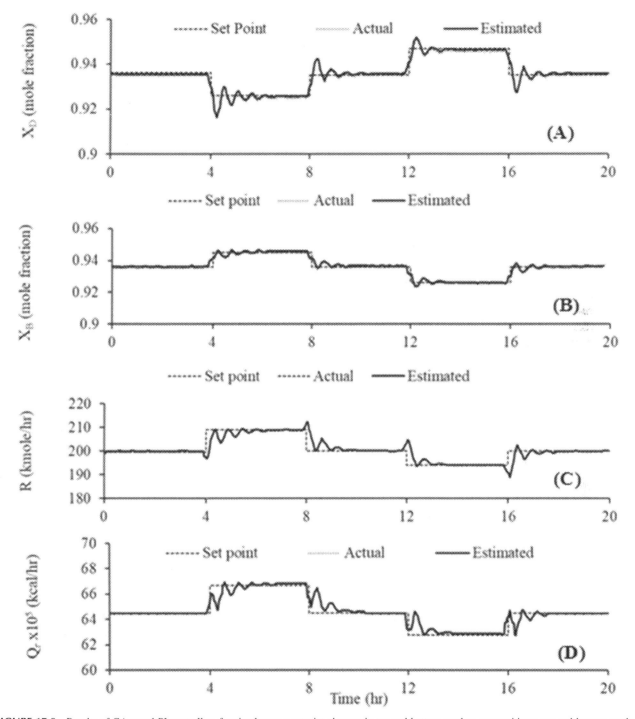

FIGURE 17.8 Results of GA tuned PI controllers for simultaneous setpoint changes in top and bottom product compositions: top and bottom product responses (A, B); manipulated variables (C, D).

responses for a step-change in feed flow rate, whereas, in the relay feedback method, the values of K_u and P_u for each loop are obtained from the sustained oscillations caused by open-loop relay input disturbance. These values of K_u and P_u are then used to evaluate the Ziegler Nichol's settings (K_{zn}, T_{zn}) for each control loop. These settings are further detuned by trial and error by analyzing the process responses to eliminate the loop interactions. The values of K_u, P_u, K_{zn}, and T_{zn} for both these methods and the detuning factors and controller parameters are given in Table 17.1.

TABLE 17.1 Controller settings.

Parameters	Step test		Relay feedback test	
	Top loop	Bottom loop	Top loop	Bottom loop
K_u	1850	1.0×10^{11}	1845.3	5.47×10^9
P_u	0.3833	0.1825	0.3881	0.1610
K_{zn}	832.50	45.0×10^9	830.37	2.46×10^9
T_{zn}	0.3194	0.1521	0.3221	0.1336
Detuning factor	1.4	300	1.1	20
Proportional	594.64	150000000	754.88	123185000
Integral	1329.79	3277778	2113.88	46092500

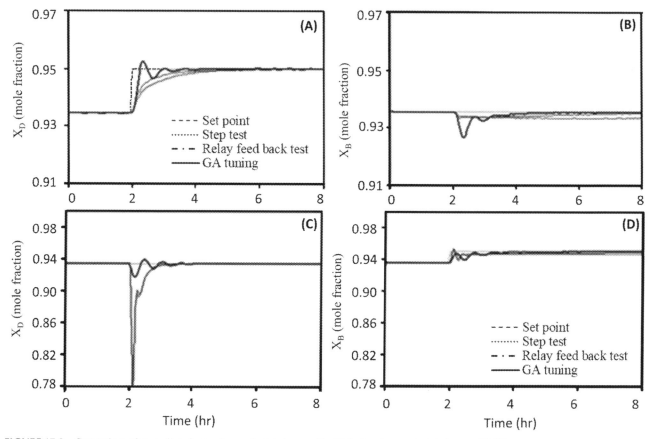

FIGURE 17.9 Comparison of controllers for top loop setpoint changes (A, B); bottom loop setpoint changes (C, D).

Fig. 17.9 compares the servo performance of the estimator based GA tuned controllers with the conventionally tuned controllers for both top and bottom loops. The subplots (A) and (B) corresponding to x_D setpoint change in Fig. 17.9 show that the GA tuned PI controllers have exhibited faster convergence than the conventionally tuned controllers. The subplots (C) and (D) corresponding to x_B setpoint change in Fig. 17.9 indicate the inferior performance of the conventional controllers in tackling the multivariable nonlinear interactions. In contrast, the GA tuned controllers have exhibited much better performance. Fig. 17.10 compares the regulatory performance of the GA tuned and conventionally tuned controllers for both the top and bottom loops for different disturbance conditions such as step increase and step decrease in feed flow rate, and step decrease in feed composition. In all these disturbance cases, the

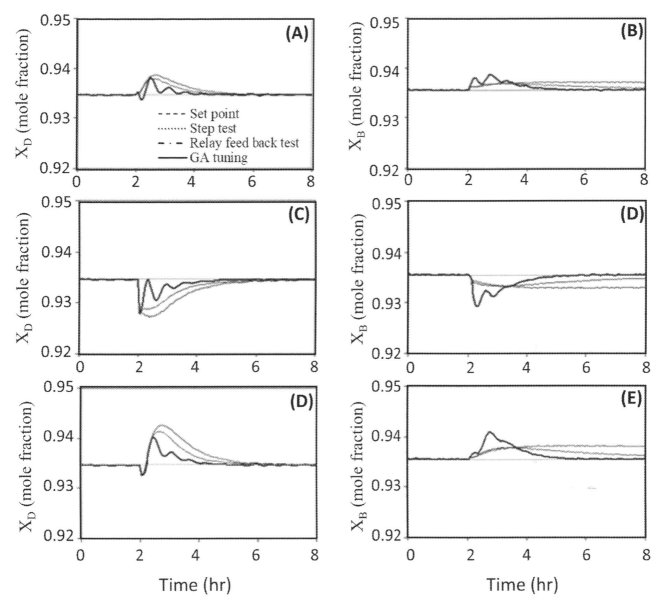

FIGURE 17.10 Comparison of controllers for load disturbances: 3% increase in F_L (A, B); 3% decrease in F_L (C, D); 5% of each x_{F2}, x_{F3} as impurities in feed (E, F).

top loop results show the better performance of the GA tuned PI controller by means of less overshoot/undershoot and faster convergence compared to the conventional controllers. The bottom loop responses show that even though the GA tuned controller has resulted in a small overshoot/undershoot, it provides much faster convergence than the conventional controllers. From these results, it is observed that the controller tuned by relay feedback test has shown very slow convergence, whereas the controller tuned by step test has resulted in a large offset. The integral absolute error (IAE) results evaluated for different disturbance cases also indicate the better performance of the GA tuned PI controllers.

The robustness of the controllers is evaluated by considering uncertainties in kinetic and thermodynamic model parameters. The uncertainties considered in thermodynamic parameters include step increase and decrease of the Antoine model coefficients, whereas ramp changes are considered as uncertainties in kinetic model parameters. These results have shown the robustness of the GA tuned PI controller in the presence of uncertainties in model parameters. The better performance of the estimator based GA tuned PI controllers can be attributed to the reasoning that these controllers are designed by accounting for the dynamic process information concerning the nonlinearities and process interactions of both the loops for different disturbances.

17.9 Summary

This chapter presents an optimal state estimator-based GA tuned PI control strategy for inferential control of continuous reactive distillation column. The controller tuning problem of this multivariable process is resolved as an optimization problem. Multiloop PI controllers are designed by exploiting the powerful global search features of GA. The GA tuning accounts for the multivariable interactions and nonlinear dynamics of the process and provides a unique set of parameters for the control system that is robust to all kinds of disturbances. An estimator is designed to provide the compositions which serve as inferential measurements to the multiloop PI controllers. The performance of the state estimator-based GA tuned decentralized control scheme is evaluated by applying to a metathesis reactive distillation column, and the results are compared with conventionally tuned PI controllers. The results demonstrate the better regulatory, and servo performance of the GA tuned inferential control scheme for composition control of reactive distillation column.

References

[1]   R. Baratti, A. Bertucco, A. Da Rold, M. Morbidelli, A composition estimator for multicomponent columns—development and experimental test on ternary mixtures, Chem. Eng. Sci. 53 (1995) 3601–3612.

[2]   N. Regnier, G. Defaye, L. Carlap, C. Vidal, Software sensor based control of exothermic batch reactors, Chem. Eng. Sci. 51 (1996) 5125–5136.

[3]   D.R. Yang, K.S. Lee, Monitoring of a distillation column using modified extended Kalman filter and a reduced order model, Comput. Chem. Eng. 21 (1997) 565–570.

[4]   M. Soroush, State and parameter estimations and their applications in process control, Comput. Chem. Eng. 23 (1998) 229–245.

[5]   R.M. Oisiovici, S.L. Cruz, State estimation of batch distillation columns using an extended Kalman filter, Chem. Eng. Sci. 55 (2000) 4667–4680.

[6]   V.M. Becerra, P.D. Roberts, G.W. Griffiths, Applying the extended Kalman filter to systems described by nonlinear differential-algebraic equations, Control Eng. Pract. 9 (2001) 267–281.

[7]   C. Venkateswarlu, S. Avantika, Optimal state estimation of batch distillation, Chem. Eng. Sci. 56 (2001) 5771–5786.

[8]   C. Sumana, C. Venkateswarlu, Genetically tuned decentralized PI controllers for inferential control of reactive distillation, Ind. Eng. Chem. Res. 49 (2010) 1297–1311.

[9]   M.J. Okasinski, M.F. Doherty, Design method for kinetically controlled, staged reactive distillation columns, Ind. Eng. Chem. Res. 37 (1998) 2821–2834.

[10] K. Alejski, F. Duprat, Dynamic simulation of the multicomponent reactive distillation, Chem. Eng. Sci. 51 (1996) 4237–4252.

[11] V.F. Boublik, E. Hala, The Vapor Pressures of Pure Substances, Elsevier, New York, 1973.

[12] R.H. Perry, D.W. Green, Perry's Chemical Engineers' Handbook, 7th ed., McGraw-Hill, New York, 1997.

[13] C.L. Yaws, Calculate liquid heat capacity, Hydrocarbon Process. 70 (1991) 73.

[14] A. Niderlinski, A. Heuristic, Approach to the design of linear multivariable control systems, Automatica 9 (1971) 691–701.

[15] H.P. Huang, J.C. Jeng, C.H. Chiang, W. Pan, A direct method for multi-loop PI/PID controller design, J. Proc. Control 13 (2003) 769A.

[16] T.N.L. Vu, M. Lee, Multi-loop PI controller design for enhanced disturbance rejection in multi-delay processes, Int. J. Math. Comput. Simulation 2 (2008) 89.

[17] W.L. Luyben, Practical Distillation Control, Van Nostrand Reinhold, 1992.

[18] M.A. Al-Arfaj, W.L. Luyben, Design and control of an olefin metathesis reactive distillation column, Chem. Eng. Sci. 57 (2002) 715.

[19] M.G. Sneesby, M.O. Tade, T.N. Smith, Two-point control of a reactive distillation column for composition and conversion, J. Proc. Control 9 (1999) 19–31.

[20] S.H. Hung, M.J. Lee, Y.T. Tang, Y.W. Chen, I.K. Lai, W.J. Hung, et al., Control of different reactive distillation configurations, AIChE J 52 (2006) 1423–1440.

[21] M.J. Olanrewaju, M.A. Al-Arfaj, Estimator-based control of reactive distillation system: application of an extended Kalman filtering, Chem. Eng. Sci. 61 (2006) 3386–3399.

[22] Y.D. Lin, H.P. Huang, C.C. Yu, Relay feedback tests for highly nonlinear processes: Reactive distillation, Ind. Eng. Chem. Res. 45 (2006) 4081–4092.

[23] D.E. Goldberg, Genetic Algorithms in Search, Optimization and Machine Learning, Reading, MA, Addison-Wesley, 1989.

[24] L. Chen, S.K. Nguang, X.D. Chen, X.M. Li, Modeling and optimization of fed-batch fermentation processes using dynamical neural networks and genetic algorithms, Biochem. Eng. J. 22 (2004) 51–61.

[25] C. Onnen, R. Babuska, J.M. Kaymark, J.M. Sousa, H.B. Verbruggen, R. Isermann, Genetic algorithms for optimization in predictive control, Control. Eng. Pract. 5 (1997) 1363–1372.

[26] C. Venkateswarlu, J.S. Eswari, Stochastic Global Optimization Methods and Applications to Chemical, Biochemical, Pharmaceutical and Environmental Processes, Elsevier, 2020.

[27] P. Wang, D.P. Kwok, Optimal design of PID process controllers based on genetic algorithms, Control Eng. Pract. 2 (1994) 641–648.

[28] C. Venkateswarlu, A.D. Reddy, Nonlinear model predictive control of reactive distillation based on stochastic optimization, Ind. Eng. Chem. Res. 47 (2008) 6949–6960.

[29] D.R. Lewin, A. Parag, A constrained genetic algorithm for decentralized control system structure selection and optimization, Automatica 39 (2003) 1801−1807.

[30] S. Jaiswal, C.S. Kumar, M.M. Seepana, G.U.B. Babu, Design of fractional order PID controller using genetic algorithm optimization technique for nonlinear system, Chemical Product and Process Modeling 15 (2) (2020) 20190072.

[31] C. Venkateswarlu, B. Jeevan Kumar, Composition estimation of multicomponent reactive batch distillation with optimal sensor configuration, Chem. Eng. Sci. 61 (2006) 55605574.

[32] C. Sumana, C. Venkateswarlu, Optimal selection of sensors for state estimation in a reactive distillation process, J. Proc. Control 19 (2009) 1024−1035.

Chapter 18

Optimal state estimation for nonlinear control of complex dynamic systems

18.1 Introduction

Optimal state estimation and estimator-based control have received significance in the control of complex nonlinear dynamical systems that are characterized by input−output multiplicities, parametric sensitivity, nonlinear oscillations, and chaos. Optimal state estimation plays a prominent role for the control of such complex dynamical systems due to the nonavailability of certain key process variables as measurable quantities.

The objective of this chapter is to present a state estimator-supported nonlinear controller, namely, inferential nonlinear internal model controller (NIMC), for the control of complex nonlinear dynamical systems. Nonlinear control is a class of advanced control in which the nonlinear process model serves as the basis for controller design. This type of control is expected to provide improved control performance since the control structure preserves the nonlinearities of the real process in the form of its mathematical model. The generic model control (GMC) introduced by Lee and Sullivan [1], the globally linearizing control (GLC) proposed by Kravaris and Chung [2], and the NIMC proposed by Henson and Seborg [3] are the prominent nonlinear model-based controllers. The NIMC proposed by Henson and Seborg is different from the GMC of Lee and Sullivan and the GLC of Kravaris and Chung in that it includes implicit integral action in the control structure by using the difference between the process output and model output as a feedback signal. A nonlinear filter employed in NIMC provides a tuning parameter that can be adjusted for process/model mismatch. The NIMC approach-based controllers have been reported earlier for nonlinear systems, whose dynamics are not so complicated [4,5]. Since reasonably accurate mathematical models have been developed to predict the complicated dynamics of nonlinear systems, the design of nonlinear controllers based on those models is a useful alternative for the control of complex dynamical systems.

The NIMC controller incorporates the nonlinear model structure and the estimator dynamics in its formulation. The estimator uses the mathematical model of the process in conjunction with the known process measurements to estimate the states. The unmeasured process states that capture the fast-changing nonlinear dynamics of the process provided by the estimator are incorporated in the controller. The design and implementation of the estimator-based NIMC strategy are investigated by choosing two typical continuous chemical and polymerization reactors that present challenging operational and control problems due to their complex open-loop dynamics, such as input−output multiplicities, parametric sensitivity, nonlinear oscillations, and chaos. Further, the comparison of the estimator-based NIMC strategy is made with conventional controllers.

18.2 Optimal state estimation and estimator-based control of chaotic chemical reactor

The design and implementation of the estimator-supported NIMC strategy are carried out by considering a continuous nonisothermal nonlinear chemical reactor.

18.2.1 Chaotic chemical reactor and its mathematical model

The reactor system and its mathematical model, along with the data and notation, are given in Section 9.2.1 of Chapter 9. The reactor system described by Eqs. (9.1)−(9.3) exhibits multistationary behavior, oscillations, and chaos for the parameter values given in Table 9.1. More details can be referred in Section 9.2 of Chapter 9.

Optimal State Estimation for Process Monitoring, Fault Diagnosis and Control. DOI: https://doi.org/10.1016/B978-0-323-85878-6.00015-4

18.2.2 Estimator design and estimation results

Optimal state estimation of nonlinear dynamical continous stirred tank reactor (CSTR) using extended Kalman filter (EKF) is detailed in Section 9.3. Sections 9.3.1 and 9.3.2 of Chapter 2 present the design of the state estimator and state estimation results for the chaotic chemical reactor.

18.2.3 Controller algorithm

The general form of a single input single output system with state space description is:

$$\dot{x} = f(x) + g(x)u \tag{18.1}$$

$$y = h(x) \tag{18.2}$$

where x is the vector of states, u is the manipulated input, y is the measured output, $f(x)$ and $g(x)$ are vector functions and $h(x)$ is the scalar function. The relative order of the system defined by Eqs. (18.1) and (18.2) is expressed by the following relations:

$$L_g L_f^k h(x) = 0, \ \ k < r - 1 \tag{18.3}$$

$$L_g L_f^{r-1} h(x) \neq 0 \tag{18.4}$$

where r represents the relative order of the system and $L_f h(x)$ is the Lie derivative of the scalar function $h(x)$ with respect to the vector function $f(x)$ with $L_f^0 h(x) = h(x)$. Similarly, higher order Lie derivatives as well as the Lie derivative of the scalar function $h(x)$ with respect to the vector function $f(x)$, and then with respect to the vector function $g(x)$ can be defined. The relative order defined by Eqs. (18.3) and (18.4) represents the number of times the output y must be differentiated with respect to time so that the input u appears explicitly.

A general form of the control law for NIMC is written as

$$u = p(x) + q(x)v \tag{18.5}$$

where v is the new input. The NIMC approach includes an implicit integral action by using the difference between the plant and model outputs as a feedback signal:

$$e = y_d - (y - \tilde{y}) \tag{18.6}$$

where $\tilde{y}$ is the process model output. The feedback signal simplifies to $e = y_d$ with perfect model assumption. A nonlinear filter is employed in NIMC which provides a tuning parameter that can be adjusted for process/model mismatch. The new input v for NIMC is defined as

$$v = -\tau_r L_f^{r-1} h(x) - \tau_{r-1} L_f^{r-2} h(x) - \ldots - \tau_1 h(x) + \tau_1 e \tag{18.7}$$

where τ_i are controller tuning parameters, r is the relative order and $e = y_d$. According to this approach, the control law for the system is given by [3]:

$$u = \frac{v - L_f^r h(x)}{L_g L_f^{r-1} h(x)} = p(x) + q(x)v \tag{18.8}$$

18.2.4 Controller design

The NIMC design for the complex dynamic chemical reactor considered in this study is briefed as follows:

The $f(x)$ and $g(x)$ in Eq. (18.1) can be written by omitting the load disturbances from the dimensionless mass and energy balances in Eqs. (9.1)–(9.3) given in Section 9.2.1 of Chapter 9:

$$f(x) = \begin{bmatrix} 1 - x_1 - Dax_1 \exp\left(\dfrac{x_3}{1 + \varepsilon_A x_3}\right) \\ -x_2 + Dax_1 \exp\left(\dfrac{x_3}{1 + \varepsilon_A x_3}\right) - DaSx_2 \exp\left(\dfrac{kx_3}{1 + \varepsilon_A x_3}\right) \\ -x_3 + BDax_1 \exp\left(\dfrac{x_3}{1 + \varepsilon_A x_3}\right) - BDa\alpha Sx_2 \exp\left(\dfrac{kx_3}{1 + \varepsilon_A x_3}\right) - \beta x_3 \end{bmatrix} \tag{18.9}$$

$$g(x) = \begin{bmatrix} 0 \\ 0 \\ \beta \end{bmatrix} \tag{18.10}$$

The controlled output in Eq. (18.2) is given by

$$y = h(x) = x_3 \tag{18.11}$$

From Eqs. (18.3) and (18.4), the relative order of the system is evaluated as one. Thus, for $r = 1$, the new input v in Eq. (18.7) is obtained as

$$v = \tau_1(x_3^s - x_3) \tag{18.12}$$

On computing the Lie derivatives and substituting them along with the new input in Eq. (18.8) leads to the following control law

$$u_t = \frac{\tau_1(x_3^s - x_3) - \left(-x_3 + BDax_1\exp\left(\frac{x_3}{1+\varepsilon_A x_3}\right) - BDa\alpha Sx_2\exp\left(\frac{kx_3}{1+\varepsilon_A x_3}\right) - \beta x_3\right)}{\beta} \tag{18.13}$$

where τ_1 is the NIMC tuning parameter. The notation and terminology for various terms of the reaction system are explained in Section 9.2.1 of Chapter 9. The schematic of inferential nonlinear NIMC is shown in Fig. 18.1.

18.2.5 Analysis of results

The temperature data used for state estimation are obtained through numerical integration of nonlinear dynamical reactor model equations, Eqs. (9.1)–(9.3) of Chapter 9, using Gear's method with a sampling time of 0.0001 units. A model-based state estimator, EKF, is designed as explained in Section 9.3.1, and applied to estimate the states x_1, x_2, and x_3 using the temperature measurement of the reactor. The performance of the EKF estimator is evaluated by considering different cases of process parameter values given in Table 9.1 of Chapter 9. The state estimation results are discussed in Section 9.3.2 of Chapter 9. The estimated states are found in close resemblance with the actual states for all the cases studied.

The state estimator-supported NIMC strategy is designed and applied to control the chemical reactor for the cases corresponding to an unstable steady state in the multiplicity region, an unique unstable steady state responsible for limit cycle oscillations, and an unique unstable steady-state responsible for chaotic motion [6,7]. The controller parameter involved in the NIMC is tuned as $\tau_1 = 14.0$. The results are also compared with those of a modified proportional integral derivative (PID) controller with a feedback mechanism, which has the form given by [8],

$$\frac{du_t}{dt} = \varepsilon_0 f'(t)e + \varepsilon_0 f(t)\frac{de}{dt} + \varepsilon_1 \int edt \tag{18.14}$$

where $f(t) = t$ and $f'(t) = 1$. The tuning parameters of the PID controller $(\varepsilon_0, \varepsilon_1)$ are selected as unity. A sampling time of 1 s is used for the implementation of the estimator and controller. For parameter values in Set I (Table 9.1 of Chapter 9), the system exhibits multiple steady-state behaviors. The controller goal is to shift the process operating at an arbitrary point $(x_1^0 = 0.04$,

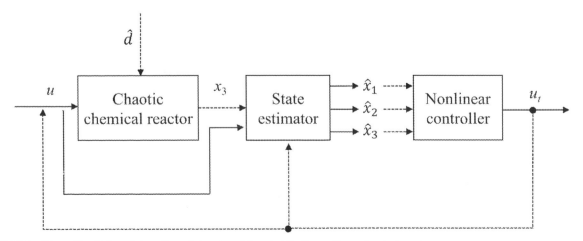

FIGURE 18.1 Schematic of inferential nonlinear internal model control for the nonlinear chemical reactor.

$x_2^0 = 0.9$, $x_3^0 = 5.75$) to the set condition in Table 9.1 and maintain it at that state. Fig. 18.2 shows the process output and controller output plots of NIMC and PID controller, respectively, for this condition. For parameters in Set II (Table 9.1 of Chapter 9), the system exhibits sustained oscillatory behavior (stable limit cycle) and the controllers are also required to regulate the trajectory at the unique unstable steady state, which is the desired condition. Fig. 18.3 shows the results of the NIMC and PID controller that realize the set condition in Table 9.1 from an arbitrarily selected initial state ($x_1^0 = 0.08$, $x_2^0 = 0.103$ and $x_3^0 = 3.654$). For parameters in Set III (Table 9.1 of Chapter 9), the system exhibits chaotic behavior. In this objective, the controller has to stabilize the chaotic trajectory exactly at the corresponding unique unstable steady state. To realize this objective, the controllers are employed with the same initial condition as used for Set II to satisfy the desired condition of Set III given in Table 9.1. The process output and controller output plots of NIMC and modified PID controller corresponding to this objective

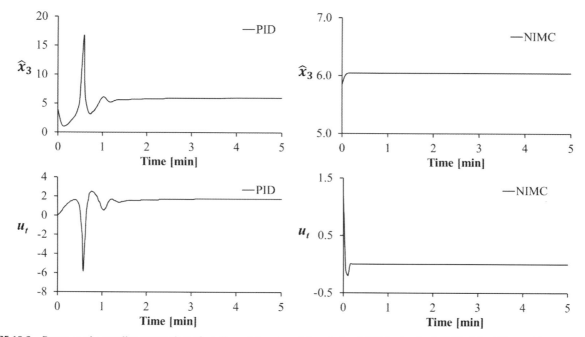

FIGURE 18.2   Process and controller output plots of nonlinear internal model control and PID for the Set I (Table 9.1 of Chapter 9).

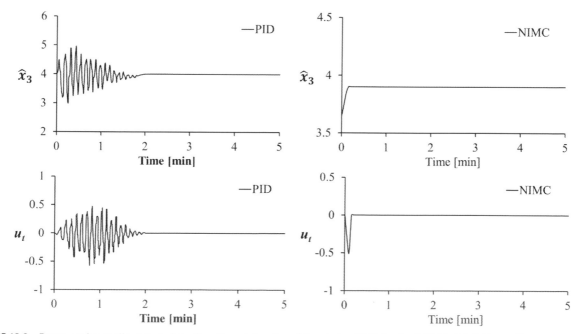

FIGURE 18.3   Process and controller output plots of nonlinear internal model control and PID for Set II (Table 9.1 of Chapter 9).

have shown the better performance of NIMC. These results indicate that the NIMC satisfies the desired objectives without any offset. The performance of the controllers is also evaluated by applying them for controlling the system at the desired conditions in the presence of deterministic and stochastic load disturbances. A deterministic disturbance of 1.0 and a stochastic load disturbance generated through random Gaussian noise of zero mean and a standard deviation of 0.25 are considered to represent the d_3 in temperature measurement. The results of the controllers for stabilizing the system at an unstable steady state responsible for chaotic motion for different disturbance conditions are shown in Fig. 18.4. These results indicate the better performance of the NIMC strategy in the presence of either type of load disturbances. The results of the controllers are also evaluated for a series of step changes in the setpoints of the controlled variables. The results in Fig. 18.5 show the

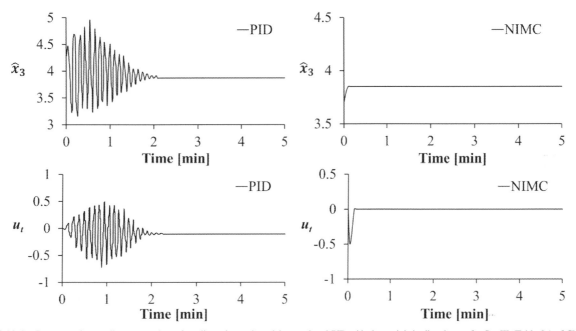

FIGURE 18.4 Process and controller output plots of nonlinear internal model control and PID with deterministic disturbance for Set III (Table 9.1 of Chapter 9).

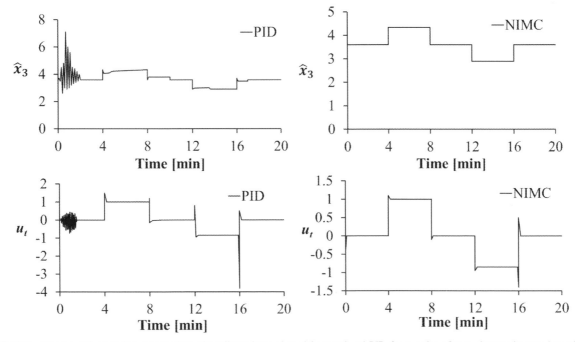

FIGURE 18.5 Process and controller output plots of nonlinear internal model control and PID for a series of step changes in setpoints of the controlled variables for Set III.

process output and controller output plots of NIMC and modified PID controller for parameters in Set III for a series of $\pm 20\%$ step changes in the set condition in Table 9.1. These results show the better performance of the inferential NIMC for chaotic chemical reactor.

18.3 Optimal state estimation and estimator-based control of homopolymerization reactor

The design and implementation of the estimator-supported NIMC strategy are carried out by considering a continuous nonlinear dynamical homopolymerization reactor.

18.3.1 Homopolymerization reactor and its mathematical model

The nonlinear dynamical homopolymerization reactor and its mathematical model, along with the data and notation, are given in Section 9.4.1 of Chapter 9.

18.3.2 Estimator design and estimation results

Optimal state estimation of nonlinear dynamical homopolymerization reactor using EKF is explained in Section 9.5 of Chapter 9. Sections 9.5.1 and 9.5.2 of Chapter 9 present the design and results of the state estimator. The estimator is employed to estimate the states, monomer concentration (v_m), solvent concentration (v_s), initiator concentration (c_i), and temperature (T) using the temperature measurement data of the reactor. The performance of the estimator is evaluated under different conditions as discussed in Section 9.5.2 of Chapter 9.

18.3.3 Controller design

The $f(x)$ and $g(x)$ in Eq. (18.1) can be defined from the mass and energy balances of Eqs. (9.16)–(9.19) given in Chapter 9:

$$f(x) = \begin{bmatrix} \dfrac{\rho_{mf}v_{mf}}{\rho_m\theta} - \dfrac{q_oq_iv_m}{\theta} - \dfrac{(MW)_mR_m}{\rho_m} + v_m\rho_m\dfrac{d(1/\rho_m)}{dT}\left(\dfrac{\rho_f(T_f-T)}{\rho} + \dfrac{\theta}{\rho C_p}\dfrac{\Delta H}{(MW)_m}v_m\rho_mk_pP - \dfrac{\theta}{\rho C_p}\dfrac{UA(T-T_c)}{V}\right) \\[3mm] \dfrac{\rho_{sf}v_{sf}}{\rho_s\theta} - \dfrac{q_oq_iv_s}{\theta} + \rho_sv_s\dfrac{d(1/\rho_s)}{dT}\left(\dfrac{\rho_f(T_f-T)}{\rho} + \dfrac{\theta}{\rho C_p}\dfrac{\Delta H}{(MW)_m}v_m\rho_mk_pP - \dfrac{\theta}{\rho C_p}\dfrac{UA(T-T_c)}{V}\right) \\[3mm] \dfrac{c_{if}}{\theta} - \dfrac{q_oq_i}{\theta}c_i - k_dc_i \\[3mm] \dfrac{\rho_f(T_f-T)}{\theta\rho} + \dfrac{\Delta Hv_m\rho_mk_pP}{\rho C_p(MW)_m} - \dfrac{UA}{V}\dfrac{(T-T_c)}{\rho C_p} \end{bmatrix}$$

(18.15)

$$g(x) = \begin{bmatrix} 0 \\ 0 \\ 0 \\ \beta \end{bmatrix}$$

(18.16)

The controlled output in Eq. (18.2) is obtained as

$$y = h(x) = T$$

(18.17)

The conditions in Eqs. (18.3) and (18.4) define the polymer reactor system is of relative order one. The new input v in Eq. (18.7) is computed as

$$v = \tau_1(T_s - T)$$

(18.18)

On computing the Lie derivatives and substituting them along with the new input in Eq. (18.8) leads to the following control law,

$$u_t = \frac{\tau_1(T_s - T) - \left(\frac{\rho_f(T_f - T)}{\theta\rho} + \frac{\Delta H v_m k_p \rho_m P}{\rho C_p (MW)_m} - \frac{UA}{V}\frac{(T - T_c)}{\rho C_p}\right)}{\beta} \qquad (18.19)$$

The notation and terminology for various terms of the reaction system are explained in Section 9.4.1 of Chapter 9.

18.3.4 Analysis of results

The mathematical model of the nonlinear dynamical homopolymerization reactor is given in Section 9.4.1 of Chapter 9. The data corresponding to this system is given in Table 9.2 of Chapter 9. The details concerning to stability analysis, steady-state solution, and bifurcation analysis of the reactor are given in Sections 9.4.2–9.4.4 of Chapter 9. The mathematical model of the homopolymerization reactor is solved by numerical integration using Gear's method with a sampling time of 0.0001 units. The details concerning optimal state estimation in nonlinear dynamical homopolymerization reactor are given in Section 9.5 of Chapter 9. A model-based state estimator, EKF is designed (Section 9.5.1) and applied to estimate the states of the polymerization reactor, monomer concentration (v_m), solvent concentration (v_s), initiator concentration (c_i), and temperature (T) using the temperature measurement data of the reactor. The performance of the EKF estimator is evaluated by considering different cases of process parameters for the dynamical situations illustrated in Section 9.4.2. The estimated states are found to be in close resemblance with the actual states for all the cases studied.

The EKF estimator-supported NIMC strategy is applied with the objective of controlling the homopolymerization reactor under different conditions, and the results are presented in terms of three cases [6]. In the first case, the process is initially operating under stable conditions. Then changes are introduced in the temperature setpoint and the controller is applied to maintain the desired operation under different input disturbance conditions introduced at different timings. In the second case, the process is operating under oscillatory conditions. The controller is applied to stabilize the process at the desired condition at a time during the open-loop oscillatory dynamics. In the third case, the process is operating under chaotic behavior. The controller is applied to suppress the chaotic dynamics and to maintain the desired operation. The results of the NIMC strategy are also compared with those of a conventional PI controller. The tuning parameter in the NIMC strategy is chosen as $\tau_1 = 16.5$.

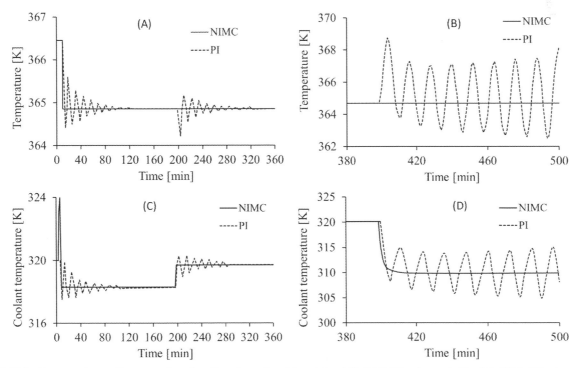

FIGURE 18.6 Process and controller output plots of nonlinear internal model control and PI: (A) & (B) - changes in T and T_f; (C) & (D) - changes in vm_f.

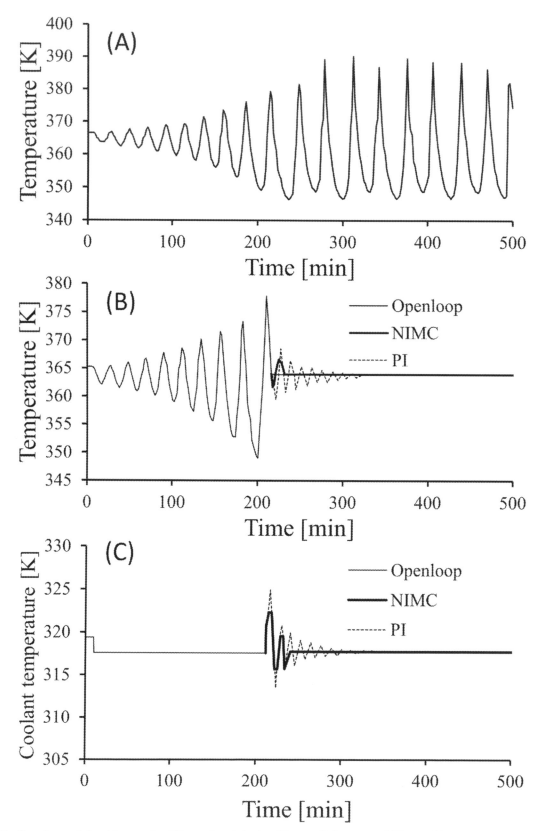

FIGURE 18.7   Open-loop and closed-loop results: (A) open-loop behavior, (B) controlled output, and (C) control input.

The PI controller parameters are evaluated by using Ziegler and Nichols method and further tuned and set as $K_c = 1.76$ and $\tau_I = 4.0$. A sampling time of 1 s is used for implementing the estimator and controller. For the first case, the process is initially at the stable steady state with $T = 366.4K$ and $T_c = 320K$. At time $t = 8.3$ min, the temperature setpoint is changed from 366.4 to 364.87K, and at $t = 200$ min, T_f is changed from 315.0 to 313.0K while keeping the setpoint as 364.87K. The input and output profiles corresponding to these conditions are shown in Fig. 18.6A and B, respectively. In the same case, at $t = 400$ min, the monomer volume fraction, v_{mf} is changed from 0.3 to 0.33, while keeping the setpoint at 364.87K and T_f at 313K. The input and output profiles corresponding to these conditions are shown in Fig. 18.6C and D, respectively. Though the changes are introduced at different timings in one time scale, the results are represented in separate figures with two time scales for the sake of clarity. From these results, it is observed that the NIMC tracks the desired condition very quickly, whereas the PI controller exhibits oscillatory

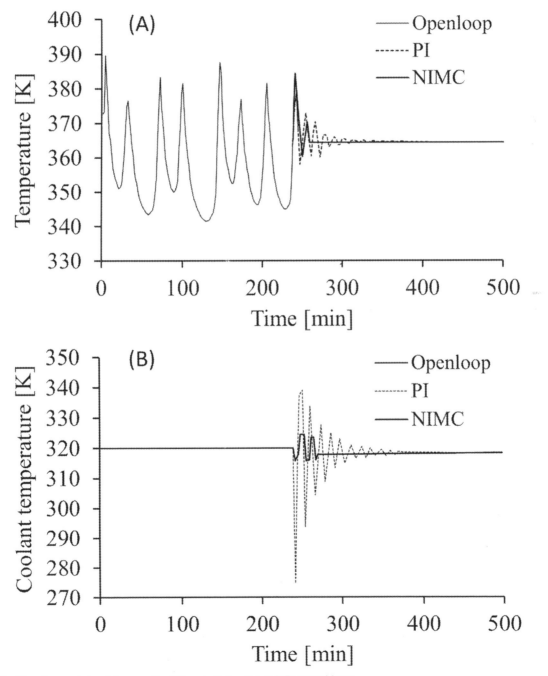

FIGURE 18.8   Open-loop and closed-loop results: (A) controlled output and (B) control input.

behavior for a longer duration. The PI controller could even fail for change in v_{mf} as shown in Fig. 18.6C. The NIMC effectively rejects the disturbance conditions and leads the process back to its desired operation in a very conservative fashion.

In the second case, the process is disturbed from its initial steady state by changing the T_c from 320K to 318K at $t = 8.3$ min. This condition makes the process exhibit oscillatory behavior. This open-loop behavior of the system is shown in Fig. 18.7A. During the oscillatory operation at $t = 217.4$ min, the controllers are applied to stabilize the operation and maintain the desired response condition, which is set as 364.87K for this case. The process output and controller output results for both the controllers are shown in Fig. 18.7B and C, respectively. From these results, it is observed that both the controllers are able to track the process to the desired condition, but the NIMC has shown better performance by quickly suppressing the oscillations. In the third case, the process is exhibiting chaotic behavior for $T = 347.6$K and $T_c = 318$K. During the chaotic operation at $t = 240$ min, the controllers are applied to stabilize the operation and maintain the desired response, which is set as 364.87K for this case. The process output and controller output results for both the controllers are shown in Fig. 18.8A and B, respectively. Again, the NIMC has performed better by quickly suppressing the chaotic dynamics with less stringent control actions. The performances of the controllers for different conditions are also expressed in terms of integral absolute error (IAE) values [6]. The results evaluated for both the chemical and polymerization reactors show the superior performance of the estimator-based NIMC strategy over the conventional controllers.

18.4 Summary

Optimal state estimation plays a vital role in estimator-based control of complex nonlinear dynamical systems that are characterized by input–output multiplicities, parametric sensitivity, nonlinear oscillations, and chaos. A NIMC strategy that incorporates the nonlinear model structure and the estimator dynamics in the control law is presented for the control of nonlinear dynamical systems. The estimator is designed to provide the unmeasured process states that capture the fast-changing nonlinear dynamics of the process to incorporate in the controller. The performance of the estimator-supported NIMC strategy is evaluated by applying it for the control of a nonisothermal nonlinear chemical reactor and a homopolymerization reactor, which exhibit rich dynamical behavior ranging from stable situations to chaos. The results evaluated under different conditions show the superior performance of the estimator-based NIMC strategy over the conventional controllers.

References

[1]   P.L. Lee, G.R. Sullivan, Generic model control (GMC), Comput. Chem. Eng. 12 (1988) 573–580.

[2]   C. Kravaris, C.B. Chung, Nonlinear state feedback synthesis by global input/output linearization, AIChE J. 33 (1987) (1987) 592–603.

[3]   M.A. Henson, D.E. Seborg, An internal model control strategy for nonlinear systems, AIChE J. 37 (1991) 1065–1081.

[4]   Ch Venkateswarlu, K. Gangiah, Comparison of nonlinear controllers for distillation startup and operation, Ind. Eng. Chem. Res. 36 (1997) (1997) 5531–5536.

[5]   M.J. Kurtz, M.A. Henson, Input-output linearizing control of constrained nonlinear processes, J. Process Control. 7 (1997) 3–17.

[6]   K.R. Rao, D.P. Rao, Ch Venkateswarlu, Soft sensor based nonlinear control of a chaotic reactor, IFAC Proc. 42 (19) (2009) 537–543.

[7]   R.R. Karri, Ch Venkateswarlu, Nonlinear model based control of complex dynamic chemical systems, Adv. Chem. Eng. Res. 2 (2013) 1–19.

[8]   J.K. Bandyopadhyay, S.S. Tambe, V.K. Jayaraman, P.B. Deshpande, B.D. Kulkarni, On control of nonlinear system dynamics at unstable steady state, Chem. Eng. J. 67 (1997) 103–114.

Chapter 19

Optimal state estimator based control of an exothermic batch chemical reactor

19.1 Introduction

State and parameter estimation plays a prominent role in the operation and control of batch reactors, which are highly nonlinear and characterized by intrinsically unsteady operating conditions. Batch reactors occupy a special position in chemical processing industries, because they are widely used to produce various fine and specialty chemicals. Lack of complete state and parameter measurements in batch reactors necessitates estimating techniques to infer the unmeasured process information and use it in their operation and control strategies. Many batch processes involve complex reaction mechanisms, and in certain cases, reaction kinetics and mechanisms are also unknown. Control of batch reactors is difficult due to their intrinsic characteristics such as finite duration of operation, lack of steady state, and high nonlinearity. When proportional integral and derivative (PID) controllers are used to control such highly nonlinear and time varying processes, the controller must be tuned very conservatively to provide stable behavior over the entire range of operating conditions. But conservative controller tuning can result in serious degradation of control system performance. As a result, these controllers cannot cope with highly nonlinear behaviors in the process and also with large disturbances. In such situations, advanced controllers can offer opportunities for more efficient operation. Model-based control is one class of advanced control that uses mathematical models of the process to compute control action. A number of model-based control strategies based on rigorous process models have been reported for batch reactors [1−5]. Although rigorous process model-based controllers are efficient, developing and validating the models that involve component balances and reaction kinetics require considerable time and effort. Moreover, it may not be possible to incorporate such rigorous model-based strategies directly for real-time applications due to the large computational requirements. Therefore, the control strategies based on models that avoid reaction kinetics and instantaneous composition measurements but preserve the rigorousness via instantaneous estimation of unknown process states and parameters are found to be useful. Different controllers that avoid the use of complete rigorous models, but employ only energy balance models with the support of an estimator have been reported for batch reactors [6−10]. Thus, most of these reported controllers have been evaluated through simulation with few experimental investigations. However, more experimental studies of energy model-based controllers with online identification of unknown process information are needed to evaluate controllers' performance in real time.

This chapter aims at deriving simple energy model-based nonlinear controllers for temperature control of batch reactor with the online inference of unknown process information. The control strategies derived in this chapter are based on an extended Kalman filter (EKF) supported by the energy balance model of the batch reactor that incorporates an unknown parameters representing the net effect of heat release due to reaction and heat loss to the environment. The control strategies derived in this study for real-time temperature control of exothermic batch reactor include EKF control (EKFC), generic model control (GMC), and model predictive control (MPC). In EKFC, EKF is designed to filter the process sensor measurements, estimate the unknown model parameter, and derive the control law. In GMC and MPC, measurement filtering and unknown parameter estimation are performed by EKF. The efficiency of these controllers is evaluated in real time by applying to an exothermic batch chemical rector involving an esterification reaction. The results are also compared to an auto-tuned proportional integral (PI) controller.

19.2 Experimental system and its mathematical model

A brief description of the experimental system, analysis method for the reaction mixture, and energy balance model of the esterification reactor, which is used to derive the control laws, is given below.

Optimal State Estimation for Process Monitoring, Fault Diagnosis and Control. DOI: https://doi.org/10.1016/B978-0-323-85878-6.00014-2

19.2.1 Experimental batch reactor

The experimental reactor setup consists of a stainless steel pressure reactor of 2 L working capacity with a sampler at the bottom. The vessel is equipped with an external jacket with an electric coil for heating and an internal cooling coil. The electric coil can supply a maximum heating power of 1.32 kW. The internal cooling coil has 20 turns. The inner and outer diameters of the cooling coil are 0.5 cm and 0.7 cm, and the surface area of the coil is 0.1314 m². The reactor is provided with an agitator, whose speed can be regulated. A thermo-well is provided for temperature measurement in the reactor, and a resistance temperature detector (RTD) is used to measure the reactor temperature. Two additional RTD's are used to measure the temperatures of the coolant inlet and outlet streams. The reactor is also equipped with a pressure gauge, gas feed, and liquid feed ports. A liquid metering pump with remote operation is employed for controlled pumping of the coolant stream with the help of a direct current (DC) drive. The power input to the heating jacket is regulated with the help of a thyristor power pack. A schematic representation of the experimental setup is shown in Fig. 19.1.

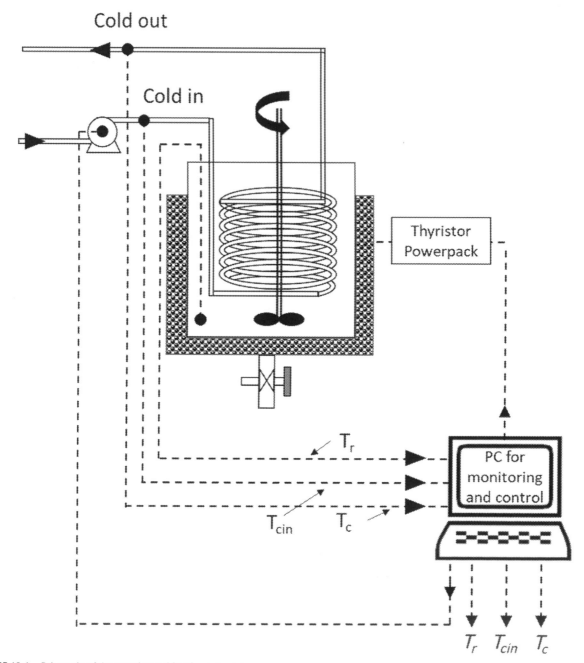

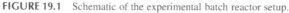

FIGURE 19.1 Schematic of the experimental batch reactor setup.

19.2.2 Analysis of reaction components

The reaction considered in this study is the esterification of maleic anhydride and n-hexanol to produce n-hexyl monoester of maleic acid. There is a possibility of the formation of diester at temperatures higher than 100°C [11], and therefore, the reaction temperature in this study is limited to 75°C. The titrimetric analysis method is employed to analyze the reaction mixture based on acid value. The acid value is defined as the milligrams of potassium hydroxide needed to neutralize the free acidity present in 1 g of the sample. This is calculated by titrating with a solution of sodium hydroxide as follows:

$$AV = \frac{(A)(N)(56)}{w} \tag{19.1}$$

where AV = acid value, A = volume of sodium hydroxide solution used for titration, mL, N = normality of sodium hydroxide solution, w = sample weight, g.

The theoretical acid value of a pure compound can be obtained from its molecular weight (MW) and basicity as expressed by

$$AV = \frac{56000 \times \text{basicity}}{MW} \tag{19.2}$$

Maleic anhydride is a dehydrated compound of maleic acid (MW = 116), which is dibasic and has an acid value of 965. On the other hand, the product of the present reaction under consideration, namely hexyl monoester of maleic acid (MW = 200), is monobasic with an acid value of 280. Therefore, the acid value has been employed to detect the completion of this reaction [12].

In this study, the difference in acid value between the reactants and products is employed to estimate the conversion achieved during the reaction. A calibration chart is prepared for conversion as a function of acid value by considering reaction mixtures at different stages. Pure hexyl monoester product is obtained by conducting a complete reaction and confirming the nuclear magnetic resonance (NMR) spectrum. On the basis of stoichiometry, and starting with the initial reaction mixture composition, several synthetic mixtures are prepared using molten maleic anhydride, hexanol, and hexyl monoester of maleic acid corresponding to different conversion levels. The reaction progress is negligible at room temperature, and therefore this method can be employed. To a known sample weight of each mixture, 25 mL of methanol and 100 mL of water are added, and the mixture is titrated against standard sodium hydroxide solution. The acid value for each mixture is calculated using Eq. (19.2), and a calibration chart is prepared by plotting % conversion, X as a function of acid value, as shown in Fig. 19.2. The relation is found to satisfy the straight line equation in the figure with reasonable accuracy.

$$X = 256.51 - 0.61(AV) \tag{19.3}$$

19.2.3 Esterification reactor model

Synthesis of hexyl monoester of maleic acid based on maleic anhydride and n-hexyl alcohol is considered in this study.

$$\text{maleic anhydride} + n\text{-hexyl alcohol} \rightarrow \text{hexyl monoester of maleic acid}$$

Maleic anhydride is a solid at room temperature and melts in the range of 53°C−55°C. The reactor is charged with maleic anhydride and hexanol, and the experiment is initiated in closed-loop operation under constant agitation by specifying a constant setpoint of 75°C. The control strategy starts with full heating to heat the reactor's contents from room temperature to the desired setpoint. Initially, part of the heat is utilized for melting the solid maleic anhydride. After melting, the concentrations of maleic anhydride and hexanol in the reactor are 4.55 kmol m^{-3} and 5.34 kmol m^{-3}, respectively. This is an exothermic reaction (ΔH = −33.5 MJ kmol^{-1}). Because of the reversibility of the reaction and formation of diester, reaction temperature exceeding 100°C should be avoided [13]. The mathematical model of the reactor consists of mass balances for maleic anhydride and n-hexyl alcohol [11], and energy balances can be defined for the reactor, heating jacket and cooling coil [14]. However, for this study of the experimental reactor with the online measurement of temperatures corresponding to the reactor and coolant, a reduced order energy balance model with the following equation:

$$\frac{dT_r}{dt} = \theta + \frac{P_w}{\rho C_p V} + \frac{U_c A_c (\overline{T}_c - T_r)}{\rho C_p V} \quad T_r(0) = T_{r0} \tag{19.4}$$

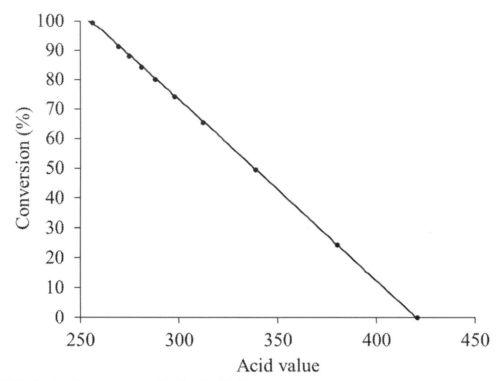

FIGURE 19.2 Calibration chart for conversion as a function of acid value.

where

$$\overline{T}_c = \frac{T_{ci} + T_c}{2}$$

Here, T_r is the reactor temperature, T_{ci} and T_c are the coolant inlet and outlet temperatures, $\overline{T}_c$ is the average coolant temperature, P_w is the energy added to the system due to heating in the jacket, U_c is the coolant side heat transfer coefficient, A_c is the surface area of the cooling coil, and θ is the unknown parameter. The parameter θ considers the net effect of heat release due to reaction and heat losses to the environment. The initial conditions for the experimentation of this study are $T_{ci} = 24°C$, $T_c = 24°C$, and $T_{ro} = 27°C$, respectively. A normalized coordinated control input u bounded between u_{min} and u_{max}, is employed. In this study, the values of $u_{max} = 5$ and $u_{min} = -5$ are used. The power input (P_w) and coolant side heat transfer coefficient (U_c) are computed as

$$P_w = (P_{wmax} - P_{wmin})\overline{u} + P_{wmin} \tag{19.5}$$

$$U_c = (U_{cmax} - U_{cmin})(1 - \overline{u}) + U_{cmin} \tag{19.6}$$

where P_{wmax} and P_{wmin} are the maximum and minimum heating powers, and U_{cmax} and U_{cmin} are the maximum and minimum coolant side heat transfer coefficents. The values of P_{wmax} and U_{cmax} considered in this experimental study are 1.32 kW and 0.7143 kJ^{-1} m^{-2} s^{-1}, respectively, whereas the values of P_{wmin} and U_{cmin} are zero. In Eqs. (19.5) and (19.6), the normalized coordinated control input is defined by

$$\overline{u} = \frac{u - u_{min}}{u_{max} - u_{min}} \tag{19.7}$$

The relation established between the heat transfer coefficient U_c and coolant flowrate F_c for this experimental system is

$$\frac{1}{U_c} = \frac{1}{4555F_c^{0.8}} + \frac{1}{54.52} \tag{19.8}$$

The maximum and minimum coolant flowrates are, $F_{cmax} = 2.03 \times 10^{-5}$ m^3 s^{-1} and $F_{cmin} = 0$ m^3 s^{-1}. Certain parameters in the above equations are ascertained with the help of experimental data generated using PI control. The power signal P_w and the coolant flowrate F_c are converted into 4–20 mA signals u_h and u_c respectively through linear transformations:

$$u_h = 4 + \frac{P_w}{P_{wmax}} \times 16 \tag{19.9}$$

$$u_c = 4 + \frac{F_c}{F_{cmax}} \times 16 \tag{19.10}$$

19.3   State/parameter estimation

The EKF computes the optimal state and/or parameter estimates of a nonlinear system at each sampling time using a linearized model and linear estimation principles. The estimation is based on the assumption that a model describing the system is available. The mean and covariances of measurement errors, inputs and outputs are considered to be known. The usefulness of EKF for state and parameter estimation in batch processes has been reported in the literature [9,15–18]. In this study, EKF is applied to filter the measured temperatures and estimate the unknown model parameter involved in the closed-loop control law.

The procedure for estimation of states, parameters, and control input by EKF is described as follows: Consider a system described by the following equations:

$$\dot{x} = f(x(t), u(t), \theta(t), t) + w(t) \quad x(0) = x_0$$
$$y = g(x(t), \theta(t)) + v(t) \tag{19.11}$$

where x represents the state variable vector of dimension n, u is the manipulated input vector of dimension m, θ is the parameter vector of dimension p, t denotes the present time, x_0 represents the initial state vector, y represents the output variable vector of dimension m, and f and g are vector functions. The process noise vector $w(t)$ and the measurement error vector $v(t)$ are assumed to be white Gaussian random processes with zero mean and covariances Q and R, respectively.

The equations for implementation of an EKF for state and parameter estimation with the help of discrete measurements can be summarized as follows:

State/parameter propagation:

$$\hat{x}_k^- = \hat{x}_{k-1}^+ + \int_{(k-1)\Delta t}^{k\Delta t} f(\hat{x}_{k-1}^+, u_{k-1}, \hat{\theta}_{k-1}^+, \tau) d\tau$$
$$\hat{\theta}_k^- = \hat{\theta}_{k-1}^+ \tag{19.12}$$

where Δt represents the sampling time.

Combined covariance propagation:

$$P_k^- = P_{k-1}^+ + \int_{(k-1)\Delta t}^{k\Delta t} \left(F(\tau) P_{k-1}^+ + P_{k-1}^+ F^T(\tau) + Q \right) d\tau \tag{19.13}$$

where

$$F(\tau) = \begin{bmatrix} \partial f/\partial x & \partial f/\partial \theta \\ 0_{p \times n} & 0_{p \times p} \end{bmatrix} \Bigg|_{\hat{x}_{k-1}^+, \hat{\theta}_{k-1}^+} \tag{19.14}$$

where $0_{p \times n}$ and $0_{p \times p}$ are matrices of dimensions $(p \times n)$ and $(p \times p)$ respectively, containing zeros as their elements. In the above equations, the superscript "−"indicates the values before measurement update, whereas the superscript " + " indicates the values after measurement update; subscript k denotes the present sampling instant,

Kalman gain:

$$K_k = \frac{P_k^- G_k^T}{[G_k P_k^- G_k^T + R]} \tag{19.15}$$

where

$$G_k = \begin{bmatrix} \partial g/\partial x & \partial g/\partial \theta \end{bmatrix} \Big|_{\hat{x}_{k-1}^+, \hat{\theta}_{k-1}^+} \tag{19.16}$$

State/parameter update:

$$\begin{bmatrix} \hat{\mathbf{x}}_k^+ \\ \hat{\boldsymbol{\theta}}_k^+ \end{bmatrix} = \begin{bmatrix} \hat{\mathbf{x}}_k^- \\ \hat{\boldsymbol{\theta}}_k^- \end{bmatrix} + \mathbf{K}_k \left[\mathbf{y}_k - \mathbf{g}(\hat{\mathbf{x}}_k^-, \hat{\boldsymbol{\theta}}_k^-) \right] \tag{19.17}$$

where $\mathbf{y}_k$ is the measured output vector at the kth sampling instant.

Covariance update:

$$\mathbf{P}_k^+ = \left[\mathbf{I}_{n+p} - \mathbf{K}_k \mathbf{G}_k \right] \mathbf{P}_k^- \tag{19.18}$$

where $\mathbf{I}_{n+p}$ denotes the identity matrix of dimension $(n+p) \times (n+p)$, $\mathbf{K}$ is the Kalman gain, and $\mathbf{P}$ denotes the combined state and parameter covariance matrix.

19.4 Control algorithms

The control algorithms employed for the esterification reactor are presented as follows.

19.4.1 Extended Kalman filter control

This strategy involves EKF for state and parameter estimation as well as control law computation. The basis of this strategy is that the input vector, $\boldsymbol{u}(t)$ is considered as an additional state and augmented to the existing states. The objective in EKF based state and parameter estimation is to reconstruct the states and parameters optimally by minimizing the deviation between the current measurements and their estimates. On the other hand, the objective of EKF for control estimation is to find $\boldsymbol{u}(t)$ optimally by minimizing the deviation between the one-step ahead predictions of the output values and their desired set values.

Consider the continuous-time system represented by Eq. (19.11). The convention followed in the present strategy is that the measurement, $\mathbf{y}_k$ is obtained after implementing the control input, $\mathbf{u}_{k-1}$. Therefore, in order to achieve an output, $\mathbf{y}_{k+1}$ close to the setpoint, $\mathbf{y}^{\mathbf{d}}_{k+1}$, the control input vector, $\mathbf{u}_k$ is determined by the following EKF formulation in an iterative manner.

State/control propagation:

$$\begin{aligned} &\text{For the first iteration,} \\ &\mathbf{u}_k^- = \mathbf{u}_{k-1}^+ \\ &\text{From the second iteration,} \\ &\mathbf{u}_k^- = \mathbf{u}_k^+ \\ &\hat{x}_{k+1}^- = \hat{x}_k^+ + \int_{k\Delta t}^{(k+1)\Delta t} f(\hat{x}_k^+, u_k^-, \hat{\theta}_k^+, \tau) d\tau \end{aligned} \tag{19.19}$$

Combined covariance propagation:

$$\mathbf{Pc}_{k+1}^- = \mathbf{Pc}_k^+ + \int_{k\Delta t}^{(k+1)\Delta t} \left(\mathbf{Fc}(\tau)\mathbf{Pc}_k^+ + \mathbf{Pc}_k^+ \mathbf{Fc}^T(\tau) + \mathbf{Qc} \right) d\tau \tag{19.20}$$

where

$$\mathbf{Fc}(\tau) = \begin{bmatrix} \partial \mathbf{f}/\partial \mathbf{x} & \partial \mathbf{f}/\partial \mathbf{u} \\ \mathbf{0}_{mxn} & \mathbf{0}_{mxm} \end{bmatrix} \Big|_{\hat{\mathbf{x}}_k^+, \hat{\boldsymbol{\theta}}_k^+, \mathbf{u}_k^-} \tag{19.21}$$

Kalman gain:

$$\mathbf{Kc}_{k+1} = \mathbf{Pc}_{k+1}^- \mathbf{Gc}_{k+1}^T \left[\mathbf{Gc}_{k+1} \mathbf{Pc}_{k+1}^- \mathbf{Gc}_{k+1}^T + \mathbf{Rc} \right]^{-1} \tag{19.22}$$

where

$$\mathbf{Gc}_{k+1} = \begin{bmatrix} \partial \mathbf{g}/\partial \mathbf{x} & \partial \mathbf{g}/\partial \mathbf{u} \end{bmatrix} \Big|_{\hat{\mathbf{x}}_k^+, \hat{\boldsymbol{\theta}}_k^+, \mathbf{u}_k^-} \tag{19.23}$$

$\mathbf{Kc}_{k+1}$ is partitioned as

$$\mathbf{Kc}_{k+1} = \begin{bmatrix} \mathbf{Kcx}_{k+1} \\ \mathbf{Kcu}_{k+1} \end{bmatrix} \Big|_{\hat{\mathbf{x}}_k^+, \hat{\boldsymbol{\theta}}_k^+, \mathbf{u}_k^-} \tag{19.24}$$

where the upper submatrix of dimension $(n \times m)$ corresponds to the gain for the states, and the lower submatrix of dimension $(m \times m)$ represents the gain for control variables.

Control update:

$$\mathbf{u}_k^+ = \mathbf{u}_k^- + \mathbf{Kcu}_{k+1}\left[\mathbf{yd}_{k+1} - \mathbf{g}(\hat{\mathbf{x}}_{k+1}^-, \hat{\boldsymbol{\theta}}_k^+, \mathbf{u}_k^-)\right] \tag{19.25}$$

where $\mathbf{yd}_{k+1}$ is the setpoint for the output at the $(k+1)$th sampling instant.

Covariance update:

$$\mathbf{Pc}_{k+1}^+ = [\mathbf{I}_{n+m} - \mathbf{Kc}_{k+1}\mathbf{Gc}_{k+1}]\mathbf{Pc}_{k+1}^- \tag{19.26}$$

where $\mathbf{I}_{n+m}$ is the identity matrix of dimension $(n+m) \times (n+m)$.

The algorithm of EKFC is summarized as follows:

1. Initialize the covariance matrices for state and parameter estimation ($\mathbf{P}_0$, $\mathbf{Q}$, and $\mathbf{R}$) as diagonal matrices and the covariance matrices for control estimation ($\mathbf{Pc}_0$, $\mathbf{Qc}$, and $\mathbf{Rc}$) as diagonal matrices.
2. With the help of the measurements of the output variables, perform EKF based state and parameter estimation according to Eqs. (19.12)–(19.18) to obtain the estimates of the states and parameters at the current sample.
3. Starting with the estimated values of states and parameters, and using the existing control vector as the initial guess, perform iterative EKF based control estimation so that the one-step ahead prediction for the outputs approach the one-step ahead desired set values using Eqs. (19.19)–(19.26). Convergence criteria employed for termination of the iterative procedure are: (1) achieving the desired performance, that is, the tracking error, defined as the squared error between the set value and the output prediction, is less than a predefined threshold value; (2) reaching the input bounds; and (3) crossing the maximum number of iterations.
4. Implement the control action, and shift to the next sampling instant, and go to Step 2.

19.4.2 Generic model control

GMC is a simple nonlinear control approach, which allows the incorporation of the nonlinear process model directly into the control law [19]. Consider a single-input single-output process described by Eq. (19.11). GMC is an optimal control approach to force the process output rate to match a reference rate defined as

$$r^* = k_1(y_d - y) + k_2 \int (y_d - y) \, dt \tag{19.27}$$

where r^* is the desired process output "rate of change," y_d is the setpoint, and k_1 and k_2 are the controller tuning parameters. The reference rate is proportional to the distance from the setpoint and includes integral action to eliminate offset. The control law is obtained from Eqs. (19.11)–(19.27) as

$$r^* = \dot{y} = \dot{g}(\mathbf{x}, \dot{\mathbf{x}}) \tag{19.28}$$

which can be solved explicitly for u in the case of control-affine systems. The tuning parameters k_1 and k_2 can be tuned to achieve the desired shape and speed of the closed-loop response given by the following transfer function

$$\frac{y}{y_d} = \frac{k_1 s + k_2}{s^2 + k_1 s + k_2} \tag{19.29}$$

The time-domain expressions for different choices of k_1 and k_2 are reported elsewhere [19]. In this study, GMC is formulated based on the energy balance model by handling the issue of process/model mismatch with the help of EKF defined by Eqs. (19.12)–(19.18) for state and parameter estimation.

19.4.3 Model predictive control

MPC has found wide applications in the control of chemical processes [20–22]. The main attractive features of this approach have been the handling of time-delays inherent in chemical processes, inverse response behavior, and easy incorporation of constraints explicitly or implicitly. Furthermore, several extensions have been proposed over the basic algorithm, incorporating nonlinear models, parameter estimation techniques, etc., to enhance performance.

The basic procedure for MPC can be summarized as follows: (1) with the help of a model, predict the output response over the prediction horizon, N; (2) find the control input over another horizon called control horizon, N_u which minimizes the deviation between the predicted output trajectory and the setpoint trajectory over the prediction horizon; and (3) implement the first of the calculated control inputs on the process, shift by a sampling time, and go back to Step (1). The control law is generally obtained by minimizing an objective function of the form

$$J = \sum_{j=1}^{Ny} (y_d(k+j) - y(k+j))^2 + \lambda \sum_{j=1}^{Nu} (\Delta u(j))^2 \tag{19.30}$$

where Δu is the change in input value, and λ is the move suppression factor. Depending on the model employed and computational techniques used, several versions of MPC schemes arise. In this study, a simplified energy balance model of the process is considered, and the unknown parameters/states are estimated with the help of EKF defined by Eqs. (19.12)–(19.18).

19.5 Design of estimator based controllers for the esterification reactor

The design details of the estimator based EKFC, GMC, and MPC for the esterification batch reactor are given as follows.

19.5.1 Extended Kalman filter control

In order to implement the EKF for state and parameter estimation, the model dependent expressions F and G of Eqs. (19.14) and (19.16) are computed based on the energy balance model defined by Eq. (19.4). The reactor temperature T_r is the measured state and θ is the unknown model parameter to be estimated. Thus F and G are obtained as

$$\mathbf{F} = \begin{bmatrix} \dfrac{\partial f}{\partial T_r} & \dfrac{\partial f}{\partial \theta} \\ 0 & 0 \end{bmatrix} = \begin{bmatrix} \dfrac{-U_c A_c}{\rho C_p V} & 1 \\ 0 & 0 \end{bmatrix} \tag{19.31}$$

$$\mathbf{G} = \begin{bmatrix} \dfrac{\partial g}{\partial T_r} & \dfrac{\partial g}{\partial \theta} \end{bmatrix} = \begin{bmatrix} 1 & 0 \end{bmatrix} \tag{19.32}$$

The energy balance model defined by Eq. (19.4) is further used in the control law derivation. However, in the computation of EKF controller involving the equations defined by Eqs. (19.19)–(19.26), the unknown model parameter θ is replaced by the control input u. The reactor temperature T_r is the state variable which is estimated along with the control input u. The coordinated control input u is related to P_w and U_c which are computed in terms of Eqs. (19.5) and (19.6). The model dependent expressions Fc and Gc involved in the EKF controller are obtained as

$$\mathbf{Fc} = \begin{bmatrix} \dfrac{\partial f}{\partial T_r} & \dfrac{\partial f}{\partial u} \\ 0 & 0 \end{bmatrix} = \begin{bmatrix} \dfrac{-U_c A_c}{\rho C_p V} & \dfrac{P_{wmax} - U_{cmax} A_c (\overline{T}_c - T_r)}{(u_{max} - u_{min}) \rho C_p V} \\ 0 & 0 \end{bmatrix} \tag{19.33}$$

$$\mathbf{Gc} = \begin{bmatrix} \dfrac{\partial g}{\partial T_r} & \dfrac{\partial g}{\partial u} \end{bmatrix} = \begin{bmatrix} 1 & 0 \end{bmatrix} \tag{19.34}$$

19.5.2 Generic model control

For batch reactor, the reference trajectory $r*$ of Eq. (19.27) is assigned as

$$r* = k_{c1}(T_r^* - T_r) + k_{c2} \int (T_r^* - T_r) dt = \dfrac{dT_r}{dt} \tag{19.35}$$

where T_r^* is the setpoint for temperature. By combining the energy balance model defined by Eq. (19.4) with Eq. (19.35), the control law is derived as

$$u = \dfrac{\left[(r* - \theta)(u_{max} - u_{min}) \rho C_p V + P_{wmax} u_{min} - U_{cmax} u_{max} A_c (\overline{T}_c - T_r) \right]}{\left[P_{wmax} - U_{cmax} A_c (\overline{T}_c - T_r) \right]} \tag{19.36}$$

19.5.3 Model predictive control

By combining Eqs. (19.5) and (19.6), the energy balance model of Eq. (19.4) can be rewritten as

$$\frac{dT_r}{dt} = \left(\theta + \frac{\left(U_{cmax}u_{max}A_c\left(\overline{T}_c - T_r\right) - P_{wmax}u_{min}\right)}{\rho C_p V(u_{max} - u_{min})}\right) + \left(\frac{\left(P_{wmax} - U_{cmax}A_c\left(\overline{T}_c - T_r\right)\right)}{\rho C_p V(u_{max} - u_{min})}\right)u \tag{19.37}$$

$$= A + Bu$$

where

$$A = \theta + \frac{\left(U_{cmax}u_{max}A_c\left(\overline{T}_c - T_r\right) - P_{wmax}u_{min}\right)}{\rho C_p V(u_{max} - u_{min})} \tag{19.38}$$

$$B = \frac{\left(P_{wmax} - U_{cmax}A_c\left(\overline{T}_c - T_r\right)\right)}{\rho C_p V(u_{max} - u_{min})} \tag{19.39}$$

Converting the above equation into discrete form with the help of Euler integration gives

$$T_r(k+1) = T_r(k) + (A + Bu(k))\Delta t \tag{19.40}$$

where Δt is the sampling time. The current input $u(k)$ is rewritten as

$$u(k) = u(k-1) + \Delta u(k) \tag{19.41}$$

where $\Delta u(k)$ is the change in control action incorporated at the kth sampling instant and $u(k-1)$ is the control action at the $(k-1)$th sampling instant. Substitution of Eq. (19.41) into Eq. (19.40) results in the following equation,

$$T_r(k+1) = T_r(k) + (A + Bu(k-1))\Delta t + B\Delta u(k)\Delta t \tag{19.42}$$

The following assumptions are made: (1) the parameters A and B are assumed to be constant over the prediction horizon, N for control calculations; however, their values are updated at the next sampling instant; (2) the control horizon is considered as a unity; (3) no explicit disturbance correction term is included in the formulation since the process state and model parameters are updated at every sampling instant using EKF; and (4) a constant setpoint is considered over the prediction horizon.

Based on the above assumptions, the prediction model derived from Eq. (19.42) is:

$$\hat{T}_r(k+i) = \hat{T}_r(k+i-1) + (A + Bu(k+i-2))\Delta t + B\Delta u(k+i-1)\Delta t \tag{19.43}$$

Since the control horizon is unity, all the future input values over the prediction horizon are equal to $u(k)$ only. Therefore, Eq. (19.43) can be rearranged as

$$\hat{T}_r(k+i) = \hat{T}_r(k) + i(A + Bu(k-1))\Delta t + iB\Delta u(k)\Delta t, \quad i = 1, 2, \ldots, N \tag{19.44}$$

The objective function to be minimized at every sampling instant is defined as

$$MinJ = \left[\sum_{j=1}^{N} \left(T_r^*(k+j) - \hat{T}_r(k+j)\right)^2 + \lambda(\Delta u(k))^2\right] \tag{19.45}$$

where λ is the move suppression factor.

Substituting Eq. (19.44) into Eq. (19.45) and differentiating the resulting expression concerning $\Delta u(k)$, and equating it to zero gives

$$\Delta u(k) = \frac{\left(\sum_{j=1}^{N} j\right)(T_r^* - \hat{T}_r(k))B\Delta t - \left(\sum_{j=1}^{N} j^2\right)(A\Delta t + B\Delta tu(k-1))B\Delta t}{\lambda + \left(\sum_{j=1}^{N} j^2\right)(B\Delta t)^2} \tag{19.46}$$

Constraints on the input are incorporated as hard bounds on the computed control action. The parameters A and B of Eqs. (19.38) and (19.39) are calculated by substituting the EKF estimated parameter θ and state T_r based on the measured temperature information.

19.6 Analysis of results

The estimator supported EKFC, GMC, and MPC are evaluated in real time by applying them for temperature control of the esterification batch reactor. The results are also compared with an auto-tuned PI controller. These controllers are implemented by considering a sampling time of 1 s. The experiments are carried out by considering a batch duration of 5 h. A temperature setpoint maintenance at 75°C with this batch time yields maximum reactant conversion.

The PI controller tuning parameters are determined by carrying out relay tests using the auto-tuning method [23]. To capture the batch reactor dynamics near the desired temperature for the auto-tuning, relay signals are introduced in the coordinated input signal in the neighborhood of the setpoint. The relay height and hysteresis are chosen as 0.8 and 0.2, respectively. The heating and cooling input current signals during the relay test duration together with the corresponding temperature response are plotted in Fig. 19.3. The tuning parameters for the PI controller are calculated from the ultimate gain and the ultimate period of oscillation. The values of these parameters are given in Table 19.1.

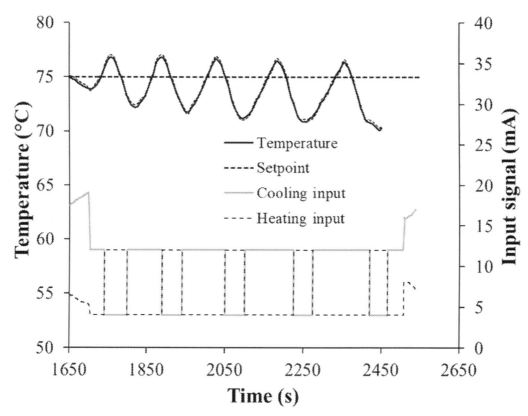

FIGURE 19.3 Relay test for auto-tuning of PI controller.

TABLE 19.1 Tuning parameters for estimator and controllers.

EKF parameters for state/parameter estimation:
$$P_0 = \begin{bmatrix} 0.1 & 0 \\ 0 & 10^{-5} \end{bmatrix}, Q = \begin{bmatrix} 100 & 0 \\ 0 & 10^{-2} \end{bmatrix}, R = \begin{bmatrix} 1 & 0 \\ 0 & 1 \end{bmatrix}, \alpha = 0.3$$

EKF controller parameters:
$$P_{c0} = \begin{bmatrix} 10^{-4} & 0 \\ 0 & 10^{-4} \end{bmatrix}, Q_c = \begin{bmatrix} 10^2 & 0 \\ 0 & 10^{-2} \end{bmatrix}, R_c = \begin{bmatrix} 1 & 0 \\ 0 & 1 \end{bmatrix}, k_1 = 0.007$$

GMC controller parameters:
$k_{c1} = 0.05, k_{c2} = 0.0$

MPC controller parameters:
$N = 30, \lambda = 0.01$

Auto-tuning parameters for PI controller:
Number of cycles = 2, Hysteresis = 0.2, Relay height = 0.8

The EKFC results have shown excessive overshoots in temperature before settling at the desired setpoint. The reason for excessive overshoots in temperature is that the control action does not change until the time at which the predicted error changes its direction. Otherwise, the cooling action starts only after the temperature crosses the setpoint. In a simulation environment, overshoot free control can be achieved with this type of controller in the absence of time delay. The control performance achieved will be similar to that of a deadbeat control algorithm with a small variation. However, in practice, this type of control action has led to large overshoots in temperature because it takes several sampling instants for receding the heating action to respond the system to cooling action. In order to overcome this problem, a small modification is introduced in the EKF based control estimation method. The basis for the proposed modification is to have the control law similar to a PI control law rather than integral-only control law. Therefore, Eq. (19.25) is modified as

$$u_k^+ = u_k^- + K_c\, u_{k+1}[K_1 e_{k+1} + (e_{k+1} - e_k)] \qquad (19.47)$$

where

$$e_k = T_r^* - T_r(k) \qquad (19.48)$$

$$e(k+1) = T_r^* - \hat{T}_r(k+1) \qquad (19.49)$$

where T_r^* is the desired temperature, $\hat{T}_r$ is predicted temperature and K_1 is additional tuning parameter. This tuning parameter can be appropriately chosen to achieve the early start of the cooling action in order to eliminate the overshoot. The control law Eq. (19.47) can be viewed as an alternate velocity form of a predictive PI controller. The temperature measurement signal is filtered using a simple exponential filter

$$T_r(k) = aT_r(k-1) + (1-a)T_r(k) \qquad (19.50)$$

where α is the filter constant. This filtered temperature is used in EKF for state and parameter estimation. The model that supported EKF for state and parameter estimation does not consider the latent heat of fusion for maleic anhydride during energy balance. Therefore, during the initial time period when the reactor temperature is less than 55°C, the estimated parameter is ignored and taken to be zero in all the controllers. This does not affect the system performance since the cooling action starts only beyond this temperature.

The tuning parameters employed for different control strategies are reported in Table 19.1. The EKF parameters for state and parameter estimation (Table 19.1) are selected initially based on process and noise information and further tuned. The tuning parameters for improving the performance of different controllers are selected heuristically. In EKFC, the study of the effect of the diagonal elements of Pc_o, Qc, and Rc shows that the controller is more sensitive to Qc. The higher and lower values of Qc than those specified in Table 19.1, have shown to result in more oscillations of larger amplitude in the temperature response. It is also observed that the increase in parameter k_1 of EKFC results to start the cooling action at a higher temperature. In GMC, low values of k_{c1} have resulted in a sluggish temperature response, and high values of k_{c1} have resulted in oscillatory temperature response. k_{c1} is also found to influence the temperature at which the cooling action starts. The effect of a nonzero value of k_{c2} on this system response is found to be detrimental. Any nonzero value of k_{c2} is found to result in minor oscillations around the major oscillations in the temperature response. Therefore, k_{c2} is chosen as zero in this study. In MPC, high values of prediction horizon are found to result in a sluggish temperature response and to regulate the control moves, move suppression factor is selected with reasonable care.

The performance of EKFC, GMC, MPC, and PI is compared in Table 19.2 for a normal batch run and the case of the presence of a disturbance in stirrer speed. Fig. 19.4 shows the temperature responses and the corresponding heating and cooling input signals for the four controllers. The temperature responses of these four controllers plotted over a

TABLE 19.2 Comparison of control performance.

Controllers case	PI	EKFC	GMC	MPC
IAE for normal run (5 h)	0.507	0.386	0.393	0.464
IAE for run with load disturbance	0.762	0.436	0.503	0.519

reduced time span are shown in Fig. 19.5. The normal batch run consists of charging the reactor with the mixture of reactants at room temperature and starting the control strategies. The batch duration is 5 h. The conversion obtained at the end of the reaction in all the four cases is around 98%—99%. This can be ascertained from the calibration chart based on an acid value obtained. These results show that the three model-based controllers have exhibited better performance over the auto-tuned PI controller. The input, as well as output responses using PI controller, are found to be more oscillatory. The integrated absolute error (IAE) values of the controllers reported in Table 19.2 also confirm the inferior performance of the PI controller. Another observation made from the manipulated input profiles for all the controllers is that the average magnitude of heat input required with a PI controller is slightly larger than the model-based controllers, whereas the cooling requirement is found to be approximately the same. This difference is due to the nonlinear nature of the three model-based controllers against the linear PI control law. The results of Figs. 19.4 and 19.5 show that EKFC has resulted in oscillations of very small amplitude before reaching the setpoint and has a very short rise time. GMC and MPC have resulted in oscillations of slightly larger amplitude before reaching the setpoint with a larger rise time. The IAE of EKFC is found to be minimum, closely followed by GMC and MPC.

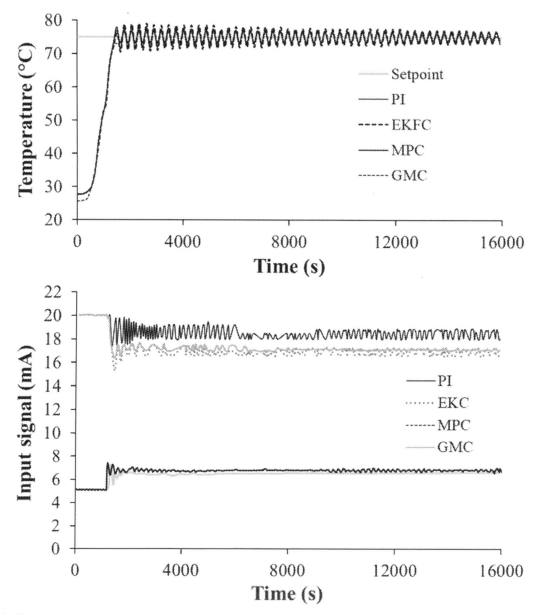

FIGURE 19.4 Temperature responses and manipulated input profiles with different controllers for setpoint tracking over the entire batch duration.

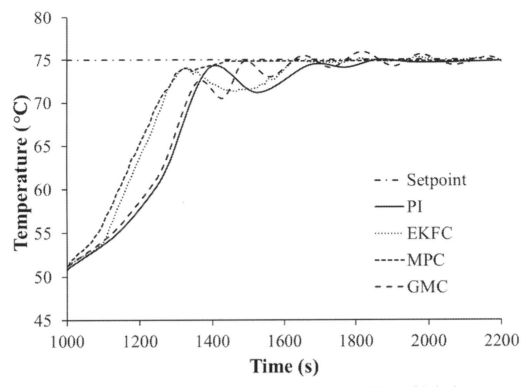

FIGURE 19.5   Temperature responses with different controllers for setpoint tracking during the initial part of the batch.

The performances of the controllers are also evaluated in the presence of an unmeasured disturbance. After the constant temperature setpoint of 75°C is attained, the stirrer speed is reduced to zero, and after leaving it for 2 min, the actual speed is restored. The disturbance rejection capability of different controllers is illustrated in Fig. 19.6. It is observed that lack of mixing has reduced the measured temperature for the same control action. All the controllers have responded to this change by increasing the heat input and decreasing the cooling input. However, on restoring the stirrer speed, the response of a PI controller is found to be totally different from that of the model-based controllers. PI controller is designed so that the integral action is active only when the controller operates within the 4−20 mA range and is reset every time it reaches the bound. The reason for the difference in the behavior of the PI controller is because of two reasons. Firstly, the manipulated input signal for heating is higher than the other controllers and therefore, the margin to reach the 20 mA limit is smaller. Secondly, as soon as the upper bound of 20 mA is reached, there is an integral reset due to which the control action changes drastically, resulting in further cooling.

The effect of the disturbance in PI control has resulted in an error in the temperature totally in the lower side of the setpoint only. On the other hand, in the case of the three model-based controllers, the first effect of the disturbance is in the lower direction, and after the restoration of stirrer speed, the response is moved above the setpoint, which is corrected in two oscillations with GMC and MPC, and within a single oscillation with the EKF controller. The controllers take the approximate time to bring the temperature to ± 0.5°C range of the setpoint is 8 min with PI control, 5 min with GMC and MPC and 4 min with EKFC, where the first 2 min represents the duration of the disturbance itself. Among the three model-based controllers, the temperature response oscillation amplitude is found to be minimum for EKFC followed by MPC and GMC. The results of the real-time control of the exothermic batch reactor of this study clearly indicate that all the three model-based controllers have exhibited better performance over the auto-tuned PI controller in a normal batch run and the presence of a disturbance. Among these state and parameter inferential model-based controllers, EKFC has shown better performance over GMC and MPC with lower IAE values, shorter rise times and lower overshoot values in setpoint tracking and disturbance rejection. Thus the EKFC proposed in this study can be effectively applied for controlling the exothermic batch reactor.

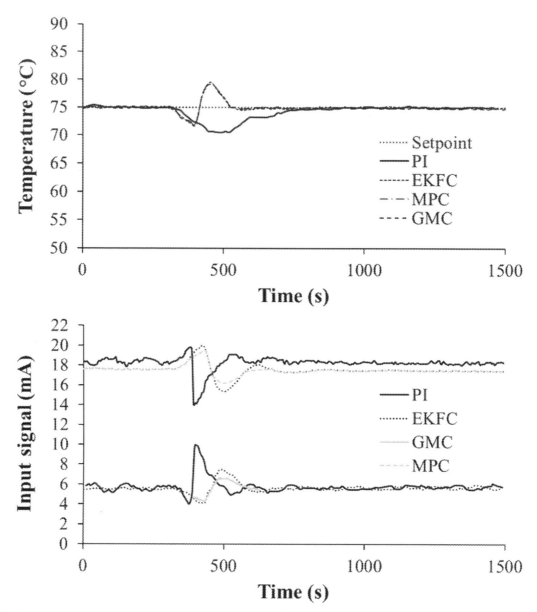

FIGURE 19.6 Temperature responses and manipulated input profiles with different controllers for load disturbance.

19.7 Summary

State and parameter estimation plays a vital role in the operation and control of batch reactors, which are highly non-linear and characterized by intrinsically unsteady operating conditions. Lack of complete state and parameter measurements in batch reactors necessitates the need for estimating techniques to infer the unmeasured process information and use it in the control and operational strategies of such reactors. Many batch processes involve complex reaction mechanisms, and in certain cases, reaction kinetics and mechanisms are also unknown. Control of batch reactors is difficult due to their intrinsic characteristics such as finite duration of operation, lack of steady state, and high nonlinearity. This study presents three energy model-based controllers, namely, EKFC, GMC, and MPC for the temperature control of an exothermic monoester formation batch chemical reactor. The state and unknown parameter information that is needed to implement the EKFC, GMC, and MPC are estimated online using an EKF. The real-time implementation results show the better performance of the model-based controllers over the auto-tuned PI controller. In addition, the results illustrate the better suitability of EKFC for real-time control of the exothermic batch chemical reactor.

References

[1] C. Kravaris, C.B. Chung, Nonlinear state feedback synthesis by global input/output linearization, AIChE J. 33 (1987) (1987) 592−603.

[2] M. Soroush, C. Kravaris, Optimal design and operation of batch reactors. 2. A case study, Ind. Eng. Chem. Res. 32 (1993) 882−893.

[3] R. Berber, Control of batch reactors: a review, Methods of Model Based Process Control, Springer, Netherlands, 1995, pp. 459−494.

[4] J.S. Chang, J.S. Hsu, Y.T. Sung, Trajectory tracking of an optimizing path in a batch reactor: experimental study, Ind. Eng. Chem. Res. 35 (1996) 2247−2260.

[5] V. Sampath, S. Palanki, J.C. Cockburn, J.P. Corriou, Robust controller design for temperature tracking problems in jacketed batch reactors, J. Process. Control. 12 (2002) 27−38.

[6] A. Jutan, A. Uppal, Combined feedforward-feedback servo control scheme for an exothermic batch reactor, Ind. Eng. Chem. Process Des. Dev. 23 (1984) 597−602.

[7] B.J. Cott, S. Macchietto, Temperature control of exothermic batch reactors using generic model control, Ind. Eng. Chem. Res. 28 (1989) 1177−1184.

[8] N. Régnier, G. Defaye, L. Caralp, C. Vidal, Software sensor based control of exothermic batch reactors, Chem. Eng. Sci. 51 (1996) 5125−5136.

[9] T. Clarke-Pringle, J.F. MacGregor, Nonlinear adaptive temperature control of multi-product, semi-batch polymerization reactors, Comput. Chem. Eng. 21 (1997) 1395−1409.

[10] X. Hua, S. Rohani, A. Jutan, Cascade closed-loop optimization and control of batch reactors, Chem. Eng. Sci. 59 (2004) 5695−5708.

[11] K.R. Westerterp, W.P. van Swaaij, A. Beenackers, H. Kramers, Chemical Reactor Design and Operation, Wiley, New York, 1984.

[12] T. Matynia, J. Ksiezopolski, Synthesis and properties of unsaturated epoxyfumarate resins, J. Appl. Polym. Sci. 77 (2000) 3077−3084.

[13] P. Hugo, Start-up and operation of exothermic batch processes, Ger. Chem. Eng. 4 (1981) 161−173.

[14] C. Venkateswarlu, K.V.S. Naidu, Adaptive fuzzy model based predictive control of an exothermic batch chemical reactor, Chem. Eng. Commun. 186 (2001) 1−23.

[15] C. Venkateswarlu, K. Gangiah, Dynamic modeling and optimal state estimation using extended Kalman filter for a kraft pulping digester, Ind. Eng. Chem. Res. 31 (1992) 848−855.

[16] B.Y. Lopez-Zapata, M. Adam-Medina, P.E. Alvarez-Cutierrez, J.P. Castillo-Gonazalez, H.X. Hernandez-DeLeon, L.G. Vera-Valdes, Virtual sensors for biodiesel production in a batch reactor, Sustainability, 9(2017)1-11.

[17] A.M. Nair, A. Fanta, F.A. Haugen, H. Ratnaweera, Implementing an Extended Kalman Filter for estimating nutrient composition in a sequential batch MBBR pilot plant, Water Science and Technology 80 (2) (2019) 317−328.

[18] G. Yang, X. Li, Y. Qian, A real-time updated model predictive control control strategy for batch processes based on state estimation, Chinese J. Chem. Eng. 22 (3) (2014) 318−329.

[19] P.L. Lee, P.L. Sullivan, Generic model control (GMC), Comput. Chem. Eng. 12 (1988) 573−580.

[20] M.A. Henson, Nonlinear model predictive control: current status and future directions, Comput. Chem. Eng. 23 (1988) 187−202.

[21] Z.K. Nagy, R.D. Braatz, Robust nonlinear model predictive control of batch processes, AIChE J. 49 (2003) 1776−1786.

[22] P.V.S. Ravichandra, C. Venkateswarlu, Multistep model predictive control of ethyl acetate reactive distillation column, Ind. J. Chem. Technol. 14 (2007) 333−340.

[23] K.J. Åström, K.J.T. Hägglund, Automatic tuning of simple regulators with specifications on phase and amplitude margins, Automatica 20 (1984) 645−651.

Part V

Optimal state estimation for online optimization

Chapter 20

Optimal state and parameter estimation for online optimization of an uncertain biochemical reactor

20.1 Introduction

Online optimization is a valuable tool for optimal control of bioprocesses involved in the production of various commodity and fine chemicals. The determination of the open-loop time varying control policies that maximize or minimize a given performance index is referred as optimal control/dynamic optimization. The optimal control policies that ensure the satisfaction of the product property requirements and the operational constraints can be calculated offline, then implemented online such that the system is operated according to these control policies. Optimal control is a widely used approach, and various techniques have been reported for chemical processes, including bioreactors [1−7]. Optimizing control is a promising tool that deals with changing conditions of a dynamic process online to achieve economic optimum. Various optimizing control techniques have been reported for different applications, including bioreactors [8−12]. The operating performance of a bioprocess is affected by several factors, such as the complex nature of microorganism growth, disturbance dynamics, parameter uncertainties, and noisy process variables. The optimizing control scheme has to consider such changes and continuously reevaluate the process to maximize its economic production. To account for the changing conditions of a bioprocess, the optimization algorithm should incorporate a model, whose parameters are to be continuously identified online. It is usual to view nonlinear systems as linear systems and compensate for nonlinearities by adapting linear model parameters. Online optimization involving adaptive process models has been widely employed in biochemical processes [13−17]. In most of those applications, a linear process model with a simple structure is employed, and convergence in parameters and efficiency in operation is achieved. The above studies are based on steady state optimization involving an identification period between optimization steps to compute steady state gain to predict process output. When process parameters change widely, methods based on steady state/linear models fail to achieve convergence. Process models based on fundamental physical and chemical laws are preferred for online optimization than the empirical input−output models because of their wide range of validity and physically more meaningful variables to identify. Since some of the bioprocess variables are not available as direct measurements, incorporation of state/parameter estimation scheme has become an integral part of online optimization. Online optimization based on a physical process model involves continuously revising the dynamic model, thus characterizing the process during transients and eliminating the need to wait for steady states. During steady state operation, the dynamic model acts as a steady state model to deduce the optimum. Earlier, a two-phase approach based on a physical process model has been presented [15], where the first phase involves identification of dynamic process model by state and parameter estimation, and the second phase deals with the determination of optimum operating strategy over a selected future time horizon based on the identified process model.

In this chapter, an online, optimizing control strategy involving a computationally efficient two-level extended Kalman filter (EKF) is presented for optimizing control of a bioreactor. In this approach, dynamic model identification is carried out using a two-level EKF for separate estimation of states and parameters, and a functional conjugate gradient method is employed for determining optimal operating conditions. The dynamic model involved in the estimation module represents the true dynamics of the system incorporating physical, chemical, and biological parameters. The two-level EKF employed in the estimation module is computationally efficient, as it provides the estimates of states and uncertain process parameters separately. In the estimation module, measurements are filtered, measured/unmeasured states are estimated, and process parameters are identified at every sampling instant and incorporated in the

Optimal State Estimation for Process Monitoring, Fault Diagnosis and Control. DOI: https://doi.org/10.1016/B978-0-323-85878-6.00018-X

optimizer to compute the optimal operating condition that maximizes the process performance. The performance of the online optimizing control strategy is evaluated through simulation by applying it to a biochemical reactor.

20.2 The process and its mathematical model

The bioreactor involves the production of biomass. The feed is a substrate of specified concentration. The manipulated variable is the dilution rate, and the process uncertainties are the feed concentration and the specific growth rate.

A dynamic model of the chemostat is described by the following differential equations [16]:

$$\frac{dc}{dt} = \frac{\mu_m w}{k_s + w} c - Dc \tag{20.1}$$

$$\frac{ds}{dt} = -\frac{1}{Y} \frac{\mu_m s}{k_s + s} c + D(s_f - s) \tag{20.2}$$

$$\frac{dw}{dt} = a(s - w) \tag{20.3}$$

where c is the biomass concentration, s is the substrate concentration, w is the weighted average of previous substrate concentrations, s_f is the substrate feed concentration, D is the dilution rate, μ_m is the maximum specific growth rate, k_s is the monad constant, and Y is the yield. The parameter a is the delay term which is a measure of the organism's ability to adjust its growth rate when a change in the condition of the chemostat occurs. For a constant volume fermentation, the product Dc is a productivity measure, where D is the dilution rate and c is the biomass concentration. The nominal process parameter values used in this simulation are: $\mu_m = 0.7 \text{ h}^{-1}$, $k_s = 22 \text{ g L}^{-1}$, $Y = 0.5$, $a = 3 \text{ h}^{-1}$, $s_f = 30 \text{ g L}^{-1}$, $c(0) = 14.153 \text{ g L}^{-1}$, $s(0) = w(0) = 1.6923 \text{ g L}^{-1}$.

20.3 State and parameter estimation using extended Kalman filter

The EKF is a heuristic filter based on the linearized dynamics of the system, and has become the standard for state and parameter estimation of nonlinear systems. Many successful applications of EKF for state and parameter estimation have been reported [18−25]. EKF provides a combined estimation of states and parameters involving full size matrix operations. The general process and measurement model representations for state and parameter estimation by EKF are given by Eqs. (4.23) and (4.24) in Section 4.4.2. The EKF algorithm given in Section 4.4.3 is designed and implemented for nonlinear state and parameter estimation of a bioreactor.

20.4 State and parameter estimation using two-level extended Kalman filter

EKF provides combined estimation of states and parameters involving full size matrix operations. If the state and parameter estimation is carried out separately, the computational effort can be reduced considerably due to the involvement of matrix calculations in reduced dimensions. Thus, in this study, a method of two-level EKF is presented for separate estimation of states and parameters. The estimation scheme is similar to that employed for process fault diagnosis via state and parameter estimation [26]. In this method, states are estimated separately in the first level by a state EKF, and in the second level, uncertain process parameters are identified separately by a parameter EKF. States and parameters are exchanged between the estimators for each new value of measurement. Process representation for state and parameter estimation by the method of two-level EKF is given in Section 4.4.2.

By considering time varying parameters in the process model, the expressions for states and parameters are given by:

$$\dot{x}(t) = f_x(x, \theta, t) + w_x(t), \quad x(o) = x_0 \tag{20.4}$$

$$\dot{\theta}(t) = p_\theta(x, \theta, t) + w_\theta(t), \quad \theta(o) = \theta_0 \tag{20.5}$$

where f_x and p_θ are nonlinear functions of states x, parameters θ, and also input u. w_x, and w_θ are process noises with covariance matrices Q_x and Q_θ. The nonlinear observation model can be expressed as

$$y(t_k) = h_k(x, \theta, t) + v(t_k) \tag{20.6}$$

where h is a nonlinear function of states and parameters, and v is observation noise with zero mean. The linear measurement relation is given by

$$y(t_k) = Hx(t_k) + v(t_k) \tag{20.7}$$

The states $x(t)$ and parameters $\theta(t)$ of Eqs. (20.4) and (20.5) can be estimated online using the known measurements $y(t_k)$, in conjunction with the process model. The two-level EKF algorithm for state and parameter estimation is given in Section 4.6.

20.5   Online optimization problem

The online optimizing control problem can be stated in abstract form as follows: Given an operating point with measurements y, and a set of manipulated inputs u, determine the values of u as a function of time that maximizes the measure of the profitability of the plant while meeting the operating constraints of the process.

For the optimization problem, the nonlinear plant and the measurement model expressions can be rewritten as

$$\dot{x} = f(x, \ u, \ \theta, \ t) \tag{20.8}$$

$$y = g(x, \ u, \ \theta, \ t) \tag{20.9}$$

The operating constraints are expressed as

$$e(x, \ u, \ \theta, \ t) \leq 0 \tag{20.10}$$

In these equations, x represents state variables, u is manipulated input, and θ denotes unknown parameters and unmeasured disturbances. This type of physical process model has a wide range of applicability for identifying more meaningful variables. However, if a more detailed model is used, the computational complexity increases. So, a compromise has to be made between the level of detail used and the number of parameters estimated online. Since the optimizer incorporates the states and parameters identified through separate estimation with less computational requirement, physical process models of higher dimensions can be employed with this strategy.

The optimization problem determines the optimal operating condition, $u(t)$, by maximizing the functional or performance index

$$I = \int \varphi(x, \ u, \ \theta, \ t) \, dt \tag{20.11}$$

subject to the model equations and constraints, Eqs. (20.1)–(20.3).

20.6   Functional conjugate gradient method

The conjugate gradient method generates a set of mutually conjugate direction vectors using the gradient vector as the basis. The conjugate gradient method, when applied to functional optimization, is called the functional conjugate gradient method. This method involves successive approximation in the control domain, utilizing the gradient to compute a new control function in each iteration. The implementation of the method involves the following procedure [27].

The objective function in Eq. (20.11) is in the form

$$I = \int f_o(x, u, \theta, t) \, dt \tag{20.12}$$

The Hamiltonian is formulated as

$$H = f_o + \lambda^T f \tag{20.13}$$

where λ represents the set of adjoint variables, the relation of which is given by

$$-\lambda_i = \frac{\partial H}{\partial x_i} \tag{20.14}$$

The gradient, g, and the direction ξ are calculated as

$$g^i(t) = \frac{\partial H}{\partial u}(t) \tag{20.15}$$

and

$$\xi^{i+1}(t) = -g^{i+1}(t) + \beta^i \xi^i(t), \quad 0 \le t \le \tau \tag{20.16}$$

where $\beta^i = \dfrac{\int_0^\tau \left(g^{i+1}\right)^2 dt}{\int_0^\tau \left(g^i\right)^2 dt}$

The initial conditions are taken to be

$$g^0(t) = g(u^0) \text{ and } \xi^0(t) = -g^0(t); \ 0 \le t \le \tau \tag{20.17}$$

The performance of the above online optimizing control strategy is evaluated by applying it to the biochemical reactor.

20.7 Extended Kalman filter-assisted online optimizing control of the biochemical reactor

This approach involves an EKF estimator for dynamic model identification and a functional conjugate gradient method for online optimization. The estimator filters out noise in the measurements and provides the estimates of uncertain process parameters. The nonlinear dynamic model identified at every sampling instant is incorporated in the optimizer to compute the optimum input to maximize the profit objective. In the biochemical reactor, the objective is to maximize the productivity measure, Dc, by optimizing the manipulated input, D, while taking care of the uncertainties in feed concentration s_f, and the specific growth rate μ_m. The EKF-assisted online optimizing control scheme is shown in Fig. 20.1.

20.7.1 Designing the strategy

This strategy involves two phases. The first phase involves identifying the unknown model parameters and unmeasured disturbances entering the process by using all available process measurements. In the second phase, the identified nonlinear model is used to determine the optimum operating strategy. The two phases are repeatedly and periodically in tandem to follow continually changing input disturbances and process parameters. The measurements available from the process are concentrations of c and s. The state w is not measured online and has to be estimated along with the parameters s_f and μ_m. EKF is designed to estimate the state w, and the parameters s_f and μ_m, using the known measurements c and s. The augmented state and parameter vector is defined as

$$x = \begin{bmatrix} c & s & w & s_f & \mu_m \end{bmatrix}^T \tag{20.18}$$

The matrices P_0, $Q(t)$, and $R(t_k)$ are selected as design parameters and used to reflect errors in the initial state, process model, and process measurements. The diagonal elements of the initial state noise covariance matrix P_0, process

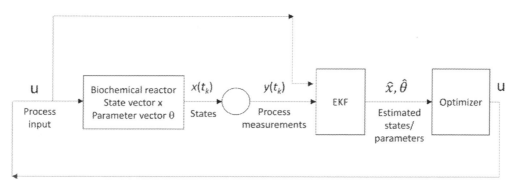

FIGURE 20.1 Extended Kalman filter-assisted online optimizing control scheme.

noise covariance matrix, Q and observation noise covariance matrix, R are selected as 0.0005, 0.2, and 1.0, respectively. The elements $f_{i,j}$ of the state transition matrix, F of Eq. (4.27) of the EKF estimation algorithm are obtained by taking the partial derivatives of the model equation, Eqs. (20.1)−(20.3) with respect to the vector of states and parameters given by Eq. (20.18). The elements of measurement covariance matrix H of Eq. (4.31) of the EKF estimation algorithm are obtained by taking the partial derivatives of the measurement equations, Eqs. (20.1) and (20.2) with respect to the augmented state vector given by Eq. (20.18).

The online optimizing control strategy has to determine the optimal dilution rate, D, that maximizes the productivity, Dc. The states and parameters estimated at each sample time are incorporated in the optimizer. The optimal operating condition, D at each sampling instant is determined by using the functional conjugate gradient method, while maximizing the objective function in Eq. (20.12).

20.7.2   Analysis of results

The state and parameter estimation and online optimization require numerical integration of the mathematical model of the biochemical reactor. The integration of model equations is performed using a step size of 0.1 h. The same step size is considered for online optimization. Since, the disturbances affect the process over time, it is necessary to revise the process model using the most recent measurements. EKF performs the model identification to estimate the states and parameters of the biochemical reactor. The identified states and parameters are used in the functional conjugate gradient method to determine the dilution rate, D, while satisfying the performance requirements.

Fig. 20.2 shows the dilution rate policy, D, when a change in substrate concentration, s_f is introduced from 30 g L^{-1} to 20 g L^{-1} at time 20 h, and a change in specific growth rate, u_m is introduced from 0.7 h^{-1} to 0.9 h^{-1} at time 60 h during the operation. The performance measure, Dc for the same changes in parameters is shown in Fig. 20.3. The change in s_f alter the Dc from 2.22 to 1.12, and the change in μ_m alter the Dc from 1.12 to 1.14. The performance of

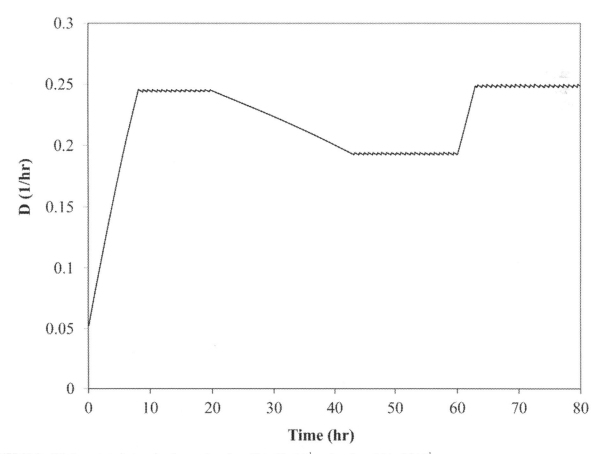

FIGURE 20.2   Dilution rate trajectory for changes in s_f from 30 to 20 g L^{-1}, and μ_m from 0.7 to 0.9 h^{-1}.

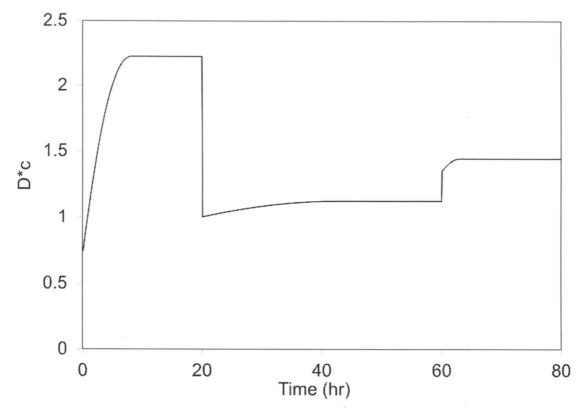

FIGURE 20.3 The plot of performance measure for changes in s_f from 30 to 20 g L^{-1}, and μ_m from 0.7 to 0.9 h^{-1}.

EKF assisted dynamic optimizer is also studied in the presence of process disturbances and noise by considering a measurement noise of 2%. The results have shown the noise filtering ability of EKF. Thus, the results show the ability of EKF-assisted dynamic optimization in achieving optimum operating conditions in a bioreactor in the presence of process disturbances and noise.

20.8 Two-level extended Kalman filter-assisted online optimizing control strategy

This strategy consists of a two-level EKF as an estimation module and a functional conjugate gradient method as an optimization module [28].

20.8.1 Designing the strategy

In this strategy, the measured/unmeasured process states and uncertain process parameters provided by the estimation module are incorporated in the optimization module to determine the optimal operating condition. Thus, the estimation and optimization modules are repeatedly in tandem to follow continually changing input disturbances and process parameters. The schematic of the online optimizing control strategy is shown in Fig. 20.4.

The online, optimizing control strategy has to determine the optimal dilution rate that maximizes productivity, Dc. The parameters s_f and μ_m are estimated at each sample time at the identification phase and incorporated in the optimizer. The nominal process parameter values used in this simulation are given in Section 20.2. The mathematical model of the process is integrated numerically using a step size of 0.1 h. The simulated process measurements of every 0.1 h are used for optimal state estimation and online optimization. The measurements are corrupted with a zero mean random Gaussian noise of about 1% of their actual values to reflect the real situation. The filter design parameters such as the initial state and parameter noise covariance matrices (P_{xo}, $P_{\theta o}$), the process noise covariance matrices (Q_x, Q_θ), and the observation noise covariance matrix (R) are initially selected using process variable and noise information and further tuned to obtain better estimator performance. These filter design parameters are given in Table 20.1. The performance index J of the two-level EKF is expressed as:

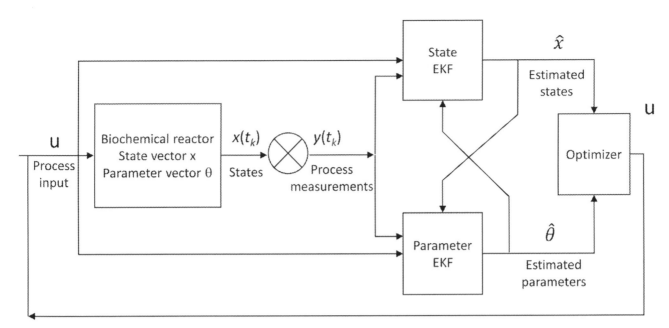

FIGURE 20.4 Two-level extended Kalman filter-assisted online optimizing control scheme.

TABLE 20.1 Filter design parameters.

$$P_{x0} = \begin{bmatrix} 0.0005 & 0 \\ 0 & 0.0005 \end{bmatrix}; \quad P_{\theta 0} = \begin{bmatrix} 0.0005 & 0.0 \\ 0.0 & 0.0005 \end{bmatrix}$$

$$Q_1 = \begin{bmatrix} 0.2 & 0.0 \\ 0.0 & 0.2 \end{bmatrix}; \quad Q_2 = \begin{bmatrix} 0.2 & 0.0 \\ 0.0 & 0.2 \end{bmatrix}$$

$$R = \begin{bmatrix} 1.0 & 0.0 \\ 0.0 & 1.0 \end{bmatrix}$$

$$J = \sum_{k=1}^{N} \left\{ (x(0) - \hat{x}(0))^T P_{x0}^{-1}(x(0) - \hat{x}(0)) + \left(\theta(0) - \hat{\theta}(0)\right)^T P_{\theta 0}^{-1}\left(\theta(0) - \hat{\theta}(0)\right) \right\}$$
$$+ \left\{ \left(y(t_k) - h_k(\hat{x}(t_k)), \hat{\theta}(t_k)\right) R^{-1}(y(t_k)) - h_k(\hat{x}(t_k), \hat{\theta}(t_k))\right) \right\} \tag{20.19}$$
$$+ \left\{ (x(t_k) - \hat{x}(t_k))^T Q_x^{-1}(x(t_k) - \hat{x}(t_k)) + \left(\theta(t_k) - \hat{\theta}(t_k)\right)^T Q_\theta^{-1}\left(\theta(t_k) - \hat{\theta}(t_k)\right) \right\}$$

The two-level EKF is thus designed and applied to estimate w, s_f and μ_m, using the known measurements c and s. At every sampling instant, the identified model is incorporated in the optimizer to determine dilution rate, D.

20.8.2 Analysis of results

The estimated state responses using normal process measurements with zero mean random Gaussian noise of about 1% to their actual values are shown in Fig. 20.5. The true values shown in this figure represent the numerical values obtained from the solution of model equations. The estimated profiles in Fig. 20.5 correspond to the biomass and substrate concentrations beginning from their initial values of 14.153 g L^{-1} and 1.6923 g L^{-1}, respectively. These results indicate the close correspondence between the estimated and true states. Fig. 20.6 shows the results of the estimated process parameters, optimal dilution rate, and process performance measure involving normal noisy measurements corresponding to change in s_f from 30 to 25 g L^{-1} at 30 h and change in u_m from 0.7 to 0.85 h^{-1} at 40 h during the operation. The identified process parameters in Fig. 20.6 show that they are in close agreement with their corresponding true values. The optimal dilution rate and process performance measure in Fig. 20.6 exhibit the fast tracking ability of the optimizer to adapt to the process disturbances. Fig. 20.7 shows the results of the estimated process parameters, optimal dilution rate, and process performance measure for change in s_f from 30 to 25 g L^{-1} at 30 h and change in u_m from 0.7

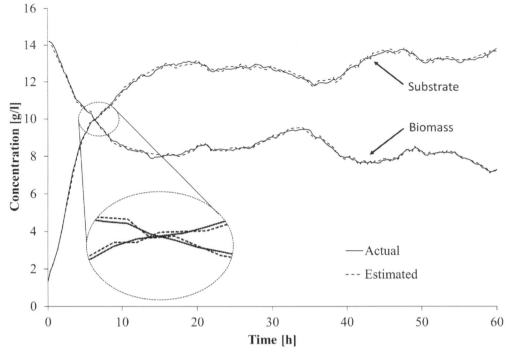

FIGURE 20.5 State estimation results of two-level extended Kalman filter-assisted online optimizer with no parameter uncertainties.

to 0.85 h^{-1} at 40 h during operation, when the measurements with zero mean random Gaussian noise of five times higher than the normal noise are employed. The results in Fig. 20.7 indicate the manifestation of process performance measure (DC) for changes in s_f and u_m in the presence of higher noise in the measurements. Because of the presence of more noise in the measurements, the specific growth rate initially deviates from the true value, and after some time, it reaches and tracks the true value quite well. These results show the noise filtering ability of the estimator and the performance of the optimizer in establishing the optimal operating conditions in the presence of uncertain parameters. The proposed strategy is also evaluated by considering measurements available at wider sampling instants. The discrete updating step of the two-level EKF is incorporated with the measurements available at wider sampling instants of 0.5 h using zero mean random Gaussian noise of five times higher than the initially set noise while iterating the continuous prediction step with lower integration time. The results evaluated indicate that the online optimizing control strategy can be employed with the noisy measurements sampled at wider discrete time instants.

The sensitivity of the estimator in conjunction with the optimizer is investigated for normal process operation by studying the effect of filter design parameters and measurement noise. Table 20.2 shows the quantitative performance of the estimator evaluated for an operating period of 60 h for normal process operation. The integral squared error (ISE) values shown in Table 20.2 are the summated squared differences between the actual states of x, s, and w obtained through numerical simulation and those corresponding to estimated states. The results shown in this table are obtained by changing each of the filter design parameters while keeping the remaining parameters unchanged. There is no significant change observed in estimator performance for 10 times increase or decrease of filter design parameters from their initially set values. Decreasing the measurement noise has shown a moderate influence on the estimator performance, whereas increasing it to a higher level has shown considerable influence on the estimator performance. These results indicate the stability of the estimator towards the effect of filter design parameters. The results thus demonstrate the better performance of the two-level EKF supported dynamic optimizer for optimal control of biochemical reactor in the presence of sudden and gradual process parameter uncertainties and noise in the measurements.

20.9 Summary

The operating performance of a bioprocess is affected by several factors such as the complex nature of microorganism growth, disturbance dynamics, parameter uncertainties, and noisy process variables. Optimizing control is a promising tool that considers process disturbances and changes in process conditions, and continuously reevaluates the process to

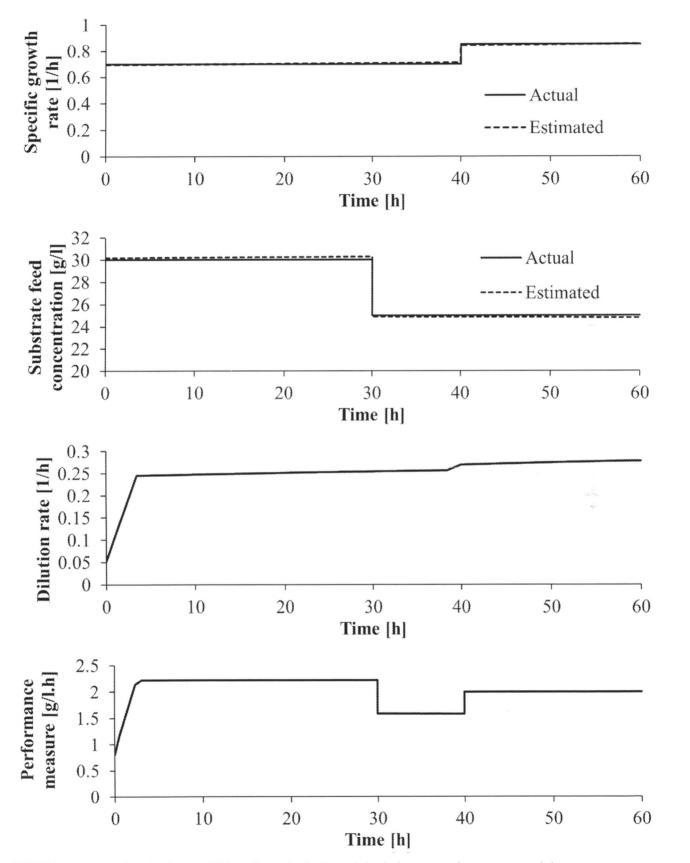

FIGURE 20.6 Results of two-level extended Kalman filter-assisted online optimizer in the presence of parameter uncertainties.

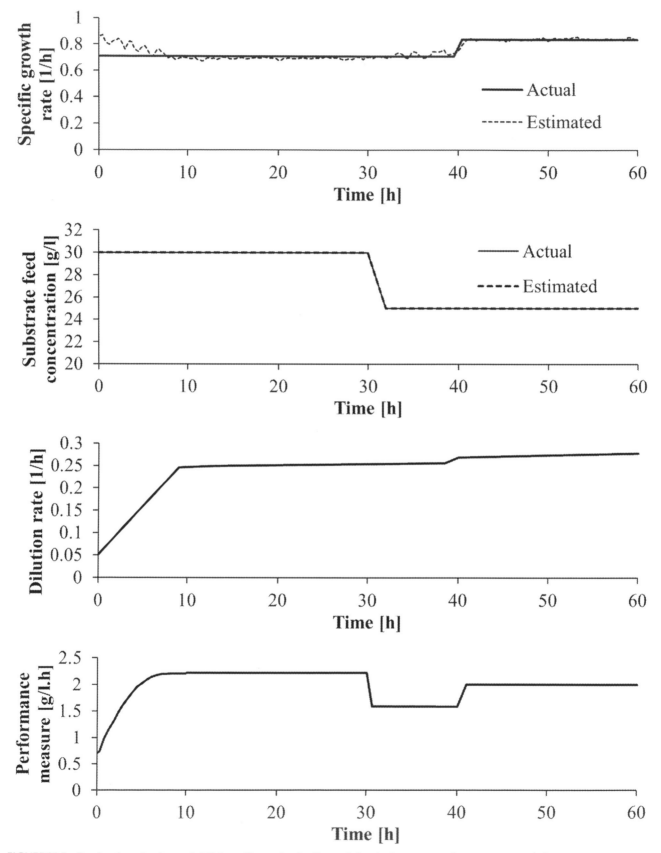

FIGURE 20.7 Results of two-level extended Kalman filter-assisted online optimizer in the presence of parameter uncertainties.

TABLE 20.2 Estimator performance for different levels of filter design parameters and noise.

Filter design parameter/noise	Level	ISE
Initial state covariance matrix	$P_{xo}/10$	3.2539
	P_{xo}	3.2463
	$10P_{xo}$	3.1763
	$P_{\theta o}/10$	3.2473
	$P_{\theta o}$	3.2463
	$10P_{\theta o}$	3.2362
Process noise covariance matrix	$Q_x/10$	3.3738
	Q_x	3.2463
	$10Q_x$	3.3660
	$Q_\theta/10$	3.2463
	Q_θ	3.2463
	$10Q_\theta$	3.2463
Observation noise covariance matrix	$R/10$	3.0188
	R	3.2463
	$10R$	3.3828
Observation noise	$v/10$	2.7995
	v	3.2463
	$10v$	12.2576

maximize its economic production. In this study, an EKF-assisted online, optimizing control strategy is presented for optimal control of a biochemical reactor. This strategy involves identifying states and parameters of the process by an EKF and determining optimal control input by functional conjugate gradient method. The simulation results demonstrate the usefulness of the strategy for online optimization of uncertain biochemical systems.

Further, an online optimizing control strategy involving a two-level EKF with separate estimation of states and parameters for dynamic model identification and a functional conjugate gradient method for determining optimal operating conditions is proposed and applied to the biochemical reactor. The optimizer incorporates the identified model and determines the optimal operating conditions, while maximizing the process performance measure. This strategy is computationally advantageous as it involves separate estimation of states and process parameters in reduced dimensions. In addition to assisting online dynamic optimization, the estimated time varying uncertain process parameter information can also be useful for monitoring the process operation. This strategy ensures that the biochemical reactor is operated at the optimal operation while taking care of the disturbances encounter during operation. The simulation results demonstrate the effectiveness of the two-level EKF-assisted dynamic optimizer for online optimizing control of uncertain nonlinear biochemical systems.

References

[1]   D.-M. Xie, D.-H. Liu, J.-A. Zhang, Temperature optimization for glycerol production by batch fermentation with *Candida krusei*, J. Chem. Technol. Biotechnol. 76 (10) (2001) 1057−1069.

[2]   J. Wei, F. Dai, L. Yan, C. Lei, G. Jia Qing, Z. Gao Ping, et al., Optimizing the feeding operation of recombinant *Escherichia coli* during fed-batch cultivation based on Pontryagins minimum principle, Afr. J. Biotechnol. 10 (64) (2011) 14143−14154.

[3]   C. Valencia, G. Espinosa, J. Giralt, F. Giralt, Optimization of invertase production in a fed-batch bioreactor using simulation based dynamic programming coupled with a neural classifier, Comput. Chem. Eng. 31 (9) (2007) 1131−1140.

[4]   I.M. Thomas, C. Kiparissides, Computation of the near-optimal temperature and initiator policies for a batch polymerization reactor, Can. J. Chem. Eng. 62 (2) (1984) 284–291.

[5]   C.V. Peroni, N.S. Kaisare, J.H. Lee, Optimal control of a fed-batch bioreactor using simulation-based approximate dynamic programming, IEEE Trans. Control Syst. Technol. 13 (5) (2005) 786–790.

[6]   A.E. Bryson, Y.-C. Ho, Applied Optimal Control, Routledge, 2018.

[7]   J.A. López, V. Bucalá, M.A. Villar, Application of dynamic optimization techniques for poly(β-hydroxybutyrate) production in a fed-batch bioreactor, Ind. Eng. Chem. Res. 49 (4) (2010) 1762–1769.

[8]   X. Chen, R.W. Pike, T.A. Hertwig, J.R. Hopper, Optimal implementation of on-line optimization, Comput. Chem. Eng. 22, 1 (1998) S435–S442.

[9]   X.-G. Zhou, L.-H. Liu, W.-K. Yuan, Optimizing control of a wall-cooled fixed-bed reactor, Chem. Eng. Sci. 54 (13–14) (1999) 2739–2744.

[10] M. Noda, T. Chida, S. Hasebe, H. Iori, On-line optimization system of pilot scale multi-effect batch distillation system, Comput. Chem. Eng. 24 (2–7) (2000) 1577–1583.

[11] Y.-H. Kim, T.-W. Ham, J.-B. Kim, On-line dynamic optimizing control of a binary distillation column, J. Chem. Eng. Jpn. 24 (1) (1991) 51–57.

[12] D. Beluhan, D. Gosak, N. Pavlović, M. Vampola, Biomass estimation and optimal control of the baker's yeast fermentation process, Comput. Chem. Eng. 19 (1) (1995) S387–S392.

[13] G. Ryhiner, I.J. Dunn, E. Heinzle, S. Rohani, Adaptive on-line optimal control of bioreactors: application to anaerobic degradation, J. Biotechnol. 22 (1–2) (1992) 89–105.

[14] M.J. Rolf, H.C. Lim, Experimental adaptive on-line optimization of cellular productivity of a continuous bakers' yeast culture, Biotechnol. Bioeng. 27 (8) (1985) 1236–1245.

[15] S.-S. Jang, B. Joseph, H. Mukai, On-line optimization of constrained multivariable chemical processes, AIChE J. 33 (1) (1987) 26–35.

[16] J. Harmon, S.A. Svoronos, G. Lyberatos, Adaptive steady-state optimization of biomass productivity in continuous fermentors, Biotechnol. Bioeng. 30 (3) (1987) 335–344.

[17] J.W. Hamer, C.B. Richenberg, On-line optimizing control of a packed-bed immobilized-cell reactor, AIChE J. 34 (4) (1988) 626–632.

[18] R.D. Gudi, S.L. Shah, M.R. Gray, Multirate state and parameter estimation in an antibiotic fermentation with delayed measurements, Biotechnol. Bioeng. 44 (11) (1994) 1271–1278.

[19] C. Venkateswarlu, K. Gangiah, Dynamic modeling and optimal state estimation using extended Kalman filter for a kraft pulping digester, Ind. Eng. Chem. Res. 31 (3) (1992) 848–855.

[20] C. Venkateswarlu, S. Avantika, Optimal state estimation of multicomponent batch distillation, Chem. Eng. Sci. 56 (20) (2001) 5771–5786.

[21] F. Crevecoeur, R.J. Sepulchre, J.L. Thonnard, P. Lefèvre, Improving the state estimation for optimal control of stochastic processes subject to multiplicative noise, Automatica 47 (3) (2011) 591–596.

[22] C. Venkateswarlu, Advances in monitoring and state estimation of bioreactors, J. Sci. Ind. Res. 64 (6) (2005) 491–498.

[23] J. Seung, F. Amir Atiya, A.G. Parlos, K. Chong, Parameter Estimation for Coupled Tank using Estimate Filtering, International Journal of Control and Automation 6 (5) (2013) 91–102.

[24] P. Mobed, S. Munusamy, D. Bhattacharyya, R. Rengaswamy, State and parameter estimation in distributed constrained systems. 1. extended Kalman filtering of a special class of differential-algebraic equation systems, Industrial & Engineering Chemistry Research 56 (1) (2017) 206–215.

[25] M. Farsi, M.D. Manshadi, A State Estimation Method Based on Integration of Linear and Extended Kalman Filters, Chemical Product and Process Modeling 14 (4) (2019).

[26] C. Venkateswarlu, K. Gangiah, M.B. Rao, Two-level methods for incipient fault diagnosis in nonlinear chemical processes, Comput. Chem. Eng. 16 (5) (1992) 463–476.

[27] R. Fletcher, Function minimization by conjugate gradients, Comput. J. 7 (2) (1964) 149–154.

[28] E.J. Satya, P. Anand, C. Venkateswarlu, Optimal state estimation and online optimization of a biochemical reactor, Chem. Process Eng. 34 (4) (2013) 449–462.

Chapter 21

Overview, opportunities, challenges, and future directions of state estimation

21.1 Overview

The technological advancements have made the process industry undergo significant changes emphasizing optimally running the plants with due consideration to safety, monitoring, and control with minimal manual intervention. Optimal estimation of the states of a plant from its model in conjunction with its measured outputs constitutes a dominant role in the process systems engineering domain. The increased degree of automation and the growing demand for higher performance, efficiency, and reliability in industrial systems makes the task of state estimation an integral part of various operational strategies, such as monitoring, diagnosis, and control concerning the field of process systems engineering. State estimation is a subject of immense interest with applications in conventional and advanced process control, supervisory control, process monitoring, online optimization, product quality monitoring, data reconciliation, fault detection and diagnosis, etc. The state estimation-based applications have become more crucial in the face of growing demand to meet process safety requirements, environmental regulations, energy efficiency, better product quality, and optimum utilization of available resources.

State estimation approaches are presented by broadly classifying them into mechanistic/first principle model-based approach and data-driven model-based approach. State estimation methods dealt under mechanistic model-based approach include various stochastic model-based filtering and observation techniques such as Kalman filter (KF), extended KF (EKF), unscented KF (UKF), square root UKF (SRUKF), particle filter (PF), ensemble KF (EnKF), and different nonlinear observers. Methods presented under data-driven model-based approach include the state estimators derived based on principal component analysis (PCA), partial least squares (PLS), nonlinear iterative PLS (NIPALS), artificial neural networks (ANN), and radial basis function networks (RBFN). Further various linear and nonlinear observers are also applied for optimal state estimation. The details regarding the description, algorithms, examples, applications and case studies concerning different approaches, and methods of state estimation are elaborated in detail in this book.

State estimation methods provide the estimates of many other variables and process parameters by strategically measuring some key variables of the process. In order to gain the maximum benefit from the state estimation methods, the sensors that provide vital information about the measured variables are to be placed at optimal locations in the process plant. Thus, optimal sensor selection is the most important issue for state estimation in dynamic systems. By means of optimal sensor configuration, it is possible to identify the minimum number of sensors to obtain the maximum amount of information of the dynamic process for reliable estimation of the states. Various methods that are discussed in the text for optimal sensor configuration include the sensitivity index, singular value decomposition, PCA, observability Grammian based quantification measures for linear systems, and empirical observability Grammians and Grammian based metrics for nonlinear systems.

The whole content of this book is divided into five parts: (1) basic details and state estimation algorithms; (2) optimal state estimation for process monitoring; (3) optimal state estimation for process fault diagnosis; (4) optimal state estimation for process control; and (5) optimal state estimation for online optimization. The details regarding the description, algorithms, examples, procedures, real system applications and case studies concerning different approaches, and methods of state estimation are elaborated in detail in each part.

21.2 Opportunities

State estimation is a contemporary discipline associated with the field of process systems engineering that emerged with the main purpose to provide the methods, tools, and human capacity that aid the industry to meet its performance

Optimal State Estimation for Process Monitoring, Fault Diagnosis and Control. DOI: https://doi.org/10.1016/B978-0-323-85878-6.00017-8

requirements. The increased degree of automation and growing demand for higher performance in industrial systems make state estimation an integral part of various operational strategies associated with process systems engineering field. State estimation has wide applications in a vast range of process industries. It plays a significant role in the design, modeling, monitoring, optimization, control, and operation of all kinds of chemical, physical, biological, and other engineering processes through the use of systematic computer-aided approaches. This discipline has grown steadily along with the developments in electronics and computing fields. The field of state estimation provides challenging opportunity to develop novel and efficient methods to meet the monitoring, diagnosis, and control requirements of process systems.

21.3 Challenges

The task of state estimation has become more challenging in the face of growing demand to meet the requirements of process safety, environmental regulations, energy efficiency, better product quality, and optimum utilization of available resources. Its major challenges are the development of concepts, methodologies, and procedures for performance improvement of an engineered system. Since mathematical model is a basis for the quantitative model-based filtering and observation oriented state estimation techniques, overcoming the mathematical modeling difficulties associated with complex processes, and real-time application of soft sensors to large scale processes require more attention. Accurate and current values of process variables are essential in order to operate manufacturing plants to maximize profits and meet all safety, health, and environmental requirements. Proper data collection is crucial for the development and validation of soft sensors. Due to the extremely large number of important process variables and measurement difficulty often associated with large plants, the capital and maintenance costs associated with the installation of additional sensors need to be accounted into the objective of sensors selection strategy. Therefore, optimization of the placement of sensors is a vital part of plant design and operation.

Data-driven soft sensors require process knowledge in the form of heuristic data to understand the inherent characteristics of process plant mechanisms and also understand the correlations between the process variables. However, the key challenges to develop a reliable and robust data-driven model-based soft sensor are as follows:

1. The quality and quantity of process data available for developing the reliable soft sensors pose hurdles. The data may be intermittent or erroneous data that are quite noisy with outliers.
2. Off-line data available from the lab measurements and analysers may not be exact due to poor calibration, measurement error, computer interface errors, etc.
3. Manual lab measurements and analyser may have different sampling intervals and available with time delays, which causes mismatch in time of measurement.
4. Large scale processes involve multiple number of variables and it may not be possible to include all of them in the development of data-driven model. Therefore, identification of significant process variables that contribute major information is a crucial issue.
5. Maintenance of data-driven model-based soft sensor is another important issue. After successful development of a soft sensor, there can be a gradual deterioration of the performance of the soft sensor, probably due to gradual changes in the process. In such a case, retraining the data-driven model, or rebuilding the soft sensor is required.

21.4 Future directions

Soft sensors play a significant role in process monitoring, diagnosis, and control, hence require continuous R&D efforts to develop novel and efficient methods. In quantitative model-based approach, EKF and its extensions have been widely applied in simulation environment with some real-time applications. However, real-time evaluation of later versions of stochastic state estimators developed based on filtering techniques such as UKF, SRUKF, EnKF, PF, and variational Bayesian filter are very limited. Experimental evaluation of state estimators built based on these algorithms can enable to identify the better methods from practicality point of view. The data-driven state estimators developed using the techniques like PLS, ANN, and RBFN with and without the combination of NIPALS have been widely employed for state estimation in various fields. The predictive models involved in data-driven soft sensors are established through training procedure resulting in fixed models and structures which are employed for state estimation using the current measured information. For highly nonlinear and uncertain systems, the state estimator performance can be improved through adaptation of the predefined data-driven models. This can become a direction for investigation towards the development of improved soft sensors. Developing novel hybrid soft sensors by integrating ANN/RBFN predictive models with

EKF/UKF techniques and exploring their usefulness in theoretical and practical environment will become a challenging research activity. Further, more theoretical and experimental research efforts are needed for the development of sensor configuration techniques and soft sensors for inferential estimation of states and parameters in highly nonlinear, high dimensional distributed dynamic systems.

Development of reliable soft sensors make the system intellectual, minimize the regular plant shut downs and industry maintenance costs as well as provides uninterrupted process outputs. This leads to better product quality control and higher productivity. Due to enormous growth in computing power and technological advancements in internet of things (IOT) has led to paradigm shifts in manufacturing and traditional manufacturing operations. Integrating the automated manufacture plant with cloud storage enables the system to store huge historical data consisting of the most of the measured variables, provides long term data for online forecasts. These historical data enable building datasets to perform predictive modeling and analyze the plant operation patterns at different operating conditions. These computational advantages and huge data processing abilities of computers can offer enormous opportunity to develop adaptive and intelligent soft sensors that can be integrated with the design and operational strategies of large plants. Manufacturing processes will increasingly benefit from intelligent soft sensing techniques and advanced machine learning algorithms and thus improve the efficiency and reduce the precious time by timely undertaking the maintenance of the failed instruments and other devices. Advanced soft sensors empower to minimize unforeseen and excessive repairs, machine downtime, and premature component replacement. The adaptive soft sensors can capture and analyze data for effective decision making. On a whole, integration of intelligent and adaptive soft sensing techniques with the plant operations provides features like tracking entire process in manufacturing, improve production performance, maintain good quality of outputs, detect faults, and regulate the process.

21.5 Summary

The technological advancements have made the process industry undergo significant changes emphasizing optimally running the plants with due consideration to safety, monitoring, and control with minimal manual intervention. Optimal estimation of the states of a plant from its model in conjunction with its measured outputs constitutes a dominant role in the process systems engineering domain. State estimation plays a significant role in the design, modeling, monitoring, optimization, control and operation of all kinds of chemical, physical, biological, and other engineering processes through the use of systematic computer-aided approaches. This chapter thus presents an overview, opportunities, challenges, and future directions of state estimation and its applications.

Index

Printed in the United States
by Baker & Taylor Publisher Services